AF877800

RECENT ADVANCES IN ROBOT KINEMATICS

RECENT ADVANCES IN ROBOT KINEMATICS

Edited by

Jadran Lenarčič
J. Stefan Institute,
University of Ljubljana,
Slovenia

and

Vincenzo Parenti-Castelli
Department of Mechanical Engineering,
University of Bologna,
Italy

KLUWER ACADEMIC PUBLISHERS
DORDRECHT / BOSTON / LONDON

A C.I.P. Catalogue record for this book is available from the Library of Congress.

ISBN 0-7923-4124-4

Published by Kluwer Academic Publishers,
P.O. Box 17, 3300 AA Dordrecht, The Netherlands.

Kluwer Academic Publishers incorporates
the publishing programmes of
D. Reidel, Martinus Nijhoff, Dr W. Junk and MTP Press.

Sold and distributed in the U.S.A. and Canada
by Kluwer Academic Publishers,
101 Philip Drive, Norwell, MA 02061, U.S.A.

In all other countries, sold and distributed
by Kluwer Academic Publishers Group,
P.O. Box 322, 3300 AH Dordrecht, The Netherlands.

Printed on acid-free paper

All Rights Reserved
© 1996 Kluwer Academic Publishers
No part of the material protected by this copyright notice may be reproduced or
utilized in any form or by any means, electronic or mechanical,
including photocopying, recording or by any information storage and
retrieval system, without written permission from the copyright owner.

Printed in the Netherlands

Preface

The articles of this book were reported and discussed at the fifth international symposium on Advances in Robot Kinematics. As is known, the first symposium of this series was organised in 1988 in Ljubljana. The following meetings took place every other year in Austria, Italy, and Slovenia (Linz, Ferrara, Ljubljana, Portorož-Bernardin). It must be emphasised that the symposia run under the patronage of the International Federation for the Theory of Machines and Mechanisms, IFToMM. In this period, Advances in Robot Kinematics has been able to attract the most outstanding authors in the area and also to create an optimum combination of a scientific pragmatism and a friendly atmosphere. Hence, it has managed to survive in a strong competition of many international conferences and meetings. In the most ancient way, robot kinematics is regarded as an application of the kinematics of rigid bodies. However, there are topics and problems that are typical for robot kinematics that cannot easily be found in any other scientific field. It is our belief that the initiative of Advances in Robot Kinematics has contributed to develop a remarkable scientific community.

The present book is of interest to researchers, doctoral students and teachers, engineers and mathematicians specialising in kinematics of robots and mechanisms, mathematical modelling, simulation, design, and control of robots. It is divided into eight sections that we found as the prevalent areas of recent kinematics research seen from the viewpoint of its implementation: kinematic control, kinematic performance of mechanisms, workspace and trajectory analysis, robot design, modelling and computation of mechanisms, analysis and simulation, performance of parallel mechanisms, and analysis of parallel mechanisms. As it can easily be noticed, an important part of the book (the last two chapters) is dedicated to parallel mechanisms. Parallel mechanisms apparently persist to be one of the most attractive areas of research in kinematics. In comparison to the previous symposia and in comparison to the quantity of articles published in scientific journals, kinematic redundancy unexpectedly lost its supremacy.

The book contains 45 contributions. The authors and co-authors are from 22 countires. It must be emphasised that the whole procedure of publishing the book (starting with draft manuscripts, reviewing, preparing final manuscripts, editing, printing, and all mailing) took just about four months. We are grateful to the authors of the articles for their valuable contributions, to the reviewers for their timely review of the articles, and to the personnel at Kluwer Academic Publishers for their (as always) excellent technical and editorial support. We are also indebted to the Slovenian Ministry of Science and Technology, the University of Ferrara (Italy), and IFToMM for their financial support, and to the staff of the Department of Automatics, Biocybernetics and Robotics of the Jožef Stefan Institute in Ljubljana (Slovenia) for their service.

Jadran Lenarčič and Vincenzo Parenti-Castelli, Editors

Table of Contents

8. Analysis of parallel mechanisms

1. Plenary

J. Nielsen, B. Roth: *The direct kinematics of the general 6-5 Stewart-Gough mechanism*

J. Angeles, F. Bulca, N. Areson, M. Areson: *The kinematic sensitivities of redundant robotic manipulators*

THE DIRECT KINEMATICS OF THE GENERAL 6-5 STEWART-GOUGH MECHANISM

J. NIELSEN AND B. ROTH
Design Division, Mechanical Engineering Department
Stanford University
Stanford, CA 94305-4021, USA
nielsen@leland.stanford.edu, roth@flamingo.stanford.edu

Abstract. This paper presents a method for solving the direct kinematics of general 6-5 Stewart-Gough platform mechanisms. The method relies on elimination techniques to reduce a system of equations to a single univariate polynomial. It is shown that, once trivial solutions are eliminated, this polynomial is of degree 40. The resulting computations are simpler than those currently required by the analogous algorithmic method for the general 6-6 configuration.

1. Introduction

Recently, an algorithmic solution for the direct kinematics of the general Stewart-Gough in-parallel platform mechanism was developed by Husty [1]. In his paper Husty also gives an extensive bibliography of papers dealing with special geometries of this mechanism.

Here we consider the case which, although a special geometry, is very close to the general configuration. This mechanism consists of a fixed base link and a moving platform link connected by six legs, Figure 1. Each of the six legs can change in length, and it is this change in length that moves the platform relative to the base. The legs are connected to the base and the platform with spherical joints. Our configuration is called a 6-5 mechanism because in one member (the base in Figure 1) the six spherical joints can be chosen completely arbitrarily, while in the other member (the moving

J. Lenarčič and V. Parenti-Castelli (eds.), Recent Advances in Robot Kinematics, 7-16.
© 1996 *Kluwer Academic Publishers. Printed in the Netherlands.*

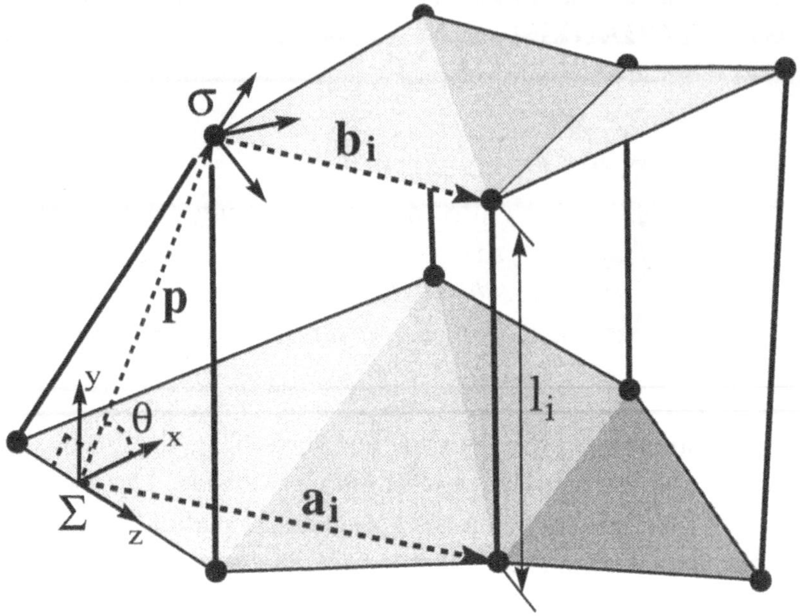

Figure 1. The 6-5 Stewart-Gough Platform.

platform in Figure 1) they need to be chosen so that two of them coincide, and there are effectively only five spherical joints.

In this paper we show that for the general 6-5 mechanism, when the six leg lengths are specified, there are at most 40 possible ways to assemble the mechanism. This number 40 is the same that exists for the most general 6-6 mechanism, however, because one of the platforms has only five joints, the algorithm which is developed in this paper is simpler than the one presently known for the 6-6 case.

2. Problem Statement

Figure 1 shows a 6-5 Stewart-Gough platform with coordinate frames Σ and σ attached to the fixed and moving platform, respectively. Note that the joints in both of these platforms are arbitrarily distributed in three-dimensional space. A triangle is formed by the two actuation legs which meet in a single point on the moving platform; the position of this triangle is determined by a single angle θ with respect to the fixed platform. This fact will prove to be critical in the solution of the forward kinematics problem.

Without loss of generality, the frame Σ is attached to the fixed platform so that the z-axis points along the line between the two fixed spherical joints on the triangle described above. The origin of this coordinate system is chosen on the z-axis at the foot of the perpendicular from the shared joint on the moving platform. The orientation of the x- and y-axes are arbitrary.

A coordinate system σ, with arbitrary orientation, is attached to the moving platform so that its origin is located at the shared spherical joint. The centers of the joints connecting the four legs, not on the triangle, are located by position vectors $\mathbf{a}_i$ in Σ and $\mathbf{b}_i$ in σ. In the forward kinematics problem, the vectors $\mathbf{a}_i$ and $\mathbf{b}_i$ are known, as well as the length l_i of each leg. Given these parameters, to solve the forward kinematics problem, one must calculate the position and orientation of the moving system σ with respect to the fixed system Σ.

The equations describing the forward kinematics problem may be derived as follows: we begin by finding the vector along each of the four legs not included in the triangle described above. The square of the magnitude of these vectors must equal the square of the specified lengths l_i. Let R represent the 3×3 orthogonal matrix describing the orientation of σ in Σ, and $\mathbf{p}$ (in Σ) the position vector of the origin of σ. Then for the ith leg we have

$$(\mathbf{p} + R\mathbf{b}_i - \mathbf{a}_i)^{\mathrm{T}} (\mathbf{p} + R\mathbf{b}_i - \mathbf{a}_i) = l_i^2 \tag{1}$$

Equation (1), with i running from 1 to 4, fully describes the 6-5 Stewart-Gough platform.

The vector $\mathbf{p}$, in Σ, lies in the xy plane at an angle θ from the x-axis. Its length p can be found given the length and location of the two legs making up the pivoting triangle. The vector $\mathbf{p}$, in the coordinate frame Σ, then, is given by

$$\mathbf{p} = \begin{bmatrix} p\cos\theta \\ p\sin\theta \\ 0 \end{bmatrix} \tag{2}$$

Using the tangent half-angle substitution, $\mathbf{p}$ may be written as

$$\mathbf{p} = \begin{bmatrix} p\frac{1-t^2}{1+t^2} \\ p\frac{2t}{1+t^2} \\ 0 \end{bmatrix} \tag{3}$$

where $t = \tan\frac{\theta}{2}$.

We express the unknown matrix elements in matrix R in terms of Rodrigues parameters [2]:

$$R = \Delta^{-1}\begin{bmatrix} 1+c_1^2-c_2^2-c_3^2 & 2(c_1c_2-c_3) & 2(c_1c_3+c_2) \\ 2(c_2c_1+c_3) & 1-c_1^2+c_2^2-c_3^2 & 2(c_2c_3-c_1) \\ 2(c_3c_1-c_2) & 2(c_3c_2+c_1) & 1-c_1^2-c_2^2+c_3^2 \end{bmatrix}$$

$$\text{where} \quad \Delta = 1+c_1^2+c_2^2+c_3^2 \tag{4}$$

This gives the matrix R explicitly in terms of the three parameters c_1, c_2, and c_3.

Expanding (1) and simplifying the result, we obtain the following four equations which represent the forward kinematics problem:

$$p^2 + 2\mathbf{p}^{\mathrm{T}}R\mathbf{b}_i - 2\mathbf{a}_i^{\mathrm{T}}R\mathbf{b}_i - 2\mathbf{p}^{\mathrm{T}}\mathbf{a}_i + \mathbf{b}_i^{\mathrm{T}}\mathbf{b}_i + \mathbf{a}_i^{\mathrm{T}}\mathbf{a}_i = l_i^2 \quad i = 1,2,3,4 \tag{5}$$

Now by substituting (4) and (3) into (5), then multiplying through by $(1+t^2)$ and $(1+c_1^2+c_2^2+c_3^2)$ to clear the denominator, we obtain:

$$\sum_{e=0,1,2;q,r=0,1,2,3} d_{eqri}t^e c_q c_r = 0 \qquad i = 1,2,3,4 \tag{6}$$

where the d_{eqri} are coefficients in the complex field, and c_0 is a homogenizing variable with respect to the variable group $\{c_1, c_2, c_3\}$ which may be considered explicitly equal to 1 (assuming the rotation represented by R is not a half-turn).

We will find that the system of equations in (6) actually has an infinite number of solutions. There is, however, only a finite set of values, for the tangent-of-the-half-angle variable t, which corresponds to solutions for the entire system. Through elimination of the remaining variables we may obtain a univariate polynomial in t of finite degree. Moreover, only 40 solutions in this finite set are not at $t = \pm i$. The number 40 has been shown previously to be the correct number of solutions for a 6-5 Stewart-Gough platform [3].

To see why this set of equations has an infinity of solutions, we begin by substituting (3) in (5), then clearing the denominator. This leaves us with the following set of equations:

$$\begin{aligned}\left(1+t^2\right)p^2 + 2p\left[\ (1-t^2) \quad 2t \quad 0\ \right]R\mathbf{b}_i - \\ 2\left(1+t^2\right)\mathbf{a}_i^{\mathrm{T}}R\mathbf{b}_i - 2p\left[\ (1-t^2) \quad 2t \quad 0\ \right]\mathbf{a}_i + \\ \left(1+t^2\right)\mathbf{b}_i^{\mathrm{T}}\mathbf{b}_i + \left(1+t^2\right)\mathbf{a}_i^{\mathrm{T}}\mathbf{a}_i = \left(1+t^2\right)l_i^2 \\ i = 1,2,3,4 \end{aligned} \tag{7}$$

If $t = i$, then this set of equations becomes:

$$2p\begin{bmatrix} 2 & 2i & 0 \end{bmatrix} R\mathbf{b}_i - 2p\begin{bmatrix} 2 & 2i & 0 \end{bmatrix}\mathbf{a}_i = 0 \quad i = 1,2,3,4 \tag{8}$$

We now introduce the Rodrigues parameters by substituting (4) in (8) above, then clearing the denominator. This leads to the following set of equations:

$$2p\begin{bmatrix} 2 & 2i & 0 \end{bmatrix}\begin{bmatrix} 1+c_1^2-c_2^2-c_3^2 & 2(c_1c_2-c_3) & 2(c_1c_3+c_2) \\ 2(c_2c_1+c_3) & 1-c_1^2+c_2^2-c_3^2 & 2(c_2c_3-c_1) \\ 2(c_3c_1-c_2) & 2(c_3c_2+c_1) & 1-c_1^2-c_2^2+c_3^2 \end{bmatrix}\mathbf{b}_i - 2p\left(1+c_1^2+c_2^2+c_3^2\right)\begin{bmatrix} 2 & 2i & 0 \end{bmatrix}\mathbf{a}_i = 0 \quad i = 1,2,3,4 \tag{9}$$

If we now set $c_1 = s$, where s is any number in the complex field, and $c_2 = si$ and $c_3 = i$, then the second term in (9) goes to zero, since $(1 + c_1^2 + c_2^2 + c_3^2) = 0$. Equation (9) now looks like:

$$2p\begin{bmatrix} 2 & 2i & 0 \end{bmatrix}\begin{bmatrix} 2\,(s^2+1) & 2i\,(s^2-1) & 4is \\ 2i\,(s^2+1) & -2\,(s^2-1) & -4s \\ 0 & 0 & 0 \end{bmatrix}\mathbf{b}_i = 0 \quad i = 1,2,3,4 \tag{10}$$

Simple expansion shows that (10) is satisfied for all values of s. We have shown, then, that an infinity of solutions exists for the set of equations in (6). Similarly, there is another infinity of solutions corresponding to $t = -i$. Experimentation with Gröbner bases and other methods, however, has shown that there is only a finite set of values for t which correspond to solutions to (6); this set includes $t = \pm i$, of course. In eliminating variables to find a solution to the system, we choose to pursue a univariate polynomial in t. This will reveal the 40 solutions which are of concern in the forward kinematics problems, in addition to solutions at $t = \pm i$.

3. Obtaining the 40th Degree Polynomial

To obtain our solution we will eliminate the c_i from (6) and solve for the non-trivial values of t. The method we use follows a technique developed by Macaulay [4]:

After suppressing the variable t, equations (6) may be rewritten as

$$f_i = A_{1_i}c_0^2 + B_{1_i}c_1 + C_{1_i}c_2 + D_{1_i}c_3 \qquad i = 1,2,3,4 \tag{11}$$

where A_{1_i} contains the suppressed variable t but none of the c_i, and B_{1_i}, C_{1_i}, D_{1_i} all contain the suppressed variable and in addition are linear in the c_i. This set of equations may be rewritten in matrix form:

$$\begin{bmatrix} A_{1_1} & B_{1_1} & C_{1_1} & D_{1_1} \\ A_{1_2} & B_{1_2} & C_{1_2} & D_{1_2} \\ A_{1_3} & B_{1_3} & C_{1_3} & D_{1_3} \\ A_{1_4} & B_{1_4} & C_{1_4} & D_{1_4} \end{bmatrix} \begin{bmatrix} c_0^2 \\ c_1 \\ c_2 \\ c_3 \end{bmatrix} = \begin{bmatrix} 0 \\ 0 \\ 0 \\ 0 \end{bmatrix} \tag{12}$$

Taking the determinant of this matrix and setting it equal to zero, we obtain the equation $g_1 = 0$, which is 8th degree in the suppressed variable t and cubic in $\{c_0, c_1, c_2, c_3\}$. Similarly, we may write:

$$f_i = A_{2_i}c_0 + B_{2_i}c_1^2 + C_{2_i}c_2 + D_{2_i}c_3 \qquad i = 1,2,3,4 \tag{13}$$

$$f_i = A_{3_i}c_0 + B_{3_i}c_1 + C_{3_i}c_2^2 + D_{3_i}c_3 \qquad i = 1,2,3,4 \tag{14}$$

$$f_i = A_{4_i}c_0 + B_{4_i}c_1 + C_{4_i}c_2 + D_{4_i}c_3^2 \qquad i = 1,2,3,4 \tag{15}$$

and, after putting each of these three sets of equations into a separate matrix form, we may obtain three other equations, $g_2 = 0$, $g_3 = 0$, $g_4 = 0$, respectively from (13), (14), and (15), which are 8th degree in the suppressed variable.

We may produce sixteen more cubic equations in $\{c_0, c_1, c_2, c_3\}$ by multiplying f_1, f_2, f_3, f_4 by each of $\{c_0, c_1, c_2, c_3\}$. With $g_1 = 0$, $g_2 = 0$, $g_3 = 0$, $g_4 = 0$ and these sixteen new equations, we have a total of twenty equations which are cubic in $\{c_0, c_1, c_2, c_3\}$. This system may be written as

$$M\mathbf{u} = 0 \tag{16}$$

where $\mathbf{u}$ is a twenty element vector containing all power products of the form $v_1v_2v_3$ where $v_i \in \{c_0, c_1, c_2, c_3\}$. This matrix contains sixteen rows which are quadratic and four rows which are of 8th degree in the suppressed variable; taking the determinant we expect at most a 64th degree polynomial in the suppressed variable.

Numerical examples show that the determinant indeed does give a 64th degree polynomial in t, and that 24 of the roots of this polynomial are at $t = \pm i$. These trivial roots may be factored out of the characteristic polynomial, leaving a 40th degree polynomial whose roots correspond to solutions of the direct kinematics problem.

Table 1 gives a set of platform parameters for a specific numerical example solved using the method described herein. After solving the characteristic polynomial for t, the c_i were calculated from the linear system, for

TABLE 1. **Platform parameters for a numerical example**

i	Base coordinates $\mathbf{a}_i$	Platform coordinates $\mathbf{b}_i$	Leg lengths l_i
1	$(2.0, -1.0, 4.0)$	$(0.0, 1.0, 3.0)$	8.621
2	$(4.0, 1.0, 2.0)$	$(4.0, -1.0, 4.0)$	3.858
3	$(5.0, 0.0, 0.0)$	$(6.0, 1.0, 0.0)$	6.870
4	$(1.0, -1.0, -3.0)$	$(4.0, 1.0, -3.0)$	9.324
5	$(0.0, 0.0, -1.0)$	$(0.0, 0.0, 0.0)$	7.071
6	$(0.0, 0.0, 2.0)$	$(0.0, 0.0, 0.0)$	7.280

$\mathbf{u}$, which results after the substitution of t into (16). (By setting $c_0 = 1$, we obtain a unique value for c_1, c_2, and c_3, corresponding to each set of ratios of the components of $\mathbf{u}$.) The final 40 solutions for this numerical example are presented in Table 2.

Four of the calculated solutions were real. These four possible assembly configurations are shown in Figure 2.

4. Conclusion

The special geometry of the 6-5 platform mechanism, where the joints in each body all lie in one plane, was studied by Liang *et al.* [5]. They showed that the direct kinematics problem has as many as 40 solutions. However, the special planar geometry, of the fixed and moving platforms, allows the solution to be expressed in terms of a 20th degree polynomial. Our results indicate that when the spherical joints are arbitrarily located the resulting polynomial remains of degree 40, just as for the general 6-6 geometry.

One of the interesting aspects of this problem is the fact, that in terms of some of the variables, there exists an infinite set of solutions. This implies that in the projective 5-space formed by t, c_0, c_1, c_2, and c_3, there exists two subspaces of solutions (one for $t = +i$, and one for $t = -i$) normal to the t-axis. Because of this, choosing t as the suppressed variable allows for determining the finite solution set that is of practical interest in this problem.

TABLE 2. **The 40 solutions to a 6-5 Stewart-Gough Platform Numerical Example.** *Each number is represented by a real and imaginary part. The 24 trivial solutions with* $t = \pm i$ *are not shown.*

t	c_1	c_2	c_3
(0.4169, 0.0000)	(2.5581, 0.0000)	(-9.7716, 0.0000)	(3.6308, 0.0000)
(1.0365, 0.0000)	(-6.6302, 0.0000)	(2.6104, 0.0000)	(-5.0815, 0.0000)
(1.0000, 0.0000)	(0.1000, 0.0000)	(0.0000, 0.0000)	(-0.1000, 0.0000)
(1.1900, 0.0000)	(0.1548, 0.0000)	(-0.0195, 0.0000)	(-0.0810, 0.0000)
(-1.7341, 0.9491)	(-1.5792, 0.1092)	(-0.0092, 0.2384)	(-0.3275, 0.4269)
(-1.7341,-0.9491)	(-1.5792,-0.1092)	(-0.0092,-0.2384)	(-0.3275,-0.4269)
(0.4849, 1.4560)	(-0.0309, 0.1386)	(0.2545,-0.2918)	(0.0023, 0.6912)
(0.4849,-1.4560)	(-0.0309,-0.1386)	(0.2545, 0.2918)	(0.0023,-0.6912)
(-0.5277, 1.5459)	(-0.3553, 0.0665)	(0.4556, 0.1727)	(-0.1429, 0.8919)
(-0.5277,-1.5459)	(-0.3553,-0.0665)	(0.4556,-0.1727)	(-0.1429,-0.8919)
(-0.7316, 1.3720)	(0.5323, 1.1996)	(-0.5162, 0.7398)	(0.1668, 0.6052)
(-0.7316,-1.3720)	(0.5323,-1.1996)	(-0.5162,-0.7398)	(0.1668,-0.6052)
(0.0052, 1.5560)	(-1.7913, 0.9728)	(-0.7069,-2.2327)	(1.4482, 0.1169)
(0.0052,-1.5560)	(-1.7913,-0.9728)	(-0.7069, 2.2327)	(1.4482,-0.1169)
(-0.9036, 1.0417)	(5.2987, 1.3911)	(1.3806, 0.0300)	(2.2292, 1.2834)
(-0.9036,-1.0417)	(5.2987,-1.3911)	(1.3806,-0.0300)	(2.2292,-1.2834)
(-0.5854, 1.0091)	(-0.1158,-0.5082)	(-0.0371,-0.5070)	(0.0873, 0.6493)
(-0.5854,-1.0091)	(-0.1158, 0.5082)	(-0.0371, 0.5070)	(0.0873,-0.6493)
(-0.0190, 1.1092)	(0.1623,-0.7895)	(0.3048,-0.0629)	(0.0542, 0.7777)
(-0.0190,-1.1092)	(0.1623, 0.7895)	(0.3048, 0.0629)	(0.0542,-0.7777)
(-0.5658, 0.2570)	(-0.5485,-0.6295)	(0.4886, 0.4447)	(-0.2814, 0.8401)
(-0.5658,-0.2570)	(-0.5485, 0.6295)	(0.4886,-0.4447)	(-0.2814,-0.8401)
(-0.0631, 0.9579)	(-0.2661, 2.0180)	(-1.9901,-0.4968)	(-0.5711, 0.9954)
(-0.0631,-0.9579)	(-0.2661,-2.0180)	(-1.9901, 0.4968)	(-0.5711,-0.9954)
(0.6968, 0.1028)	(0.3243, 0.2596)	(0.4446, 0.6775)	(0.1352,-0.4875)
(0.6968,-0.1028)	(0.3243,-0.2596)	(0.4446,-0.6775)	(0.1352, 0.4875)
(0.0627, 0.5707)	(-0.2758, 1.1304)	(-0.8992,-0.2741)	(-0.5250, 0.3782)
(0.0627,-0.5707)	(-0.2758,-1.1304)	(-0.8992, 0.2741)	(-0.5250,-0.3782)
(0.1218, 0.3088)	(1.8197, 0.5092)	(-1.7703, 3.9456)	(-1.6574, 0.0469)
(0.1218,-0.3088)	(1.8197,-0.5092)	(-1.7703,-3.9456)	(-1.6574,-0.0469)
(-0.0435, 0.2699)	(-0.4173, 1.2824)	(-0.7677, 0.4665)	(-0.7660, 1.0197)
(-0.0435,-0.2699)	(-0.4173,-1.2824)	(-0.7677,-0.4665)	(-0.7660,-1.0197)
(-0.0354, 0.1464)	(-0.3135,-0.6313)	(0.1533,-0.0012)	(-0.0022, 0.6681)
(-0.0354,-0.1464)	(-0.3135, 0.6313)	(0.1533, 0.0012)	(-0.0022,-0.6681)
(0.0861, 0.1735)	(0.8731, 2.4187)	(-4.3534, 1.6078)	(0.1849,-1.5173)
(0.0861,-0.1735)	(0.8731,-2.4187)	(-4.3534,-1.6078)	(0.1849, 1.5173)
(0.1384, 0.0159)	(-1.1740, 0.3509)	(-0.2157, 1.7706)	(-1.4819, 1.0033)
(0.1384,-0.0159)	(-1.1740,-0.3509)	(-0.2157,-1.7706)	(-1.4819,-1.0033)
(0.0806, 0.0183)	(-1.0129, 1.1973)	(0.7387, 1.5584)	(-1.0587, 1.7385)
(0.0806,-0.0183)	(-1.0129,-1.1973)	(0.7387,-1.5584)	(-1.0587,-1.7385)

Figure 2. Four Real Solutions to the Numerical Example.

5. Acknowledgment

The financial support of the National Science Foundation is gratefully acknowledged.

References

1. M.L. Husty, "An Algorithm for Solving the Direct Kinematics of General Stewart-Gough Platforms," *Mech. Mach. Theory,* vol. 31, no. 4, pp. 365-379, 1996.
2. O. Bottema and B. Roth, *Theoretical Kinematics*, pp. 147–148, North Holland, 1979 (reprinted by Dover Publications, NY, 1990).
3. D. Lazard and J.C. Faugère, "The combinatorial classes of parallel manipulators," 1994, to appear in *Mech. Mach. Theory.*
4. F.S. Macaulay, "On Some Formulae in Elimination," *Proc. London Mathematical Society,* pp. 3-27, May 1902.
5. C. Liang, L. Han, and F. Wen, "Forward Displacement Analysis of the 5-6 Stewart Platforms," *Proceedings of the Ninth World Congress on the Theory of Machines and Mechanisms,* Milan, Italy, vol. 1, pp. 184-187, 1995.

THE KINEMATIC SENSITIVITIES OF REDUNDANT ROBOTIC MANIPULATORS

Department of Mechanical Engineering & Centre for Intelligent Machines
McGill University, 817 Sherbrooke Street West, Montreal, Quebec, Canada H3A 2K6
angeles@cim.mcgill.ca, bulca@cim.mcgill.edu, noga@cim.mcgill.ca,
mordi@cim.mcgill.ca

Abstract. The kinematic sensitivity of a manipulator is the effect that a 'small' change in the required task produces on the posture of the manipulator. Redundant manipulators being capable of performing both a primary and a secondary task admit, correspondingly, two kinematic sensitivities. We refer to them as the primary and the secondary sensitivities, each being associated with a *sensitivity matrix*. The primary sensitivity matrix turns out to be the partial derivative of the joint rates with respect to the twist of the end-effector, the secondary sensitivity matrix being defined as the partial derivative of the joint rates with respect to the secondary task. It is shown that the primary sensitivity can be minimized by design, while the secondary sensitivity is constant.

1 Introduction

The concept of *kinematic sensitivity* has been introduced in the robotics literature to evaluate and compare the performance of manipulators under small changes in the task at hand. The kinematic sensitivity of non-redundant manipulators depends on both their design parameters and their posture, as has been shown in the literature. However, redundant manipulators admit two sensitivities, which are associated with the primary and the secondary tasks. We will refer to the former as the primary sensitivity, to the latter as the secondary sensitivity. One should pay attention to the sensitivity of a manipulator under development at the early stages of the design process in order to design a kinematically robust manipulator.

Most of the research on sensitivity reported in the literature has dealt with the numerical stability of the system $\mathbf{A}\mathbf{x} = \mathbf{b}$, where $\mathbf{A}$ is an $n \times n$ matrix and $\mathbf{x}$ and $\mathbf{b}$ are n-dimensional vectors. A literature review on the sensitivity of the foregoing equations reveals some recent publications in connection with the solution of the Poisson equation (Chan and Foulser, 1988) and with polynomial regressions (Winkler, 1994). The research in the area of robotics has been divided so far into two groups: kinematic sensitivity and dynamic sensitivity. Togai (1985) proposed qualitative measures for manipulability and kinematic sensitivity. Kaizerman et al. (1994) introduced an inverse kinematics calibration procedure using a sensitivity analysis. Vuskovic (1989) reported on two approaches for the compensation of the position and orientation errors due to variations of the geometric parameters of manipulators. Moreover, Wiens et al. (1989 & 1992), and Vesovic and Vukobratovic (1994) dealt with the dynamic sensitivity of manipulators.

Manipulators that have more joints than the dimension of their task space are called kinematically redundant, in order to distinguish them from redundantly actuated manipulators. In this paper we will not be concerned with the latter, and so, we will refer to the for-

J. Lenarčič and V. Parenti-Castelli (eds.), Recent Advances in Robot Kinematics, 17-26.
© 1996 *Kluwer Academic Publishers. Printed in the Netherlands.*

mer as *redundant manipulators* for brevity. Redundant manipulators are attractive because of the advantages that they offer in singularity and obstacle avoidance, which makes them more dextrous than their nonredundant counterparts. This being the case, it is noteworthy that no reference has been made to redundant manipulators in the sensitivity analysis reported so far. The motivation behind this paper lies in the need of sound design criteria for redundant manipulators envisioned for space applications. Canada is in charge of the development and commissioning of the Special-Purpose Dextrous Manipulator (SPDM) and the Remote Manipulator System (RMS), to operate on the International Space Station. As such, Canada has embarked on a major project, the Strategic Technologies for Autonomous Robotics, or STEAR for brevity. This study was motivated to a great extent by the STEAR Project.

In this paper, we introduce two kinematic-sensitivity measures, the primary sensitivity and the secondary sensitivity. After the background of Section 2, we derive the sensitivity measures in Section 3. Then, we give a few examples of different redundant and nonredundant manipulators in Section 4 and, finally, draw some conclusions.

2 Preliminaries

The Jacobian matrix of a general n-axis manipulator takes the form (Whitney, 1972)

$$\mathbf{J} = [\ \mathbf{j}_1 \quad \mathbf{j}_2 \quad \cdots \quad \mathbf{j}_n\ ] \tag{1}$$

For revolute (R) and prismatic (P) joints, the 6-dimensional ith column of $\mathbf{J}$, $\mathbf{j}_i$ ($i = 1, \ldots, n$), is given, correspondingly, as

$$R: \quad \mathbf{j}_i \equiv \begin{bmatrix} \mathbf{e}_i \\ \mathbf{e}_i \times \mathbf{r}_i \end{bmatrix} \qquad P: \quad \mathbf{j}_i \equiv \begin{bmatrix} \mathbf{0} \\ \mathbf{e}_i \end{bmatrix} \tag{2}$$

where $\mathbf{e}_i$ is the unit vector parallel to the axis of the ith joint and $\mathbf{r}_i$ is the vector directed from any point on the joint axis to the operation point of the end-effector (EE). Henceforth, we limit ourselves to revolute joints.

As is commonly known, the Jacobian matrix $\mathbf{J}$ of a manipulator maps the joint rates, grouped in the n-dimensional vector $\dot{\boldsymbol{\theta}}$, into the 6-dimensional EE twist array $\mathbf{t}$, namely,

$$\mathbf{J}\dot{\boldsymbol{\theta}} = \mathbf{t} \tag{3}$$

with $\mathbf{t}$ defined as

$$\mathbf{t} \equiv \begin{bmatrix} \boldsymbol{\omega} \\ \mathbf{v} \end{bmatrix} \tag{4}$$

where $\boldsymbol{\omega}$ and $\mathbf{v}$ denote the angular velocity of the EE and the velocity of a point of the same body, termed the *operation point*, respectively.

Now, let us recall that the ith column of $\mathbf{J}$ represents the Plücker coordinates of the ith joint axis. Thus, the first three coordinates are dimensionless, while the last three have units of length. This dimensional inhomogeneity prevents us from evaluating the condition number of the Jacobian. In order to achieve a dimensionally homogeneous $\mathbf{J}$, and be able to evaluate its condition number, the *characteristic length* was introduced by Ranjbaran, Angeles, González-Palacios, and Patel (1995). Upon dividing the last three entries of each column of **J** by the characteristic length, one obtains a dimensionally homogeneous Jacobian, and hence, the associated condition number becomes meaningful. It is recalled that the *condition number* of **J** is defined as the product of the norm of **J** by the norm of $\mathbf{J}^{-1}$ and is, hence, norm-dependent. Based on the 2-norm (maximum singular value), then, the condition number is the ratio of the largest to the smallest singular values of **J** or, correspondingly, the square root of the ratio of the largest to the smallest eigenvalues of $\mathbf{J}\mathbf{J}^T$, where the superscript T denotes transposition. If, on the other hand, we use the Frobenius norm (square root of $1/m$ times the trace of $\mathbf{J}\mathbf{J}^T$), then the condition number of **J** becomes

$$\kappa_F(\mathbf{J}) = \frac{1}{m}\sqrt{\mathrm{tr}(\mathbf{J}\mathbf{J}^T)\mathrm{tr}[(\mathbf{J}\mathbf{J}^T)^{-1}]} \tag{5}$$

were m stands for the dimension of the task space, which is 3 for position tasks and 6 for position and orientation tasks, but other dimensions are possible.

A matrix is said to be *isotropic* when all its singular values are nonzero and identical. Furthermore, a manipulator whose associated Jacobian matrix is isotropic at certain postures is called *isotropic*. When the Jacobian matrix **J** is isotropic, then

$$\mathbf{J}\mathbf{J}^T = \alpha^2\mathbf{1} \tag{6}$$

in which α is the multiple nonzero singular value of **J** and **1** is the 6×6 identity matrix.

3 Sensitivity Analysis

Our understanding of sensitivity is the effect that a 'small' change in the assigned task produces on the posture of the manipulator, i.e, on its joint variables. Therefore, we define the sensitivity matrix as *the ratio, in a large sense, of the change in the joint variables to a small change in the given task.*

Let us multiply both sides of eq.(3) by a small time-increment Δt and substitute

$$\Delta\boldsymbol{\theta} = \dot{\boldsymbol{\theta}}\Delta t, \quad \Delta\mathbf{s} = \mathbf{t}\Delta t \tag{7}$$

where $\Delta\mathbf{s}$ denotes a 'small' change in the given task, i.e., in the EE pose. We have, therefore,

$$\mathbf{J}\Delta\boldsymbol{\theta} = \Delta\mathbf{s} \tag{8}$$

3.1 NONREDUNDANT MANIPULATORS

Now, let us consider a nonredundant manipulator with $n = m$. Under the assumption that **J** is nonsingular, the sensitivity matrix **S** of such a manipulator is now formally defined as

$$\mathbf{S} \equiv \frac{\partial\Delta\boldsymbol{\theta}}{\partial\Delta\mathbf{s}} = \mathbf{J}^{-1} \tag{9}$$

which can also be expressed as

$$\mathbf{S} \equiv \frac{\partial\dot{\boldsymbol{\theta}}}{\partial\mathbf{t}} \tag{10}$$

We thus consider $\mathbf{J}^{-1}$ as the kinematic sensitivity matrix of a given manipulator. Any norm of $\mathbf{J}^{-1}$ gives a scalar that we term the sensitivity σ of the manipulator. Note that $\mathbf{J}$ has been assumed dimensionless by using the concept of characteristic length, and hence, a norm of $\mathbf{J}$ is physically meaningful.

If we use the Frobenius norm, for example, then,

$$\sigma \equiv \|\mathbf{J}^{-1}\|_F \equiv \sqrt{\frac{1}{m}\operatorname{tr}(\mathbf{J}^{-1}\mathbf{J}^{-T})} = \sqrt{\frac{1}{m}\operatorname{tr}[(\mathbf{J}\mathbf{J}^T)^{-1}]} = \sqrt{\frac{1}{m}\sum_{i=1}^{m}\frac{1}{{\alpha_i}^2}} \tag{11}$$

where $\mathbf{J}^{-T}$ stands for the inverse of the transpose of $\mathbf{J}$ and α_i denotes the i-th singular value of $\mathbf{J}$. Therefore, under the Frobenius norm, the sensitivity is simply the root-mean square value of the reciprocals of the singular values of $\mathbf{J}$.

3.2 REDUNDANT MANIPULATORS

In the case of redundant manipulators, the EE pose is not sufficient to determine the manipulator posture, and hence, a secondary task is introduced. Usually, this task derives from a minimality condition imposed in addition to the primary task, associated with the EE twist. The process of calculating a specific value of the joint variables (posture), or their time-rates of change, is referred to as redundancy resolution. Under the assumption that $\mathbf{J}$ is of full rank (the manipulator is at a nonsingular configuration), the redundancy-resolution algorithm minimizes an objective function $z(\Delta\boldsymbol{\theta})$ subject to the condition of eq.(8), i.e., $\Delta\boldsymbol{\theta}$ is calculated as the solution of an optimization problem, namely,

$$\min_{\Delta\boldsymbol{\theta}} z(\Delta\boldsymbol{\theta}) \tag{12}$$

subject to condition (8).

The objective function is most commonly given in the form

$$z(\Delta\boldsymbol{\theta}) = \frac{1}{2}\Delta\boldsymbol{\theta}^T\mathbf{W}\Delta\boldsymbol{\theta} \tag{13}$$

where $\mathbf{W}$ is a positive-definite weighting matrix. In the case at hand, $\Delta\boldsymbol{\theta}$ consists of two parts, namely, a part that depends on the pose change, and one that depends on the gradient of the objective function z, i.e.,

$$\Delta\boldsymbol{\theta} = \mathbf{J}^{\dagger}\Delta\mathbf{s} + \mathbf{J}^{\ddagger}\nabla z \tag{14}$$

where $\mathbf{J}^{\dagger}$ is the applicable generalized inverse of $\mathbf{J}$ and $\mathbf{J}^{\ddagger}$ is an orthogonal complement of $\mathbf{J}$ and, moreover, is a *projection* of ∇z onto the nullspace of $\mathbf{J}$. Here, we term $\Delta\mathbf{s}$ and ∇z the primary task and the secondary task, respectively.

The first term of eq.(14) thus depends on the design parameters, the manipulator posture, and the primary task, while the second term is apparently dependent on the design parameters, the manipulator posture and the secondary task. We thus have two sensitivity matrices, the *primary sensitivity matrix* $\mathbf{S}_p$ and the *secondary sensitivity matrix* $\mathbf{S}_s$, namely,

$$\mathbf{S}_p \equiv \frac{\partial\Delta\boldsymbol{\theta}}{\partial\Delta\mathbf{s}} = \mathbf{J}^{\dagger} \tag{15}$$

$$\mathbf{S}_s \equiv \frac{\partial \Delta \theta}{\partial \nabla z} = \mathbf{J}^{\ddagger} \tag{16}$$

Similar to the sensitivity definition given in eq.(11), the primary sensitivity σ_p and the secondary sensitivity σ_s of a redundant manipulator are defined as

$$\sigma_p = ||\mathbf{S}_p|| = ||\mathbf{J}^{\dagger}|| \tag{17a}$$

and

$$\sigma_s = ||\mathbf{S}_s|| \equiv ||\mathbf{1} - \mathbf{J}^T(\mathbf{J}\mathbf{J}^T)^{-1}\mathbf{J}|| \tag{17b}$$

with $\mathbf{1}$ denoting the $n \times n$ identity matrix. Now, in order to evaluate σ_s from the above expression, we find the norm of a *projection* $\mathbf{P}$ of a vector space $\mathcal{U}$ of dimension p onto a vector space $\mathcal{V}$ of dimension $q < p$. That is, if $\mathbf{x} \in \mathcal{U}$, then $\mathbf{y} \equiv \mathbf{P}\mathbf{x} \in \mathcal{V}$, but with $\mathcal{V}$ being a subspace of $\mathcal{U}$. Since a projection is represented by a square matrix, $\mathbf{P}$ is a $p \times p$ matrix and $\mathbf{x}$ and $\mathbf{y}$ are p-dimensional, but $\mathbf{y}$ lies entirely in the subspace $\mathcal{V}$ of $\mathcal{U}$. If we recall the basic definition of a matrix norm (Golub and Van Loan, 1989), then we can write

$$||\mathbf{P}|| \equiv \max_{\mathbf{x} \neq \mathbf{0}} \frac{||\mathbf{P}\mathbf{x}||}{||\mathbf{x}||} \tag{18}$$

Furthermore, $\mathbf{P}\mathbf{x}$ is restricted to a subspace of $\mathcal{U}$, its norm, any norm in fact, thus being bound to be not greater than the corresponding norm of $\mathbf{x}$. Clearly, the maximum of the above norm ratio is attained when $\mathbf{x}$ itself lies in $\mathcal{V}$, in which case $\mathbf{P}\mathbf{x} = \mathbf{x}$, and hence,

$$||\mathbf{P}|| = 1 \tag{19}$$

As a consequence, then

$$\sigma_s = 1 \tag{20}$$

i.e., the secondary sensitivity is equal to unity, *regardless of the manipulator posture*. Therefore, the secondary sensitivity needs no further discussion.

On the other hand, the condition number of $\mathbf{J}$ is defined, in our case, as

$$\kappa(\mathbf{J}) = ||\mathbf{J}|| ||\mathbf{J}^{\dagger}|| \tag{21}$$

Hence,

$$\sigma_p \equiv ||\mathbf{J}^{\dagger}|| = \frac{\kappa(\mathbf{J})}{||\mathbf{J}||} \tag{22}$$

In eq.(22), note that $||\mathbf{J}|| > 0$, and hence, the division is always possible. Now, the denominator of σ_p undergoes small changes over the whole workspace of the manipulator, and attains its maximum where the manipulator is postured at its maximum reach. However, κ varies from 1 to ∞; as a consequence, σ_p is dominated by κ. Therefore, σ_p reaches its minimum value at postures of minimum condition number. In other words, the manipulator is least sensitive to changes in the EE pose at low values of its condition number.

3.2.1 *Isotropic Manipulators*

When an isotropic manipulator (Ranjbaran, Angeles, González-Palacios and Patel) is in its isotropic posture, $\kappa = 1$, and hence,

$$\sigma_p = \frac{1}{||\mathbf{J}||} \tag{23}$$

If we use the Frobenius norm in eq.(23), then,

$$||\mathbf{J}||_F = \sqrt{\frac{1}{m}\,\mathrm{tr}(\mathbf{J}\mathbf{J}^T)} \tag{24}$$

For position and orientation tasks, let us partition $\mathbf{J}$ into the two $3 \times n$ blocks $\mathbf{A}$ and $\mathbf{B}$:

$$\mathbf{J} \equiv \begin{bmatrix} \mathbf{A} \\ \mathbf{B} \end{bmatrix} \tag{25}$$

Then,

$$\begin{aligned} \frac{1}{6}\,\mathrm{tr}(\mathbf{J}\mathbf{J}^T) &= \frac{1}{6}\,\mathrm{tr}\left(\begin{bmatrix} \mathbf{A}\mathbf{A}^T & 0 \\ 0 & \mathbf{B}\mathbf{B}^T \end{bmatrix}\right) \\ &= \frac{1}{6}\left[\mathrm{tr}(\mathbf{A}\mathbf{A}^T) + \mathrm{tr}(\mathbf{B}\mathbf{B}^T)\right] = \frac{1}{3}\,\mathrm{tr}(\mathbf{A}\mathbf{A}^T) \end{aligned} \tag{26}$$

where we have used the equality of the two traces at the isotropic posture. Moreover, $\mathrm{tr}(\mathbf{A}\mathbf{A}^T) = n$ at this posture, and hence,

$$||\mathbf{J}||_F = \sqrt{\frac{n}{3}} \tag{27}$$

Thus, at the isotropic posture,

$$\sigma_p = \sqrt{\frac{3}{n}} \tag{28}$$

Note that eq.(28) holds also for the case of positioning tasks only. If the 2-norm is used instead, then,

$$||\mathbf{J}||_2 = \alpha \tag{29}$$

were α stands for the common singular value of $\mathbf{J}$. Moreover (Ranjbaran, Angeles, González-Palacios and Patel, 1995),

$$\alpha = \sqrt{\frac{n}{3}} \tag{30}$$

From the above discussion, it turns out that, at the isotropic posture, both the Frobenius norm and the 2-norm give the same sensitivity, which depends only on the number of joints of the manipulator.

3.2.2 *Nonisotropic Manipulators*

When the manipulator is not isotropic, or when an isotropic manipulator is not at its isotropic posture, then eq.(22) is used. If we use the Frobenius norm, then,

$$\sigma_p = \sqrt{\frac{1}{m}\operatorname{tr}[(\mathbf{J}\mathbf{J}^T)^{-1}]} = \sqrt{\frac{1}{m}\sum_{i=1}^{m}\frac{1}{\alpha_i{}^2}} \tag{31}$$

If we use the 2-norm instead, then,

$$\sigma_p = \frac{1}{\alpha_m} \tag{32}$$

where α_m denotes the smallest singular value of $\mathbf{J}$. These results hold also for nonredundant manipulators. Note that the primary sensitivity obtained with the 2-norm is greater than that obtained with the Frobenius norm and, therefore, it is important to specify the norm that has been adopted. In practice, it does not matter which norm is used, as long as the same is used throughout the whole study. However, notice that the Frobenius norm is simpler to calculate, as it requires only the trace of $(\mathbf{J}\mathbf{J}^T)^{-1}$; on the contrary, the 2-norm requires singular-value calculations. As a consequence, the Frobenius norm is to be preferred over the 2-norm in real-time calculations, e.g., for control purposes.

4 Sensitivities of Some Robotic Manipulators

Now we calculate the primary sensitivities of three redundant manipulators. Two of these manipulators, REDIESTRO 1 and the C3 Arm, were developed at McGill University's Centre for Intelligent Machines (CIM). Moreover, REDIESTRO 1 is supplied with seven joints, while the C3 Arm with only four, which renders the latter redundant for positioning tasks of its EE. For positioning and orientation tasks, the EE is supplied with a spherical wrist. The third robot is the Advanced Research Manipulator II (ARMII) with eight revolute joints (Williams, 1994). Moreover, REDIESTRO 1 and the C3 Arm are designed to be isotropic, their primary sensitivities being calculated at their isotropic posture in Table 1. The posture where ARMII attains the lowest condition number has been found and its primary sensitivity is calculated at this posture. It can be seen that the isotropic manipulators given below have the lowest primary sensitivity values, i.e., the effect of small changes in the EE pose leads to minimal effects on the joint variables. The sensitivity measure reported in this paper can be directly applied to nonredundant manipulators. The nonredundant manipulators used here are the popular Puma 560, the Yaskawa manipulator, and DIESTRO, a six-axis isotropic robot designed and manufactured at CIM. Again, the configurations at which the Puma and Yaskawa robots attain their lowest condition numbers are found and used in sensitivity calculations. The sensitivity of DIESTRO is found at its isotropic posture. It is worthwhile to note that isotropic manipulators achieve lower primary sensitivities as compared with their nonisotropic counterparts.

5 Conclusions

The concept of kinematic sensitivity, as pertaining to redundant manipulators, was discussed in this paper. Two sensitivities of redundant manipulators are found, which are

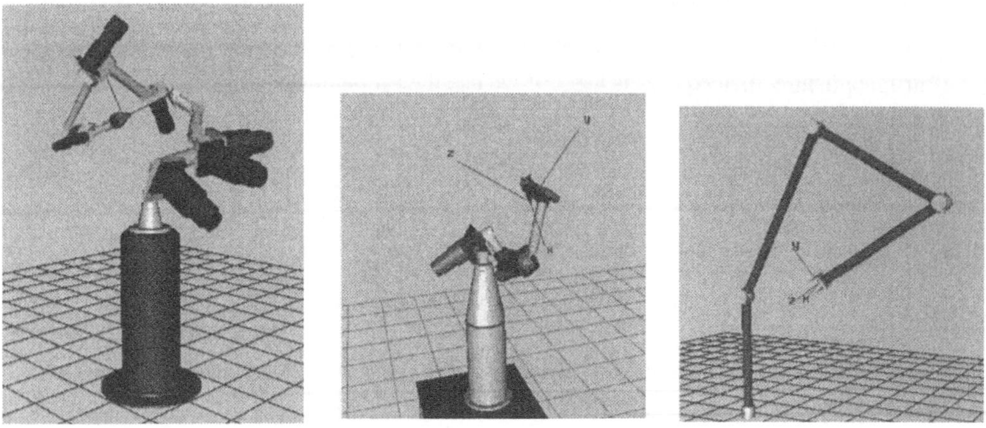

(a) (b) (c)

Figure 1: Examples of redundant manipulators: (a) REDIESTRO 1 (b) C3 Arm (c) ARMII

TABLE 1. Primary Sensitivities

Example	κ_2	$\|\|\mathbf{J}\|\|_2$	2-norm σ_d
Nonredundant manipulators:			
Puma 560 (Fig. 2a)	3.0629	2.3513	1.3026
Yaskawa (Fig. 2b)	3.1196	2.5700	1.2139
DIESTRO (Fig. 2c)	1.0000	1.4142	0.7071
Redundant manipulators:			
ARMII (Fig. 1c)	2.9632	2.2864	1.2960
C3 Arm (Fig. 1b)	1.0000	1.1547	0.8660
REDIESTRO 1 (Fig. 1a)	1.0000	1.5275	0.6547

referred to as primary and secondary, because redundant manipulators are capable of performing both a primary and a secondary task simultaneously. It is shown that the primary sensitivity depends on the design parameters and the posture of the manipulator, while the secondary sensitivity is constantly equal to unity, and hence, independent of design parameters and posture. Focusing on the primary sensitivity of redundant manipulators, this paper shows that isotropic manipulators have minimum sensitivities to small changes in the pose of the EE, i.e., the changes in the joint variables induced by changes in the pose of the EE are minimal. The primary sensitivity was obtained by using both the Frobenius norm and the 2-norm. It was shown that the former is simpler to compute than the latter. Moreover, at the isotropic posture, the primary sensitivity depends only on the number of joints of the manipulator. As expected, the primary sensitivity decreases as the number of joints increases. The kinematic sensitivities of six redundant and nonredundant manipulators were calculated.

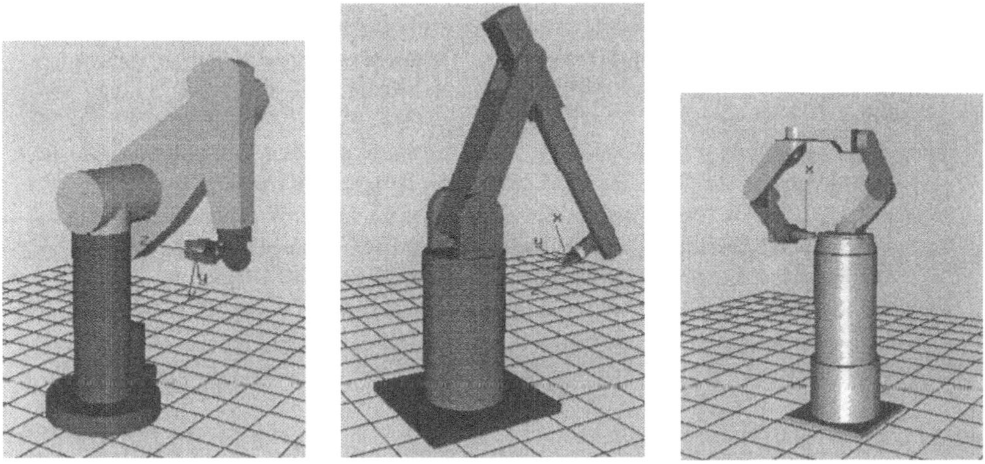

(a) (b) (c)

Figure 2: Examples of non-redundant manipulators: (a) Puma 560 (b) Yaskawa (c) DIESTRO

6 Acknowledgements

The research work reported here was made possible by a Collaborative Research and Development Grant from both the Canadian Natural Sciences and Engineering Research Council and the Canadair Division of Bombardier Inc. Dr. Alan Robins, of Bombardier's Canadair Division, is duly acknowledged for his insightful criticism to a first draft of this paper. The third and fourth authors acknowledge the support of McGill University through Eileen Peters and David Stewart McGill Major Fellowships.

References

Chan, T. F. and Foulser, D. E., November, 1988, "Effectively well-conditioned linear systems," *SIAM J. Sci. Stat. Comput.*, Vol. 9, No. 6, pp. 963–969.

Golub G.H. and Van Loan C.F., 1989, *Matrix Computations*, The Johns Hopkins University Press, Baltimore and London.

Kaizerman, S., Zak, G., Benhabib, B. and Fenton, R. G., 1994, "A sensitivity analysis based method for robot calibration," *ASME J. of Mechanical Design*, Vol. 116, pp. 607–613.

Ranjbaran, F., Angeles, J., González-Palacios, M.A and Patel, R.V, September 1995, " The mechanical design of a seven-axes manipulator with kinematic isotropy," *J. Intelligent and Robotic Systems*, Vol. 14, No.1, pp. 21–41.

Togai, M., 1985, "Manipulability and sensitivity for design and evaluation of industrial robots: kinematic consideration," *Proc. 15th International Symposium on Industrial Robots,* Tokyo, Sept. 11–13, Vol. 2, pp. 865–872.

Vesovic, N. and Vukobratovic, M., 1994, "Accuracy of manipulation robot dynamics," *Mechanism and Machine Theory*, Vol. 29, No. 3, pp. 415–425.

Vuskovic, M. I., 1989, "Compensation of kinematic errors using kinematic sensitivities," *Proc. IEEE Int. Conf. on Robotics and Automation*, May 14–19, Vol. 2, pp. 745–750.

Whitney, D. E., 1972, "The mathematics of coordinated control of prosthetic arms and manipulators," *ASME J. Dyn. Syst., Meas., and Control*, Series G, Vol. 94, No. 4, pp. 303–309.

Wiens, G. J., Scott, R. A. and Zarrugh, M. Y., 1989, "The role of inertia sensitivity in the evaluation of manipulator performance," *ASME J. Dyn. Syst., Meas., and Control*, Vol. 111, pp. 194–199.

Wiens, G. J., Zarrugh, M. Y. and Scott, R. A., 1992, "Effects of geometric parameters on manipulator dynamic performance," *ASME J. Mechanical Design*, Vol. 114, pp. 137–142.

Williams II, R. L., 1994, "Local performance optimization for a class of redundant eight-degree-of-freedom manipulatorms," NASA Technical Paper 3417, National Aeronautics and Space Administration, Langley Research Center, Hampton, Virginia 23681-0001.

Winkler J. R., 1994, "The Sensitivity of Linear Algebraic Equations," *Proc. 5th SIAM Conf. on Applied Linear Algebra*, Snowbird, Utah, June 15–18, pp. 279–283.

2. Control and optimisation

MODULAR DECOMPOSITION FOR OPTIMAL DYNAMIC DESIGN OF REDUNDANT MACRO/MINI MANIPULATORS

ALAN BOWLING AND OUSSAMA KHATIB
Robotics Laboratory
Department of Computer Science
Stanford University, Stanford, CA 94086

Abstract. This article investigates the design of redundant mechanisms for improved dynamic performance. Our study has resulted in a new methodology which allows the decomposition of the design problem into smaller, more manageable problems. With this method, a redundant mechanism is subdivided into a chain of non-redundant subsystems which can be designed separately. The result is a highly efficient design process.

1. Introduction

The study of manipulator dynamic performance involves issues dealing with how quickly the system responds to actuator commands and how it reacts to contact forces and moments. Our previous work in this area focused on the optimal design of non-redundant manipulators [2], [3] and resulted in an effective approach for optimizing their dynamic performance.

These studies were based on analysis of end-effector inertial and acceleration properties. For redundant manipulators however, the effect of the redundant degrees of freedom on the overall mechanism performance must be analyzed.

In addition to the performance requirements at the end-effector, redundant systems must also exhibit adequate dynamic performance in the manipulator null-space. The article discusses the effect of poor null-space performance on degrading the overall performance at the end-effector, and presents a new design methodology for optimizing the dynamics of redundant manipulators.

J. Lenarčič and V. Parenti-Castelli (eds.), Recent Advances in Robot Kinematics, 29-38.
© 1996 *Kluwer Academic Publishers. Printed in the Netherlands.*

The approach provides a decomposition of the overall design optimization problem into a set of smaller optimization sub-problems. Each of these problems corresponds to a structure obtained by subdivision of the mechanism into non-redundant mini and macro(s) subsystems.

The total search space explored when solving the sub-problems is greatly reduced from that required by a single overall optimization. The goal of the design is to obtain a mechanism with the smallest, most isotropic inertial properties and the largest, most isotropic acceleration capability at its end-effector, over the workspace.

2. Overview

Our analysis of inertial properties for macro/mini manipulator systems has shown that for all directions and configurations, the effective mass/inertia of a macro/mini manipulator is smaller than or equal to the inertia associated with the mini-manipulator structure, considered alone [1]. A more general statement of this result is that *the inertial properties of a redundant manipulator are bounded above by the inertial properties of the structure formed by the smallest distal set of degrees of freedom (DOF) that span the operational space.*

In this article we present an analysis of acceleration properties for macro/mini manipulator systems which shows that the acceleration capability of the mechanism, described by the end-effector *isotropic acceleration*, must be smaller than or equal to twice the *isotropic acceleration* of the mini's end-effector, considered alone. Here the *isotropic acceleration* is defined as the maximum acceleration achievable at the end-effector in or about every direction.

These results show the significant impact the incorporation of a fast lightweight mini-manipulator structure can have on the dynamic performance of a manipulator. It also shows that the dynamic performance of a redundant manipulator will be determined essentially by the performance of its last set of joints that provide full freedom for end-effector linear/angular motions. The structure formed by these degrees of freedom, act as a mini-manipulator for the redundant mechanism and will be referred to as such. If the structure formed by remaining links is non-redundant, it will be considered as a macro-manipulator. If this structure is redundant, then it will be again treated as macro/mini manipulator. The last link of each subsystem is then considered

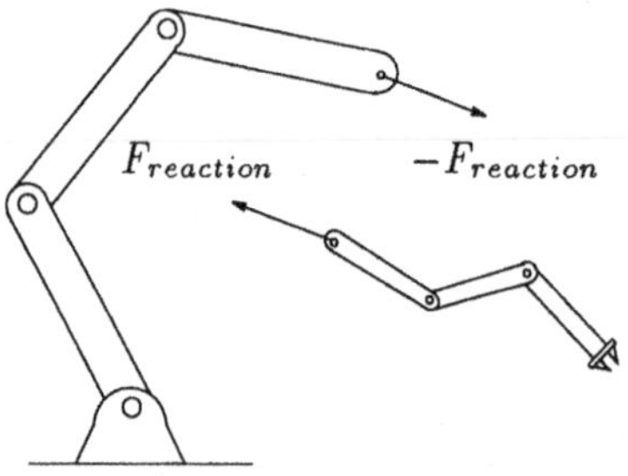

Figure 1. Decomposition.

to be that subsystem's end-effector. An example of this is shown for a six DOF planar manipulator in Figure 1.

With this decomposition, the mini structure will be designed separately to provide a desired load capacity and to achieve the desired inertial and acceleration properties. Once the mini design is accomplished the proceeding macro structure can be designed to complement that mini. This procedure is repeated to the final macro.

To design each of these structures separately, load requirements and desirable inertial and acceleration properties must be specified. Determining these specifications for the design(s) of the successive structures requires an analysis of the interactions between the different subsystems. The interactions should be analyzed in both directions, going from tip to base and from base to tip. To analyze the interactions from tip to base, i.e. mini to macro, the reaction forces at the base of each system are analyzed and the maximum reaction forces are passed to the next subsystem in the chain. These maximum reaction forces can be interpreted as the load on the macro end-effector. At this point a design constraint is imposed requiring the macro(s) to be capable of stabilizing its(their) end-effector(s) against the motion of the subsystems that follow in the chain.

For instance, a macro should be able to maintain zero acceleration at it's end-effector when the mini is accelerating at its isotropic accelerations. This process results in a lower bound on the acceleration capability required at the macros' end-effectors.

The interactions from base to tip can be analyzed from the kinematics of the overall mechanism. Here it is assumed that the macro(s) have been designed to handle the loads passed to them from subsystems later in the chain. Thus the acceleration capability of the connection points between subsystems is known. This analysis results in a second design constraint. This constraint requires each subsystem to be able to stabilize its end-effector against the motion of subsystems that proceed in the chain, i.e. the mini must compensate for macro motion. This constraint results in an upper bound on the acceleration capability of the macros' end-effectors.

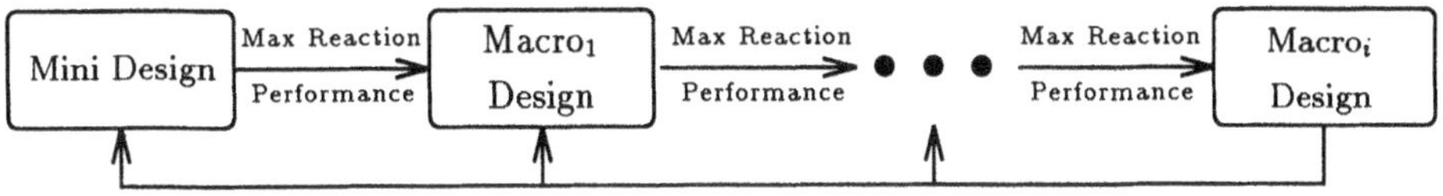

Figure 2. Design Decomposition.

The above analysis suggests a modular process where each design is conducted independently and where the upper and lower bounds on performance are passed between the successive stages of the design, as illustrated in Figure 2. The next sections detail this process for a redundant system consisting of two structures: one mini and one macro. This analysis is applied to the design of the 6 DOF planar manipulator shown in Figure 1. The optimization of the mini is discussed first and applied to the last three links of the manipulator shown in Figure 1. The study of the connection between the mini and macro is then analyzed. These results are then applied to the design of the macro manipulator, the first three links of the mechanism shown in Figure 1.

3. Mini Manipulator Optimization

The process for optimizing the mini-manipulator is outlined below. Further details on this process can be found in [4]. We present first, the models used to characterize inertial and acceleration properties. We will then briefly discuss the optimization scheme.

3.1. INERTIAL PROPERTIES

The inertial properties as perceived at the end-effector are described by the pseudo-kinetic energy matrix $\Lambda(\mathbf{q})$,

$$\Lambda^{-1}(\mathbf{q}) = J(\mathbf{q})A^{-1}(\mathbf{q})J^T(\mathbf{q}) \tag{1}$$

where A(q) is the joint space kinetic energy matrix and the Jacobean, $J(\mathbf{q})$, is defined as,

$$\vartheta \triangleq \begin{bmatrix} v \\ \omega \end{bmatrix} = J(\mathbf{q})\dot{\mathbf{q}}, \tag{2}$$

where v and ω are the end-effector linear and angular velocities and $\dot{\mathbf{q}}$ is a vector of n joint velocities.

Analysis of the matrix $\Lambda(\mathbf{q})$ is based on a decomposition of the Jacobian matrix into its linear and angular sub-matrices:

$$J(\mathbf{q}) = \begin{bmatrix} J_v(\mathbf{q}) \\ J_\omega(\mathbf{q}) \end{bmatrix} \tag{3}$$

where the matrix $J_v(\mathbf{q})$ transforms joint velocities into end-effector linear velocities and $J_\omega(\mathbf{q})$ does likewise for end-effector angular velocities. From these relationships the mass and inertia properties are described by the matrices, Λ_v and Λ_ω;

$$\Lambda_v = (J_v(\mathbf{q})A^{-1}(\mathbf{q})J_v^T(\mathbf{q}))^{-1} \tag{4}$$

and

$$\Lambda_\omega = (J_\omega(\mathbf{q})A^{-1}(\mathbf{q})J_\omega^T(\mathbf{q}))^{-1}. \tag{5}$$

The end-effector acceleration capability is characterized by the isotropic acceleration. In the development of this characterization we consider the bounds on joint torques, determined by the maximum torque of the actuators,

$$-\tau_{bound} \leq \tau \leq \tau_{bound}. \tag{6}$$

The end-effector behavior is described by,

$$\Lambda\dot{\vartheta} + \mu + \mathbf{p} = \mathbf{F} \tag{7}$$

where μ, $\mathbf{p}$, and $\mathbf{F}$ are respectively the centrifugal and Coriolis force vector, gravity force vector, and generalized force vector acting in operational space. Using the above equations the following relationship is obtained,

$$\tau_{lower} \leq E_v\dot{v} + E_\omega\dot{\omega} + \bar{\mathbf{b}} \leq \tau_{upper} \tag{8}$$

where

$$\begin{aligned} [E_v\ E_\omega] &= NJ^T\Lambda & \tau_{upper} &= \mathbf{1} - NJ^T\mathbf{p} \\ \bar{\mathbf{b}} &= NJ^T\mu & \tau_{lower} &= -\mathbf{1} - NJ^T\mathbf{p}. \end{aligned} \tag{9}$$

In obtaining this equation the bounds, τ_{bound}, have been normalized using a diagonal matrix N with elements $N_{ii} = \frac{1}{\tau_{bound_i}}$.

Equation (8) is analyzed by visualizing each component as a geometric object. The torque bounds are visualized as an n-dimensional hypercube whose center is shifted from the origin by the gravity effect, i.e. $NJ^T\mathbf{p}$ from equation (9). An isotropic acceleration is visualized as a sphere with radius, a,

$$\dot{v}^T\dot{v} = a^2. \tag{10}$$

The vectors comprising this sphere are transformed by matrix E_v in equation (8) into a torque ellipsoid of dimension three or less, using the relationship $\tau_v = E_v\dot{v}$,

$$\tau_v^T(E_vE_v^T)^+\tau_v = a^2, \tag{11}$$

where $(M)^+$ is the left pseudo-inverse of M. This surface is mapped inside the torque bounds. The isotropic acceleration is then determined by expanding/contracting the ellipsoid, changing a, until it lies completely within and is tangent to one of the torque bounds, Figures 3 and Figure 4. The dashed ellipses in both figures correspond to $a = 1$.

Similar visualizations are performed on the other terms of equation (8). The resulting surfaces are added together within the bounds. Note that if the bounds are not normalized and desired macro-manipulator isotropic accelerations are known, the required actuator capacity can be estimated by circumscribing a hyper-cube around the composite surface.

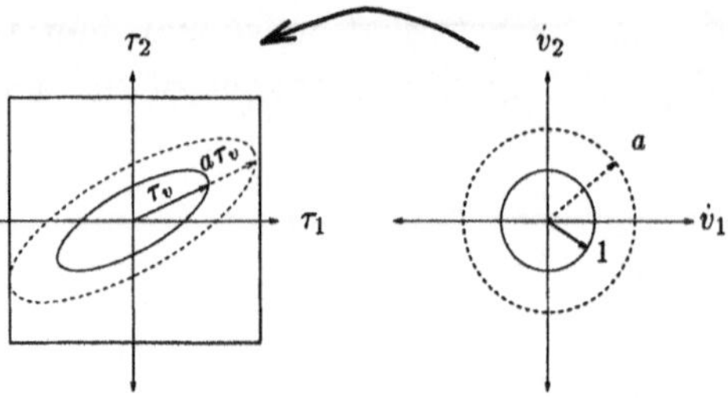

Figure 3. Ellipse τ_v.

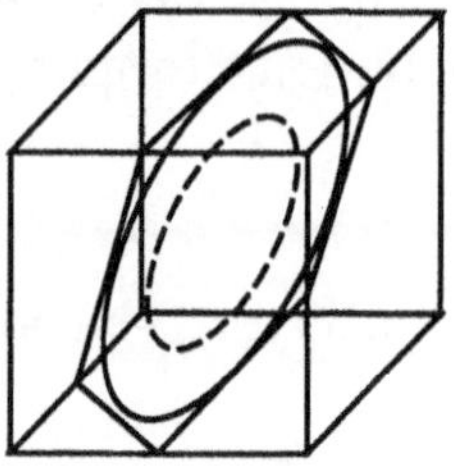

Figure 4. Ellipsoid in 3D.

Analysis of equation (8) yields a set of inequalities describing the relationship between the isotropic accelerations, $||\dot{v}||$ and $||\dot{\omega}||$, isotropic velocities, $||v||$ and $||\omega||$, and actuator torque capacities. Figure 5 shows a plot of these relationships for the 3 DOF mini-manipulator, where $||\omega|| = 0$. The three points where the curves intersect the coordinate axes will be of interest in later sections.

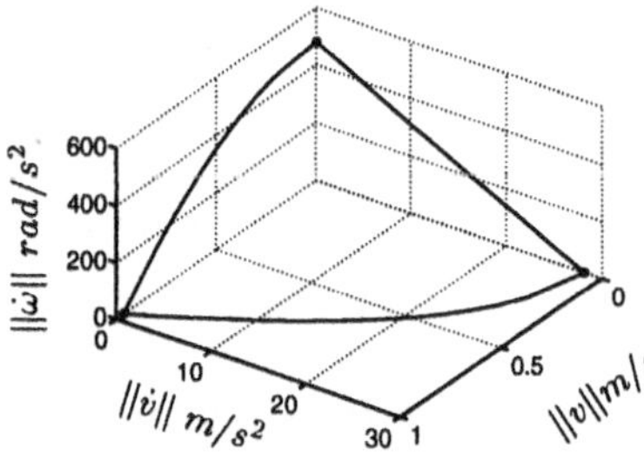

Figure 5. Isotropicity Curves.

3.3. OPTIMIZATION

The cost function at a given configuration is defined as;

$$\begin{aligned} cost_i \ = \ & w_1\,\kappa(\Lambda_v) + w_2\,\kappa(\Lambda_\omega) + w_3\,||\Lambda_v|| + w_4\,||\Lambda_\omega|| + \\ & w_5\,\kappa(E_v) + w_6\,\kappa(E_\omega) + w_7\,||E_v|| + w_8\,||E_\omega|| \end{aligned} \tag{12}$$

where w_i is a weight, $\kappa(M)$ is the condition number of M, and $||M||$ is the norm of M. This function is summed over a set of i representative configurations to obtain the total cost.

The steps in the optimization are shown in Figure 6.

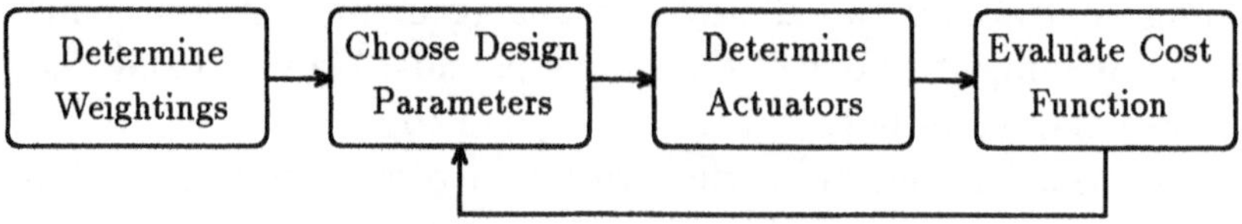

Figure 6. Single Subsystem Design.

The results for the optimization of the mini are shown in Figure 7.

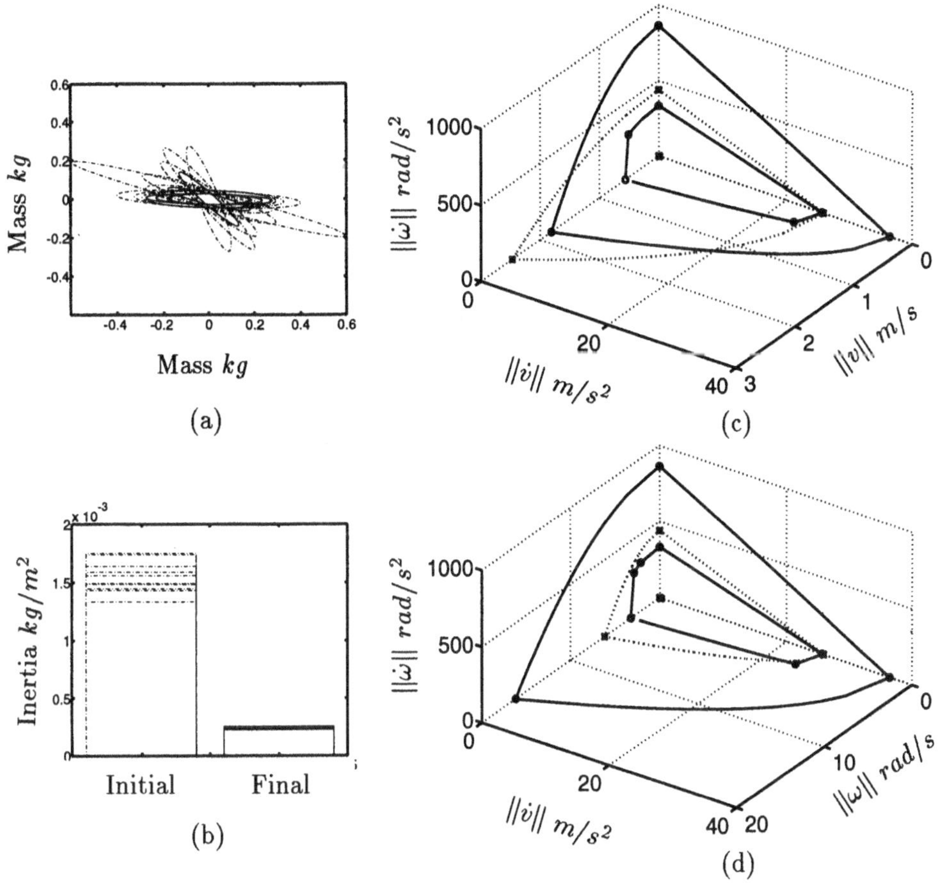

Figure 7. Mini-Manipulator Properties

The mini-manipulator was optimized for twelve different configurations in the workspace. The dotted and solid lines represent the initial and final designs respectively. The effective masses of the initial and final designs at each of the twelve positions in the workspace are shown as overlayed ellipses in Figure 7a. The effective inertia of the initial and final designs are shown as overlayed bars on Figure 7b. The figures show a clear improvement of the inertial properties.

Figures 7c and 7d show the isotropicity curves for the initial and final designs. In Figures 7c and 7d the dotted lines represent the minimum and maximum envelopes for the isotropicity curves of the initial design at the twelve configurations in the workspace. The minimum envelope for the initial design is very near the origin. In Figure 7c and 7d the solid lines represent the minimum and maximum envelopes for the isotropicity curves of the final design at the twelve configurations in the workspace. Figure

7c shows the curves for $||\omega|| = 0$ while 7d shows the curves for $||v|| = 0$. The figures show overall improvement in the mini's end-effector acceleration capability.

4. Macro Manipulator Design

In order to design the macro, the bounds on acceleration capability required of the macro in order to satisfy the design constraints are determined. A discussion of this analysis follows.

4.1. TIP TO BASE ANALYSIS

The first design constraint requires that the macro manipulator be able to hold its end-effector fixed in the workspace while the mini is in motion. For this case the interaction forces are the reaction forces and moments at the base of the mini when it is attached to the ground. The end-effector behavior of the macro is then described by,

$$\Lambda_{mac}\dot{\vartheta}_{mac} + \mu_{mac} + \mathbf{p}_{mac} = \mathbf{F}_{mac} - \mathbf{F}_{reaction}, \tag{13}$$

where the reaction forces, $\mathbf{F}_{reaction}$, are a function of the mini configuration, velocity, and acceleration. We can obtain a relation similar to equation (9) using the macro dynamic equations and reaction forces,

$$\begin{array}{llll} \tau_{lower} \leq & E_{v_{mac}}\dot{v}_{mac} + E_{\omega_{mac}}\dot{\omega}_{mac} + \bar{\mathbf{b}}_{mac} + & \\ & E_{v_r}\dot{v}_{min} + E_{\omega_r}\dot{\omega}_{min} + \bar{\mathbf{b}}_r & \leq \tau_{upper} \end{array} \tag{14}$$

where the terms with subscript r are obtained from the reaction forces. The subscripts *mac* and *min* relate to the macro and mini manipulators. This equation is derived and analyzed in the same manner as equation (9) discussed in section 3.2. However, the surfaces formed from the reaction force terms have a constant size and represent torque required to compensate for the maximum reaction forces from the mini. They determine a lower bound on the acceleration required of the macro manipulator.

4.2. BASE TO TIP ANALYSIS

Just as for the case of the non-redundant systems, the acceleration characteristics of a redundant mechanism, at a given configuration, can be described by isotropic curves, as illustrated in Figure 8. The isotropic accelerations of the overall redundant mechanism depend on the acceleration characteristics of the macro and mini. This dependence can be analyzed by determining the end-effector acceleration in terms of the macro and mini end-effector accelerations,

$$\dot{v} = \dot{v}_{mac} + \dot{v}_{min} + \dot{\omega}_{mac} \times r + \omega_{mac} \times (\omega_{mac} \times r) + 2\omega_{mac} \times v_{min} \quad (15)$$

$$\dot{\omega} = \dot{\omega}_{mac} + \dot{\omega}_{min} + \omega_{mac} \times \omega_{min} \quad (16)$$

where r is the vector connecting the extremity of the macro to the end-effector of the mini.

First consider the intersection point between the isotropic surface and the linear acceleration axis, as shown in Figure 8. From equation (15) this point is described by

$$||\dot{v}|| = ||\dot{v}_{mac}|| + ||\dot{v}_{min}||. \quad (17)$$

The second design constraint requires that the mini to stabilize the end-effector against macro motion. This means that the mini must be capable of cancelling macro motions so that $||\dot{v}|| = 0$. Thus the isotropic acceleration capability of the mini must be greater than or equal to that of the macro. Omitting the details, we can also find that this holds for the intersection point between the isotropicity curves and the angular acceleration axis as well as the accelerations described by the curves in the $||\dot{v}||$-$||\dot{\omega}||$ plane, Figure 8;

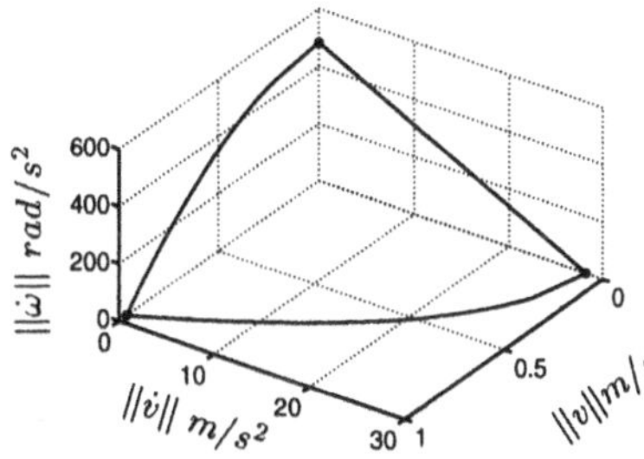

Figure 8. Example of Isotropicity Curves.

$$||\dot{v}_{mac}|| \leq ||\dot{v}_{min}|| \quad (18)$$

$$||\dot{\omega}_{mac}|| \leq ||\dot{\omega}_{min}||. \quad (19)$$

If these constraints are violated, at least some of the subsystems will have been overdesigned. If the mechanism has overdesigned components the amount of controllable end-effector acceleration will be limited to twice the acceleration capability of the mini.

4.3. APPLICATION

In this example the macro was simply designed to compensate for mini motions at twelve points in its workspace; making the design quite simple. The results of this design are given in Figure 9. The maximum and minimum envelopes for the isotropicity curves for the macro and mini designs are shown. The broken lines represent the macro and the solid ones correspond to the mini. At all the above configurations, the maximum macro accelerations are less than the maximum mini accelerations. Thus both design constraints are satisfied.

5. Overall Properties

Finally we examine the overall acceleration properties. Figure 10 shows the envelopes for the overall system acceleration capability in the $||\dot{v}||$-$||\dot{\omega}||$ plane. These curves are produced by summing the macro and mini isotropicity curves.

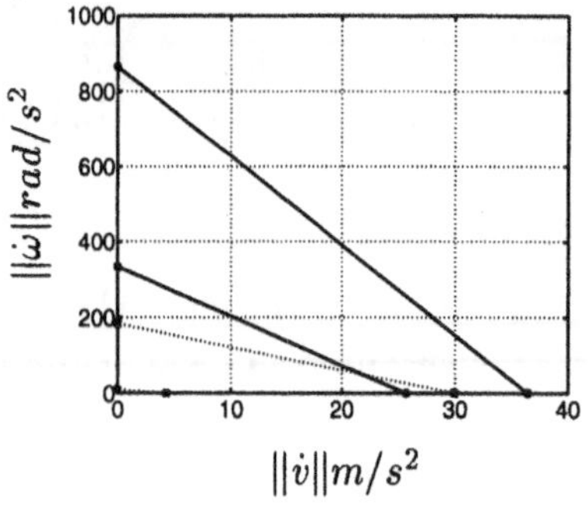

Figure 9. Macro vs Mini Accelerations

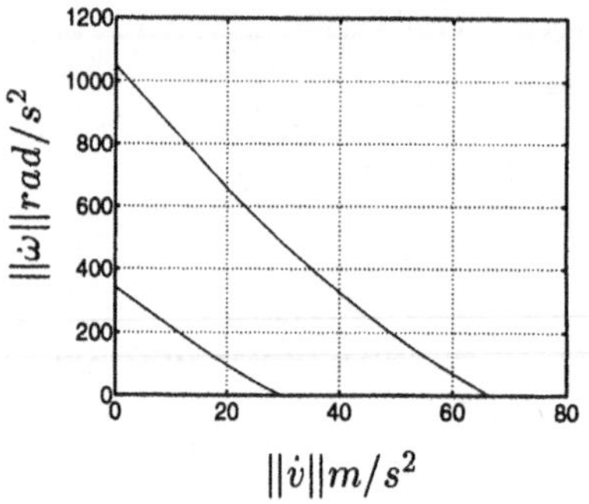

Figure 10. Overall Accelerations

6. Conclusion

We have presented a general methodology for the design of redundant manipulators for high dynamic performance. This methodology provides a modular decomposition of the design process into separate stages. This greatly reduces the search space for an optimization, since each of these stages depends on an independent subset of the design parameters. The design process was illustrated on an example of a six DOF planar manipulator. These results can be easily extended to systems involving several macro/mini structures.

Acknowledgments

The financial support of NASA/JPL, Boeing, Hitachi Construction Machinery, and NSF, grant IRI-9320017, are gratefully acknowledged.

References

1. Khatib, O.: Inertial Properties in Robotic Manipulation: An Object-Level Framework, *The International Journal of Robotics Research*, vol. 13, no. 1, February 1995, pp. 19-36.
2. Khatib, O., and Burdick, J.: Dynamic Optimization in Manipulator Design: The Operational Space Formulation, *The International Journal of Robotics and Automation*, vol.2, no.2, 1987, pp. 90-98.
3. Khatib, O. and Agrawal, S.: Isotropic and Uniform Inertial and Acceleration Characteristics: Issues in the Design of Redundant Manipulators, *IUTAM/IFAC Symposium on Dynamics of Controlled Mechanical Systems*, Zurich, Switzerland, 1989.
4. Khatib, O. and Bowling, A.: Optimization of the Inertial and Acceleration Characteristics of Manipulators, *IEEE International Conference on Robotics and Automation*, Minneapolis, Minnesota, April 1996.

SINGULARITY-ROBUST SECOND-ORDER KINEMATIC CONTROL OF ROBOT MANIPULATORS

F. CACCAVALE, S. CHIAVERINI AND B. SICILIANO
Dipartimento di Informatica e Sistemistica
Università degli Studi di Napoli Federico II
Via Claudio 21, 80125 Napoli, Italy

Abstract. The goal of this work is to report an experimental study of second-order kinematic control of robot manipulators with robustness to the occurrence of kinematic singularities. A damped least-squares inverse of the Jacobian is used within a closed-loop inverse kinematics algorithm to compute joint accelerations, as well as velocities and positions by integration. These are input to a joint space computed torque control. Results are presented for a six-joint industrial robot with open control architecture.

1. Introduction

The cascade of an inverse kinematics scheme and a joint space control scheme is typically referred to as kinematic control (Siciliano, 1990). This approach is different from operational space control (Khatib, 1988) where manipulator kinematics is handled inside the control loop.

An efficient technique to transform a given end-effector motion into a corresponding joint motion is the closed-loop inverse kinematics (CLIK) algorithm which is based on the (pseudo-)inverse of the manipulator Jacobian and makes use of a feedback correction term on the end-effector error (Chiacchio *et al.*, 1991). Such algorithm can be formulated either at the first-order level to solve for joint velocities and positions or at the second-order level to solve for joint accelerations too (Siciliano, 1990; Novaković and Siciliano, 1991). In order to cope with occurrence of kinematic singularities, a damped least-squares (DLS) inverse of the Jacobian can be used to gain numerical robustness (Nakamura and Hanafusa, 1986; Wampler, 1986). This has been embedded into a first-order CLIK algorithm (Chiaverini *et al.*, 1991; Chiaverini *et al.*, 1994).

J. Lenarčič and V. Parenti-Castelli (eds.), Recent Advances in Robot Kinematics, 39-46.
© 1996 *Kluwer Academic Publishers. Printed in the Netherlands.*

Industrial robots are typically endowed with independent joint PID controllers. In order to enhance their tracking performance, a model-based compensation can be provided, e.g., in a feedforward fashion (computed torque method) (Sciavicco and Siciliano, 1996).

This work describes the application of a singularity-robust second-order CLIK algorithm to kinematic control of robot manipulators with adoption of a joint space computed torque control scheme. Experiments on a six-joint industrial robot with open control architecture have been performed, and the results for an end-effector path involving a near-singular configuration are presented.

2. Second-order CLIK algorithm with DLS inverse

The direct kinematics equation of a robot manipulator with open kinematic chain can be written in the form

$$\boldsymbol{x} = \boldsymbol{k}(\boldsymbol{q}), \tag{1}$$

where $\boldsymbol{q}$ is the $(n \times 1)$ vector of joint position variables, $\boldsymbol{x}$ is the $(m \times 1)$ vector of end-effector variables (including position variables and a minimum number of orientation variables), and $\boldsymbol{k}(\cdot)$ denotes the nonlinear direct kinematics function. For the purpose of the present work, we consider the case of a nonredundant six-joint manipulator, i.e., $m = n = 6$.

The differential kinematics equation can be obtained by differentiating (1) with respect to time, leading to

$$\dot{\boldsymbol{x}} = \boldsymbol{J}_A(\boldsymbol{q})\dot{\boldsymbol{q}} \tag{2}$$

where $\boldsymbol{J}_A = \partial\boldsymbol{k}/\partial\boldsymbol{q}$ is the analytical Jacobian (Sciavicco and Siciliano, 1996). Further differentiation of (2) with respect to time gives

$$\ddot{\boldsymbol{x}} = \boldsymbol{J}_A(\boldsymbol{q})\ddot{\boldsymbol{q}} + \dot{\boldsymbol{J}}_A(\boldsymbol{q}, \dot{\boldsymbol{q}})\dot{\boldsymbol{q}}. \tag{3}$$

Hereafter, the dependence on $\boldsymbol{q}$ and $\dot{\boldsymbol{q}}$ is omitted for notation compactness.

Given a desired end-effector trajectory in terms of position, velocity and acceleration $(\boldsymbol{x}_d(t), \dot{\boldsymbol{x}}_d(t), \ddot{\boldsymbol{x}}_d(t))$, the second-order CLIK algorithm giving the joint acceleration vector is (Sciavicco and Siciliano, 1988)

$$\ddot{\boldsymbol{q}} = \boldsymbol{J}_A^{-1}\left(\ddot{\boldsymbol{x}}_d - \dot{\boldsymbol{J}}_A\dot{\boldsymbol{q}} + \boldsymbol{K}_D\dot{\boldsymbol{e}} + \boldsymbol{K}_P\boldsymbol{e}\right) \tag{4}$$

where

$$\boldsymbol{e} = \boldsymbol{x}_d - \boldsymbol{x} \tag{5}$$

is the end-effector tracking algorithmic error, and $\boldsymbol{K}_P$, $\boldsymbol{K}_D$ are suitable diagonal positive definite matrix gains. The joint position and velocity vectors are then computed by integrating (4), where the feedback correction terms are aimed at counteracting the drift caused by numerical integration. It can be easily recognized that substitution of (4) into (3) leads to a second-order exponentially stable linear and decoupled time-invariant system of the error, i.e.,

$$\ddot{e} + K_D\dot{e} + K_P e = \mathbf{0}. \tag{6}$$

In the above algorithm, the inversion of the Jacobian is ill-conditioned in the neighbourhood of kinematic singularities. A remedy for such occurrence is the adoption of a DLS inverse of the Jacobian, i.e.,

$$J_A^{\#} = J_A^{\mathrm{T}}\left(J_A J_A^{\mathrm{T}} + \lambda^2 I\right)^{-1}, \tag{7}$$

where I is the identity matrix of proper dimensions and λ is the damping factor (Nakamura and Hanafusa, 1986; Wampler, 1986). Using (7) in lieu of the inverse in (4) yields

$$\ddot{q} = J_A^{\#}\left(\ddot{x}_d - \dot{J}_A\dot{q} + K_D\dot{e} + K_P e\right). \tag{8}$$

Substituting (8) into (3) gives, after some algebraic manipulation,

$$\ddot{e} + K_D\dot{e} + K_P e = \Delta\left(\ddot{x}_d - \dot{J}_A\dot{q} + K_D\dot{e} + K_P e\right) \tag{9}$$

where the matrix

$$\Delta = I - J_A J_A^{\#} \tag{10}$$

characterizes the discrepancy of using the DLS inverse versus the inverse.

For system (9), it can be shown that it is possible to choose suitable values of the damping factor λ and feedback gains K_P, K_D so as to ensure ultimate boundedness of the end-effector error. The details of the stability proof are not included here for brevity and are reported in (Caccavale *et al.*, 1996).

Far from singularities, no damping is needed and the error system (9) reduces to (6). The damping factor λ in (7) is aimed at achieving a trade-off between solution accuracy (small end-effector errors) and solution feasibility (limited joint accelerations), and thus it can be suitably tuned by defining a singular region in the neighbourhood of the singularity on the basis of the estimate of the smallest singular value of J_A, e.g., (Chiaverini, 1993),

$$\lambda^2 = \begin{cases} 0 & \widehat{\sigma}_6 \geq \varepsilon \\ \left(1 - \left(\frac{\widehat{\sigma}_6}{\varepsilon}\right)^2\right)\lambda_M^2 & \widehat{\sigma}_6 < \varepsilon \end{cases} \tag{11}$$

where $\widehat{\sigma}_6$ is the estimate of the smallest singular value, and ε defines the size of the singular region; the value of λ_M serves to suitably shape the solution in the neighbourhood of a singularity.

3. Computed torque control

The above second-order CLIK algorithm gives the solution in terms of joint position, velocity and acceleration variables, which can be denoted by $\boldsymbol{q}_r$, $\dot{\boldsymbol{q}}_r$ and $\ddot{\boldsymbol{q}}_r$, respectively. These can be used as the reference inputs to a model-based joint space control scheme.

The dynamic model of a robot manipulator can be written in the form

$$\boldsymbol{B}(\boldsymbol{q})\ddot{\boldsymbol{q}} + \boldsymbol{c}(\boldsymbol{q}, \dot{\boldsymbol{q}}) + \boldsymbol{f}(\boldsymbol{q}, \dot{\boldsymbol{q}}) + \boldsymbol{g}(\boldsymbol{q}) = \boldsymbol{u}, \tag{12}$$

where $\boldsymbol{B}$ is the (6×6) symmetric positive definite inertia matrix, $\boldsymbol{c}$ is the (6×1) vector of Coriolis and centrifugal torques, $\boldsymbol{f}$ is the (6×1) vector of friction torques, $\boldsymbol{g}$ is the (6×1) vector of gravity torques, and $\boldsymbol{u}$ is the (6×1) vector of joint driving torques.

By adopting a computed torque control strategy, the joint driving torques can be chosen as the sum of a feedforward centralized nonlinear control action and a feedback decentralized linear control action, i.e.,

$$\boldsymbol{u} = \boldsymbol{u}_{FF}(\boldsymbol{q}_r) + \boldsymbol{u}_{PID}(\boldsymbol{q}, \boldsymbol{q}_r) \tag{13}$$

where

$$\boldsymbol{u}_{FF} = \widehat{\boldsymbol{B}}(\boldsymbol{q}_r)\ddot{\boldsymbol{q}}_r + \widehat{\boldsymbol{c}}(\boldsymbol{q}_r, \dot{\boldsymbol{q}}_r) + \widehat{\boldsymbol{f}}(\boldsymbol{q}_r, \dot{\boldsymbol{q}}_r) + \widehat{\boldsymbol{g}}(\boldsymbol{q}_r) \tag{14}$$

in which $\widehat{\boldsymbol{B}}$, $\widehat{\boldsymbol{c}}$, $\widehat{\boldsymbol{f}}$, $\widehat{\boldsymbol{g}}$ denote the estimates of the respective model terms, and

$$\boldsymbol{u}_{PID} = \boldsymbol{G}_P(\boldsymbol{q}_r - \boldsymbol{q}) + \boldsymbol{G}_I \int_0^t (\boldsymbol{q}_r - \boldsymbol{q})d\tau + \boldsymbol{G}_D(\dot{\boldsymbol{q}}_r - \dot{\boldsymbol{q}}) \tag{15}$$

in which $\boldsymbol{q}$, $\dot{\boldsymbol{q}}$ denote the actual joint positions and velocities, and $\boldsymbol{G}_P$, $\boldsymbol{G}_I$, $\boldsymbol{G}_D$ are suitable diagonal positive definite matrix gains characterizing a PID controller in the joint space.

The joint PID servos of a conventional industrial robot controller are typically tuned for average constant inertias at the joints while the nonlinear terms depending on the configuration are treated as disturbances. Some intentional simplifications aimed at reducing the computational burden of the model-based feedforward action can be made for compensation of nonlinear terms. In particular, it is customary to compensate at least the terms on the diagonal of the inertia matrix as well as the gravity term, while coupling inertias and velocity-dependent terms can be partially compensated or even neglected. On the other hand, if the complete dynamic model of the robot manipulator is available, e.g., as the result of an identification process, it is possible to achieve accurate estimates of the model terms and compute a full model-based feedforward action; this ensures better tracking performance than the above partial compensation.

4. Experimental results

The proposed second-order kinematic control scheme has been implemented on the industrial robot Comau SMART-3 S available in the laboratory. This is a six-joint manipulator with nonnull shoulder and elbow offsets and a nonspherical wrist whose axes intersect two-by-two. The joints are actuated by brushless motors via gear trains; shaft absolute resolvers provide motor position measurements, and thus velocities have to be reconstructed via numerical filtering. The robot is controlled by an open version of the C3G 9000 control unit which has a VME-based architecture with a bus-to-bus communication link to a standard PC (Dogliani *et al.*, 1993).

The second-order CLIK algorithm and computed torque control have been implemented as C modules running on a 486DX2/66 PC. During each sampling interval, the PC computes the joint reference positions, velocities and accelerations by means of Euler integration of (8), which are then used to compute the feedforward action (14) on the basis of an identified dynamic model of the robot manipulator, as in (Caccavale and Chiacchio, 1994). The torque references are then computed by summing this action to the PID action (15) available from the manufacturer as a C module on the PC. The overall computation allows the control to be run at 2 ms sampling time.

The reference currents are passed to the servos through the communication link. For logging purposes, resolver readings and motor currents are downloaded to the PC at each sampling instant.

With reference to the manipulator base frame, the case study consists of taking the end effector from the initial location described by the position vector

$$\boldsymbol{p}_{di} = \begin{bmatrix} 0.113 \\ 1.193 \\ 0.078 \end{bmatrix}$$

and the rotation matrix

$$\boldsymbol{R}_{di} = \begin{bmatrix} 0 & 1 & 0 \\ 0 & 0 & 1 \\ 1 & 0 & 0 \end{bmatrix}$$

to the final location described by the position vector

$$\boldsymbol{p}_{df} = \begin{bmatrix} 0.513 \\ 0.993 \\ 0.078 \end{bmatrix}$$

and the rotation matrix

$$\boldsymbol{R}_{df} = \begin{bmatrix} 0 & 1 & 0 \\ \sqrt{3}/2 & 0 & -1/2 \\ 1/2 & 0 & \sqrt{3}/2 \end{bmatrix}.$$

The end-effector path for position is a segment in the Cartesian space, whereas the equivalent axis/angle method has been used to find an end-effector path for orientation. This results into a displacement of 0.447 m in the xy plane with a rotation of $-\pi/3$ rad about the x axis. The initial and final path coordinates have been interpolated using 5th-order polynomial time laws such that null initial and final velocities and accelerations are obtained. The duration of the motion is 5 s.

In order to tune the parameters of the CLIK algorithm, the software has been run on the PC without activating the communication with the robot control unit. As a result of these, the gains in (8) have been set to

$$\begin{aligned} \boldsymbol{K}_P &= 10^5 \text{diag}\{2, 2, 2, 8, 8, 8\} \\ \boldsymbol{K}_D &= 10^3 \text{diag}\{1.5, 1.5, 1.5, 1.5, 1.5, 1.5\}. \end{aligned}$$

The parameters characterizing the singular region in (11) have been set to $\varepsilon = 0.1\sqrt{5}$ and $\lambda_M = \sqrt{0.4}$.

The results are illustrated in Fig. 1 in terms of the time history of the minimum singular value, the damping factor, the joint accelerations, the end-effector position and orientation error norms, and the reference currents. Notice that the errors have been computed by comparing the desired end-effector variables with the end-effector variables as reconstructed from the actual joint measurements. It can be recognized that the motion of the manipulator is quite demanding as far as the wrist is concerned, and involves a near-singular configuration. Thanks to the intervention of the damping factor after about 2 s, the wrist joint accelerations are kept limited. The overall tracking performance is very good and the steady-state errors are practically null. Finally, it can be observed that the reference currents are limited too.

5. Conclusion

A kinematic control scheme for industrial robot manipulators has been presented in this work. The desired end-effector trajectory specified in terms of position, velocity and acceleration is transformed into joint trajectories by using a second-order CLIK algorithm with DLS inverse to ensure smooth motion in the neighbourhood of kinematic singularities. The joint trajectories are then used as references to a computed torque control scheme which consists of a complete model-based feedforward action and an independent joint control PID action. The results of experiments conducted on a conventional industrial robot with open control architecture have shown the effectiveness of the scheme for a motion passing near a singularity.

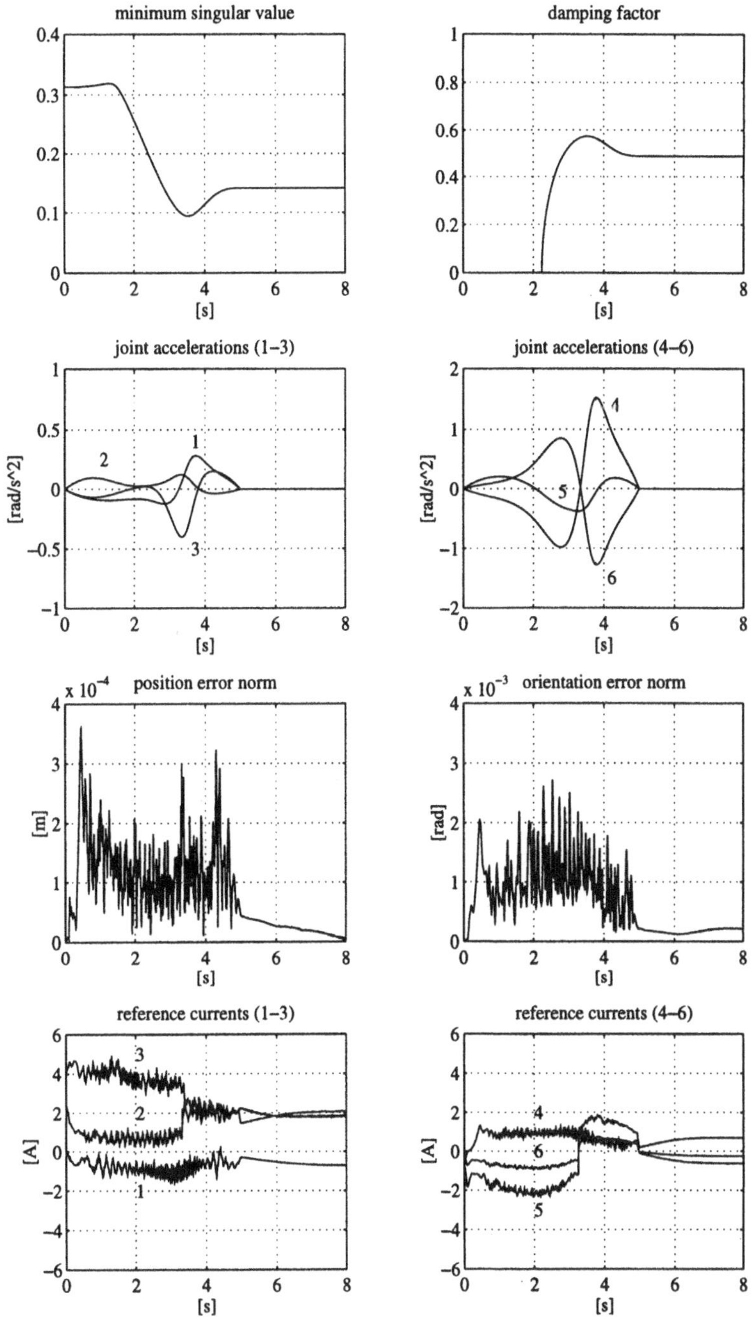

Figure 1. Performance of kinematic control scheme in the case study.

Acknowledgements

This work was partly supported by *Ministero dell'Università e della Ricerca Scientifica e Tecnologica* under *40%* and *60%* funds and partly by *Agenzia Spaziale Italiana* under contract *ASI-94-RS-71*.

References

Caccavale, F., Chiacchio, P. (1994) Identification of dynamic parameters and feedforward control for a conventional industrial manipulator, *Control Engineering Practice*, vol. 2, pp. 1039–1050.

Caccavale, F., Chiaverini S., Siciliano, B. (1996) Second-order kinematic control of robot manipulators with Jacobian damped least-squares inverse: Theory and experiments, *IEEE/ASME Trans. on Mechatronics*, submitted.

Chiacchio, P., Chiaverini, S., Sciavicco, L., Siciliano, B. (1991) Closed-loop inverse kinematics schemes for constrained redundant manipulators with task space augmentation and task priority strategy, *Int. J. of Robotics Research*, vol. 10, pp. 410–425.

Chiaverini, S. (1993) Estimate of the two smallest singular values of the Jacobian matrix: Application to damped least-squares inverse kinematics, *J. of Robotic Systems*, vol. 10, pp. 991–1008.

Chiaverini, S., Sciavicco, L., Siciliano, B. (1991) Control of robotic systems through singularities, *Int. Work. on Adaptive and Nonlinear Control: Issues in Robotics*, Grenoble, F, in *Advanced Robot Control*, Lecture Notes in Control and Information Sciences 162, C. Canudas de Wit (Ed.), Springer-Verlag, Berlin, D, pp. 285–295.

Chiaverini, S., Siciliano, B., Egeland, O. (1994) Review of the damped least-squares inverse kinematics with experiments on an industrial robot manipulator, *IEEE Trans. on Control Systems Technology*, vol. 2, pp. 123–134.

Dogliani, F., Magnani, G., Sciavicco, L. (1993) An open architecture industrial controller, *Newsl. of IEEE Robotics and Automation Soc.*, vol. 7, n. 3, pp. 19–21.

Khatib, O. (1988) A unified approach for motion and force control of robot manipulators: The operational space formulation, *IEEE J. of Robotics and Automation*, vol. 4, pp. 43–53.

Nakamura, Y., Hanafusa, H. (1986) Inverse kinematic solution with singularity robustness for robot manipulator control, *ASME J. of Dynamic Systems, Measurements, and Control*, vol. 108, pp. 163–171.

Novaković, Z.R., Siciliano, B. (1991) A new second-order inverse kinematics solution for redundant manipulators, *2nd Int. Work. on Advances in Robot Kinematics*, Linz, A, in *Advances in Robot Kinematics*, S. Stifter and J. Lenarčič (Eds.), Springer-Verlag, Wien, A, pp. 408–415.

Sciavicco L., Siciliano, B. (1988) A computational technique for solving robot end-effector trajectories into joint trajectories, *Proc. 1988 American Control Conf.*, Atlanta, GA, pp. 535–536.

Sciavicco, L., Siciliano, B. (1996) *Modeling and Control of Robot Manipulators*, McGraw-Hill, New York, NY.

Siciliano, B. (1990) A closed-loop inverse kinematic scheme for on-line joint-based robot control, *Robotica*, vol. 8, pp. 231–243.

Wampler, C.W. (1986) Manipulator inverse kinematic solutions based on vector formulations and damped least-squares methods, *IEEE Trans. on Systems, Man, and Cybernetics*, vol. 16, pp. 93–101.

DYNAMICS AND CONTROL OF ELASTIC JOINT ROBOTS ON A GROUP CONFIGURATIONAL MANIFOLD

CLEMENTINA D. MLADENOVA

Institute of Mechanics, Bulgarian Academy of Sciences
Acad.G.Bonchev Str., Bl.4, 1113 Sofia, Bulgaria

Abstract. The present paper treats the problems of modeling and control of manipulators with elastic joints in terms of vector-parametrization of the **SO(3)** group. Since the computational effectiveness of the vector-parameter approach increases with the increasing number of the revolute degrees of freedom, this fact is successfully used in the problems of elastic joint manipulators, where except the real n links, n fictious links are included as well as additional n revolute degrees of freedom. A dynamic model in "pure" vector-parameter form is developed and the inverse dynamic control is discussed.

1. Introduction

The vector-parametrization of **SO(3)** may be recognized in the literature as "vector of finite rotations" or "Rodrigues'parameters" but it is not used till now as an element of a Lie group with its nice properties and advantages. Studying the different group parametrizations of the rotational motions and their after-effects some quite interesting theoretical results were obtained by Mladenova 1990, 1991,1992. As an example, the trivialization of the **SO(3)** tangent bundle in different ways gives the possibility to consider kinematic and dynamic problems in different configurational, respectively state manifolds, and this makes many theoretical investigations more flexible for practical applications. Especially, the comparison of computational costs within vector-parameter formalism with other methods for simulation and control used till now shows its efficiency and feasibility for real time applications. This fact is heavily used in the present paper for modeling and control of elastic joint manipulators.

J. Lenarčič and V. Parenti-Castelli (eds.), Recent Advances in Robot Kinematics, 47-56.
© 1996 *Kluwer Academic Publishers. Printed in the Netherlands.*

2. Basic Notation

We consider a flexible joint manipulator with n rotational degrees of freedom, and links $B(0)$ (fixed), $B(1), B(2), \ldots, B(n)$ (Fig.1). A frame $R(i)$ centered in $C(i)$ is referred to the link $B(i)$ (and joint i) from the chain. The frame $R(n+1)$ is built in the gravity center $G(n)$ of the link $B(n)$ (gripper), which may be coincident with the gripper characteristic point C. We denote by $Rm(i)$ the frame referred to the motor $M(i)$ that moves the link $B(i)$, with $Cm(i)$ as its origin. We introduce the following notation:

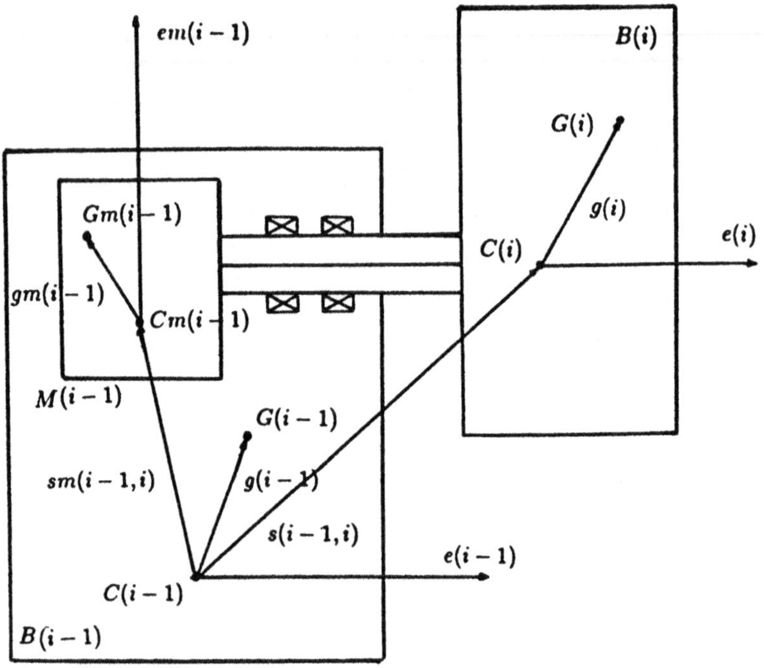

Figure 1. Scheme of a manipulator

$q := [q(1) \ldots q(n)]^T$ is $n \times 1$ vector of the generalized coordinates (joint displacements) of the manipulator system (MS) in the usual sense, $q \in Q \subset R^n$ and Q is the configurational manifold. $qm := [qm(1) \ldots qm(n)]^T$ is $n \times 1$ vector of the motor rotation angles, $qm \in Qm \subset R^n$ and Qm is the space of motor angles, and x is m-dimensional vector of end-effector position and orientation. In the configurational space Q the dynamic equations for rigid body manipulator look like

$$H(q)\ddot{q} + h(q, \dot{q}) = P \tag{1}$$

where $H := H(q)$ is $n \times n$ inertia matrix, the $n \times 1$ vector $h := h(q, \dot{q})$ takes into account Coriolis, centrifugal and gravitational forces, and P is $n \times 1$

vector of the generalized forces and moments. Further we denote by :
$\times$ - the cross product of vectors, $a \times b := a^{\times} b$, $a^{\times}$ is the skew-symmetric matrix whose components are the components of vector a.
$G(i)$ - mass centre of link $B(i)$ and $Gm(i)$ - mass centre of motor $M(i)$;
$s(i, i+1)$ – 3×1 vector from $C(i)$ to $C(i+1)$, written in $R(i)$,
$sm(i-1, i)$ – 3×1 vector from $C(i-1)$ to $Cm(i)$, written in $R(i-1)$;
$g(i)$ - the 3×1 vector from $C(i)$ to the mass center $G(i)$, written in the frame $R(i)$, which is constant in this frame, $gm(i)$ - the 3×1 vector from $Cm(i)$ to the mass center $Gm(i)$, written in the frame $Rm(i)$, which is constant in this frame;
$e(i) := [0\ 0\ 1]^T$ is 3×1 vector aligned with the i-th rotation axis,
$em(i) := [0\ 0\ 1]^T$ is 3×1 vector of rotation for the motor $M(i)$ written in the frame $Rm(i)$;
$p(i) := s(0,1) + s(1,2) + \ldots + s(i-1, i) \quad (s(0,1) := 0) \qquad pg(i) = p(i) + g(i)$;
$c(i)$ - the vector-parameter of rotation of $R(i-1)$ according to $R(i)$.

Vector c is directed along the rotation axis and its module is equal to to the tangent of the half rotation angle. The vector-parameters form a group with a simple composition law: $c' = < c_1, c_2 > = (c_1 + c_2 + c_1 \times c_2)/(1 - c_1 c_2)$.

After this remark we continue the notation:
$cc(i) := < c(1), c(2), \ldots, c(i) >$,
$tr(i)$ - the translation vector along the axis of i-th prismatic joint;
$O(i, i+1) = O(c(i+1))$ is the rotation matrix that relates the orientation of frame i to $i+1$ and the following identities are valid:
$O(1, i) = O(cc(i)), O(i, k) = O(cc(k))O(-cc(i))$;

$\omega(i)$ – 3×1 vector of angular velocity of link i at $C(i)$ written in the frame $R(i)$,
$\omega m(i)$ - the 3×1 vector of angular velocity of motor i at $C(i)$ written in the frame $R(i)$;
$v(i)$ - the 3×1 vector of linear velocity of link $B(i)$ written in the frame $R(i)$,
$vm(i)$ - the 3×1 vector of linear velocity of the motor $M(i)$ at $Cm(i)$ written in $Rm(i)$;
$V(i)$–6×1 vector of spatial velocity of link i in $R(i)$, $V(i) := [\omega(i)^T\ v(i)^T]^T$;
$Vm(i)$ – 6×1 vector of spatial velocity of the motor $M(i)$ in $Rm(i)$,
$Vm(i) := [\omega(i)^T\ vm(i)^T]^T$;
$W(i)$ – 6×1 vector of spatial acceleration of $C(i)$ in $R(i)$,
$W(i) := [\dot{\omega}(i)^T\ \dot{v}(i)^T]^T$,
$Wm(i)$ – 6×1 vector of spatial acceleration of $Cm(i)$ in $Rm(i)$,
$Wm(i) := [\omega \dot{m}(i)^T\ v\dot{m}(i)^T]^T$;

$F(i)$ – 3×1 vector of inertial forces acting upon link i at point $G(i)$,
$N(i)$ – 3×1 vector of inertial moments acting upon link i at $G(i)$,
$f(i)$ – 3×1 vectors of forces acting upon link i at point $C(i)$, written in the frame $R(i)$,
$n(i)$ – 3×1 vectors of moments acting upon link i at point $C(i)$, written in $R(i)$;
$t(i)$ – 6×1 vector of spatial force acting upon link i at point $C(i)$ written in $Rm(i)$, $t(i) := [f(i)^T \, n(i)^T]^T$,
$P(i)$ - the torque acting on the link i, and $Pm(i)$ - the i-th motor torque.

Let $(c(1), tr(1))$, ..., $(c(n), tr(n))$ are couples of vectors describing the link movements of the MS. They define the group configurational space Q_{ct}. The entries of the i-th pair $((c(i), \; tr(i))$ coincide either with the vector-parameter $c(i)$ or translational vector $tr(i)$ in dependence on the type of the i-th joint. Further we assume that all joints are rotational and $c := [c(1) \dots c(n)]^T$, $c\prime := g(n) :=< c(1), \dots, c(n) >$ and $Q_{ct} \equiv Q_c$, where $c(i) = (cx(i), \, cy(i), \, cz(i))$, $g(i) = (gx(i), \, gy(i), \, gz(i))$.

Remark: Only one of the components of the vector-parameters contains the whole information for the joint rotation, and the superfluous coordinates are discarded by proper geometrical considerations. So in reality the dimensions of the kinematic and dynamic models do not increase.
The dynamic equations for a rigid body MS on the group configurational space Q_c look like

$$H_V(c)\ddot{c} + h_V(c, \dot{c}) = P_V, \tag{2}$$

where $H_V := H_V(c)$ is $3n \times 3n$ inertia matrix, the $3n \times 1$ vector $h_V := h_V(\dot{c}, c)$ takes into account Coriolis, centrifugal and gravitational forces, and P_V is $3n \times 1$ vector of the generalized forces and moments. Analogically the vector cm with components $cm(i)$ $(i = 1, \dots .n)$ which describe the motor rotations is introduced, $cm \in Q_{cm} \subset R^{3n}$.
We continue with the appropriate notation for spatial transformations [8]:
$\Phi(i+1, i)$ – 6×6 matrix which translates a spatial velocity at point $C(i)$ written in the frame $R(i)$ to a spatial velocity at $C(i+1)$ written in $R(i+1)$

$$\Phi(i+1, i) := \begin{bmatrix} O(i, i+1) & \emptyset \\ -O(i, i+1)s^{\times}(i, i+1) & O(i, i+1) \end{bmatrix},$$

where $\emptyset$ means 3×3 zero matrix.
$\Phi^T(i+1, i)$ - transpose of $\Phi(i+1, i)$ matrix. Φ^T translates the spatial force acting at $C(i+1)$ written in the frame $R(i+1)$ to a spatial force acting at

$C(i)$ written in $R(i)$;
$Om(i-1,i)$ – 3×3 rotation matrix which relates the orientation of the frame $Rm(i)$ at $Cm(i)$ to the frame $R(i-1)$. The corresponding vector-parameter is $cm(i)$ and $Om(i-1,i) := O(cm(i))$;

$\Phi m(i,i-1)$ - 6×6 matrix which translates a spatial velocity at point $C(i-1)$ written in the frame $R(i-1)$ to a spatial velocity at $Cm(i)$ written in $Rm(i)$;

$mb(i)$ - mass of link i, $mm(i)$ - mass of motor i, that drives link i, $I(i) - 3\times 3$ inertia matrix of the link i at $C(i)$, written in $R(i)$, $Im(i) - 3 \times 3$ inertia matrix of the motor/gear at $Cm(i)$, written in the frame $Rm(i)$;

$Mb(i) - 6 \times 6$ spatial mass matrix of link i at $C(i)$ written in the in frame $R(i)$. It connects the inertia matrix of link i, link mass and the location of the mass centre $g(i)$

$$Mb(i) := \begin{bmatrix} I(i) & m(i)g^{\times}(i) \\ -m(i)g^{\times}(i) & m(i)I \end{bmatrix}$$

where I is 3×3 identity matrix,
$Mm(i) - 6 \times 6$ spatial mass matrix of the motor and gear at $Cm(i)$ written in $R(i)$;
$a(i) - 6 \times 1$ vector of spatial accelerations of link i, written in frame $R(i)$, which in case of a rotational joint,

$$a(i) := \begin{bmatrix} O(i-1,i)\omega(i-1) \times \omega(i) \\ O(i-1,i)(\omega(i-1) \times (\omega(i-1) \times s(i-1,i))) \end{bmatrix},$$

$am(i)$ - the 6×1 vector of spatial accelerations of the motor(gear) driving link i, written in frame $Rm(i)$;
$b(i)$ - 6×1 vector of spatial forces acting on link i at point $C(i)$, written in frame $R(i)$, which for a rotational joint look as

$$b(i) := \begin{bmatrix} \omega(i) \times I(i)\omega(i) \\ mb(i)\omega(i) \times (\omega(i) \times g(i)) \end{bmatrix},$$

$bm(i)$ - the 6×1 vector of spatial forces acting on the motor/gear at point $C(i)$, written in frame $Rm(i)$;
$E(i)$ - the 6×1 vector for the spatial axis of motion for link i, written in $R(i)$, which for a rotational joint is $E(i) := [e^T(i)\ 0\ 0]^T$ or equivalently $E(i) := [0\ 0\ 1\ 0\ 0\ 0]^T$ and
$E^T(i)$ - transpose of $E(i)$, $Em(i)$ - the 6×1 vector for the spatial axis of motion for the motor driving link i, written in $R(i)$, $Em(i) := [em^T(i)\ 0\ 0\ 0]^T$,

or $Em(i) := [0\,0\,1\,0\,0\,0]^T$ and $Em^T(i)$ - transpose of $Em(i)$.

The mathematical models of kinematics, dynamics and control of rigid body manipulators are treated in details elsewhere by Mladenova 1990, 1991,1992 'through" and in "pure" vector-parameter language. The idea of the second approach is that the geometry, kinematics and dynamics of an open loop kinematic chain should be described in a new extended configurational space with a group structure - the space of vectors describing the joints displacements (vector-parameters c and translational vectors tr) as it is presented above in the paper. In this manner the transition from vector-parameters to generalized coordinates is saved and computational efficiency proved in the first approach becomes here even a stronger one. In this way the kinematic and dynamic equations in are pure algebraic and the differential equations of motion "feel" the Lie group structure. In this frame the merge of the powerful theory of Lie groups and Lie algebras with fundamental problems of controllability and observability of manipulator systems is quite natural and promissing.

3. Inverse Dynamic Problem (IDP) for Elastic Joint Manipulators.

The Inverse Dynamic Problem (IDP) plays an important role in real time simulation and control. The problem is treated here "through" vector-parameter and spatial operator technique as developed by Rodrigues et al., 1992 is also used (see also Murphy, 1992).

3.1. EXTERNAL ITERATIVE PROCEDURE (FORWARD RECURSION)

This procedure is realized from the first link to the last one. It contains the propagation of the velocities and the accelerations. The velocity of each link is the sum of the velocity of the previous link and the link's own relative rotational velocity. The velocity of each motor is the velocity of the link upon which it is mounted plus the local rotational velocity of the motor. For the flexible joint, the motor velocity is independent of the link velocity. For a geared joint, the motor velocity and link velocity are related through the gear ratio, Γ.
Similar equations hold for the accelerations but with additional terms added to account the centrifugal accelerations. The accelerations are contained in $a(i)$ and $am(i)$ terms. The base is assumed to be fixed with respect to an inertial frame. The recursive equations in an algorithmic form seem like:

$$V(0) := \text{Base Spatial Velocity} \qquad W(0) := \text{Base Spatial Acceleration} \quad (3)$$

For $i = 1, 2, \ldots, n$ we have

$$\begin{aligned} \text{link}: \ V(i) &:= \Phi(i,i-1)V(i-1) + E(i)\dot{q}(i) \\ W(i) &:= \Phi(i,i-1)W(i-1) + E(i)\dot{q}(i) + a(i) \\ \text{motor}: \ Vm(i) &:= \Phi m(i,i-1)Vm(i-1) + E(i)\dot{q}(i) \\ Wm(i) &:= \Phi m(i,i-1)Wm(i-1) + E(i)q\ddot{m}(i) + am(i) \\ \text{Geared}: \ q\dot{m}(i) &:= \Gamma(i)\dot{q}(i); \ \ q\ddot{m}(i) = \Gamma(i)\ddot{q}(i). \end{aligned} \tag{4}$$

3.2. INTERNAL ITERATIVE PROCEDURE (BACKWARD RECURSION)

It is realized from the end-effector to the base. Independent of the geared or flexible joint case, the force acting over every link is a sum of the link inertial forces, the forces from the outer link and the motor inertial forces

$$t(n+1) := \text{vector of external torques(forces) applied on the end-effector} \tag{5}$$

$$tm(n+1) := 0. \tag{6}$$

Besides, for $i = n, \ldots, 1$ we have that

$$\begin{aligned} \text{link}: t(i) &:= \Phi^T(i+1,i)t(i+1) + \Phi m^T(i+1,i)tm(i+1) \\ &+ \ M(i)W(i) + b(i) \\ \text{motor}: tm(i) &:= Mm(i)Wm(i) + bm(i). \end{aligned} \tag{7}$$

The next step aims obtaining of motor and link torques, namely: The torque acting on the joint axis i is the projection of the force acting on the link i along the axis of the joint,

$$P(i) = E^T(i)t(i). \tag{8}$$

This torque is also a function of the state-dependent coupling between the motor and the link. For example, for a simple torsional spring model, the torque is a function of the angular displacement between the motor and the link

$$P(i) = K(i)(qm(i) - q(i)), \tag{9}$$

where $K(i)$ is the torsional spring constant for *ith* axis. Through a force balance on the motor and the link, the motor torque can be expressed as the sum of the projection of the forces along the motor axis and the link torque

$$Pm(i) = E^T(i)tm(i) + P(i). \tag{10}$$

4. Dynamic Problem and Control

4.1. DIRECT DYNAMIC PROBLEM

The goal of simulation, or forward dynamics, is to find the motor and link accelerations when the motor(link) angular positions, velocities and motor torques are given. A lot of algorithms exist. They are also recursive and their main idea is that for any link, the force on the link may be written as

$$t(i) = A(i)W(i) + B(i). \tag{11}$$

In Mladenova 1990, 1991, we have presented Eqn (1) as

$$H(q)\ddot{q} + h_1(q,\dot{q})\dot{q} + h_2(q) = P, \tag{12}$$

where in $h_1(q,\dot{q})\dot{q}$ enter centrifugal terms (these which contain $\dot{q}^2$) and Coriolis terms (these which contain $\dot{q}\dot{q}$), and $h_2(q)$ is the gravity vector. We consider now a model of an n-link manipulator with joint flexibility. For simplicity we assume (as it is done by Spong, 1987) that the joints are revolute, they are actuated by DC motors and the flexibility of i-th joint is modeled as a linear torsional spring with spring constants $K(i)$ for $i = 1, \ldots, n$. We note that due to the joint flexibility, there are now twice as many degrees of freedoms as compared with the rigid joint case. We denote

$$\bar{q} := [q_1 \ldots q_{2n}]^T \tag{13}$$

the set of generalized coordinates of the system, where

$$q_{2i-1} := q(i) \quad \text{(the angle of link i)}, \qquad q_{2i} := -\frac{1}{\Gamma(i)} qm(i). \tag{14}$$

In this case $q_{2i} - q_{2i-1}$ is the elastic displacement of joint i. If we define the n-dimensional vectors $\bar{q}_1 := [q_1\, q_3\, \ldots\, q_{2n-1}]^T$, $\bar{q}_2 := [q_2\, q_4\, \ldots\, q_{2n}]^T$ then the dynamics of elastic joint manipulator can be described by

$$H(\bar{q}_1)\ddot{\bar{q}}_1 + h_1(\bar{q}_1, \dot{\bar{q}}_1) + h_2(\bar{q}_1) + K(\bar{q}_1 - \bar{q}_2) = 0 \tag{15}$$

$$Jm\ddot{\bar{q}}_2 - K(\bar{q}_1 - \bar{q}_2) = u. \tag{16}$$

Jm is a $n \times n$ diagonal rotors inertia matrix.
The input vector u in (16) (input torque applied to the motor shaft) which has units of torque is a function of the generalized forces and moments of the rigid mechanical system and some motor characteristics and constants. The Eqs (15) and (16) may be combined in the form

$$(H(\bar{q}) + Jm)\ddot{\bar{q}} + h_1(\bar{q}, \dot{\bar{q}}) + h_2(\bar{q}) = u. \tag{17}$$

Viscous friction terms acting both on link and on the motor sides of the elastic joints could be easily included in the dynamic model. This model describes system dynamics "through" vector-parameter coordinates.
How does the dynamic model looks in "pure" vector-parameter notation? Analogically we can define the $3n-$dimensional vectors:

$$\bar{c}_1 := [c_1\, c_3 \ldots c_{2n-1}]^T, \quad \bar{c}_2 := [c_2\, c_4 \ldots c_{2n}]^T$$

and using the formalism described already in Mladenova, 1991 we obtain

$$H_V(\bar{c}_1)\ddot{\bar{c}}_1 + h_{V1}(\bar{c}_1, \bar{c}_1) + h_{V2}(\bar{c}_1) + K_V(\bar{c}_1 - \bar{c}_2) = 0 \tag{18}$$

$$Jm_V\ddot{\bar{c}}_2 - K_V(\bar{c}_1 - \bar{c}_2) = u_V \tag{19}$$

$$(H_V(\bar{c}) + Jm_V)\ddot{\bar{c}} + h_{V1}(\bar{c}, \dot{\bar{c}}) + h_{V2}(\bar{c}) = u_V. \tag{20}$$

However, the real dimension of the differential equations is just $2n$ because only one of the components of the vectors $c(i)$ is essential and informative. Here the subindex V means again that the model is considered in "pure" vector-parameter approach.
We note that the gyroscopic forces between each rotor and the other links are not included in the last equations since we assume that the kinetic energy of the rotors is due mainly to its own rotation, equivalently, the motion of the rotor is a pure rotation with respect to an inertial frame.

4.2. INVERSE DYNAMICS CONTROL AND FEEDBACK LINEARIZATION

It is well known that the rigid robot Eqn (17) may be globally linearized and decoupled by nonlinear feedback. This is just the scheme of the inverse dynamic control. The key idea of the inverse dynamics is to seek a nonlinear feedback control law $u = f(\bar{q}, \dot{\bar{q}}) \quad u_V = f_V(\bar{c}, \dot{\bar{c}})$, which after entering in the Eqn (17) or (20) results in a linear closed loop system. For example applying the nonlinear control law to the system described by Eqn (17)

$$u = A(\bar{q})v + B(\bar{q}, \dot{\bar{q}}), \quad A(\bar{q}) = H(\bar{q}) + Jm, \quad B(\bar{q}, \dot{\bar{q}}) = h_1(\bar{q}, \dot{\bar{q}}) + h_2(\bar{q}) \tag{21}$$

which is also called the inverse dynamic control, the so called double integrator system is obtained $\ddot{\bar{q}} = v$. The new system is linear and decoupled and it can be controlled by adding an "outer loop" control. We have to note that here v does not mean velocity vector but control variable. The technique of the inverse dynamic control may be considered as a special case of a more general procedure for transforming a nonlinear system to a linear one, known as external or Feedback Linearization (see Isidori,1985; Nicosia et al., 1988; Nijmeijer et al., 1990; Spong, 1987, and etc.). In the case of rigid manipulators the inverse dynamic problem and feedback linearizations are covered each other. The power of the techniques of the feedback linearization may be really seen when manipulator systems with elastic joints are considered. In the general case of n-link manipulator, the dynamic equations represent a multi-input nonlinear system.

5. Conclusion

The considerations concerning rigid body manipulators are extended here for the elastic joint manipulators. Because of additional degrees of freedom introduced by the elastic coupling of the motors to the links we model the rotor of each actuator as a fictious link, i.e. an additional rigid body in the chain with its own inertia. The resulting nonlinear equations of motion are globally linearizable by smooth invertible coordinate transformation and inner control low.

Acknowledgments.

This paper is partially supported by Bulgarian NSF - Grant No. MM 445/94.

References

Isidori A. (1985) *Nonlinear Control Systems: An Introduction, In: Lecture Notes in Control and Inf. Sciences*, **Vol. 72**, Springer-Verlag, Berlin.

Mladenova C. (1990) A contribution to the modelling and control of manipulators, *J. Intell. Robot. Syst.*,**no. 3**, pp. 349-363.

Mladenova C.(1991) Mathematical modelling and control of manipulator systems, *Robot. Comput. Integr. Mfg.*, **no. 8**, pp. 233-242.

Mladenova C. (1992) Robot Modeling over Extended Configurational Manifold, *Proc. 1992 IEEE Int. Conf. on Robot & Automation, Nice, France*, **Vol. 1**, pp. 326-331.

Murphy S.H. (1992) Simulation of space manipulators,*In Intelligent Robotic Systems for Space Exploration (Edited by Desrochers A.A.)*, Kluwer Acad. Publishers, Boston, pp.257-295.

Nicosia S. and Tomei P. (1988) On the feedback linearization of robots with elastic joints, *Proc. 27th Conf. on Decision and Control, Austin, Texas*, pp. 180-185.

Nijmeijer H. and Schaft A. (1990) *Nonlinear Dynamical Control Problems*, Springer-Verlag.

Rodrigues G.,Jain A., Kreutz-Delgado K. (1992) Spatial operator algebra for multibody system dynamics, *J. Astronautical Sci.*, **Vol. 40, no. 1**, pp. 27-50.

Spong M.W.(1987) Modeling and control of elastic joint robots, *Trans. ASME, J. Dyn. Syst. Meas. Cont.*, **no. 109**, pp. 310-319.

SIMULATION OF GRASPING WITH A FLEXIBLE-JOINT ROBOT

J. KÖVECSES
Department of Mechanics, University of Miskolc
Miskolc-Egyetemváros, Hungary, H-3515

AND

R.G. FENTON AND W.L. CLEGHORN
Department of Mechanical Engineering, University of Toronto
5 King's College Rd., Toronto, Ontario, Canada M5S 1A4

Abstract. In this paper, the motion of a spatial two-arm robot manipulator undergoing grasping and lifting a three dimensonal object is investigated. The manipulator has rigid links. The joints are modeled as flexible, continuous shafts. The object is travelling on a conveyor. A mechanical model is derived for the problem. The effects of grasping, when the end effector of the manipulator approaches, suddenly grasps and lifts are analysed. Simulation results are presented.

1. Introduction

One of the causes of vibration during motion of a robot is that the transmission systems between the driving motors and the robot links contain flexible elements. Even in the case of a direct-drive robot (investigated by Asada and Youcef-Toumi, 1987 in detail), there is at least one element in the driving system (the motor shaft) which is much more flexible than other parts of the robot, i.e. robotic joints are generally flexible. Furthermore, the robot's flexibility is necessary in a number of robotic operations, to prevent the fracture of the robot's components. In order to model joint flexibility and investigate its effects on the robot's operation, many papers have been published (for instance Anderson, 1985, Dado and Soni, 1986, Spong, 1987, Potkonjak, 1988, Readman, 1994, Xi and Fenton, 1994). In these papers, joint flexibility is modeled by discrete springs and the mass of the joints is

J. Lenarčič and V. Parenti-Castelli (eds.), Recent Advances in Robot Kinematics, 57-66.
© 1996 *Kluwer Academic Publishers. Printed in the Netherlands.*

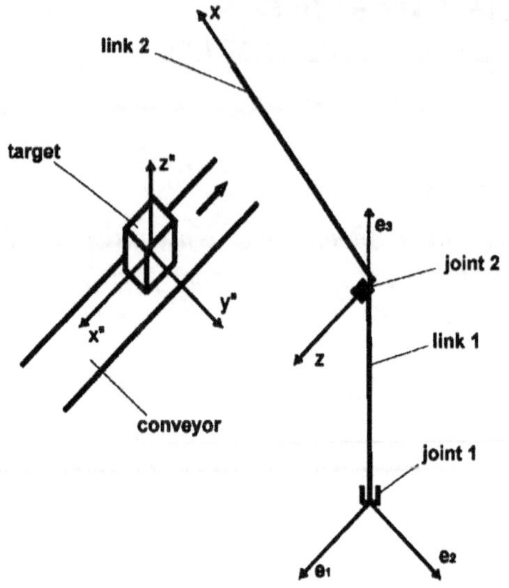

Figure 1. A sketch of the system.

normally neglected. Due to this assumption, the flexibility properties have been completely described by using the stiffness coefficients of the joint or the transmission system (Potkonjak, 1988).

This paper investigates a robot model in which each robotic arm is modeled as an assembly of a rigid rotor, a flexible joint and a rigid link. Moreover, the flexible joint is modeled as a continuous circular shaft with torsional flexibility. In this model, other transmission elements are not taken into consideration, (the driving torques directly act on the rotors). However, the investigation may easily be extended to include more elements.

Another topic discussed in this paper is the investigation of a robotic operation in which a two-arm manipulator approaches, suddenly grasps and lifts a target moving on a conveyor. A sketch of the system is shown in Figure 1. At the time of grasping, the target suddenly becomes part of the manipulator. This process causes impulsive effects on the system (robot + target) which will be investigated. The reduction of the impulsive effects is also discussed.

2. Modeling the Joints

In order to describe the motion of a flexible joint, the concept of fictitious rigid joints is used. A fictitious rigid joint is attached to each of the joints

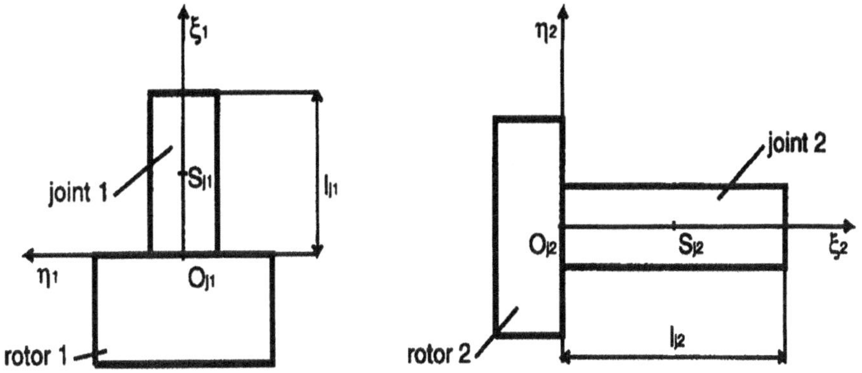

Figure 2. Models for the flexible joints.

in theory. Fictitious rigid joint i can be identified as the undeformed state of joint i having the same mass distribution as the real joint does (rigid counterpart of the real joint). Using this concept, motion of the joints may be divided into rigid body motion described by the fictitious rigid joints and flexible body motion (deflection). For describing joint deflections, coordinate system ξ_i, η_i, ζ_i is fixed to rotor i according to Figure 2 ($i = 1, 2$). Point S_{ji} denotes the center of mass of fictitious rigid joint i. Only torsional flexibility of the joints is considered. The time- and space dependent coordinate $\varphi_i(\xi_i, t)$ describes the deflections of joint i with respect to the coordinate system attached to rotor i. Let J_i be an arbitrary point of joint i, and I_i be the corresponding point in the undeformed state indicated by coordinates ξ_i, η_i, ζ_i. Then the position, velocity and acceleration of point J_i with respect to fictitious rigid joint i are described as

$$\vec{r}_{OJ_i} = \vec{r}_{OI_i} - \varphi_i \zeta_i \vec{e}_{\eta i} + \varphi_i \eta_i \vec{e}_{\zeta i} = \vec{r}_{OI_i} + \vec{u}_i, \tag{1}$$

$$\overset{*}{\vec{r}}_{OJ_i} = -\dot{\varphi}_i \zeta_i \vec{e}_{\eta i} + \dot{\varphi}_i \eta_i \vec{e}_{\zeta i} = \overset{*}{\vec{u}}_i, \tag{2}$$

$$\overset{**}{\vec{r}}_{OJ_i} = -\ddot{\varphi}_i \zeta_i \vec{e}_{\eta i} + \ddot{\varphi}_i \eta_i \vec{e}_{\zeta i} = \overset{**}{\vec{u}}_i, \tag{3}$$

where $\vec{r}_{OI_i} = \xi_i \vec{e}_{\xi i} + \eta_i \vec{e}_{\eta i} + \zeta_i \vec{e}_{\zeta i}$ is the position vector of point I_i with respect to origin O_{ji} and $\vec{u}_i$ is the deflection vector. The position, velocity and acceleration of point J_i with respect to the base (inertial) frame can be described as

$$\vec{r}_{J_i} = \vec{r}_{O_i} + \vec{r}_{OI_i} + \vec{u}_i = \vec{r}_{I_i} + \vec{u}_i, \tag{4}$$

$$\vec{v}_{J_i} = \dot{\vec{r}}_{J_i} = \vec{v}_{I_i} + \dot{\vec{u}}_i = \vec{v}_{I_i} + \vec{\omega}_{irot} \times \vec{u}_i + \overset{*}{\vec{u}}_i, \tag{5}$$

$$\vec{a}_{J_i} = \ddot{\vec{r}}_{J_i} = \vec{a}_{I_i} + \ddot{\vec{u}}_i = \vec{a}_{I_i} + \vec{\varepsilon}_{irot} \times \vec{u}_i + \vec{\omega}_{irot} \times (\vec{\omega}_{irot} \times \vec{u}_i) + 2\vec{\omega}_{irot} \times \overset{*}{\vec{u}}_i + \overset{**}{\vec{u}}_i, \tag{6}$$

where $\vec{\omega}_{irot}$, $\vec{\varepsilon}_{irot}$ are the angular velocity and acceleration vectors of rotor i which are equal to the angular velocity and acceleration vectors of fictitious rigid joint i. Using these quantities the kinetic and acceleration energy functions of the joints can be obtained as

$$T_{ji} = \frac{1}{2}\left(m_{ji}\vec{v}_{S_{ji}} \cdot \vec{v}_{S_{ji}} + \vec{\omega}_{irot} \cdot \hat{J}_{S_{ji}} \cdot \vec{\omega}_{irot}\right) +$$

$$+\frac{1}{2}\int_0^{l_{ji}}\int_{A_i} \rho_i \left(2\vec{v}_{I_i} \cdot \dot{\vec{u}}_i + \dot{\vec{u}}_i^2\right) dA_i d\xi_i \tag{7}$$

and

$$G_{ji} = \frac{1}{2}\int_0^{l_{ji}}\int_{A_i} \rho_i \left(2\vec{a}_{I_i} \cdot \ddot{\vec{u}}_i + \ddot{\vec{u}}_i^2\right) dA_i d\xi_i +$$

$$+\frac{1}{2}\left[m_{ji}\vec{a}_{S_{ji}} \cdot \vec{a}_{S_{ji}} + \vec{\varepsilon}_{irot} \cdot \hat{J}_{S_{ji}} \cdot \vec{\varepsilon}_{irot} + 2\vec{\varepsilon}_{irot} \cdot \left(\vec{\omega}_{irot} \times \left(\hat{J}_{S_{ji}} \cdot \vec{\omega}_{irot}\right)\right)\right] + \ldots \tag{8}$$

where m_{ji} is the mass of joint i, $\vec{v}_{S_{ji}}$ and $\vec{a}_{S_{ji}}$ are the velocity and acceleration of the center of mass of fictitious rigid joint i, $\hat{J}_{S_{ji}}$ is the tensor of inertia of fictitious rigid joint i around its center of mass. In the acceleration energy functions, only those parts are indicated which contain acceleration vectors. The developed joint model is not limited to the special manipulator configuration investigated in this paper. It may be adopted for any robot configuration with arbitrary degrees of freedom.

3. Constraint Equations

At time t_1, the manipulator suddenly grasps and starts to lift the target. Before grasping, the manipulator moves freely and the motion of the target is prescribed, it travels on a conveyor with velocity $\dot{q}_c$. After grasping, the manipulator lifts the target. The six constraint equations which are imposed on the system may be derived from vector equations $\vec{v}_{S_o} = \vec{v}_{O_r}$ and $\vec{\omega}_o = \vec{\omega}_{tool}$, where $\vec{v}_{S_o}$ is the velocity of the center of mass of the target, $\vec{v}_{O_r}$ is the velocity of the reference point of the end effector, $\vec{\omega}_o$ is the angular velocity of the target, and $\vec{\omega}_{tool}$ is the angular velocity of the end effector. At the time of grasping, t_1, the object becomes part of the manipulator and these constraint equations suddenly emerge. This is the point of discontinuity of the constraint equations because at t_1-0, all the coefficients are zero, and at t_1+0, the coefficients are not zero. As a result, impulsive effects are induced, and the velocity set becomes discontinuous. For further investigations, the time interval of the motion is divided into two continuous domains in which the constraint equations are continuous, and a discontinuous domain in the small neighbourhood of t_1, time interval $[t_1 - 0, t_1 + 0]$. The discontinuous domain separates the continuous domains. The continuous domains may be identified as follows: in the first, the robot approaches the target, and in the second, lifts it. The discontinuous domain is the short time interval when the robot grasps the target and the system's degrees of freedom are suddenly decreased. Since in this interval the position of the system does not change, the constraint equations mentioned above have to be considered as nonholonomic equations in the vicinity of grasping. In the continuous domains, the dynamic equations can be derived as the equations of motion. However, at the time of grasping, in the discontinuous domain, special dynamic equations, impulsive equations are needed to connect the systems of equations of motion before and after t_1. A detailed derivation and analysis of the equations of motion and the impulsive equations have been performed by the authors.

4. Computational Investigation of Grasping

Let us consider a numerical example. The reference position of the manipulator is when its second link is horizontal, and perpendicular to the conveyor. q_1, q_2 and k_1, k_2 will denote the relative rotations of the rotors and the links about the joint axes. The kinematic and dynamic parameters of the model investigated can be found in (Kövecses *et al.*, 1995). The task of the manipulator is to approach, grasp and lift the target travelling on the conveyor. The desired angular veloctiy profile of the second link with respect to the first link is shown in Figure 3. At t_1 ($t_1 = 1.5s$), the manipulator grasps the target. In the first step of the investigations, it is assumed that the driving torque of the second rotor has been determined according to this velocity profile, without considering the effects of joint flexibility, the impulsive motion and the motion of the previous link. First let us consider that the initial positions of the first rotor, link and the second rotor are given by coordinates $q_1 = k_1 = 0$ and $q_2 = -1\ rad$, furthermore $\dot{q}_1$ is zero during the robotic operation. In other words, the first rotor is held in a fix position, the motion of the first link is caused only by joint flexibility. This fact imposes an additional constraint equation on the system as

$$\dot{q}_1 = 0, \qquad \dot{q}_1|_{t_1+0} = \dot{q}_1|_{t_1-0}. \tag{9}$$

Simulation results derived in this case are presented in Figures 4 and 5. Before grasping, in time interval $(0, 1.5s)$, when the manipulator approaches the target, the motion of the system is almost the same as desired with additional small vibrations. However, the grasping induces relatively large vibrations in the first joint and hence in the motion of the first link. After grasping these vibrations parametrically excite the vibration of the second joint, and hence the motion of the second rotor and link are also excited. The vibratory motion of the second link is strongly increased in the domain $1.5s$ to $3s$. In the mechanical modeling of this robotic operation, it is necessary to model at least the first joint as a flexible component, because the impulsive motion caused by the robotic grasping of the moving target mainly influences the vibration of this element. Otherwise, if all parts of the manipulator are rigid, the most important property of the robotic grasping that the target moves in the direction of axis $-e_1$ (Figure 1), cannot be taken into consideration in the presence of constraint Equation (9).

In order to reduce the effects discussed above, consider first a possibility concerning the relative motion of the second manipulator arm. Constraint Equation (9) concerning the motion of the first rotor is further valid. Let us prescribe an additional constraint equation

$$\dot{q}_2 = f(t), \qquad \dot{q}_2|_{t_1+0} = \dot{q}_2|_{t_1-0}, \tag{10}$$

where $f(t)$ is the desired velocity profile as shown in Figure 3. This means that we prescribe the relative motion of the second rotor, rather than the motion of the second link, to be as desired. The ideal solution would be if we prescribed the relative motion of the second link, however, it generally does not lead to useful results, since the control input acts on the rotor. In this case, the simulation results obtained for the motion of the first joint and link are almost the same as shown in Figure 4. The simulation result for the motion of the second link is shown in Figure 6. It can be seen that the motion of the second link approaches the desired profile better than in the previous case. The motion of the system can be acceptable if the displacements caused by the vibrations of the first joint can be neglected from the point of view of the robotic operation.

It can be seen from Figure 4 that holding the first rotor in a fix position causes large vibrations of the first joint. Consider now another possible design of the robotic operation in which the desired motion of the second link with respect to the first link is the same as before, but the first rotor is not held in a fix position. The motion of this robotic part is prescribed to reduce the effects of grasping. Let us investigate the case when the motion of the first rotor is designed in such a way that, at the time of grasping, the velocity of the end effector of the manipulator would be the same as the velocity of the target if all the parts of the manipulator are modeled as rigid members. A possible solution of this problem is outlined here. Let the initial position of the first rotor be $q_1 = k_1 = 1.5\ rad$. The first rotor, joint and link start to move from the position given by these values. The constraint Equation (9) can be reformulated, in order to satisfy the above

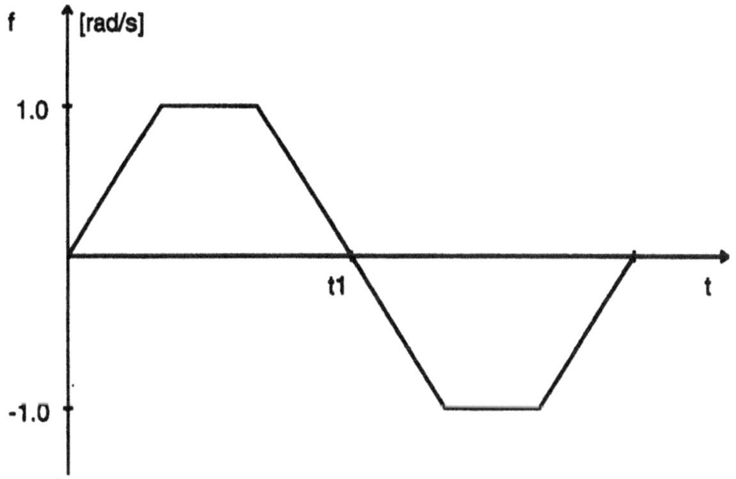

Figure 3. Desired relative angular velocity of the second link.

condition, as

$$\dot{q}_1 = \frac{\dot{q}_c}{\ell_2 t_1} t, \ \ if \ \ t < t_1; \ \ \ \dot{q}_1 = \frac{\dot{q}_c}{\ell_2} e^{-\lambda(t-t_1)}, \ \ if \ \ t > t_1; \ \ \ \dot{q}_1|_{t_1+0} = \dot{q}_1|_{t_1-0}, \tag{11}$$

where ℓ_2 is the length of the second link, $\lambda = -\frac{\dot{q}_c}{t_1}$ and $\dot{q}_c$ is the velocity of the conveyor. The specifications for the motion of the second rotor and link remain the same as before. Simulation results computed in this case concerning to the motion of the second link are almost the same as it is shown in Figure 6. However, motion of the first link is significiantly different from the previous results. These values are shown in Figures 7. In this case, the vibrations are reduced in the first joint too.

5. Conclusions

In this paper, the modeling of a flexible joint robot manipulator and the analysis of an important robotic operation have been discussed. For modeling the joint flexibility, a new approach is presented in which the joints are modeled as continuous shafts. During the investigated operation the robot approaches, grasps and lifts a moving target. This process imposes discontinuous constraint equations on the system consisting of the manipulator and the target. Impulsive effects are caused by these discontinuities. It is shown that relatively small vibrations are induced by the motion of the manipulator before grasping, but the impulsive effects of grasping and the parametrically excited joint vibrations together cause fast and relatively

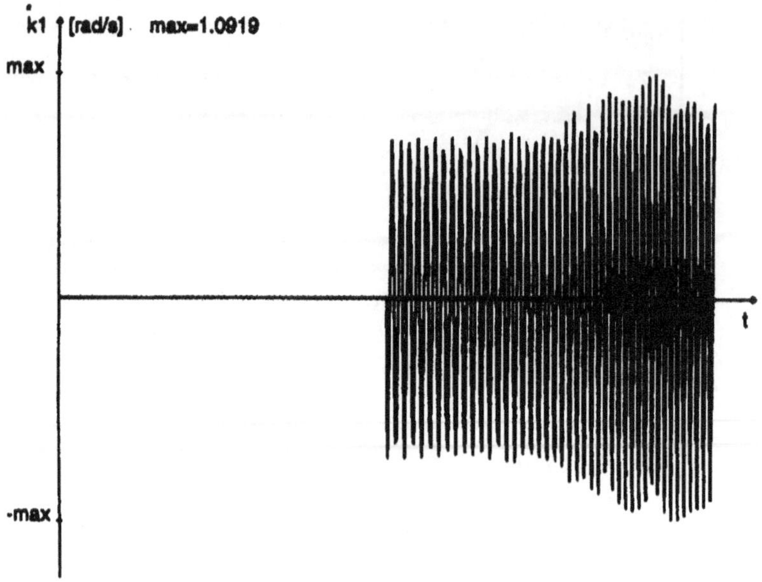

Figure 4. Angular velocity response of the first link with respect to the base, when the manipulator motion is constrained by Equation (9).

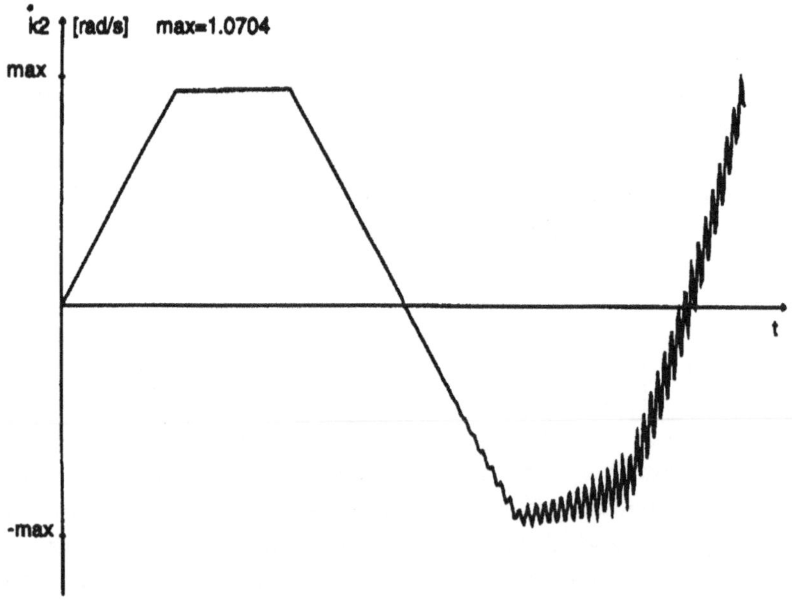

Figure 5. Angular velocity response of the second link with respect to the first link, when the manipulator motion is constrained by Equation (9).

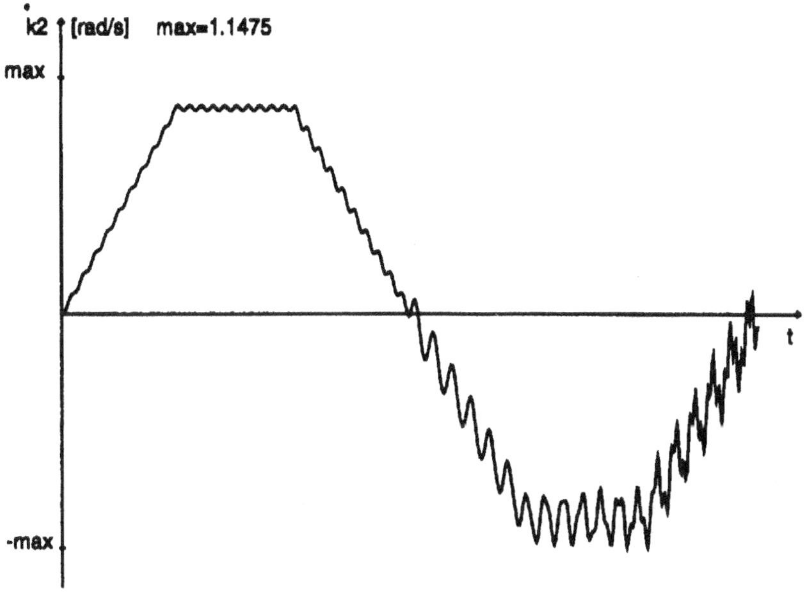

Figure 6. Angular velocity response of the second link with respect to the first link, when the manipulator motion is constrained by Equations (9) and (10).

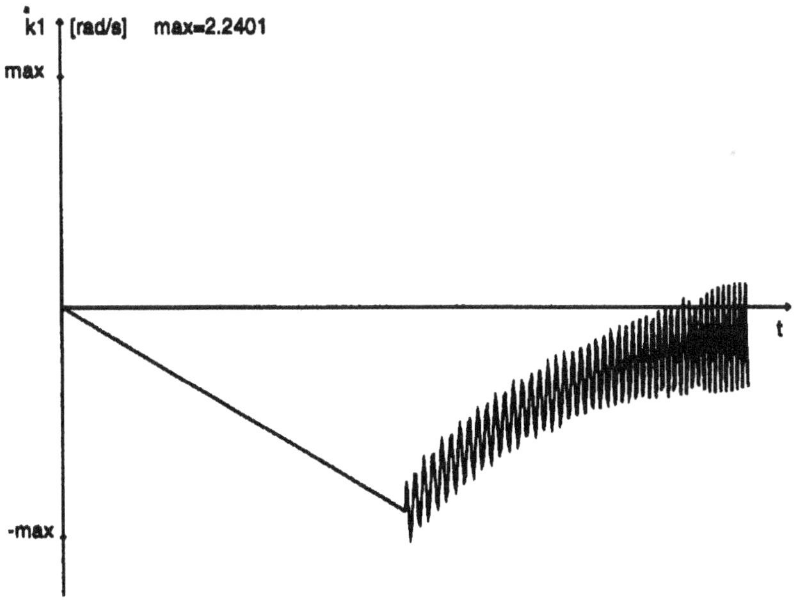

Figure 7. Angular velocity response of the first link with respect to the base, when the manipulator motion is constrained by Equations (10) and (11).

large vibrations in the system after grasping. The vibrations of the joints have small displacements, but these vibrations are very fast. The high velocities and accelerations significiantly influence the rigid body motion of the manipulator elements. Procedures are suggested and investigated to reduce these effects.

6. Acknowledgement

This work was supported in part by the Hungarian National Research Foundation (OTKA), under grant T 016 836.

References

Anderson, G.L. (1985) Stability of a Manipulator with Resilient Joints, *J. Sound and Vibration* Vol. 101, No. 4, pp. 463-480.

Asada, H. and Youcef-Toumi, K. (1987) Direct Drive Robots, Theory and Practice, The MIT Press, Cambridge.

Dado, M. and Soni, A.H. (1986) A Generalized Approach for Forward and Inverse Dynamics of Elastic Manipulators, *Proc. 1986 IEEE Conf. on Robotics and Automation* pp. 359-364.

Kövecses, J., Fenton, R.G. and Cleghorn, W.L. 1995 The Dynamic Effects of Robotic Grasping Considering Flexible Joints, *Proc. 1995 Design Engineering Technical Conferences of the ASME DE-Vol. 84-3* pp. 1375-1383.

Potkonjak, V. (1988) Contribution to the Dynamics and Control of Robots Having Elastic Transmissions, *Robotica* vol. 6, pp. 63-69.

Readman, M.C. (1994) Flexible Joint Robots, CRC Press.

Spong, M.W., (1987) Modeling and Control of Elastic Joint Robots, *ASME J. Dynamic Systems, Measurement, and Control* Vol. 109, pp. 310-319.

Xi, F. and Fenton, R.G. (1994) Coupling Effect of A Flexible Link and A Flexible Joint, *J. Robotics Research* Vol. 13, No. 5, pp. 443-453.

ON-LINE COLLISION-RECOGNITION AND COLLISION-AVOIDANCE CONTROL FOR REDUNDANT ARTICULATED ROBOTS

N. RAHMANIAN-SHAHRI AND I. TROCH
Technical University Vienna
Wiedner Hauptstrasse 8-10, A-1040 Wien, Austria
Tel: 0043-1-58801-5367, Fax: 0043-1-5868093
email: nrahman@ws1.atv.tuwien.ac.at, itroch@email.tuwien.ac.at

Abstract. This work [1] presents a suitable mathematical formulation of robot and obstacles such that for on-line collision recognition only robot joint positions in the workspace are required. This reduces calculation time essentially, because joint positions in workspace can be computed at every time from the joint variables through robot geometry. It is supposed that the obstacles in the workspace of the manipulator are represented by convex polygons in 2D. The recognition of collisions of the links of the manipulator with obstacles results on-line through a nonsensory method. For every link of the redundant robot and every obstacle a boundary ellipse in 2D is defined in workspace such that there is no collision if the robot joints are outside these ellipses. In addition to this, two collision avoidance methods are presented which allow the use of redundant degrees of freedom such, that a manipulator can avoid obstacles in workspace while tracking the desired end-effector trajectory. Moreover, the new methods are applied directly in workspace. This paper gives also a comparison of these two methods. The effectiveness of the porposed methods is discussed by theoretical considerations and illustrated by simulation of the motion of a planar four-link manipulator between obstacles.

1. Introduction

A robot manipulator is defined to be kinematically redundant if it possesses more degrees of freedom than are required to achieve the desired position and orientation of the end-effector. One of the advantages of robot redundancy is the potential to use the extra degrees of freedom to manoeuver in a congested workspace and to avoid collisions with obstacles, where obstacles are defined as objects in the robot workspace.

Generally, there are two kinds of obstacle-avoidance strategies for robot manipulators. One deals with high-level path planning, in which the end-effector path is planned off-line such that collisions with workspace obstacles are avoided. This path planning problem has been studied by Lozano-Perez [1], Lumelsky [2],

[1] Supported by the Austrian Funds for the Promotion of the Scientific Research, grant Nr. P10413-MAT

J. Lenarčič and V. Parenti-Castelli (eds.), Recent Advances in Robot Kinematics, 67-76.
© 1996 *Kluwer Academic Publishers. Printed in the Netherlands.*

Freund and Hoyer [3], Gilbert and Johnson [4], and others, and is not discussed in this article.

Alternatively, the obstacle avoidance problem of redundant robots can be solved on-line by the robot controller at the low level such that a redundant robot closely tracks the desired end-effector trajectory and simultaneously avoids workspace obstacles. In the field of kinematic control of redundant manipulators many algorithms have been proposed, mainly based on using a pseudoinverse matrix as reviewed by Klein and Huang [5]. The obstacle avoidance problem has been studied by many researchers, including Maciejewski and Klein [6], Nakamura et al. [7], Anthimopoulou and Aspragathos [8],Baillieul [9], Chen and Vidyasagar [10], Oh et al [11], Sciavicco and Siciliano [12], Das et al. [13], Khatib [14], Colbaugh et al [15], and Bagchi and Hatwal [16].

This article reviews a new method for on-line collision-recognition for robot manipulators. It is supposed that the obstacles in the workspace of the manipulator are represented by convex polygons. The recognition of collisions of the links of the manipulator with obstacles results on-line by a nonsensory method. For every link of the redundant robot and every obstacle a boundary ellipse is defined in workspace such that there is no collision if the robot joints are outside these ellipses. In addition to this, two collision avoidance methods are presented which allow the use of redundant degrees of freedom such that a manipulator can avoid obstacles in the workspace while tracking the desired end-effector trajectory. Moreover, the new methods are applied directly in workspace. The optimization method is based on the generalized inverse with boundary ellipse functions as optimization criteria. The method permits the tip of the hand to approach any arbitrary point in the free space while the kinematic control algorithm maximizes the boundary ellipse function of the critical link. The self-motion-control method is based on coordinate transformation and inverse kinematics and leads to the favorable use of the abilities of redundant robots to collision-avoidance. In case a collision is imminent, the collision-avoidance algorithm computes the self-motion movements necessary to avoid the collision. This method has the advantage that the configuration of the manipulator after collision-avoidance can be influenced by further requirements such as avoidance of singularities, joint limits etc. The algorithms are developed for a planar manipulator with rotational joints moving on a plane cluttered with obstacles. Future work will extend the algorithm to the kinematic control of a general manipulator moving on a three dimensional space cluttered with obstacles.

2. Collision recognition

It is supposed that the obstacles in the workspace of the manipulator are represented by convex polygons. One way to prevent collisions between the manipulators and the obstacles is to maintain nonzero distances between the links of the manipulators and the obstacles. Unfortunately, closed-form expressions for these distances are in general quite complicated to determine, even when the obstacles are modeled by primitive objects. A simple way to keep the links of a robot away form the obstacles is to set up a barrier between them. We illustrate this idea through the following example.

Example 1

Consider link l of a planar revolute robotic manipulator and the obstacle $ABCD$ shown in Figure 1.a. Assume that joint $i+1$ of link l moves from point a to point b on the line k which has an arbitrary distance d from the edge AB of the obstacle such that link l moves tangential to the corner point A of the obstacle as shown in Figure 1.a. Joint i of link l moves then along a parabolic curve as

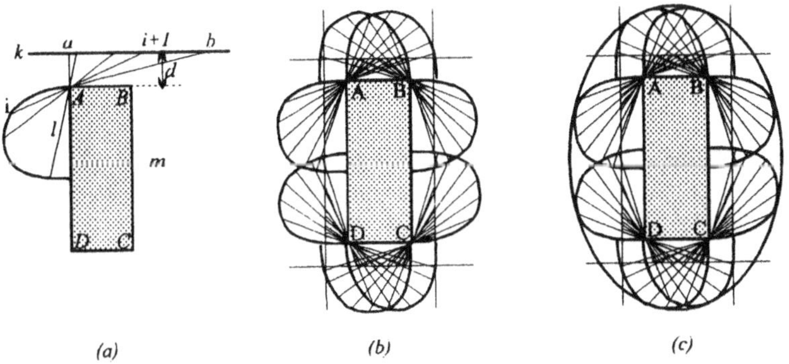

Figure 1. Construction of boundary ellipse relative to link l
(a) Tangential moving of the link l to the corner point A of the obstacle
(b) Parabolic curves at any corner point of the obstacle
(c) Resulting boundary ellipse

shown in Figure 1.a. If joint $i+1$ moves above line k and joint i moves out of this parabolic curve at the left side of the obstacle, there is no collision between link l and the obstacle. This can be repeated for each edge and each corner point of the obstacle such that we get two parabolic curves at any corner point of the obstacle as shown in Figure 1.b. It makes sense to approximate these parabolic curves by an closed algebraic curve in workspace. A rather obvious candidate is an ellipse because of the simplicity of its analytical expression. Moreover, an ellipse can be easily extended to higher dimensions. To prevent collision between link l of the manipulator and the obstacle, it is sufficient that joint i of link l is kept away from this boundary ellipse and joint $i+1$ of link l moves "outside" the lines which are parallel to each edge of the obstacle as shown in Figure 1.c. Joint $i+1$ of link l moves certainly "outside" such lines, if we calculate the boundary ellipse for the next link of the manipulator.

The boundary ellipse function has a negative value for points which lie inside the boundary ellipse, a positive value for points outside and it vanishes for points on the boundary ellipse. If we calculate now the boundary ellipse function of obstacle m relative to link l for the workspace coordinates of joint i, we can decide whether joint i lies inside, outside or on the boundary ellipse. The joint variables of an articulated planar robot can be represented by the vector $\underline{q} = [q_1, q_2, \ldots, q_n]^T$ and world coordinates of the tip of the hand by the vector $\underline{x} = [x, y]^T$. The vectors $\underline{x}$ and $\underline{q}$ are connected by the geometric relation

$$\underline{x} = \underline{f}(\underline{q}) \tag{1}$$

The workspace coordinates of the joints can be computed from the joint variables at any time as

$$x_i = l_1 \cos(q_1) + l_2 \cos(q_1 + q_2) + \ldots + l_i \cos(q_1 + q_2 + \ldots + q_i) \quad (2)$$

$$y_i = l_1 \sin(q_1) + l_2 \sin(q_1 + q_2) + \ldots + l_i \sin(q_1 + q_2 + \ldots + q_i), \quad (3)$$

where $i = 1, 2, \ldots, n$.
The boundary ellipse function of obstacle k can be computed for the joint space coordinate of link i as follows:

$$h_{ik}(x_i, y_i) = h_{ik}(q_1, q_2, \ldots, q_i) \quad (4)$$

Let $h_{lm}(x, y)$ be the boundary ellipse function of obstacle m relative to link l and (x_i, y_i) the position of joint i in workspace. Then, collision recognition of link l relative to obstacle m can be represented as follows:

$$h_{lm}(x_i, y_i) = h_{lm}(\underline{q}) \quad \begin{cases} > 0 & \ldots \text{ there is no collision} \\ = 0 & \ldots \text{ there is possibility of collision} \\ < 0 & \ldots \text{ possibility of collision is high} \end{cases}$$

Nevertheless, there is no boundary ellipse defined for the first link of the manipulator, because, the first joint is hold in its position. Collision-avoidance of the first link can be performed via constraints for this joint, respectively. Therefore, for a manipulator with n links and a workspace, with m obstacles there are $(n-1) * m$ boundary ellipse functions to be established which guarantee that there is no collision if all robot joints are outside their respective ellipses.

3. Collision-avoidance control

3.1. Optimization method

Now, a kinematic control algorithm is developed for a planar redundant manipulator with n degrees of freedom, which allows the use of redundant degrees of freedom such that a manipulator can avoid obstacles in workspace while tracking the desired end-effector trajectory. The links of the manipulator are represented by straight line segments connecting two adjacent joints of each link. The joint variables are represented by the vector $\underline{q} = [q_1, q_2, \ldots, q_n]^T$ and world coordinates of the tip of the hand are represented by the vector $\underline{x} = [x, y]^T$. The vectors $\underline{x}$ and $\underline{q}$ are connected by the geometric relation (1). Linearization leads to the various types of resolved motion rate control (RMRC) algorithms. Liegeois [17] used for redundant manipulators the following equation for the derivatives of the joint variables:

$$\dot{\underline{q}} = \mathbf{J}^+ \dot{\underline{x}} + (\mathbf{I} - \mathbf{J}^+ \mathbf{J}) \alpha \nabla h(\underline{q}), \quad (5)$$

with the Moore-Penrose inverse $\mathbf{J}^+ = \mathbf{J}^T (\mathbf{J}\mathbf{J}^T)^{-1}$, where $\mathbf{J}$ and $\mathbf{J}^T$ denote the Jacobian of $\underline{f}(\underline{q})$ and its transpose respectively, $\mathbf{I}_n$ the identity, α a gain constant, $h(\underline{q})$ an optimization criterion, and $\nabla h(\underline{q})$ its gradient.

The optimization criterion consists in the minimization of the possibility of collision through maximization of the boundary ellipse function of the critical

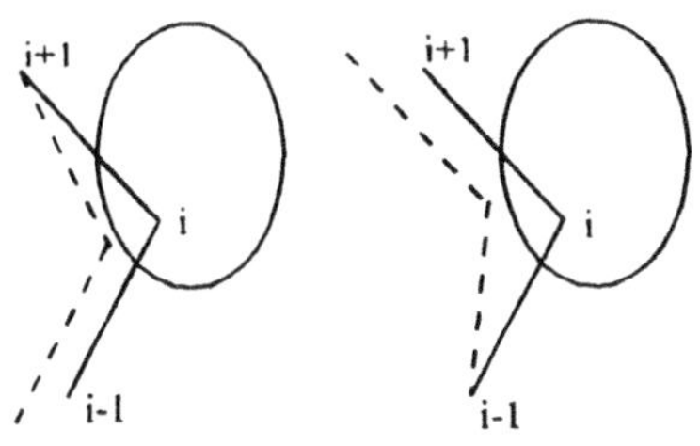

Figure 2. Reconfiguration of links connected with the joint i due to obstacle-avoidance

link. As we mentioned earlier, workspace coordinates of joints can be computed from joint variables at any time from (2) and (3). The boundary ellipse function of any obstacle can be computed for the joint space coordinate of any link by (4). If a joint u lies outside of the danger zone of an obstacle w, then $h_{uw}(x_u,y_u) > 0$ must hold. The $h_{ik}(x_i,y_i)$-values can be used in this case as a measure for the danger of the collision with obstacles. One way to avoid obstacles is now to increase h_{ik} with minimum value. By comparison of the h_{ik}-values we can obtain which link of the manipulator is most exposed to collision with which obstacle:

$$h_{uw}(x_u,y_u) = \min_{ik} h_{ik}(x_i,y_i) \tag{6}$$

To avoid the possibility of collision of link u with obstacle w we can use the fore-mentioned optimization method with $h_{uw}(\underline{q})$ as optimization function instead of $h(\underline{q})$ in (5). The method permits the tip of the hand to track the end effector trajectory while the kinematic control algorithm minimizes the possibility of collision via maximization of the boundary ellipse function of the critical link. A detailed description of the method together with simulation results is given in Rahmanian-Shahri and Troch [18].

3.2 Self-motion-control method

Now, we present a kinematic control algorithm to avoid collision of the links of redundant robots with obstacles through control of the self-motion of the manipulator. In case a collision is imminent, the collision-avoidance algorithm computes the self-motion movements necessary to avoid the collision.

As mentioned earlier the boundary ellipse function of any obstacle can be computed from the joint space coordinates of any link. If a joint i lies outside of the danger zone of an obstacle k, then $h_{ik}(x_i,y_i) > 0$ must hold. The $h_{ik}(x_i,y_i)$-values can be used in this case as a measure for the danger of the collision with obstacles. By comparison of the h_{ik}-values we can obtain which link of the manipulator is most exposed to collision with which obstacle. One way to avoid obstacles is now to reconfigurate the links of the manipulator to shift joint i, with negative h_{ik}-value, away from the boundary ellipse of obstacle k. Because of robot redundancy there is a possibility for reconfiguration while the end-effector of the manipulator is hold fixed. A simple way to shift the joint i away form the boundary ellipse of the obstacle k is to reconfigurate the links connected with this joint by rotating one of this links as shown in Figure 2.

The kinematic control establishes a suitable joint as rotating point and the configuration of the rest of the manipulator belonging to it. We illustrate this method by the following example.

Consider a three-link planar revolute robotic manipulator. In the case there are no constraints of the orientation of the end-effector trajectory, we have one redundant degree of freedom. We define a new ($\hat{x}$,$\hat{y}$)-coordinate system on the end-effector parallel to the (x,y)-coordinate system of the base as shown in Figure 3.a. These frames are related according

$$\hat{x} = x - x_3, \qquad \hat{y} = y - y_3, \tag{7}$$

where (x_3,y_3) is the position of the origin of the ($\hat{x}$, $\hat{y}$)-frame in the (x,y)-coordinate system as shown in Figure 3.a.

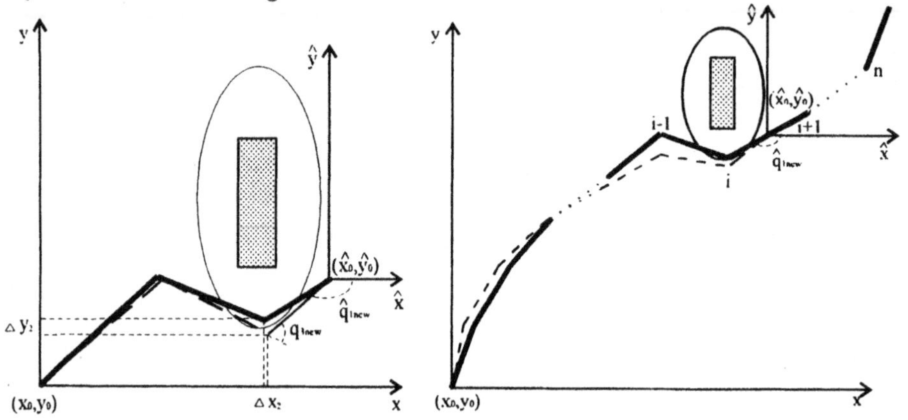

(a) *(b)*

Figure 3. Collision-avoidance through self-motion-control

Thus we obtain new joint coordinates ($\hat{q}_1$, $\hat{q}_2$, $\hat{q}_3$) with respect to the ($\hat{x}$,$\hat{y}$)-coordinate system.

Assume that the third joint of the manipulator lies inside of the boundary ellipse of a rectangular obstacle as shown in Figure 3.a. The boundary ellipses are determined as described in the previous section. To avoid the possibility of collision with the obstacle, we rotate the third link of the manipulator about ($\hat{x}_0$, $\hat{y}_0$) by a small amount Δq such that the third joint of the manipulator moves outside the boundary ellipse belonging to it. For this reason we enlarge or reduce the $\hat{q}_1$-value by a multiple of a Δq-value as long as the third joint of the manipulator lies inside the boundary ellipse. Thus we determine new $\hat{q}_1$-value as

$$\hat{q}_{1new} = \hat{q}_1 \pm k\Delta q \tag{8}$$

where k is the number of the required steps. After each step of changing the $\hat{q}_1$-value we can calculate the work space coordinates of the third joint relative to the (x,y)-coordinate system

$$x_{2new} = x_3 + l_3 \cos(\hat{q}_{1new}), \qquad y_{2new} = y_3 + l_3 \sin(\hat{q}_{1new}) \tag{9}$$

Calculating the boundary ellipse function relative to the third link for the workspace coordinates of the third joint, allows to establish after which step and in which direction of the changing, the third joint lies outside the boundary ellipse or, $h_2(x_{2new}, y_{2new}) > 0$ holds, respectively. Now we determine the change of coordinates of the third joint in world-coordinate

$$\Delta x_2 = x_{2new} - x_2, \qquad \Delta y_2 = y_{2new} - y_2 \tag{10}$$

The necessary rotation of the remaining links of the manipulator is based on the following consideration:
The remaining links of the manipulator build a two-link manipulator. The end-effector of this manipulator is moved about $(\Delta x_2 , \Delta y_2)$ by rotating the third link of the base manipulator as shown in Figure 3.a. The generalized joint displacement increments, Δq_1 and Δq_2, which enable the remaining two-link robot to move its end-effector about $(\Delta x_2 , \Delta y_2)$, can be determined by making use of the inverse kinematics [5]. q_{3new} can be determined from coordinate transformation.

Generally, the reconfiguration strategy to avoid the collision of link i $(i > 2)$ of a n-link robot manipulator with an obstacle works as follows:
We suppose that joints $i+1,\ldots,n$ and the end-effector are held stationary in their positions and define the $(\hat{x},\hat{y})$-coordinate system parallel to the base (x,y)-coordinate system fixed on joint $i+1$ of the manipulator as shown in Figure 3.b. Coordinate transformation yields world- and joint-coordinates of links $1,2,\ldots,i$ of the manipulator with respect to the $(\hat{x},\hat{y})$-coordinate system. To avoid the possibility of collision with the obstacle we rotate link i of the manipulator about $(\hat{x}_0 , \hat{y}_0)$ by a small amount Δq such that the i-th joint of the manipulator is outside the boundary ellipse of the obstacle belonging to it.

To determine the necessary reconfiguration of the remaining links, we suppose that the remaining links of the manipulator build a $(i-1)$-link manipulator with (x_0 , y_0) as base coordinates. The end-effector of this manipulator is moved about $(\Delta x_{i-1} , \Delta y_{i-1})$ by rotating link i of the base manipulator. The generalized joint displacement increment, $(\Delta q_1, \Delta q_2, \ldots, \Delta q_{i-1})$, which enables the remaining $(i-1)$-link robot to move its end-effector by $(\Delta x_{i-1} , \Delta y_{i-1})$, can be determined by making use of the inverse kinematics. Determination of q_{inew} is performed via coordinate transformation.

To avoid the collision of the second link of the manipulator with the obstacle, we suppose that joints $4,5,\ldots,n$ and the end-effector of the manipulator are held stationary and define the $(\hat{x},\hat{y})$-coordinate system on the fourth joint of the manipulator. We rotate the first link about (x_0 , y_0) by a small amount Δq such that the second joint of the manipulator is outside the boundary ellipse of the obstacle belonging to it. The reconfiguration of the remaining links results by making use of the inverse kinematics as explained in the example. Thus, we obtain the joint displacement increments, $\Delta \underline{q}_r$

$$\Delta \underline{q}_r = \underline{q}_{new} - \underline{q} \tag{11}$$

The joint displacement increments, $\Delta \underline{q}_m$, which enable the robot manipulator to move its end-effector about $\Delta \underline{x}$, can be determined now by solving the equation

$$\Delta \underline{x} = \mathbf{J} \cdot \Delta \underline{q}_m \tag{12}$$

by making use of the pseudoinverse. Thus, the generalized joint displacement increments, $\Delta\underline{q}_g$ result in this step of path tracking from

$$\Delta\underline{q}_g = \Delta\underline{q}_r + \Delta\underline{q}_m \tag{13}$$

If more than one joint lies inside the boundary ellipses of some obstacles, the collision-avoidance results step by step. At each step, the reconfiguration shifts exactly one critical joint away from a boundary ellipse. The total reconfiguration results then as sum of the individual reconfigurations. It is important to note, that the total number of critical joints must not be greater than the number of degrees of redundancy. If the number of critical joints is greater than the number of degrees of redundancy, a collision is not avoidable and our policy is to abort the task. In general, there is no safe trajectory solution to the tracking problem in this case. Observe that this does not prevent us from successfully manoeuvering in a workspace containing more obstacles than the number of degrees of redundancy, provided that only so many critical joints as the number of degrees of redundancy are handled at one instant of time.

This method has the advantage that the configuration of the manipulator after collision-avoidance can be influenced by further requirements such as avoidance of singularities, joint limits etc. The activation of obstacle-avoidance control is simply performed by detecting that one of the joints of the robot enters the boundary ellipse of an obstacle. If all joints of the manipulator lie outside the boundary ellipses of the obstacles, than the obstacles have no influence on the manipulator and the redundant degrees of freedom can be used to satisfy some other desirable criterion. A detailed description of the method together with simulation results is given in Rahmanian-Shahri and Troch [19].

3.3. Simulation results

The preceding obstacle recognition and avoidance schemes were implemented and tested by computer simulations. The specified end-effector trajectory and mathematical description of the world are program inputs. The methods were tested for several cases with different numbers of obstacles for three- and four-link manipulators. The results presented here are snapshoots from samples selected from a comprehensive computer simulation study that was carried out to test the performance of the proposed method. The boundary ellipse functions are determined as described in section 2. Figure 4.a and 5.a show a planar four-link manipulator, two rectangular obstacles in the workspace and the desired path. The movement of the manipulator between the obstacles by tracking a part of the desired path using the optimization method is shown in Figures 4.b and 4.c and, via self-motion-control method is shown in Figures 5.b and 5.c.

In these examples, we have demonstrated the obstacle avoidance capability in a variety of simulations by using a simple control scheme. Moreover, these methods are applied directly in workspace. When using the optimization method we don't have any information about the configuration of the manipulator after collision-avoidance. The self-motion-control method has the advantage that the configuration of the manipulator after collision-avoidance can be influenced by

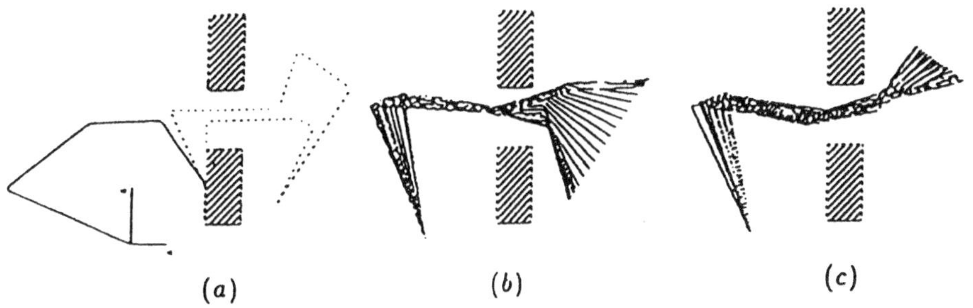

Figure 4. Snapshoots from the simulation of obstacle avoidance through optimization

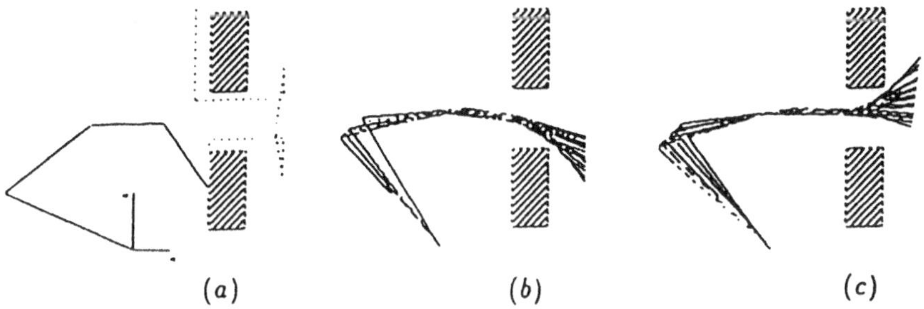

Figure 5. Snapshoots from the simulation of obstacle avoidance through self-motion-control

further requirements such as avoidance of singularities, joint limits etc. Nevertheless there is no optimization parameter which influences the collision-avoidance. The self-motion-control method leads to the favorable use of redundant degrees of freedom to obstacle-avoidance.

4. Conclusions

A comparision between two new collision avoidance methods for redundant robots is presented which allows the use of redundant degrees of freedom such that a manipulator can avoid obstacles in workspace while tracking the desired end-effector trajectory. The methods use a geometric strategy to reconfigurate the manipulator due to collision-avoidance. The collision recognition results on-line by defining a boundary ellipse for every link of the manipulator and every obstacles in workspace. The collision-avoidance methods are applied directly in workspace. But the optimization method does not yield any information about the configuration of the manipulator after collision-avoidance. The self-motion-control method has the advantage that the configuration of the manipulator after collision-avoidance can be influenced by further requirements such as avoidance of singularities, joint limits etc. Nevertheless, there is no optimization parameter which influences the collision-avoidance. The self-motion-control method leads to the favorable use of redundant degrees of freedom to obstacle-avoidance. Similar algorithms for the three dimensional case are under development.

References

1. T. Lozano-Perez, Automatic Planning of Manipulator Transfer Movements, *IEEE Trans. on SMC*, **SMC-11**, 681-698 (1981).
2. V. J. Lumelsky, Effect of kinematics on motion planning for planar robot arms moving admist unknown obstacles, *IEEE J. of Robotics and Automation*, **RA-3**, 207-223 (1987).
3. E. Freund and H. Hoyer, Collision avoidance for industrial robots with arbitrary motion, *Journal of Robotic Systems*, **1**, 317-329 (1984).
4. E. G. Gilbert and D. W. Johnson, Distance functions and their application to robot path planning in the presence of obstacles, *IEEE Journal of Robotics and Automation*, **RA-1**, 21-29 (1985).
5. C. Klein and C. Huang, Review of Pseudoinverse Control for Use with Kinematically Redundant Manipulators. *IEEE Transactions on Systems, Man, and Cybernetics*, **SMC-13**, No 3, (1983).
6. A. A. Maciejewski and C. A. Klein, Obstacle avoidance for kinematically redundant manipulators in dynamically varying environments, *The Intern. Journal of Robotics Research*, 4, 109-117 (1985).
7. Y. Nakamura, H. H. Hanafusa, and T. Yoshikawa, Task-priority based redundancy control of robot manipulators, *The Intern. J. of Robotics Research*, **6**, 3-15 (1987).
8. M. Anthimopoulou and N. Aspragathos, Kinematic Control of Planar Redundant Manipulators Moving Between Obstacles, *Advances in Robot Kinematics*, Springer Verlag, 380-391 (1990).
9. J. Baillieul, Avoiding obstacles and resolving kinematic redundancy, *Proc. IEEE Int. Conf. on Robotics and Automation*, San Francisco, CA, April 1986, 1698-1704.
10. Y. Chen and M. Vidyasagar, Optimal Control of Robotic Manipulators in the Presence of Obstacles, *Journal of Robotic Systems*, 7(5), 721-740 (1990).
11. S. Oh, D. Orin, and M. Bach, An inverse kinematic solution for kinematically redundant robot manipulators, *Journal of Robotic Systems*, **1**, 235-249 (1987).
12. L. Sciavicco and B. Siciliano, A solution algorithm to the inverse kinematic problem for redundant manipulators, *IEEE J. of Robotics and Automation*, **4**, 403-410 (1988).
13. H. Das, J-J. Slotine, and T. Sheridan, Inverse kinematic algorithms for redundant systems, *Proc. IEEE Intern. Conf. on Robotics and Automation*, Philadelphia, PA. April 1988, pp. 43-48.
14. O. Khatib, Real-time obstacle avoidance for manipulators and mobile robots, *The Intern. Journal of Robotics Research*, **5**, 90-98 (1986).
15. R. Colbaugh, H. Seraji, and K. L. Glass, Obstacle Avoidance for Redundant Robots Using Configuration Control, *Journal of Robotic Systems*, 6(6), 721-744 (1989).
16. A. Bagchi and H. Hatwal, Fuzzy logic-based techniques for motion planning of a robot manipulator amongst unknown moving obstacles, *Robotica*, Volume 10, pp 563-573 (1992).
17. A. Liegeois, Automatic Supervisory Control of the Configuration and Behavior of the Multibody Mechanisms, *IEEE Transaction on systems, Man, and Cybernetics*, Vol SMC-7, No 12, December 1977, pp. 868-871.
18. N. Rahmanian-Shahri and I. Troch, Collision-avoidance control for redundant articulated robots , *Robotica*, volume 13, pp 159-168 (1995).
19. N. Rahmanian-Shahri and I. Troch, Collision-avoidance for redundant robots through control of the self-motion of the manipulator , (to appear in *Journal of Intelligent and Robotic Systems)*.

GRADIENT-BASED KINEMATIC CONTROL OF NONREDUNDANT SERIAL ROBOTS IN SPECIAL CONFIGURATIONS

T. F. PARIKIAN
Department of Mechanics and Machine Science
State Engineering University of Armenia
105 Terian Street, Yerevan, 375009, Armenia
Fax: +374-2-151068 E-mail: politech@arminco.com

Abstract. The paper presents a reliable method of bringing any nonredundant serial robot out of a singular configuration the shortest way. The method implies the combination of motion performed in direction of gradient (or anti-gradient) of Jacobian or Gramian with approximation of the twist of the target end-effector motion.

1. Introduction

The motion planning algorithms for the path-controlled nonredundant serial robots may fail in special (or singular) configurations, when the inverse kinematic problem becomes unsolvable. The number of instantaneous degrees of freedom of the end-effector of the serial robot in such configurations becomes less than the number of independently controlled joints. As a result, the twist of the target end-effector motion may become irreproducible by the linkwork of the robot.

To prevent the robot from losing control, the following remedies are proposed in the existing works on singularity avoidance, known to the author: a) performing a screw motion of the end-effector which is the nearest to the target one [1]; b) determining the conditions (e.g., the minimal number of joints to be actuated) for escaping from specific types of singular configurations, based on theorems formulated using the theory of screw systems [2]; c) finding out the numerical solution of the motion planning task in terms of optimal control problem with penalty-function constraints representing the singularity avoidance condition [3]. The method (a) gives no guarantee for getting out of singular or sub-singular area in the course of motion, (b) doesn't consider the special configurations of general type, (c) lacks visuality inherent in geometric methods and gives no analytic solution.

In [4] the author proposed another method based on screw theory, which shows how to bring any 6-dof serial robot out of singularity region the shortest way, in order to rehabilitate the kinematic capabilities of the robot as quick as possible.

A modified method is proposed here for synthesizing a motion which is the nearest to the target one among those which provide the fastest escape from the singularity region. This method is the extension of the approach described in [4] to the case of any nonredundant serial robots.

J. Lenarčič and V. Parenti-Castelli (eds.), Recent Advances in Robot Kinematics, 77-86.
© 1996 *Kluwer Academic Publishers. Printed in the Netherlands.*

2. Instantaneous kinematics of serial robots: screws, singularities and relevant matrices

Let us consider a typical nonredundant serial robot having $n \leq 6$ revolute and/or prismatic pairs, or joints (Fig.1). The configuration of the robot is uniquely defined by the vector $\mathbf{Q} = [q_1, \dots, q_n]^T$ of its joint coordinates.

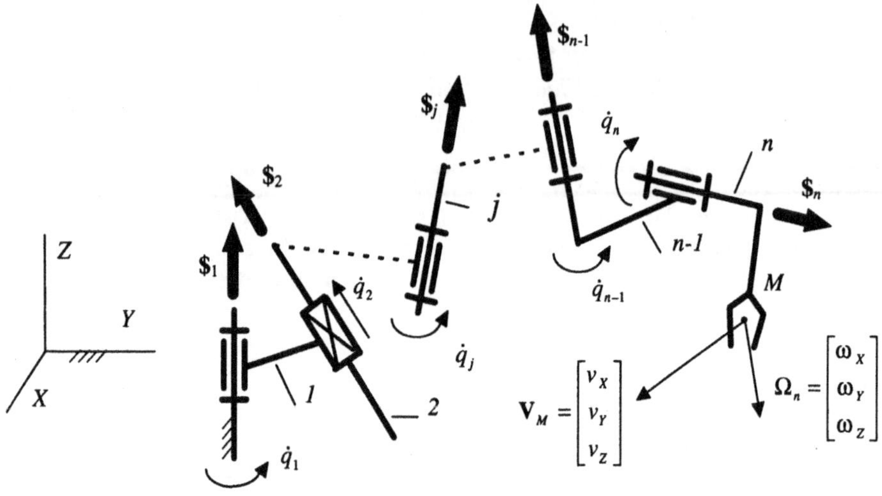

Figure 1. A general *n*-dof serial robot.

One of the main problems in the motion planning of the path-controlled serial robots is the inverse velocity problem, which is reduced to the solution of the following linear system:

$$\mathbf{T} = [J]\,\dot{\mathbf{Q}} \tag{2.1}$$

where $\mathbf{T} = [\omega_x\ \omega_y\ \omega_z\ v_x\ v_y\ v_z]^T$ is the twist of the target end-effector motion;
$\dot{\mathbf{Q}} = [\dot{q}_1, \dots, \dot{q}_n]^T$ is the vector of joint velocities of the robot;
$[J]$ is the *Jacobian matrix* of transformation which maps $\dot{\mathbf{Q}}$ into $\mathbf{T}$.

Let us associate kinematic screw $\$_j$ with the *j*th kinematic pair of the robot (Fig.1). Each screw can be represented by means of its Plücker coordinates:

$$\$_j = \left[s_j^{(1)}\ \ s_j^{(2)}\ \ \cdots\ \ s_j^{(6)}\right]^T \tag{2.2}$$

The components of the Jacobian matrix are nothing but the Plücker coordinates of these kinematic screws. The Jacobian matrix can therefore be written in the following compact form:

$$[J] = [\ \$_1\ \ \$_2\ \dots\ \$_n\ ] \tag{2.3}$$

While the direct transformation of $\dot{\mathbf{Q}}$ into $\mathbf{T}$ always exists for serial robots, the inverse one is not free from singularities. For n=6 the inverse mapping degenerates for those points of the configuration space $\{\mathbf{Q}\}$, for which the rank of the Jacobian matrix becomes less than 6. For robots with $n \le 5$ system (2.1) has no solutions for a general configuration of robot, and can be simultaneous only if the twist $\mathbf{T}$ is the linear combination of the kinematic screws associated with the joint axes.
Zero values of Jacobian $|J|$ -- the determinant of the Jacobian matrix $[J]$ -- can serve as indicators of singular configurations for 6-dof robots. For less-than-6-dof robots the Jacobian $|J|$ is not defined, as the Jacobian matrix $[J]$ is not a square one. To identify the singularities in these cases the *Gramian matrix* can be used. It is a symmetric square matrix

$$[G] = [J]^T\,[J] = \begin{pmatrix} \$_1 * \$_1 & \$_1 * \$_2 & \dots & \$_1 * \$_n \\ \$_2 * \$_1 & \$_2 * \$_2 & \dots & \$_2 * \$_n \\ \vdots & \vdots & \ddots & \vdots \\ \$_n * \$_1 & \$_n * \$_2 & \dots & \$_n * \$_n \end{pmatrix} \tag{2.4}$$

composed of all possible inner products of screws as defined in [5]:

$$\$_i * \$_j = \sum_{k=1}^{6} s_i^{(k)} s_j^{(k)} \tag{2.5}$$

As was proved in [5], the zero values of Gramian $|G|$ -- the determinant of the Gramian matrix $[G]$ -- correspond to special configurations of less-than-6-dof robots. For n=6 the Gramian can also be expressed through the Jacobian [6]:

$$|G| = |J|^T\,|J| = |J|^2 \tag{2.6}$$

3. The basic concept of escape from singularity

The basic idea of this paper is to provide the fastest escape of serial robots from special configurations. Although different motions can take the robot out of the singularity region, "minimal" motions are of interest.
Case 1 (n=6). If $|J|$ is considered as a scalar function in the joint space $\{\mathbf{Q}\}$, then the fastest way out of the singularity region (where $|J|$=0) will be the direction of gradient (or anti-gradient, as $|J|$ can be negative) of that function, provided that we consider the joint space $\{\mathbf{Q}\}$ as having a Euclidean norm $\Delta\mathbf{Q} = \sqrt{\sum_j (\Delta q_j)^2}$.

As $|J|$ is invariant with respect to coordinate transformations [7], then no motions in the first and the last kinematic pairs will change the value of $|J|$. It is natural therefore to consider the subspace of four joint coordinates $\{\mathbf{Q}'\}=\{q_2, \dots, q_5\}$ instead of the full joint space $\{\mathbf{Q}\}=\{q_1, \dots, q_6\}$.

Case 2 ($n \leq 5$). All the above speculations for $|J|$ are applicable to $|G|$, although the latter is not invariant with respect to all coordinate transformations [6].

4. Gradient of Jacobian for 6-dof serial robots

The results of this paragraph are not new as they have been already derived in [4]. However, the new method leading to the same results is simpler and shorter than one previously used. The main difference of obtaining the components of gradient of Jacobian, used in this paper, is in applying the theorem on derivative of determinant [8][1].

It is obvious that in any serial robot the position of the jth joint axis depends on coordinates in the previous (j-1) joints only. This means that for the derivatives of the kinematic screw $\$_j$ with respect to the joint coordinates q_i we can write:

$$\frac{\partial \$_j}{\partial q_i} = 0, \quad i \geq j \tag{4.1}$$

$$\frac{\partial \$_j}{\partial q_i} = \mathbf{R}_{ij}, \quad i<j \tag{4.2}$$

where

$$\mathbf{R}_{ij} = \$_i \times \$_j \tag{4.3}$$

is the *screw-product* of the screws $\$_i$ and $\$_j$ as defined by F. M. Dimentberg [9][2].
Now we can find the gradient of Jacobian:

$$\mathbf{grad}\,|J| = \nabla|J| = \left[\frac{\partial |J|}{\partial q_2} \quad \frac{\partial |J|}{\partial q_3} \quad \frac{\partial |J|}{\partial q_4} \quad \frac{\partial |J|}{\partial q_5}\right]^T \tag{4.4}$$

According to the theorem on derivative of determinant [8] we have:

$$\frac{\partial |J|}{\partial q_i} = \left| \frac{\partial \$_1}{\partial q_i} \; \$_2 \; \dots \; \$_6 \right| + \left| \$_1 \; \frac{\partial \$_2}{\partial q_i} \; \dots \; \$_6 \right| + \dots + \left| \$_1 \; \$_2 \; \dots \; \frac{\partial \$_6}{\partial q_i} \right| \tag{4.5}$$

Taking into account equations (4.1)-(4.3) we come to the following expressions:

---

[1]The statement of the theorem is proposed in [8] as an exercise without a proof, which therefore must be well known, so we do not prove it here.

[2]It was defined as the cross-product of the respective dual vectors $\$_i \times \$_j = (\mathbf{s}_i+\varepsilon\mathbf{s}_i^{\circ}) \times (\mathbf{s}_j+\varepsilon\mathbf{s}_j^{\circ}) = \mathbf{s}_i \times \mathbf{s}_j + \varepsilon\,(\mathbf{s}_i \times \mathbf{s}_j^{\circ} + \mathbf{s}_i^{\circ} \times \mathbf{s}_j)$, where ε is the Clifford's dual unity ($\varepsilon^2 = 0$).

$$\frac{\partial |J|}{\partial q_5} = |\ \$_1\ \ \$_2\ \dots\ \$_5\ \ \mathbf{R}_{56}\ | \tag{4.6}$$

$$\frac{\partial |J|}{\partial q_4} = |\ \$_1\ \ \$_2\ \dots \mathbf{R}_{45}\ \ \$_6\ | + |\ \$_1\ \ \$_2\ \dots\ \$_5\ \ \mathbf{R}_{46}\ | \tag{4.7}$$

$$\frac{\partial |J|}{\partial q_3} = |\ \$_1\ \ \$_2\ \ \$_3\ \ \mathbf{R}_{34}\ \ \$_5\ \ \$_6\ | + |\ \$_1\ \dots\ \$_4\ \ \mathbf{R}_{35}\ \ \$_6\ | + |\ \$_1\ \dots\ \$_5\ \ \mathbf{R}_{36}\ | \tag{4.8}$$

$$\frac{\partial |J|}{\partial q_2} = |\ \$_1\ \ \$_2\ \ \mathbf{R}_{23}\ \ \$_4\ \ \$_5\ \ \$_6\ | + |\ \$_1\ \ \$_2\ \ \$_3\ \ \mathbf{R}_{24}\ \ \$_5\ \ \$_6\ | + $$

$$+ |\ \$_1\ \dots\ \$_4\ \ \mathbf{R}_{25}\ \ \$_6\ | + |\ \$_1\ \dots\ \$_5\ \ \mathbf{R}_{26}\ | \tag{4.9}$$

Assuming the reference frame as being attached to the 3rd link of the robot, and keeping in mind the invariance of $|J|$ with respect to coordinate transformations [7], equations (4.8) and (4.9) can be simplified if the inverted motion is considered (like we did it in [4]):

$$\frac{\partial |J|}{\partial q_3} = |\ \mathbf{R}_{13}\ \ \$_2\ \dots\ \$_6\ | + |\ \$_1\ \ \mathbf{R}_{23}\ \ \$_3\ \dots\ \$_6\ | \tag{4.10}$$

$$\frac{\partial |J|}{\partial q_2} = |\ \mathbf{R}_{12}\ \ \$_2\ \dots\ \$_6\ | \tag{4.11}$$

5. Gradient of Gramian for nonredundant serial robots

Below we determine the gradient of Gramian

$$\mathbf{grad}\ |G| = \nabla |G| = \left[\frac{\partial |G|}{\partial q_2}\ \ \frac{\partial |G|}{\partial q_3}\ \ \frac{\partial |G|}{\partial q_4}\ \ \frac{\partial |G|}{\partial q_5} \right]^T \tag{5.1}$$

First we write the Gramian matrix (2.1) in a compact form:

$$[G] = [\ \mathbf{M}_1\ \ \mathbf{M}_2\ \dots\ \mathbf{M}_n\ ] \tag{5.2}$$

where

$$\mathbf{M}_j = \begin{pmatrix} \$_j * \$_1 \\ \$_j * \$_2 \\ \vdots \\ \$_j * \$_n \end{pmatrix} = [J]^T\ \$_j \tag{5.3}$$

According to the theorem on derivative of determinant [8] we have:

$$\frac{\partial|G|}{\partial q_i} = \left|\frac{\partial \mathbf{M}_1}{\partial q_i}\ \mathbf{M}_2\ \ldots\ \mathbf{M}_n\right| + \left|\mathbf{M}_1\ \frac{\partial \mathbf{M}_2}{\partial q_i}\ \ldots\ \mathbf{M}_n\right| + \ldots + \left|\mathbf{M}_1\ \mathbf{M}_2\ \ldots\ \frac{\partial \mathbf{M}_n}{\partial q_i}\right| \tag{5.4}$$

Taking into account equations (4.1)-(4.3) we can write:

$$\frac{\partial \mathbf{M}_j}{\partial q_i} = \begin{pmatrix} 0 \\ \vdots \\ 0 \\ \$_j * \mathbf{R}_{i,i+1} \\ \vdots \\ \$_j * \mathbf{R}_{in} \end{pmatrix},\ i \geq j \tag{5.5}$$

$$\frac{\partial \mathbf{M}_j}{\partial q_i} = \begin{pmatrix} \$_1 * \mathbf{R}_{ij} \\ \vdots \\ \$_i * \mathbf{R}_{ij} \\ \$_{i+1} * \mathbf{R}_{ij} + \$_j * \mathbf{R}_{i,i+1} \\ \vdots \\ 2\$_j * \mathbf{R}_{ij} \\ \vdots \\ \$_n * \mathbf{R}_{ij} + \$_j * \mathbf{R}_{in} \end{pmatrix},\ i < j \tag{5.6}$$

For the first particular case of $\frac{\partial|G|}{\partial q_{n-1}}$ we derive:

$$\frac{\partial \mathbf{M}_1}{\partial q_{n-1}} = \begin{pmatrix} 0 \\ \vdots \\ 0 \\ \$_1 * \mathbf{R}_{n-1,n} \end{pmatrix};\ \ldots\ ;\ \frac{\partial \mathbf{M}_{n-1}}{\partial q_{n-1}} = \begin{pmatrix} 0 \\ \vdots \\ 0 \\ \$_{n-1} * \mathbf{R}_{n-1,n} \end{pmatrix};\ \frac{\partial \mathbf{M}_n}{\partial q_{n-1}} = \begin{pmatrix} \$_1 * \mathbf{R}_{n-1,n} \\ \vdots \\ \$_{n-1} * \mathbf{R}_{n-1,n} \\ 2\$_n * \mathbf{R}_{n-1,n} \end{pmatrix} \tag{5.7}$$

Substituting expression (5.7) into (5.4) yields

$$\frac{\partial|G|}{\partial q_{n-1}} = \begin{vmatrix} 0 & & & \\ \vdots & \mathbf{M}_2 & \cdots & \mathbf{M}_n \\ 0 & & & \\ \$_1 * \mathbf{R}_{n-1,n} & & & \end{vmatrix} + \begin{vmatrix} & 0 & & & \\ \mathbf{M}_1 & \vdots & \mathbf{M}_3 & \cdots & \mathbf{M}_n \\ & 0 & & & \\ & \$_2 * \mathbf{R}_{n-1,n} & & & \end{vmatrix} + \ldots +$$

$$+\begin{vmatrix} \mathbf{M}_1 & \cdots & \mathbf{M}_{n-2} & \begin{matrix} 0 \\ \vdots \\ 0 \\ \$_{n-1}*\mathbf{R}_{n-1,n} \end{matrix} & \mathbf{M}_n \end{vmatrix} + \begin{vmatrix} \mathbf{M}_1 & \cdots & \mathbf{M}_{n-1} & \begin{matrix} \$_1*\mathbf{R}_{n-1,n} \\ \vdots \\ \$_{n-1}*\mathbf{R}_{n-1,n} \\ 2\$_n*\mathbf{R}_{n-1,n} \end{matrix} \end{vmatrix} \tag{5.8}$$

Keeping in mind that for any square matrix A we have $|A| = |A|^T$, expression (5.8) can be assembled into a single determinant taken two times:

$$\frac{\partial |G|}{\partial q_{n-1}} = 2\begin{vmatrix} \mathbf{M}_1 & \cdots & \mathbf{M}_{n-1} & \begin{matrix} \$_1*\mathbf{R}_{n-1,n} \\ \vdots \\ \$_{n-1}*\mathbf{R}_{n-1,n} \\ \$_n*\mathbf{R}_{n-1,n} \end{matrix} \end{vmatrix} = 2\begin{vmatrix} \mathbf{M}_1 & \cdots & \mathbf{M}_{n-1} & \mathbf{W}_{n-1,n} \end{vmatrix} \tag{5.9}$$

where $\mathbf{W}_{n-1,n}$ stands for the product $[J]^T\mathbf{R}_{n-1,n}$ (and $\mathbf{W}_{ik} = [J]^T\mathbf{R}_{ik}$ throughout the paper). By similarity with (5.7), for the second derivative $\frac{\partial |G|}{\partial q_{n-2}}$ we write:

$$\frac{\partial \mathbf{M}_1}{\partial q_{n-2}} = \begin{pmatrix} 0 \\ \vdots \\ 0 \\ \$_1*\mathbf{R}_{n-2,n-1} \\ \$_1*\mathbf{R}_{n-2,n} \end{pmatrix}; \; \ldots \; ; \frac{\partial \mathbf{M}_{n-2}}{\partial q_{n-2}} = \begin{pmatrix} 0 \\ \vdots \\ 0 \\ \$_{n-2}*\mathbf{R}_{n-2,n-1} \\ \$_{n-2}*\mathbf{R}_{n-2,n} \end{pmatrix};$$

$$\frac{\partial \mathbf{M}_{n-1}}{\partial q_{n-2}} = \begin{pmatrix} \$_1*\mathbf{R}_{n-2,n-1} \\ \vdots \\ \$_{n-2}*\mathbf{R}_{n-2,n-1} \\ 2\$_{n-1}*\mathbf{R}_{n-2,n-1} \\ \$_n*\mathbf{R}_{n-2,n-1} + \$_{n-1}*\mathbf{R}_{n-2,n} \end{pmatrix}; \frac{\partial \mathbf{M}_n}{\partial q_{n-2}} = \begin{pmatrix} \$_1*\mathbf{R}_{n-2,n} \\ \vdots \\ \$_{n-2}*\mathbf{R}_{n-2,n} \\ \$_n*\mathbf{R}_{n-2,n-1} + \$_{n-1}*\mathbf{R}_{n-2,n} \\ 2\$_n*\mathbf{R}_{n-2,n} \end{pmatrix} \tag{5.10}$$

Substituting expressions (5.10) into (5.4) and assembling the determinants together yields:

$$\frac{\partial |G|}{\partial q_{n-2}} = 2\left(\begin{vmatrix} \mathbf{M}_1 & \cdots & \mathbf{M}_{n-2} & \begin{matrix} \$_1*\mathbf{R}_{n-2,n-1} \\ \vdots \\ \$_{n-1}*\mathbf{R}_{n-2,n-1} \\ \$_n*\mathbf{R}_{n-2,n-1} \end{matrix} & \mathbf{M}_n \end{vmatrix} + \begin{vmatrix} \mathbf{M}_1 & \cdots & \mathbf{M}_{n-1} & \begin{matrix} \$_1*\mathbf{R}_{n-2,n} \\ \vdots \\ \$_{n-1}*\mathbf{R}_{n-2,n} \\ \$_n*\mathbf{R}_{n-2,n} \end{matrix} \end{vmatrix} \right) =$$

$$= 2\left(\left|\mathbf{M}_1 \quad \cdots \quad \mathbf{M}_{n-2} \quad \mathbf{W}_{n-2,n-1} \quad \mathbf{M}_n\right| + \left|\mathbf{M}_1 \quad \cdots \quad \mathbf{M}_{n-1} \quad \mathbf{W}_{n-2,n}\right|\right) \qquad (5.11)$$

By similarity with the first two derivatives the others can be obtained as well, using the following general formula:

$$\frac{\partial |G|}{\partial q_i} = 2\sum_{k=i+1}^{n} \left|\mathbf{M}_1 \quad \cdots \quad \mathbf{M}_{k-1} \quad \mathbf{W}_{ik} \quad \mathbf{M}_{k+1} \quad \cdots \quad \mathbf{M}_n\right| \qquad (5.12)$$

For sake of shorter expressions and hence faster calculations like in the case of Jacobian (*see* paragraph 4), it would pay to consider the reference frame as being attached to the middle link of the robot[3] (e.g., when n=6, the 3rd link is the middle one: *see* Fig.1).
By similarity with (4.11) and (5.9) we derive

$$\frac{\partial |G|}{\partial q_2} = -2\left|\mathbf{W}_{21} \quad \mathbf{M}_2 \quad \cdots \quad \mathbf{M}_n\right| = 2\left|\mathbf{W}_{12} \quad \mathbf{M}_2 \quad \cdots \quad \mathbf{M}_n\right| \qquad (5.13)$$

Following the same lines, we come to

$$\frac{\partial |G|}{\partial q_3} = 2\left(\left|\mathbf{W}_{13} \quad \mathbf{M}_2 \quad \cdots \quad \mathbf{M}_n\right| + \left|\mathbf{M}_2 \quad \mathbf{W}_{23} \quad \mathbf{M}_3 \quad \cdots \quad \mathbf{M}_n\right|\right) \qquad (5.14)$$

and to a general formula - a kind of an "inverse sum" as compared with (5.12):

$$\frac{\partial |G|}{\partial q_i} = 2\sum_{k=1}^{i-1} \left|\mathbf{M}_1 \quad \cdots \quad \mathbf{M}_{k-1} \quad \mathbf{W}_{ki} \quad \mathbf{M}_{k+1} \quad \cdots \quad \mathbf{M}_n\right| \qquad (5.15)$$

6. Kinematic control procedure

In order to take a 6-dof robot out of sigularity region one should move it in direction of gradient of Jacobian[4] in the joint subspace {**Q**'}, with the rates defined as follows:

$$(\dot{q}_2 : \dot{q}_3 : \dot{q}_4 : \dot{q}_5) = \left(\frac{\partial |J|}{\partial q_2} : \frac{\partial |J|}{\partial q_3} : \frac{\partial |J|}{\partial q_4} : \frac{\partial |J|}{\partial q_5}\right) \qquad (6.1)$$

Under this motion the end-effector's partial twist will become

$$\mathbf{T}_{2\text{-}5} = \dot{q}_2 \$_2 + \dot{q}_3 \$_3 + \dot{q}_4 \$_4 + \dot{q}_5 \$_5 \qquad (6.2)$$

[3] Or to one of two middle links, if n is odd.
[4] The same is true for anti-gradient.

for which the relevant screw $\$_{2\text{-}5}$ can be defined:

$$\$_{2\text{-}5} = \frac{1}{\dot{q}_2}\mathbf{T}_{2-5} \tag{6.3}$$

In addition, as no motion in the first and the last joints affects the position of the axes, the free joint velocities $\dot{q}_1$ and $\dot{q}_6$ can be used to approximate the target twist **T** of the end-effector motion. Thus, the screws $\$_1$, $\$_{2\text{-}5}$, $\$_6$ form in general (if they are not dependent of each other) the basis of the three-system

$$\Theta = \dot{q}_1\$_1 + \dot{q}_2\$_{2-5} + \dot{q}_6\$_6 \tag{6.4}$$

Using the Gram-Schmidt orthogonalization procedure [5], one can always build a basis $\{\mathbf{U}_1, \mathbf{U}_2, \mathbf{U}_3\}$ of the screw system $\tilde{\Theta}$ which is the orthogonal complement of the three-system Θ. The screw system $\tilde{\Theta}$ contains all the "twists of non-freedom" [10].
Decomposing now the target twist **T** by the above six basic screws $\$_1$, $\$_{2\text{-}5}$, $\$_6$, $\mathbf{U}_1$, $\mathbf{U}_2$ and $\mathbf{U}_3$, one can find the actual rates of motion in all joints by solving the linear system

$$\dot{q}_1\$_1 + \dot{q}_2\$_{2-5} + \dot{q}_6\$_6 + k_1\,\mathbf{U}_1 + k_2\,\mathbf{U}_2 + k_3\,\mathbf{U}_3 = \mathbf{T} \tag{6.5}$$

and calculating the values for $\dot{q}_3$, $\dot{q}_4$ and $\dot{q}_5$ using (6.1).
The same approach is applicable for $n \leq 5$ cases, for which the Gramian can be used.

7. Conclusion

In special configuration the end-effector of a serial robot loses one or more of its degrees of freedom. This does not allow the end-effector to perform all the instantaneous motions, which is especially undesirable for path-controlled robots. That is why a technique leading to rehabilitation of kinematic capabilities of the robot currently being in singularity region becomes a major task, and could serve as a valuable addition to the algorithms used both in on-line and off-line programming of robots.
The paper presents a reliable method of bringing any nonredundant serial robot out of the singular configuration the shortest way, in the sense of Euclidean norm measured in the joint space of the robot. Two determinants -- the Jacobian (for n=6) and the Gramian (for $n \leq 5$) -- are used as indicators of proximity of the robot to the singularity region. These determinants are considered as scalar functions in the joint space of the robot, and the direction of gradient (or anti-gradient) of Jacobian or Gramian shows the fastest way out of the special configuration. The theorem on derivative of determinant is used to determine analytically (in terms of kinematic screws) the components of vector of gradient. The invariance of determinants with respect to motions in the first and the last joints gives an additional opportunity to approximate the target twist of the end-effector motion.

References

1. Glazunov V. A., (1985), "On Manipulator Control in Special Configurations," *Izvestiya AN SSSR: Mekhanika Tverdogo Tela*, #4, 61-65 (in Russian).
2. Hunt K. H., (1986), "Special Configurations of Robot-Arms via Screw Theory. Part 1: The Jacobian and its Matrix of Cofactors," *Robotica*, **4**, 171-179.
3. Mayorga R. V., Wong A. K. C., (1987), "A Singularities Avoidance Method for the Trajectory Planning of Redundant and Nonredundant Robot Manipulators,", *Research Report Feb. 1987,* Dept. of Systems Design Engineering, University of Waterloo, Waterloo, Ontario (Canada).
4. Parikyan T. F., (1989), "Manipulator Kinematic Control Strategy Using Gradient Direction of Jacobian," *Preprints of 6th IFAC Symposium on Information Control Problems in Manufacturing Technology INCOM'89,* Madrid, **1**, 365-368.
5. Sugimoto K., Duffy J., (1982), "Application of Linear Algebra to Screw Systems," *Mechanism and Machine Theory*, **17,** #1, 73-83.
6. Soylu R., Duffy J., (1988), "Hypersurfaces of Special Configurations of Serial Manipulators and Related Concepts. Part I: Theory and Development of the Hypersurfaces of Special Configurations," *Journal of Robotic Systems*, **5,** #1, 13-30.
7. Soylu R., Duffy J., (1988), "Hypersurfaces of Special Configurations of Serial Manipulators and Related Concepts. Part II: Passive Joints, Configurations, Component Manifolds and Some Applications," *Journal of Robotic Systems*, **5,** #1, 31-53.
8. Lankaster P., (1969), *Theory of Matrices*, Academic Press, New York - London.
9. Dimentberg F. M., (1982), *Theory of Spatial Linkages*, Nauka, Moscow (in Russian).
10. Lipkin H., Duffy J., (1985), "The Elliptic Polarity of Screws," *Journal of Mechanisms, Transmissions and Automation in Design*, **107**, #3, 377-387.

3. Performance

KINEMATICS OF A THREE DEGREE-OF-FREEDOM MOTION PLATFORM FOR A LOW-COST DRIVING SIMULATOR

PO-HUA YANG, KENNETH J. WALDRON
Department of Mechanical Engineering

AND

DAVID E. ORIN
Department of Electrical Engineering
The Ohio State University
Columbus, Ohio, U.S.A.

Abstract. The objective of the project described in this paper is to provide a motion base for a low-cost driving simulator which is a research topic in the development of an automated vehicle/highway system. The motion platform is driven by a parallel mechanism with three degrees of freedom. The kinematics for the position problem of this type of mechanism was quite well developed. The velocity problem, on which there is a focus in the paper, is less stressed by other researchers. A closed form of Jacobian which could be computed efficiently by feedback data from sensors is presented. Static force analysis is carried out before the velocity problem is solved. From the result, the singular positions of the structure, which constrain the range of motions, are located and described in the paper. The design of a table-top model, which was built to assist the development of control software, is also briefly described in the paper.

1. Introduction

Following from the success of flight simulators for training and human factors research, the importance of driving simulators was identified. Some applications, including research on monitoring for driver drowsiness and measurement of driver performance, look promising. Unlike flight simulators, there is little motivation to use driving simulators for training. The main application is human factors research. A few large installations ca-

J. Lenarčič and V. Parenti-Castelli (eds.), Recent Advances in Robot Kinematics, 89-98.
© 1996 *Kluwer Academic Publishers. Printed in the Netherlands.*

pable of carrying an entire vehicle and providing six or more degrees of motion freedom have been built or are planned (Haug *et al.*, 1990) (Weir *et al.*, 1995). While the design of a low-cost driving simulator has been suggested by other researchers, motion cues were not provided (Allen *et al.*, 1995).

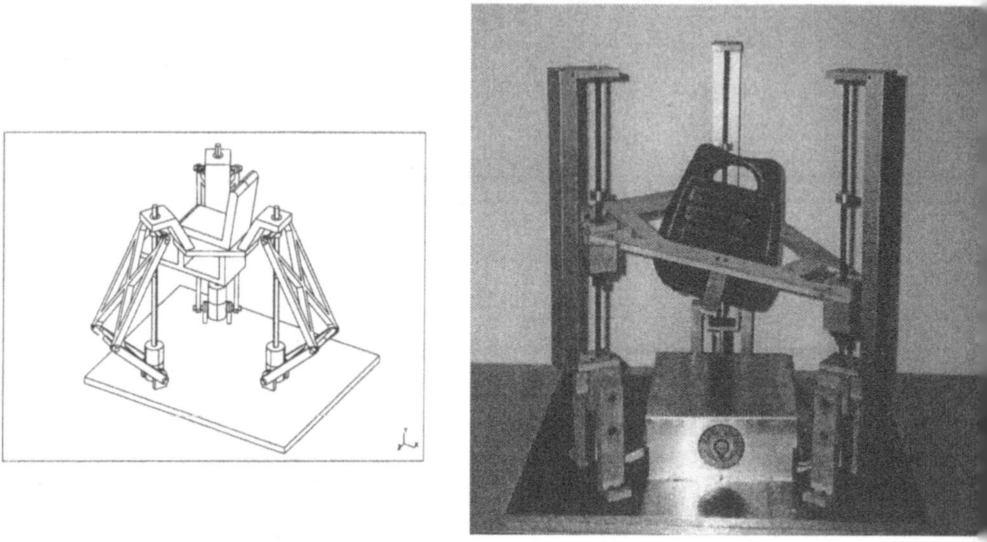

(a) (b)

Figure 1. (a) CAD Drawing of the Motion Platform (b) Photograph of the Table-Top Model

An important function of a driving simulator is to provide a simulation of the experience of the accelerations of driving, and this can be achieved by a motion platform. The Stewart Platform, a parallel mechanism with closed kinematic chains, is widely used by aircraft flight simulators (Stewart, 1965) and has also been applied to driving simulators (Salcudean *et al.*, 1994). There are several other classes of parallel mechanisms which are derived on the basis of the Stewart Platform. Given the actuator displacements, these mechanisms belonging to RRR-3S, PRR-3S, PPR-3S, and PPP-3S structures, where P,R and S stand for prismatic, revolute,and spherical pairs respectively, have been discussed by Innocenti *et al.*(1990).

Considering the sensations produced by braking, acceleration, turning, ascending and descending while driving, we can partially simulate all of these experiences by pitching, rolling and heaving the motion platform. The mechanism we designed for the motion platform is composed of 8 links, 3 revolute joints, 3 spherical joints, and 3 prismatic joints. The total number of degrees of freedom is 3, which can be determined by the Kutzbach mobility equation:

$$DF = 6\,n_L - 5\,n_J - 6$$

where n_L and n_J are, respectively, the numbers of links and elementary revolute (R) or prismatic (P) joints. The same type of parallel mechanism was used for the **ARTISAN** micro-manipulator developed in the late 1980s (Waldron *et al.*, 1989).

Compared with other parallel mechanisms, which indeed have some advantages in maneuverability, the mechanism for the motion platform is relatively simple from the point of view of kinematic computation, mechanical design and feedback control, but its capabilities can still satisfy most of the required specifications.

2. Description of the Device and Notation

In our model, prismatic and spherical joints are replaced by ball screws and universal joints. An active ball screw in conjunction with a passive universal joint on a ball nut performs the same kinematic function as a passive spherical joint and an active prismatic pair working together. Three brushless D.C. motors are used as the driving devices and the ball screws are used to transform rotating motions to translating motions. These features can be seen from the CAD model and photograph in Figure 1.

A schematic drawing of the driving simulator is shown in Figure 2(a). The device is fixed to the base by three passive revolute joints denoted by 1, 2, and 3 at the vertices of a base triangle with equilateral side length a. The joint axes directions $\mathbf{u}_1, \mathbf{u}_2$, and $\mathbf{u}_3$, lie in the plane of the base triangle and are normal to the medians of the triangle. The motion platform is connected to the ball screws by three universal joints denoted by 4, 5, and 6, which form the vertices of an upper triangle with the same geometry as the base triangle. Each ball screw can only move in the plane normal to the respective base joint axis as shown in Figure 2(b). The lengths of the ball screws, denoted by r_1, r_2, and r_3, are measured from the revolute joint to the universal joint in each case. The tilt angles of the links are denoted by ϕ_1, ϕ_2, and ϕ_3.

Reference frame {B} is fixed at the centroid of the base triangle with the z-axis pointing vertically downward. The x-axis points toward the midpoint of the line connecting the joints 2 and 3. The mechanism is initialized at the position where all the ball nuts touch the hard stops on top of the ball screws. At the initial position, the operator of the driving simulator will face in the x direction. The y-axis can be determined by the right-hand rule. Similarly, reference frame {U} has its origin at the centroid of upper triangle with its x, y, z axes parallel to those of frame {B} in the initial position.

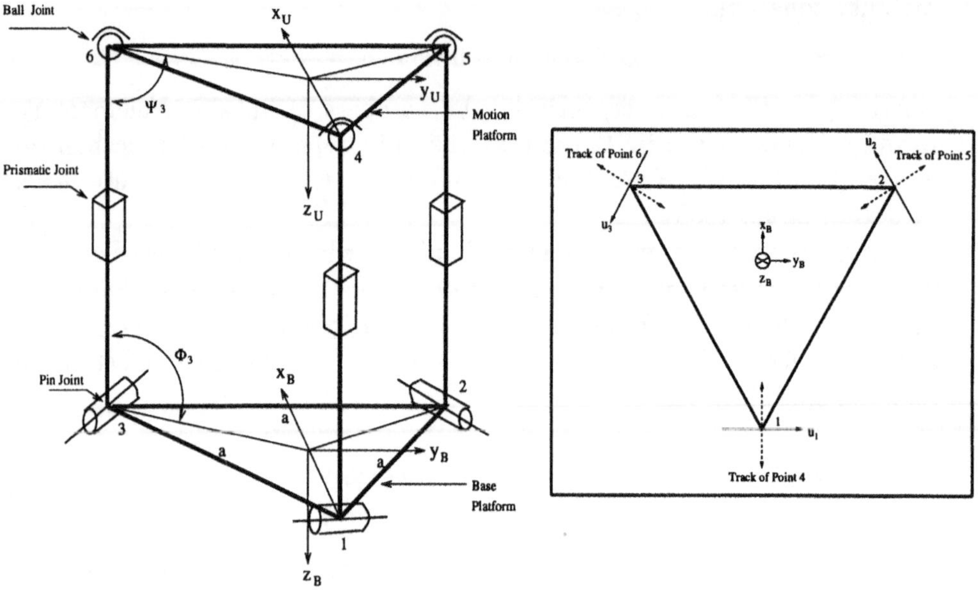

(a) (b)

Figure 2. (a) Kinematic Scheme of Motion Platform (b) Constraint Motions of Limbs

3. Position Kinematics

As a rigid body moving in space, the position of the motion platform can be described by six variables which include 3 translations: surge (${}^Bp_U^x$), sway (${}^Bp_U^y$), yaw (${}^Bp_U^z$) and three rotations: pitch (α, y), roll (β, x), yaw (γ,z). The position kinematics, which is the basis of the velocity problem, is briefly described here.

Inverse Kinematics

The inverse position kinematic problem for this device is: *Given the position of the motion platform, find the actuator position variables r_1, r_2, and r_3.*

In our application, the motion platform is rotated by the sequence of pitch, roll, and yaw which are **Y-X-Z** Euler angles by the definition of our coordinate system. Any specified points on the motion platform can be referred back to earth coordinates by multiplying by the homogeneous transformation matrix ${}^B\mathbf{T}_U$.

The lengths of the actuators can be calculated by

$$r_1 = |\ {}^B\mathbf{T}_U \cdot {}^U\mathbf{p}_4 - {}^B\mathbf{p}_1\ |, \tag{1}$$
$$r_2 = |\ {}^B\mathbf{T}_U \cdot {}^U\mathbf{p}_5 - {}^B\mathbf{p}_2\ |, \tag{2}$$

$$r_3 = \mid {}^B\mathbf{T}_U \cdot {}^U\mathbf{p}_6 - {}^B\mathbf{p}_3 \mid. \tag{3}$$

From the characteristics of the mechanism, the limbs are constrained to move in the planes: $y = 0$, $y = \sqrt{3}\ x$, and $y = -\sqrt{3}\ x$, respectively, as is seen in Figure 2(b). These constraint conditions give three constraint equations:

$${}^B\mathbf{p}_4 \cdot \mathbf{u}_1 = 0, {}^B\mathbf{p}_5 \cdot \mathbf{u}_2 = 0, {}^B\mathbf{p}_6 \cdot \mathbf{u}_3 = 0.$$

which can be written explicitly as:

$${}^Bp_U^y - \frac{a}{\sqrt{3}}\ c\beta\, s\gamma = 0, \tag{4}$$

$$\begin{aligned} &\tfrac{a}{4}(c\alpha\, c\gamma + s\alpha\, s\beta\, s\gamma) + \tfrac{a\sqrt{3}}{4}(-c\alpha\, s\gamma + s\alpha\, s\beta\, c\gamma) \\ &+\tfrac{\sqrt{3}}{2}\ {}^Bp_U^x - \tfrac{a}{4\sqrt{3}}\ (c\beta\, s\gamma) - \tfrac{a}{4}\ (c\beta\, c\gamma) - \tfrac{{}^Bp_U^y}{2} = \quad 0, \end{aligned} \tag{5}$$

$$\begin{aligned} &-\tfrac{a}{4}\ (c\alpha\, c\gamma + s\alpha\, s\beta\, s\gamma) + \tfrac{a\sqrt{3}}{4}(-c\alpha\, s\gamma + s\alpha\, s\beta\, c\gamma) \\ &-\tfrac{\sqrt{3}}{2}\ {}^Bp_U^x - \tfrac{a}{4\sqrt{3}}\ (c\beta\, s\gamma) + \tfrac{a}{4}(c\beta\, c\gamma) - \tfrac{{}^Bp_U^y}{2} = \quad 0 \end{aligned} \tag{6}$$

Since pitch(α), roll(β) ,and heave(${}^Bp_U^z$) are designed as the inputs, the other three degrees of freedom yaw(γ), surge(${}^Bp_U^x$), and sway(${}^Bp_U^y$) become dependent variables:

$$\begin{aligned} \gamma &= \tan^{-1}\left(\frac{s\alpha\, s\beta}{c\alpha + c\beta}\right), &(7)\\ {}^Bp_U^y &= \frac{a}{\sqrt{3}} c\beta\, s\gamma, &(8)\\ {}^Bp_U^x &= \frac{a(c\beta\, s\gamma - c\alpha\, c\gamma - s\alpha\, s\beta\, s\gamma)}{2\sqrt{3}}. &(9) \end{aligned}$$

Forward Kinematics

The forward kinematic problem for this device is: *Given the actuator position parameters* $\mathbf{r_1}$, $\mathbf{r_2}$, *and* $\mathbf{r_3}$, *find the position of the motion platform.* A difficult approach is solving the equations (1), (2), (3), (4), (5), and (6). A more straightforward calculation is possible if the tilt angles of the actuators (ϕ_i) are measured by optical encoders on the shafts of the revolute joints. Then the positions of points 4, 5, and 6 can be expressed as:

$${}^B\mathbf{p}_4 = \begin{bmatrix} -a/\sqrt{3} + r_1\, c\phi_1 \\ 0 \\ -r_1\, s\phi_1 \end{bmatrix}, \tag{10}$$

$$
{}^{B}\mathbf{p}_5 = \begin{bmatrix} a/2\sqrt{3} - (1/2)r_2\, c\phi_2 \\ a/2 - (\sqrt{3}/2)\, r_2\, c\phi_2 \\ -r_2\, s\phi_2 \end{bmatrix}, \tag{11}
$$

$$
{}^{B}\mathbf{p}_6 = \begin{bmatrix} a/2\sqrt{3} - (1/2)r_3\, c\phi_3 \\ -a/2 + (\sqrt{3}/2)\, r_3\, c\phi_3 \\ -r_3\, s\phi_3 \end{bmatrix}. \tag{12}
$$

The centroid of the motion platform can then be written in terms of r_i, ϕ_i as shown below

$$
{}^{B}\mathbf{p}_C = \frac{1}{3}({}^{B}\mathbf{p}_4 + {}^{B}\mathbf{p}_5 + {}^{B}\mathbf{p}_6). \tag{13}
$$

The Euler angles of α, β, γ can be solved for via the following discussion. The platform is rotated by the sequence of angles pitch(α), roll(β), and yaw(γ). The normal vector of the motion platform can be calculated by

$$
\mathbf{z}_u = \frac{{}^{B}\mathbf{p}_{6/4} \times {}^{B}\mathbf{p}_{5/4}}{|\,{}^{B}\mathbf{p}_{6/4}\,||\,{}^{B}\mathbf{p}_{5/4}\,|}. \tag{14}
$$

From the last column of the transformation matrix(${}^{B}\mathbf{T}_U$), the $\mathbf{z}_u$ vector can also be written as

$$
\mathbf{z}_u = [s\alpha\, c\beta \quad -s\beta \quad c\alpha\, c\beta]^{\mathrm{T}}, \tag{15}
$$

so we can find β, α:

$$
\beta = \sin^{-1}(-{z_u}^{y}), \tag{16}
$$

$$
\alpha = \sin^{-1}(\frac{{z_u}^{x}}{\cos\beta}). \tag{17}
$$

The angle γ can be computed from the constraint equation (7) from the inverse kinematics.

4. Velocities of Motion Platform

The required actuator speeds can be represented in two different ways. A straightforward way is to differentiate the position equations with respect to time. Another approach starts from the static force analysis. With the principle of virtual work, the actuator speed can be computed by the transpose of a Jacobian matrix which is developed by static force analysis. The result from the second approach could be computed more efficiently by feeding back the tilt angles of the actuators.

4.1. METHOD OF DIRECT DIFFERENTIATION

From inverse kinematics, the actuator lengths can be computed by three highly nonlinear equations (1), (2), and (3). By differentiating these three equations with respect to time, the results can be written as:

$$\dot{r}_i = \mathrm{f}_i({}^{B}p_U^x, {}^{B}p_U^y, {}^{B}p_U^z, \alpha, \beta, \gamma, {}^{B}\dot{p}_U^x, {}^{B}\dot{p}_U^y, {}^{B}\dot{p}_U^z, \dot{\alpha}, \dot{\beta}, \dot{\gamma})$$

where i = 1 ,2, and 3

As the pitch $(\alpha, \dot{\alpha})$, roll $(\beta, \dot{\beta})$ and heave $({}^{B}p_U^z, {}^{B}\dot{p}_U^z)$ are chosen as the inputs, the rates of yaw $(\dot{\gamma})$, surge $({}^{B}\dot{p}_U^x)$, and sway $({}^{B}\dot{p}_U^y)$ are dependent variables which can be obtained by differentiating equations (4), (5), and (6) from position kinematics as

$$\begin{aligned}
\dot{\gamma} &= (\frac{1}{1 + c\alpha\, c\beta})(s\beta\ \dot{\alpha} + s\alpha\ \dot{\beta}), && (18)\\
{}^{B}\dot{p}_U^x &= (\frac{a}{2\sqrt{3}})[(s\alpha\, c\gamma - c\alpha\, s\beta\, s\gamma)\ \dot{\alpha} \\
&- (s\beta\, c\gamma + s\alpha\, c\beta\, s\gamma)\ \dot{\beta} \\
&+ (c\alpha\, s\gamma - c\beta\, s\gamma - s\alpha\, s\beta\, c\gamma)\ \dot{\gamma}], && (19)\\
{}^{B}\dot{p}_U^y &= (\frac{a}{\sqrt{3}})[(-s\beta\, s\gamma)\ \dot{\beta} + (c\beta\, c\gamma)\ \dot{\gamma}]. && (20)
\end{aligned}$$

The above computation is quite cumbersome. However, a purely geometric relationship between the velocity of the end effector and the velocities of the actuators is possible (Waldron *et al.*, 1991). The analysis leads to a 3 by 6 Jacobian matrix, since only three actuators are involved in the computation. A square matrix is essential to perform the analysis of workspace problem. As the static forces analysis was carried out, a 6 by 6 Jacobian matrix was found, since there are three corresponding forces from revolute joints besides the forces from three actuators. The symmetry between the force/torque and velocity/angular velocity can be fully explained by the concept of virtual work.

4.2. METHOD OF VIRTUAL WORK

As the platform is subject to a external force and moment at its centroid, there will be six reaction force components. Three of them are along the ball screw axes $\mathbf{r}_1$, $\mathbf{r}_2$, $\mathbf{r}_3$ with the magnitudes of R_1, R_2, R_3. The other three are in the directions which is parallel to the passive revolute joint axes $\mathbf{u}_1$, $\mathbf{u}_2$, $\mathbf{u}_3$ with the magnitudes of U_1, U_2, U_3.

The force balance equation can be written in the form:

$$\begin{bmatrix} \mathbf{F}_{load} \\ \mathbf{N}_{load} \end{bmatrix} = \begin{bmatrix} \mathbf{J}_{6\mathrm{X}6} \end{bmatrix} \begin{bmatrix} \mathbf{R} \\ \mathbf{U} \end{bmatrix} \qquad (21)$$

where $[\ \mathbf{J}\ ]$ is the Jacobian matrix, which can be explicitly written as

$$\begin{bmatrix} \mathbf{r}_1 & \mathbf{r}_2 & \mathbf{r}_3 & \mathbf{u}_1 & \mathbf{u}_2 & \mathbf{u}_3 \\ \mathbf{c}_1 \times \mathbf{r}_1 & \mathbf{c}_2 \times \mathbf{r}_2 & \mathbf{c}_3 \times \mathbf{r}_3 & \mathbf{c}_1 \times \mathbf{u}_1 & \mathbf{c}_2 \times \mathbf{u}_2 & \mathbf{c}_3 \times \mathbf{u}_3 \end{bmatrix}.$$

The vectors $\mathbf{c}_i$ are from centroid of the platform to the points 4, 5, and 6 respectively, which could be computed from equations (10), (11), (12), and (13) from forward position kinematics. The three ball screw axes $\mathbf{r}_1, \mathbf{r}_2, \mathbf{r}_3$ can be expressed in terms of the tilt angles of ϕ_1, ϕ_2, and ϕ_3 without much computation.

We can relate the parameters in the force/torque field with the parameters in the velocity/angular velocity field by the principle of virtual work. The equation for velocities can be written as

$$\begin{bmatrix} \dot{\mathbf{r}} \\ \dot{\mathbf{u}} \end{bmatrix} = \mathbf{J}^T \begin{bmatrix} \mathbf{v} \\ \omega \end{bmatrix}. \tag{22}$$

Since motions along the axes of the revolute joints are not allowed, $\dot{u}_1$, $\dot{u}_2$, and $\dot{u}_3$ will be zero. Solving for three unknowns: $\dot{r}_1, \dot{r}_2, \dot{r}_3$, we can select three variables from ${}^B\dot{p}_U^x$, ${}^B\dot{p}_U^y$, ${}^B\dot{p}_U^z$, ω_x, ω_y, ω_z as knowns. The other three variables will be constrained.

In our application, $\omega_x, \omega_y, \omega_z$ are not designed inputs. The angular velocity for the $\mathbf{Y}-\mathbf{X}-\mathbf{Z}$ Euler angles can be converted to the angular velocity vector as (Craig, 1986):

$$\begin{bmatrix} \omega_x \\ \omega_y \\ \omega_z \end{bmatrix} = \begin{bmatrix} 0 & c\alpha & s\alpha\, c\beta \\ 1 & 0 & -s\beta \\ 0 & -s\alpha & c\alpha\, c\beta \end{bmatrix} \begin{bmatrix} \dot{\alpha} \\ \dot{\beta} \\ \dot{\gamma} \end{bmatrix}. \tag{23}$$

As we substitute equations (18), (19), and (20) into equation (22), it shows consistency with $\dot{u}_1 = \dot{u}_2 = \dot{u}_3 = 0$.

5. Workspace of the Motion Platform

In the practical design, there are physical constraints on the positions of the motion platform. These can be identified by directly analyzing the Jacobian Matrix. Another approach for identifying singular configurations of this type of mechanisms was developed by Shi and Fenton (1992). We find that both methods provide identical results.

Singular positions are determined by the condition $\det[\mathbf{J}] = 0$, where either force or speed control of the actuators will become difficult. In our application, the inputs are α, β and ${}^Bp_U^z$. The result can be plotted by

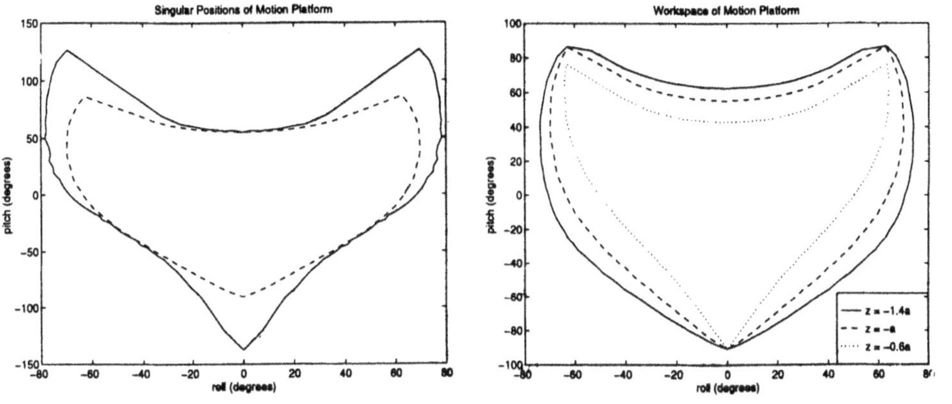

(a)Singular Positions (b)Coplanar Positions

Figure 3. Workspace of Motion Platform

varying α, β and keeping ${}^{B}p_{U}^{z}$ constant, as is seen by the solid curve in Figure 3(a).

In some of the literature, the limit of the workspace is defined as satisfying the condition that the limb is coplanar with the motion platform. These positions occur when the angle between the normal vector $\mathbf{z}_U$ of the motion platform and one of the limbs is orthogonal. The result is plotted by the dashed curve in Figure 3(a). Compared with the envelope of singularities, the coplanar workspace is smaller. Both curves consist of three segments: the segment on the top indicates the motion limit of limb 1, the segment on the right-hand side indicates the motion limit of limb 2, and the segment on the left-hand side indicates the motion limit of limb 3. These two curves are coincident at the center portions of the segments. A larger heave will increase the boundary of the workspace envelope, but it does not affect the result too much as shown in Figure 3(b).

6. Summary and Conclusion

In this paper, both position and velocity relationships between three drivers and the motion platform are presented. These computations are essential for the development of the control system of the driving simulator. The analysis of the velocity problem for a parallel mechanism can be more direct if the static force problem is solved first. The results obtained for the singularity analysis are verified by that of reference (Shi *et al.*, 1992), but the approach is totally different.

Our future work includes dynamic analysis and design of a full-scale prototype. Hopefully, these studies will be provided for an OSU Driving Simulator.

7. Acknowledgment

The work reported here is supported by a grant to The Ohio State University by Center for Intelligent Transportation Research (CITR). Thanks are due to Mr. Gary Gardner for machining the table-top model.

References

Allen, R.W., Rosenthal, T.J., and Parseghian, Z. (1995) Low Cost Driving Simulation for Research, Training and Screening Applications, *International Congress and Exposition, Detroit, MI*, **SAE Paper 950171**

Craig, John J. (1986) Introduction to Robotics Mechanics and Control, *Addison-Wesley Publishing Company, Menlo Park, California*

Dutt, V., Orin, D.E., Yang, P., Waldron, K.J. (1995) Control of a Motion Platform for a IVHS Driving Simulator, *Fourth Applied Mechanism and Robotics Conference, Cincinnati, Ohio*

Haug, E.J. (1990) Feasibility Study and Conceptual Design of a National Advanced Driving Simulator, *Final Report, U.S. Department of Transportation*, **DOT HS 807 596**

Innocenti C., Parenti-Castelli V. (1990) Direct Displacement Analysis for Some Classes of Spatial Parallel Mechanism, *Proceedings of the 8th Symposium on Theory and Practice of Robots and Manipulators, Warsaw*, pp. 126–133.

Lee, K.M., Shah D.K. (1988) Kinematic Analysis of a Three-Degree-of-Freedom In-Parallel Actuated Manipulator, *IEEE J. of Robotics and Automation*, **Vol. 4, no. 3**, pp. 354–360,

Salcudean, S.E., Drexel, P.A., Ben-Dov, D., Taylor, A.J., Lawrence, P.D. (1994) A Six Degree-of-Freedom, Hydraulic, One Person Motion Simulator, *IEEE*, pp. 2437-2442.

Shi, X., Fenton, R.G. (1992) Structural Instabilities in Platform-Type Parallel Manipulators Due to Singular Configuration, *Robotics, Spatial Mechanisms, and Mechanical Systems, ASME*, **DE-Vol 4**, pp. 347–352.

Stewart D. (1965) A Platform with Six Degrees of Freedom, *Proc. Instn. Mech. Engrs.*, **Vol. 180, Part 1, No. 15**, p. 371

Waldron, K.J., Raghavan M., and Roth B. (1989) Kinematics of Hybrid Series-Parallel Manipulation System, *Journal of Dynamics Systems, Measurement, and Control*, **Vol. 111**, pp. 211–221

Waldron, K.J., Hunt, K.H. (1991) Series-Parallel Dualities in Actively Coordinated Mechanisms, *International Journal of Robotics Research*, Vol. 10, no. 5, pp. 473–480

Weir, D.H., Clark, A.J. (1995) A Survey of Mid-Level Driving Simulators, *International Congress and Exposition, Detroit, MI*, **SAE Paper 950172**

KINEMATIC MANIPULABILITY OF CLOSED CHAINS

F. C. PARK AND J. W. KIM
Mechanical Design and Production Engineering
Seoul National University
Kwanak-Ku Shinlim-Dong San 56-1
Seoul 151-742, Korea
E-mail: fcp@robotics.snu.ac.kr

Abstract. This paper presents a coordinate-invariant differential geometric formulation of manipulability for closed kinematic chains containing active and passive joints. The formulation treats both redundant and nonredundant mechanisms, as well as overactuated and exactly actuated ones, in a uniform manner. We illustrate the formulation with several planar closed chain examples.

1. Introduction

The notion of manipulability has been particularly useful in the analysis and design of robot manipulators. Manipulability, or the ability to move and apply forces in arbitrary directions, has been used in applications ranging from the kinematic design of robotic fingers (Salisbury and Craig, 1982), to the optimal positioning of the workpiece in a robot's workspace (Klein and Blaho, 1987). Salisbury and Craig (1982) first proposed the concept in the design context mentioned above, while Yoshikawa (1985) gave one of the first comprehensive mathematical treatments of manipulability for general open chains. Since then alternate formulations and various aspects of manipulability have also been investigated in, *e.g.*, Gosselin *et. al.*, (1988), (1988b), (1990).

Nearly all of the previous work on open chain manipulability begins by parametrizing the forward kinematics in local coordinates, and measuring in some fashion the extent to which the Jacobian of the forward kinematics (i.e., the linear map relating joint rates to tip rates) preserves distances

J. Lenarčič and V. Parenti-Castelli (eds.), Recent Advances in Robot Kinematics, 99-108.
© 1996 *Kluwer Academic Publishers. Printed in the Netherlands.*

and angles; Section 2 discusses this idea in more detail. In (Park, 1995b) these and other manipulability measures are formulated under a general mathematical framework using the coordinate-free methods of Riemannian geometry. While the coordinate-free methods resolve some apparent ambiguities that arise in defining a combined position and orientation manipulability, the geometric analysis more or less confirms the intuitive approach of the coordinate-based methods. This is due in part to the fact that the configuration (joint) space for open chains can be regarded as a vector space. Moreover, since all joints are assumed actuated, a coordinate-based formulation of manipulability is particularly straightforward and easy to visualize in this case.

Closed chains, on the other hand, present a number of subtleties. Kinematically speaking, the most obvious difference is that a closed chain's configuration space will no longer be a flat space; in general it will be a curved multi-dimensional surface embedded in a higher-dimensional flat space. Also, dual to the open chain case, the forward kinematics for closed chains is generally more difficult than the inverse kinematics, with the possibility of multiple solutions. Another important difference is that only a subset of the joints may be actuated, and it is not difficult to see that manipulability will be affected by the number and choice of actuated joints. A number of coordinate-based formulations for closed chain manipulability have been proposed for specific mechanisms (Gosselin and Angeles, 1988b) and for cooperating robot systems all of whose joints are assumed actuated (Bicchi *et. al*, 1995), (Chiacchio *et. al.*, 1991), at least some of which has led to apparently paradoxical results (Melchiorri *et. al*, 1993), (Chiacchio *et. al.*, 1993). Because of the nonlinear characteristics unique to closed chain mechanisms discussed above, particular care must be exercised in formulating coordinate-based manipulability measures.

In this paper we present a differential geometric formulation of manipulability for closed chains and parallel manipulators. Manipulability, like any other performance measure, should be formulated such that it is independent of the choice of coordinates, and takes into account the nonlinearity of both the configuration space and the rigid-body displacements. We argue that the global, coordinate-free methods of differential geometry provide the most natural means of formulating manipulability for mechanisms with curved configuration spaces. In this framework actuator characteristics are reflected in the choice of a (pseudo-)Riemannian metric on the configuration space. As we show below, this geometric framework uniformly treats both redundant and nonredundant robots, as well as exactly actuated and over-actuated systems, eliminating the need for ad hoc methods.

The paper is organized as follows. In Section 2 we present a geometric formulation of manipulability for general mechanisms, briefly reviewing the essential differential geometric background. In Section 3 this formulation of manipulability is specialized to closed chain systems and parallel manipulators, while Section 4 presents an analysis of manipulability for some closed-chain planar mechanisms using our geometric framework.

2. Manipulability and Riemannian Geometry

In this section we present a Riemannian geometric formulation of manipulablity for general chains, beginning with some of the ideas presented in (Park, 1995b). Because of space limitations we refer the reader to (Park, 1995b) first, where an intuitive understanding can be developed via a simple spherical mechanism example.

NONREDUNDANT MECHANISMS

We first fix some notation. Let M denote the manifold corresponding to the joint space of the mechanism, and $\mathbf{H}$ the Riemannian metric on M. For an n-revolute open chain M will be an n-dimensional torus, and $\mathbf{H}$ a diagonal matrix (in terms of the standard coordinates for a flat torus) whose entries will be proportional to the maximum velocity of the corrresponding actuator. Let N denote the manifold corresponding to the end-effector space of the mechanism, and $\mathbf{G}$ the Riemannian metric on N. For example, if orientations are ignored, then N will be $\Re^2$ for a planar mechanism, and $\Re^3$ for a spatial mechanism. The Riemannian metric in this situation will be chosen to be the Euclidean metric, *i.e.*, $\mathbf{G} = \mathbf{I}$.

Manipulability can now be defined in a coordinate-invariant way as follows. First, any symmetric function of the roots of $\det(\mathbf{H}\lambda - \mathbf{J}^T\mathbf{G}\mathbf{J}) = 0$ (or, in coordinate-free language, the proper values of the pullback metric f^*g) is a coordinate-invariant function defined on the Riemannian manifold M. At a regular point the Jacobian of f by definition will have maximal rank, and the proper values of f^*g will have exactly m nonzero values (assuming $n \geq m$); we label these, in descending order, as $\lambda_1, \lambda_2, \ldots, \lambda_m$. Various symmetric functions of these proper values can now be constructed to reflect kinematic manipulability. For example, since singularities are marked by λ_m going to zero, maximizing λ_m is one possibility for a local kinematic manipulability measure that penalizes singular configurations. A condition number-based manipulability measure is

$$c(f) = \frac{\lambda_1}{\lambda_m} \tag{1}$$

while the generalization of Yoshikawa's manipulability measure is

$$v(f) = (\lambda_1 \cdots \lambda_m)^{\frac{1}{2}} \tag{2}$$

To obtain a global measure of manipulability one can integrate any of the various local manipulability measures over the joint space manifold M:

$$E(f) = \int_M \mu(f)\, dV \tag{3}$$

where $\mu(f)$ is an appropriate local measure, and $dV = \sqrt{\det G} dx_1 \cdots dx_m$ is the volume form on M. One interesting choice of $\mu(f)$ is given by $\mu(f) = \mathrm{Tr}(\mathbf{J}^{\mathrm{T}}\mathbf{G}\mathbf{J}\mathbf{H}^{-1})$; this leads to the integral functional for harmonic maps, whose minima can be viewed as the mapping that best approximates a global isometry, see (Park, 1995b).

REDUNDANT MECHANISMS

The astute reader will have observed that the concept of isometry only makes sense if the dimension of $\mathbf{G}$ is at least as large as that of $\mathbf{H}$ ($n \geq m$). Another way of expressing this is that the dimension of the range space be at least as large as that of the domain. While our spherical mechanism example satisfies this requirement (the dimensions are equal), redundant mechanisms do not. This apparent dilemma can be resolved by instead considering the adjoint mapping $\mathbf{J}^{\mathrm{T}}$; recall that this is the linear map relating output forces to input torques. In this case the dimensions of the domain and range are reversed, so that one can now formulate a "dual" notion of isometry. Mathematically the idea can be realized as follows.

Let $\mathcal{L} : \mathcal{U} \rightarrow \mathcal{V}$ be a linear mapping between two vector spaces. Then its adjoint map $\mathcal{L}^* : \mathcal{V}^* \rightarrow \mathcal{U}^*$ is a linear mapping between the corresponding dual vector spaces. If $\mathbf{L}$ is the matrix representation for $\mathcal{L}$ with respect to some bases on $\mathcal{U}$ and $\mathcal{V}$, then with respect to their standard dual bases the adjoint map $\mathcal{L}^*$ can be expressed in coordinates as the matrix transpose $\mathbf{L}^{\mathrm{T}}$. For the kinematic Jacobian, $\mathcal{U}$ and $\mathcal{V}$ are velocity spaces, while their dual spaces correspond to the spaces in which the dual vectors of forces and torques reside. Given metrics $\mathbf{H}$ and $\mathbf{G}$ on $\mathcal{U}$ and $\mathcal{V}$, respectively, their corresponding dual metrics on $\mathcal{U}^*$ and $\mathcal{V}^*$ are $\mathbf{H}^{-1}$ and $\mathbf{G}^{-1}$, respectively. Hence, the requirement that $\mathcal{L}^*$ be an isometry can be expressed in coordinates as $\mathbf{J}\mathbf{H}^{-1}\mathbf{J}^{\mathrm{T}} = \mathbf{G}^{-1}$, or $\mathbf{J}\mathbf{H}^{-1}\mathbf{J}^{\mathrm{T}}\mathbf{G} = \mathbf{I}$. Thus, as in the nonredundant case, a coordinate-invariant function of manipulability can be formulated as a symmetric function of the roots of $\det(\mathbf{G}^{-1}\lambda - \mathbf{J}\mathbf{H}^{-1}\mathbf{J}^{\mathrm{T}}) = 0$ (or alternatively, the eigenvalues of $\mathbf{J}\mathbf{H}^{-1}\mathbf{J}^{\mathrm{T}}\mathbf{G}$).

The ellipsoid associated with the above formulation of manipulability is obtained from the eigenvalues and eigenvectors of $\mathbf{J}\mathbf{H}^{-1}\mathbf{J}^{\mathrm{T}}\mathbf{G}$. Specifically, $\mathbf{J}\mathbf{H}^{-1}\mathbf{J}^{\mathrm{T}}\mathbf{G}$ is a quadratic form on the space of tip velocities, and the principal axes of the ellipsoid are given by its eigenvectors, while their lengths are given by the square root of the eigenvalues. The ellipsoid admits the following two physical interpretations. The more obvious interpretation is that the ellipsoid can be regarded as the image, under $\mathbf{J}$, of the unit sphere (in the space of joint velocities) into the space of tip velocities, which in this case is $\Re^m$. Note that "sphere" and "ellipsoid" here are relative to the quadratic forms $\mathbf{G}$ and $\mathbf{H}$, respectively. The other dual interpretation is that the ellipsoid is obtained by mapping, via $\mathbf{J}^T$, the unit sphere in the space of tip forces into the space of joint torques. Here "sphere" and "ellipsoid" in this context are with respect to the quadratic forms $\mathbf{G}^{-1}$ and $\mathbf{H}^{-1}$, respectively. Also, the ellipsoid in this case is of dimension n-1, residing inside $\Re^m$. The manipulability ellipsoid described here is valid for both redundant and nonredundant mechanisms.

3. Manipulability for Closed Chains

The concept of manipulability described above was formulated in the setting of general Riemannian manifolds, with a coordinate-invariant manipulability function defined on the configuration space manifold. Because of its generality, this formulation is valid for both open and closed chains; the only other issue to consider in the latter case is what the proper choices of Riemannian metric should be. We now discuss the issue of what metrics to choose, along with the general procedure for formulating closed chain manipulability.

Consider a mechanism with a total of k prismatic and revolute joints, possibly containing closed loops. Denote the m-dimensional configuration space manifold by M, where $m \leq k$; M can be viewed as an m-dimensional surface in $\Re^k$. Let $\mathbf{x} = (x_1, \ldots, x_k)$ and $\mathbf{u} = (u_1, \ldots, u_m)$ denote local coordinates on $\Re^k$ and M, respectively, so that $\mathbf{x}$ can be written as a function of $\mathbf{u}$, i.e., $\mathbf{x} = \mathbf{x}(\mathbf{u})$. Suppose a tool reference frame is attached to some point on the mechanism, and denote its configuration manifold by N. Let $\mathbf{f} = (f_1, \ldots, f_n)$ denote local coordinates for N, where $n \leq 6$. For general spatial mechanisms N will be some subset of the group of spatial displacements, or the Special Euclidean Group SE(3). For planar mechanisms N will correspond to the planar displacements SE(2), while for spherical mechanisms N will be SO(3). For certain applications one may also wish to ignore orientations and consider position only; in this case N can be taken to be a subset of $\Re^2$ or $\Re^3$.

We now discuss the choice of an appropriate Riemannian metric for M, followed by that for N. Rather than choose a metric directly on M, we first choose a metric for $\Re^k$, and project this metric onto M. The metric on the ambient space $\Re^k$ can be given the same physical interpretation as for the open chain case. That is, the metric is taken to be of the form $ds^2 = \epsilon_1 dx_1^2 + \ldots + \epsilon_k x_k^2$, where each ϵ_i can be interpreted as the gear ratio between the ith actuator and ith joint. In the event that the joint is not actuated, ϵ_i is then set to zero. This choice of metric can be justified from virtual work considerations. Recalling that work is force times distance, if a particular joint is not actuated then its contribution to the total work should be zero regardless of any displacement that the joint may have undergone. Setting ϵ_i to zero achieves the desired result by ignoring the joint displacement.

Denote the metric on $\Re^n$ by $\mathbf{E} = \text{Diag}[\epsilon_1 \ldots \epsilon_k]$. Note that because some of the ϵ_i may be zero, the metric may only be positive semi-definite (and hence strictly speaking only a pseudo-Riemannian metric). The metric on M is then obtained by projecting $\mathbf{E}$ onto M: since $\mathbf{dx} = \mathbf{x_u du}$, where the subscript denotes partial differentiation,

$$\begin{aligned} ds^2 &= \frac{1}{2}\mathbf{dx}^T\mathbf{E}\,\mathbf{dx} \\ &= \frac{1}{2}\mathbf{du}^T\mathbf{x}_\mathbf{u}^T\mathbf{E}\mathbf{x}_\mathbf{u}\mathbf{du} \end{aligned}$$

The projected metric on M, denoted $\mathbf{H}$, is therefore $\mathbf{H} = \mathbf{x}_\mathbf{u}^T\mathbf{E}\mathbf{x}_\mathbf{u}$.

The metric on N, which we denote $\mathbf{G}$, is also chosen from physical considerations. In the event that N is some proper subset of SO(3) or Euclidean space, there exists a natural choice of Riemannian metric, given by the standard Euclidean inner product for velocities and angular velocities (see (Park, 1995)). When N is taken to be either SE(2) or SE(3), however, there now exists a one-parameter family of suitable Riemannian metrics parametrized by the choice of length scale for physical space. There is a certain amount of arbitrariness involved in this choice of length scale, and (Park, 1995) suggests some ways in which to choose a scale that takes into account the given task. In general, however, this arbitrariness is an unavoidable consequence of the geometry of SE(3).

Denote the forward kinematics $\mathbf{f}: M \rightarrow N$ in local coordinates as $\mathbf{f}(\mathbf{u})$, and its Jacobian by $\mathbf{J} = \frac{\partial \mathbf{f}}{\partial \mathbf{u}} = \mathbf{f_u}$.[1] The manipulability ellipsoid is then determined by the eigenvalues and eigenvectors of $\mathbf{J}\mathbf{H}^{-1}\mathbf{J}^T\mathbf{G}$, assuming $m \geq n$. Two remarks are in order. First, because $\mathbf{H}$ is induced from only a pseudo-Riemannian metric, there may be instances in which $\mathbf{H}$ is singular. Physically this corresponds to situations where the mechanism is underactuated. Second, the points at which $\mathbf{J}\mathbf{H}^{-1}\mathbf{J}^T\mathbf{G}$ becomes singular correspond to actuator singularities; that is, with the current set of actuated joints the tip frame loses a degree of freedom of motion.

[1] For closed chains the forward kinematics may be multiple valued, but from both practical and mathematical considerations it is not unreasonable to eliminate extra solutions, and consider only one solution branch.

4. Examples

In this section we examine the manipulability of two planar closed chains. The first mechanism is a nonredundant 5-bar linkage, while the second mechanism is a redundant 6-bar linkage. In both cases we ignore orientation and consider only Cartesian position of the tip.

FIVE-BAR LINKAGE

The five-bar linkage is shown in Fig. 2. We assume all the link lengths to be equal, with the two fixed pivots also separated by a link length. Two cases are considered. In the first case we assume that the two base joints are actuated (observe that this mechanism has two degrees of freedom), while in the second case the two neighboring middle joints are actuated. In the figures the actuated joints are shaded, while the passive joints are indicated by unfilled circles. We draw the manipulablility ellipsoid at 3 points in the workspace for both cases. From a visual inspection it is clear that for the given set of workspace points, placing the actuators at the middle joints provides better kinematic manipulability.

SIX-BAR LINKAGE

To illustrate how our formalism can uniformly handle both overactuated and exactly actuated mechanisms, consider the three degree-of-freedom six-bar linkage of Fig. 3. Because we ignore orientation and consider only planar position this mechanism is kinematically redundant. Two cases are considered: in the first case the mechanism is overactuated, by placing actuators at the four joints closest to the fixed pivots (shown as filled circles). In the second case only three of the mechanism's joints are actuated. Once again we draw manipulability ellipsoids for three points in the workspace. As expected, for the overactuated case the manipulability ellipsoid is aligned along the x- and y-axes when the mechanism is in a symmetric posture. However, when only three of the joints are actuated the ellipsoid's orientation is skewed for the same posture.

5. Conclusions

In this paper we have presented a coordinate-invariant formulation of manipulability for both open and closed chain mechanisms, that is based on methods of Riemannian geometry. The actuator characteristics are reflected in the choice of a Riemannian metric for the joint and end-effector configuration spaces, and the resulting definitions of manipulability are independent of the choice of coordinate parametrization for these two spaces.

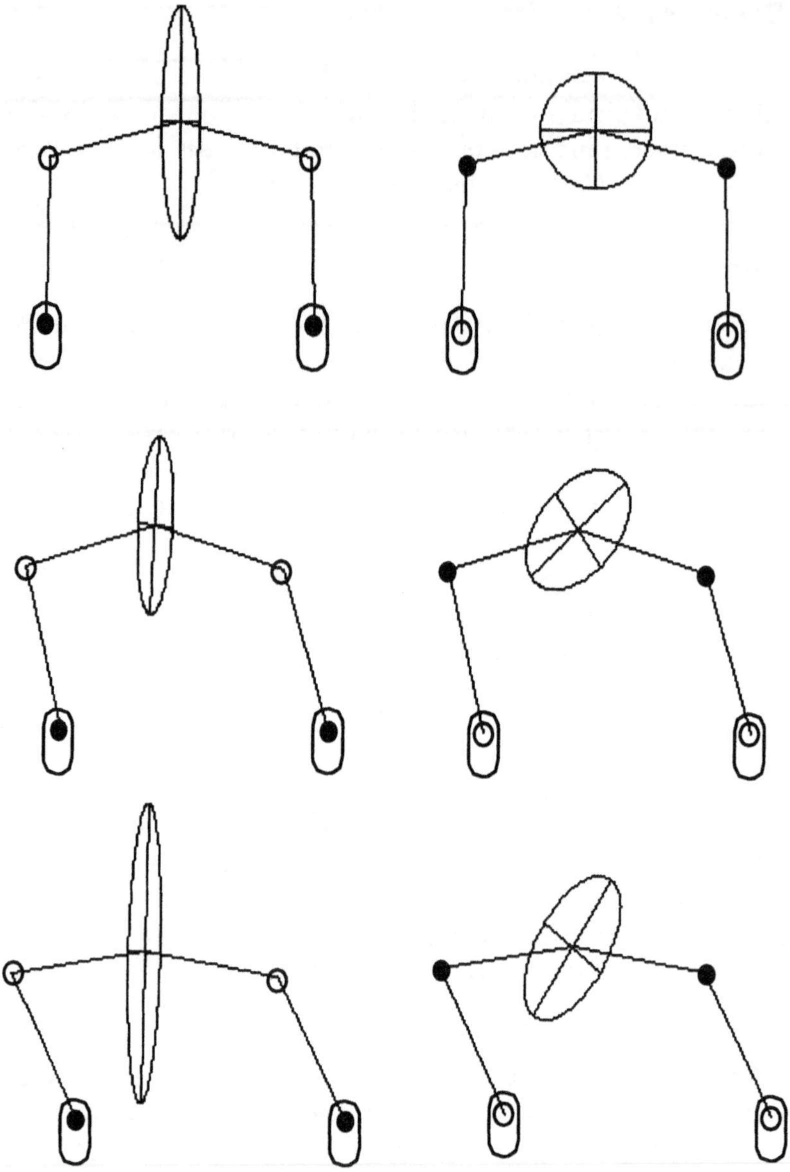

Figure 1. Manipulability ellipsoids for the *5R* linkage.

One of the appealing features of this formulation is that the dynamics of the mechanism can be naturally included; one can choose the configuration space metric $\mathbf{H}$ to be the inertia matrix, and for cooperating systems manipulating a common object, the object's generalized inertia matrix can be chosen as the metric $\mathbf{G}$. The classification of various singularities of parallel

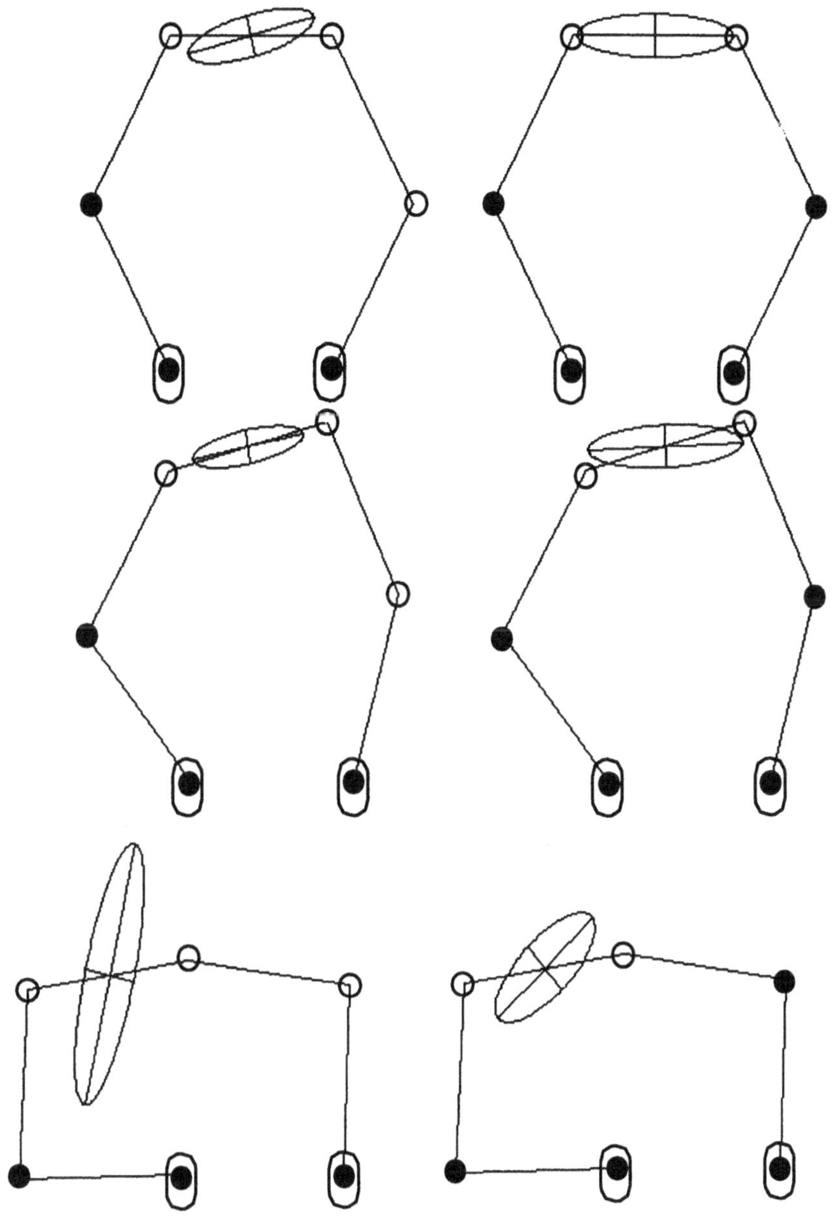

Figure 2. Manipulability ellipsoids for the *6R* linkage.

manipulators, as well as their stiffness characteristics, can all be be pursued rigorously under this mathematical framework. These are topics of current investigation.

Acknowledgements

This research was supported by the Engineering Research Center for Advanced Instrumentation and Control at Seoul National University.

References

Chiacchio, P., Chiaverini, S., Sciavicco, L., and Siciliano, B. (1991) Global task space manipulability ellipsoids for multiple-arm systems, *IEEE Trans. Robotics Autom.*, **Vol. 7, No. 5**, pp. 678-685.

Chiacchio, P., Chiaverini, S., Sciavicco, L., and Siciliano, B. (1993) Reply to "comments on 'Global task space manipulability ellipsoids for multiple-arm systems' and further considerations," *IEEE Trans. Robotics Autom.*, **vol. 9, No. 2**, pp. 235-236.

Gosselin, C. M., and Angeles, J. (1988) A new performance index for the kinematic optimization of robotic manipulators, *Proc. 20th ASME Mechanisms Conf.*, Kissimmee, FL., pp. 441-447.

Gosselin, C. M., and Angeles, J. (1988b) The optimum kinematic design of a planar three-degree-of-freedom parallel manipulator, *ASME J. Mech., Trans., and Autom. in Design*, **Vol. 110**, pp. 35-41.

Gosselin, C. M. (1990) Dexterity indices for planar and spatial robotic manipulators. *Proc. IEEE Int. Conf. Robotics & Autom.*, pp. 650-655.

Klein, C. A., and Blaho, B. (1987) Dexterity measures for the design and control of kinematically redundant manipulators, *Int. J. Robotics Research*, **Vol. 6, No. 2**, pp. 72-83.

Melchiorri, C. (1993) Comments on 'Global task space manipulability ellipsoids for multiple-arm systems' and further considerations, *IEEE Trans. Robotics Autom.*, **Vol. 9, No. 2**, pp. 232-235.

Bicchi, A., Melchiorri, C., and Balluchi, D. (1995) On the mobility and manipulability of general multiple limb robots, *IEEE Trans. Robotics Autom.*, **Vol. 11, No. 2**, pp. 232-235.

Park, F. C. (1995b) Optimal robot design and differential geometry, *Trans. ASME 50th Anniv. Special Issue*, **Vol. 117(B)**, pp. 87-92.

Park, F. C. (1995) Distance metrics on the rigid-body motions with applications to mechanism design, *ASME J. Mechanical Design*, **Vol. 117**, pp. 48-54.

Salisbury, J. K., and Craig, J. J. (1982) Articulated hands: force control and kinematic issues, *Int. J. Robotics Research*, **Vol. 1, No. 1**, pp. 4-17.

Yoshikawa, T. (1985) Manipulability of robotic mechanisms, *Int. J. Robotics Research* **Vol. 4, No. 2**, pp. 3-9.

SINGULARITY ANALYSIS AND CATEGORISATION OF GENERIC 3-R REGIONAL MANIPULATORS

Laboratoire d'Automatique de Nantes (U.A. CNRS 823)
Ecole Centrale de Nantes, University of Nantes
1, rue la Noë 44072 NANTES - FRANCE
Tel (33) 40371671 -Fax (33) 40747406 - email : wenger@ec-nantes.fr

Abstract - This paper investigates the global kinematics of generic 3-R regional manipulators through singularity analysis. Generic manipulators are those manipulators whose singularities form smooth manifolds in the jointspace. Starting from a preliminary homotopy based classification scheme recently proposed in the literature, the complete categorisation of all generic regional manipulators is established. It is shown that there are only eight classes of homotopic generic manipulators. Several illustrative examples are provided.

1. Introduction

Singularities play a central role in manipulator kinematics. Most of the huge amount of literature dealing with manipulator singularities is concerned with manipulator control and trajectory planning. On the other hand, few authors have so far addressed manipulator singularities for global analyses purposes. The global kinematics properties of a manipulator are often analysed by investigating its workspace structure (see for example [1-6]). The geometry and topology of the singular manifolds in the joint space turn out to be an interesting alternate way of globally analysing the kinematics properties of manipulators. In 1988, Burdick presented a detailed analysis of 3-R manipulator singularities [7]. One year later, Paï introduced the notion of generic manipulators [8]. A manipulator is said generic if its singularities are generic, that is, if they form smooth manifolds in the joint space. Paï showed that the set of non generic manipulators is codimension one in the space of all manipulators. That is, a manipulator is almost generic, in the sense that if the geometric parameters of a manipulator are given at random, the probability to obtain a non generic manipulator is null. At the end of his work [9], Burdick proposed a preliminary classification scheme for generic manipulators, using the number and homotopy class of their singular manifolds. However, he did not attempt a complete enumeration of the generic manipulators.

The primary goal of this paper is to enumerate all possible classes of homotopic generic regional 3-R manipulators. Regional manipulators are referred to as serial manipulators whose primary task is to reach points in the 3-D Cartesian space. One of the applications of this study is the complete categorisation of the 3-R generic cuspidal[1] and non cuspidal manipulators.

This remainder of this paper is organized as follows:

- section 2 recalls the notions of singularity, genericity and homotopy class,
- section 3 sets new results about the geometry and topology of the singular manifolds, and states that there are only 8 classes of homotopic generic manipulators
- section 4 illustrates the different manipulator classes

The last section is devoted to concluding remarks.

---

[1] A cuspidal manipulator is one that can change posture without meeting a singularity [10]

J. Lenarčič and V. Parenti-Castelli (eds.), Recent Advances in Robot Kinematics, 109-118.
© 1996 *Kluwer Academic Publishers. Printed in the Netherlands.*

2. Theoretical background

2.1. THE SINGULARITIES OF REGIONAL 3-DOF MANIPULATORS

Manipulator singularities arise when the manipulator Jacobian matrix looses rank. In a regional (resp. rotational) singularity, the end-effector cannot be instantaneously translated along (resp. rotated around) an axis. Only regional singularities will be studied here, and from now on, the word singularity will stand for regional singularity.
The singularities of a regional manipulator are characterized by the set of joint configurations q that nullify the determinant of the Jacobian matrix. They form manifolds in the joint space, and divide it into at least two singularity-free domains called "aspects" in [11] and "c-sheets" in [7] and [9]. The global kinematics properties of a manipulator are intimately related to the geometry and topology of the singular manifolds, which govern the global transformation, under the forward kinematics map, of the joint space into the workspace. Under the forward kinematics map, the singular manifolds are rearranged into manifolds in the workspace, which divide it into regions with different accessibility. The joint space of a 3-R manipulator is a 3-dimensional torus, but since the singularities are independent of the first joint axis, the singular manifolds can be more simply analysed in the (θ2-θ3)-torus, where they form closed curves. For more convenience, however, the singular manifolds are traced in a square of dimension 2π, by cutting the torus along its generators. In order to keep the topology of the torus, the opposite sides of square should be identified.

2.2 GENERIC MANIPULATORS

The genericity of manipulators is based on the notion of generic maps. A manipulator is generic when its singularities form a collection of smooth non-intersecting closed manifolds [8], that will be referred to as *branches* in this paper. Paï derived an algebraic condition for a 3-DOF regional manipulator to be generic. A 3-DOF manipulator is generic if and only if the following two conditions are satisfied :

(1) The rank deficiency of the Jacobian matrix is 1 at all singular configurations

(2) For all singular configurations q_s : $\dfrac{\partial\,[\det(J(q^{(s)}))]}{\partial\, q_i} \neq 0$ for at least i=2 or 3

It appears that non-genericity often arises from geometric simplification conditions in the manipulator structure (like two intersecting or parallel joint axes), and that most industrial manipulators, in turn, are non-generic [7], [9].
An important feature of generic manipulators is that their global kinematics properties remain stable under small changes in their kinematics parameters. This is not true for non-generic manipulators. This means that particular attention must be paid when manufacturing a non-generic manipulator, since too large manufacturing tolerances may profoundly modify the expected kinematic properties of the manipulator. Figures 1, 2 illustrate the instability with respect to small variations in the design parameters of a non-generic manipulator. Figure 1 depicts the singularities (from -π to π) of a non-generic manipulator with standard DH parameters : a1=0, a2=2, a3=1.5, d2=1, d3=0, α1=-90° and α2=90°

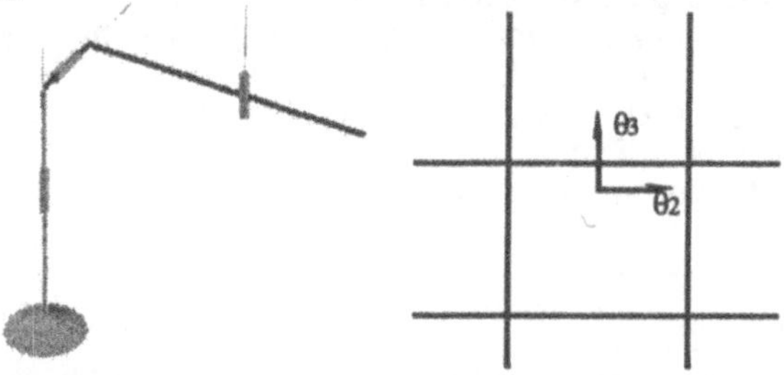

Fig. 1 : Singular manifolds of a non-generic manipulator with two intersecting joint axes

The singularities clearly intersect and divide the joint space into 4 c-sheets (do not forget to identify the opposite sides of the square), and it can be shown that there is one inverse kinematic solution in each c-sheet, thus this manipulator must pass through a singularity when changing posture [10]. Fig. 2 shows that, when the first two joint axes do not exactly intersect, the singularities form two distinct branches, yielding only two c-sheets.

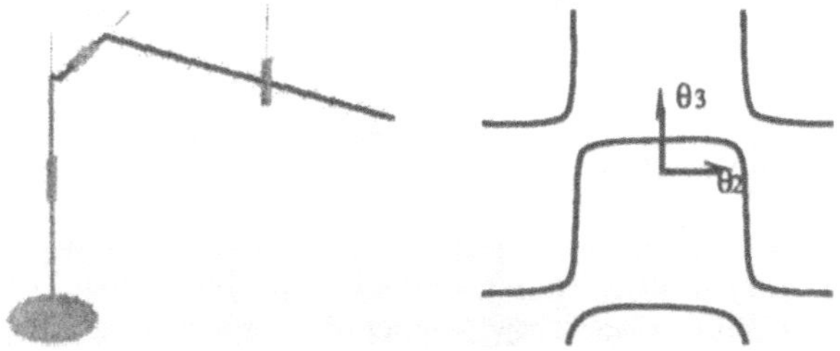

Fig. 2 : Singular manifolds when the first two joint axes do not exactly intersect (a1= 0.01)

In addition, it can be shown that there are two inverse kinematics solutions in each c-sheet, thus showing that non-singular changes of posture are possible for this robot.

2.3 CLASSIFICATION USING THE NOTION OF HOMOTOPY CLASS

Burdick proposed an interesting way of classifying 3-R generic regional manipulators using the number and homotopy class of their singularities.

Two maps f and g are said to be homotopic if f can be smoothly (i.e. through a smooth map) deformed to g. In the context of manipulator kinematics, two generic manipulators M1 and M2 are homotopic if the singular manifolds of M1 can be smoothly deformed to the singular manifolds of M2. More importantly, it was shown in [7] that the kinematic maps of two homotopic manipulators have the same multiplicity. This means that the number of inverse kinematics solutions per aspect is the same for two homotopic manipulators. Consequently, if M1 is cuspidal (resp. non-cuspidal), M2 is also cuspidal (resp. non-cuspidal). The classification scheme proposed by Burdick is the following :

- the space of all quaternary[2] 3-R regional manipulators is naturally divided by the set of non-generic manipulators, into disjoints subspaces of homotopic generic manipulators. Note that this classification does not hold for binary manipulators [10]. Only quaternary manipulators will be considered in this study.
- the homotopic generic manipulators are characterized by the respective homotopy classes of their singular manifolds branches.

For a generic manipulator, it can be shown that each singular manifold branch forms a loop (i.e. a closed curve that starts and ends at the same point) on the (q2-q3)-torus. The homotopy class of the singular manifold branches is related to the topological properties of loops on the (q2-q3)-torus. In the Cartesian space (R2 or R3), all loops can be continuously deformed to a point : they are all homotopic. This is not the case on the torus, where there are as many loop classes as ways of encircling the two generators of the torus. Figure 3 illustrates loops with three different homotopy classes. The homotopy class of L1 is a circle surrounding the q2-generator of the torus. In a square representation ($-\pi\leq\theta 2\leq\pi, -\pi\leq\theta 3\leq\pi$) of the torus, L1 appears as an horizontal line. The homotopy class of L2 is a circle that wraps around the q3-generator, and its square

---

[2] Quaternary positional manipulators have 4 inverse kinematic solutions. Binary manipulators have only two inverse solutions [10].

representation is a vertical line. Finally, L3 does not encircle any of the two generators. In both representations (torus and square), L3 is homotopic to a point.

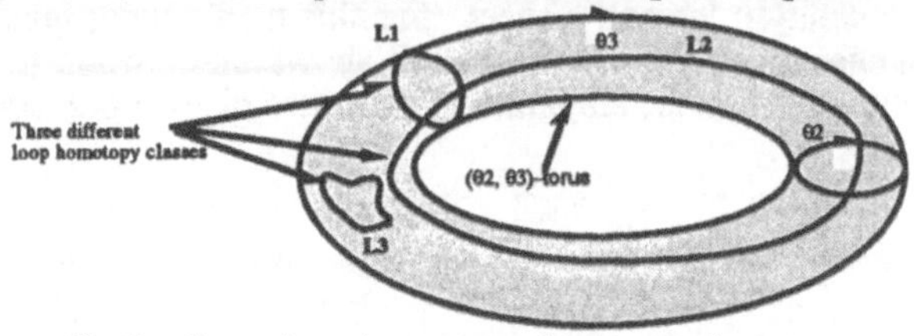

Fig.3 : Some loop homotopy classes on the torus

The homotopy class of a singular manifold branch can be quantified by a set of two integers (n2, n3). Integer n2 (resp. n3) characterizes the number of times the branch encircles the θ2-generator (resp. θ3-generator) of the torus (θ2,θ3). In figure 3, L1 has homotopy class (1,0), L2 has (0,1), and L3 has (0,0). Accordingly, the homotopy class of a generic manipulator is characterized by a series of couples (n2,n3) which define the homotopy classes of each of its branch. For example, the generic manipulator shown in figure 2 has two (1,0) singular manifolds branches. In effect, both branches can be smoothly deformed to an horizontal line (in the square representation). We say that it is a 2(1,0) manipulator. Figure 4a depicts a (0,0)-manipulator : there is one singular manifold branch with homotopy class (0,0) ($a1=1$, $a2=2$, $a3=2.5$, $d2=1$, $d3=0.2$, $\alpha1=-62°$ $\alpha2=90°$).

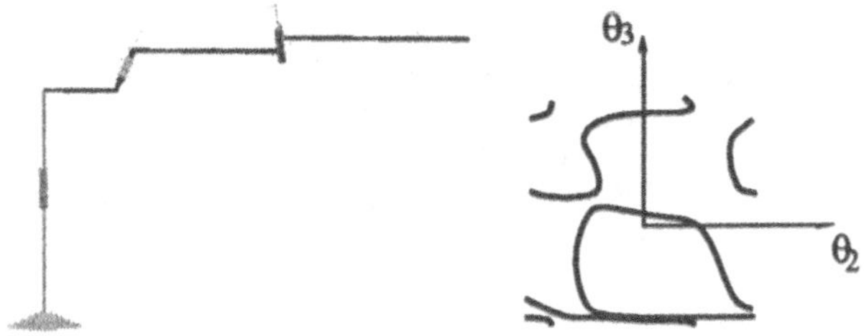

Fig.4a : Manipulator with (0,0) homotopy class

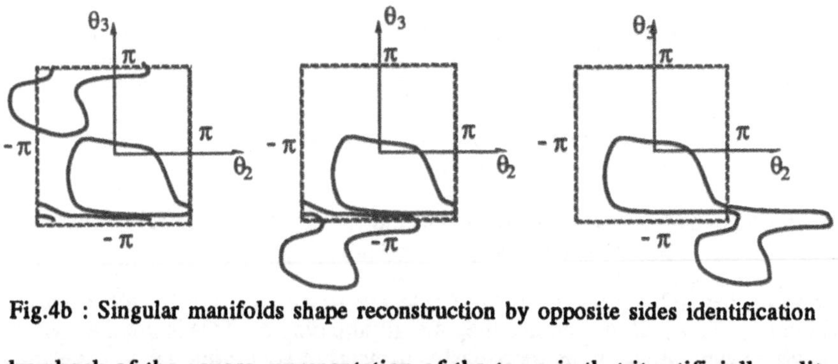

Fig.4b : Singular manifolds shape reconstruction by opposite sides identification

The drawback of the square representation of the torus is that it artificially splits the singular manifolds along the generators of the torus, and it is difficult to identify their shape, especially when we have a (0,0)-branch which does not lie within the square representation. To better understand that there is actually one single singular manifold branch with homotopy class (0,0), it is necessary to "reglue" the singular manifold by identifying the opposite sides of the square representation of the torus (see figure 4b, which shows that this singular manifold can be smoothly contracted to a point). In some cases, we may have a more recognisable (0,0)-branch that appears completely within the square. Such (0,0)-branches, referred to as isolated (0,0)-branches, are quite unusual, and it was shown in [7] that a 3-R generic manipulator cannot have several isolated (0,0)-branches. Other manipulator classes may exist. The following section states new results which will permit to enumerate all possible classes.

3. Enumeration of all singular manifold branches

3.1. SEPARATING AND NON-SEPARATING SINGULAR MANIFOLD

A singular manifold branch is said to be separating if it can divide by itself the joint space into several c-sheets [1].

Theorem 1 : *A singular manifold branch is separating iff its homotopy class is (0,0).*

Proof : This result stems from the topology of the torus, which, unlike R^2, is not simply connected. The only loop which can divide the torus is one which encircles no generators. In figure 3 it is clear that L1 and L2 do not divide the torus, while L3 does. Any other (regular, i.e. without self-intersection) loop form helicoïdal closed curves on the torus, and it is always possible to link any two points on the torus without encountering the loop. In effect, the set obtained by removing the helicoïdal loop from the torus forms an helicoïdal closed band. Fig. 5 illustrates this result with a (5,2)-loop.

Note : To understand that the homotopy class of a loop is (n2,n3), it is useful to follow this practical technique. Start from one point on the loop and set n2 and n3 to zero. Track the loop. As soon as the side $\theta2=\pm\pi$ (resp. $\theta3=\pm\pi$) is encountered, "jump" to the opposite side and increment n2 (resp. n3). Continue until you come back to the starting point. On figure 5, the "jumps" have been traced in dotted lines : the loop was completely tracked with 5 jumps through $\theta2=\pm\pi$, and 2 jumps through $\theta3=\pm\pi$. Thus its homotopy class is (5,2).

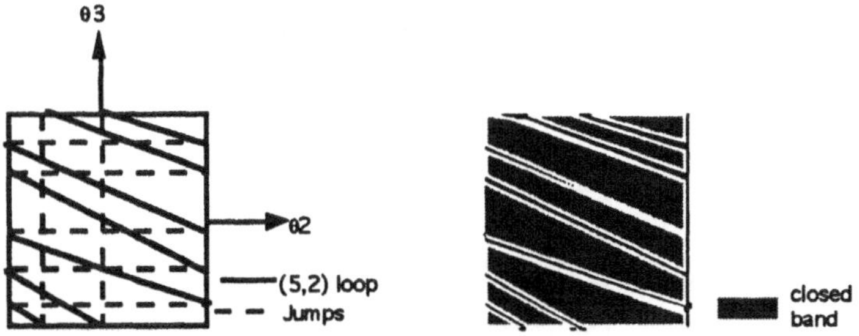

Fig. 5 : Square representation of a non-separating loop on the torus

3.2. IMPOSSIBLE LOOP HOMOTOPY CLASSES

Here, we state a series of results which will drastically reduce the possible homotopy classes of generic manipulators.

Theorem 2 : *A generic 3-DOF regional manipulator cannot have a (n2,n3)-singular manifold branch with n3>1.*

Proof : The singular manifolds of 3-R regional manipulators are analytically defined by the zero sets of the determinant of the Jacobian det(J), which can be put in a polynomial of degree 2 in t_2 and of degree 4 in t_3, where $t_2=\tan(\theta_2/2)$ and $t_3=\tan(\theta_3/2)$. Thus, det(J)=0 has at most two solutions in θ_2 when θ_3 is fixed. In other words, the singular manifolds cannot be intersected more by the θ_2-generator more than twice. Since there are n3 intersections between the θ_2-generator and a loop which wraps around the θ3-generator of the torus n3 times, n3 is necessarily less than 3. Assume that n3=2. Since a (n2,2)-singular manifold branch is not separating (from theorem 1), and since a 3-R regional manipulator has at least two c-sheets, a (n2,2)-branch cannot exist alone. But an additional singular manifold branch would leads to, either more than two intersections with the θ_2-generator, or non-generic intersecting points between the singular manifold branches (fig 6). Thus n3≤1.

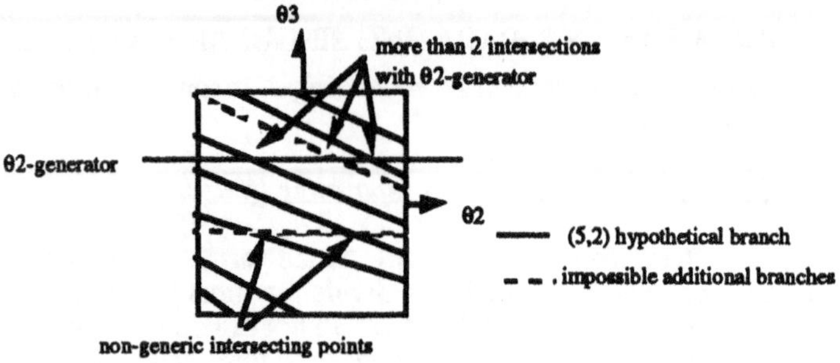

Fig. 6 Impossible singular manifold branch

Theorem 3 : A generic 3-DOF regional manipulator cannot have a (n2,n3)-singular manifold branch with n2>2.

Proof : The beginning of this proof is similar to the preceding one. Since det(J) is fourth degree in t_3, n2 cannot be more than 4. On the other hand, n2=4 is not possible since a (n3,4)-singular manifold branch cannot exist alone, and its coexistence with any other branch would yield more than 4 intersections with the θ3-generator.
Now, we show that n2=3 is not possible either. We need only to prove that (3,1) and (3,0) cannot exist (by virtue of theorem 2).
(3,1) : this branch cannot exist alone since it is not separating. However, it is not possible to add another branch. In effect, (1,0) and (0,1) additional branches would intersect the (3,1)-branch, and any other additional branch would exceed the maximum number of allowable intersections with the θ3-generator (see figure 7).

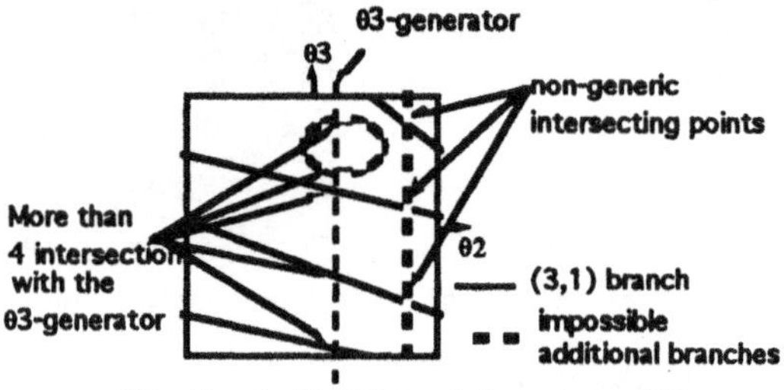

Fig. 7 : A (3,1)-branch is not possible

(3,0) : A loop which does not wrap around one of the generator of the torus, cannot wrap around the other generator more than once without self-intersection. Thus (3,0) and (2,0) are impossible.

Consequences : Theorems 1-3 show that the only possible homotopy classes of a singular manifold branch are (0,0), (0,1), (1,0), (1,1), (2,1). Numerous examples have confirmed this result [10].

3.3. IMPOSSIBLE BRANCH COMBINATIONS

A generic 3-R regional manipulator may have at most 4 singular manifold branches. In effect, it can be shown that more than 4 branches either intersect, or exceed the maximum number of intersections with the generators [10]. The branches could be, in theory, of any of the possible homotopy classes listed above.
However, it turns out that many branch combinations are not possible.

Theorem 4 : Incompatible branch homotopy classes :
Branches with homotopy class (0,1), (1,0), (1,1), and (2,1) never appear in mixed combination.

Proof : Any mixed combination would lead to intersecting branches, as shown in fig. 8.

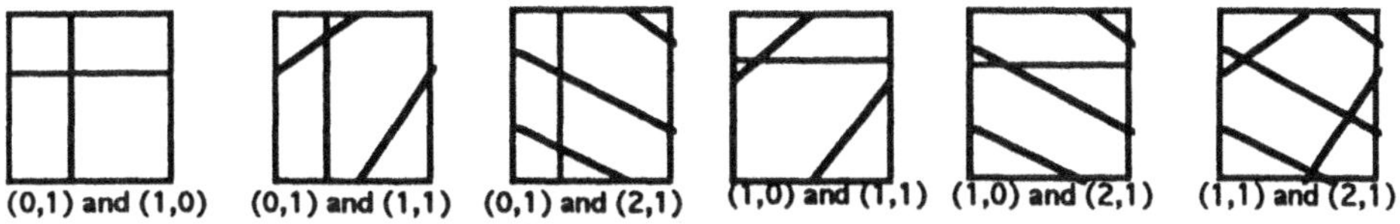

Fig. 8 : Incompatible homotopy classes

Theorem 5: Number of branches :
Branches of homotopy class (0,1), (1,1) and (2,1) always appear in pairs ; (1,0)-branches appear in pairs or in sets of four.

Proof : We know that (0,1), (1,1), (2,1), and (1,0) branches cannot appear alone (from theorem 1), and that they cannot coexist with branches of different classes (from theorem 4). On the other hand, more than two coexisting branches with homotopy class (0,1), (1,1) or (2,1) would cut the θ2-generator more than twice, which is not possible. Finally, a generic manipulator cannot have three (1,0)-branches, otherwise its singular manifolds would be cut less than four times by any θ3-generator. If so, the coefficient in t^4 of det(J) would be zero for all θ2, which was proved in [9] to be a special case of non-genericity.

Theorem 6: A (0,0)-branch either appears together with an additional (0,0)-branch, or with two (1,0)-branches.

Proof : A (0,0)-branch cannot coexist with a (0,1), (1,1) or (2,1)-branch, otherwise the singular manifolds would be cut more than twice by the θ2-generator. A (0,0)-branch with only one (1,0)-branch would yield less than 4 intersecting points with the θ3-generator, which is a non-generic case. On the other hand, since a generic manipulator cannot have more than one isolated (0,0)-branch (see §3.C), more than two coexisting (0,0)-branches would necessarily yield more than 4 intersections with the θ3-generator. For the same reason, a pair of (0,0)-branches cannot coexist with a (1,0)-branch.

3.4. SUMMARY

As a result of the preceding 6 theorems, there are only 8 classes of homotopic generic 3-R manipulators : {(0,0), 2(0,0), (0,0)+2(1,0), 2(1,0), 4(1,0), 2(0,1), 2(1,1), (2,1)}

4. Examples

In this section, we provide an example of one representative of each class. For each example, a cross section of the workspace is depicted in a plane (x, $\rho=\sqrt{x^2+y^2}$).

4.1. A (0,0)-MANIPULATOR

An example of this manipulator was already presented in the preceding section, fig. 4a, which can be shown to be cuspidal. Thus, all manipulators with one (0,0)-branches are cuspidal.

4.2. A 2(0,0)-MANIPULATOR

A 2(0,0)-manipulator is depicted in figure 9. The DH-parameters of this manipulator are : a1=1, a2=2, a3=1.5, d2=1, d3=0.1, α1=-90° and α2=90°. We can show that this manipulator is cuspidal. Thus, all manipulators with two (0,0)-branches are cuspidal.

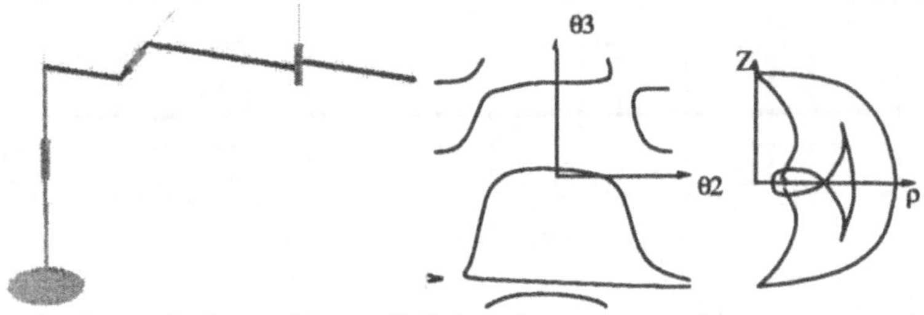

Fig. 9 : Aa manipulator with two (0,0)-branches and its workspace cross-section

4.3. MANIPULATOR WITH HOMOTOPY CLASS (0,0)+2(1,0)

Figure 10 shows a manipulator with homotopy class (0,0)+2(1,0).(DH-parameters : a1=7, a2=2, a3=6.5, d2=0.2, d3=2.5, α1=70° and α2=-56°). We can show that this manipulator is also cuspidal. Thus, all manipulators with one (0,0)-branch plus two (1,0)-branches are cuspidal.

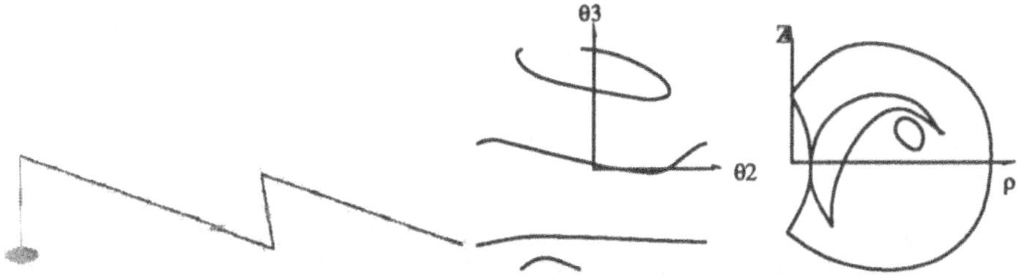

Fig. 10 : a manipulator with one (0,0)-branch and two (1,0)-branches

4.4. MANIPULATOR WITH HOMOTOPY CLASS 2(1,0)

An example of such a manipulator was shown in figure 2. We can show, again, that this manipulator is cuspidal. Thus, all manipulators with two (1,0)-branches are cuspidal.

4.5. MANIPULATOR WITH HOMOTOPY CLASS 4(1,0)

Figure 11 shows a manipulator with homotopy class 4(1,0). (a1=2.5, a2=0.5, a3=2, d2=1, d3=0.3, α1=-80° and α2=70°). We can show that this manipulator, unlike the preceding ones, is non-cuspidal. Thus, all manipulators with four (0,0)-branches are non-cuspidal (i.e. must pass through a singularity when changing posture).

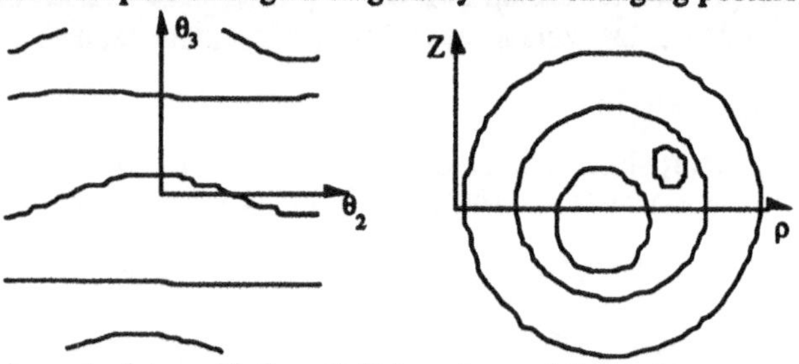

Fig. 11 : A manipulator with four (1,0)-branches and its workspace cross-section

4.6. MANIPULATOR WITH HOMOTOPY CLASS 2(0,1)

Figure 12 shows a manipulator with homotopy class 2(0,1) (DH parameters : a1=0.5, a2=0.7, a3=1.5, d2=0, d3=0, α1=80° and α2=60°). We can show that this manipulator is cuspidal. Thus, all manipulators with two (0,1)-branches are cuspidal.

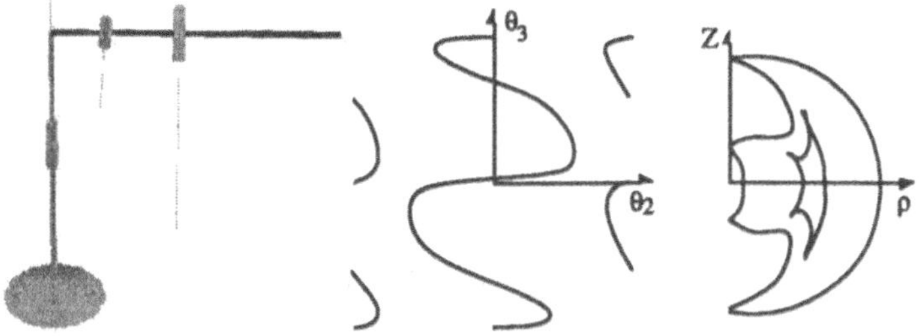

Fig. 12 : A manipulator with two (0,1)-branches and its workspace cross-section

4.7. MANIPULATOR WITH HOMOTOPY CLASS 2(1,1)

Figure 13 shows a manipulator with homotopy class 2(1,1) (DH parameters : a1=1, a2=2, a3=2.5, d2=1, d3=0.1, α1=-50° and α2=90°). We can show that this manipulator is also cuspidal. Thus, all manipulators with two (1,1)-branches are cuspidal.

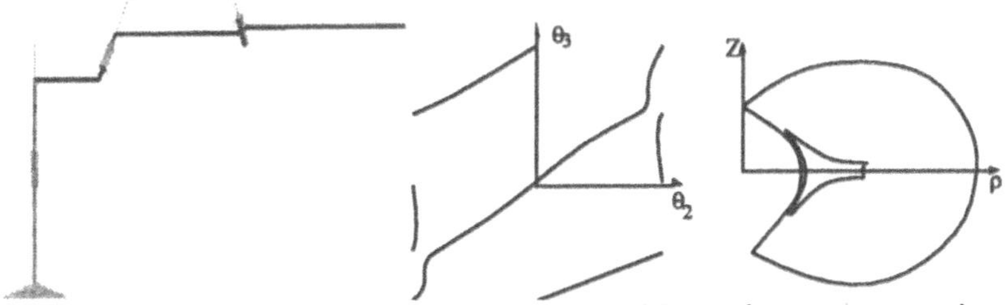

Fig. 13 : A manipulator with two (1,1)-branches and its workspace cross-section

4.8. MANIPULATOR WITH HOMOTOPY CLASS 2(2,1)

Figure 14 shows a manipulator with homotopy class 2(2,1) (DH parameters : a1=0.8, a2=1, a3=1.4, d2=0.3, d3=0.3, α1=45° et α2=-70°). We can show that this manipulator is cuspidal. Thus, all manipulators with two (2,1)-branches are cuspidal.

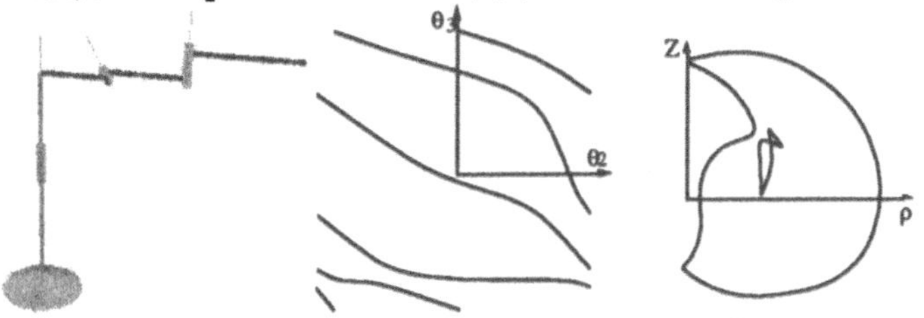

Fig. 14 : A manipulator with two (2,1)-branches and its workspace cross-section

5. Concluding remarks

This paper has established the complete classification of all generic regional 3-R manipulators into classes of homotopic manipulators. The classes are discriminated by the number and homotopy class of their singular manifold branches By examination of the geometry of the singular manifolds, it was shown that there are only eight distinct classes of homotopic manipulators. The examples show that all manipulators with homotopy class (0,0), 2(1,0), 2(0,1), 2(1,1), and 2(2,1) have two c-sheets. Manipulators with homotopy class 2(0,0) and (0,0)+2(1,0) have 3 c-sheets. Finally, all manipulators with homotopy class 4(1,0) have four c-sheets, and are the only generic 3-R manipulators which do not have the nonsingular posture changing ability. This is an interesting new result. However, it seems that 4(1,0) generic manipulators are very unlikely, since it has been difficult to find an example. On the other hand, 2(1,0) generic manipulators seem to be the most widespread. Future research work is to include the categorisation of non-generic 3-R manipulators.

References

[1] Kumar A. and Waldron K.J., "The workspace of Mechanical Manipulators", Journal of, Mechanical Design, Vol. 103, July 1981.

[2] Gupta K.C. and Roth B., "Design Considerations for Manipulator Workspaces", Journal of Mechanical Design, Vol. 104, October 1982.

[3] Kholi D. and Hsu M.S., "The Jacobian Analysis of Workspaces of Mechanical Manipulators", Journal of Mechanisms and Machine Theory, Vol 22, No 3, pp 265-275, 1987.

[4] Rastegar J., Deravi P. : "Methods to Determine Workspace with Different Numbers of Configurations and all the Possible Configurations of a Manipulator", Journal of Mechanisms and Machine Theory, Vol. 22, No 4, pp343-350, 1987.

[5] Wenger Ph. and Chedmail P. : "On the Connectivity of Manipulator free-Workspace", Journal of Robotic Systems, Vol 8(6), Dec 1991.

[6] Ceccarelli, M., "On the Workspace of General 4R Manipulators", the International Journal of Robotics Research, Vol 14, No 2, pp 152-160, April 1995.

[7] Burdick J. W., "Kinematic Analysis and Design of Redundant Manipulators", PhD Diss., Stanford, 1988.

[8] Paï D.K., "Generic Singularities of Robot Manipulators", Proc. IEEE Conf. on Rob. and Aut., pp738-744, 1989.

[9] Burdick J. W., "A classification of 3R Regional Manipulator Singularities and Geometries", Journal of Mechanisms and Machine Theory, Vol. 30, pp 71-89, 1995.

[10] El Omri J., "Analyse Géométrique et Cinématique des Mécansimes de type Manipulateur", Doctoral Thesis, Ecole Centrale de Nantes, France, 1996.

[11] Borrel P. and Liegeois A. , "A Study of Manipulator Inverse Kinematic Solutions with Application to Trajectory Planning and Workspace Determination, Proc. IEEE Int. Conf. Rob. and Aut., pp1180-1185, 1986.

COORDINATE-FREE FORMULATION OF THE CARTESIAN STIFFNESS MATRIX

MILOŠ ŽEFRAN AND VIJAY KUMAR
GRASP Laboratory, University of Pennsylvania
3401 Walnut St., Philadelphia, PA 19104-6228, USA

Abstract. In the paper we study the Cartesian stiffness matrix using methods of differential geometry. We show that the stiffness of a conservative mechanical system is described by a $\binom{0}{2}$ tensor and that components of the Cartesian stiffness matrix are given by evaluating this tensor on a pair of basis twists. Our formulation leads to three important results: (a) The stiffness matrix does not depend on the parameterization of the manifold; (b) The stiffness matrix depends on the choice of a connection on the manifold; and (c) The standard definition of the Cartesian stiffness matrix assumes an asymmetric connection and this is the reason that the matrix is, in general, asymmetric.

1. Introduction

This paper focuses on the static analysis of conservative systems in which the associated potential energy, Φ, is a function of position only. A displacement of a conservative system changes the force which acts on the system. The relation between a displacement and the change in force is given by a stiffness matrix. When the task is described in Cartesian coordinates, the linear and angular velocities of a rigid body are represented by a twist. Accordingly, forces and torques acting on the rigid body are represented as wrenches. Changes in the wrench components as the rigid body moves along the basis twists are given by a $n \times n$ Cartesian stiffness matrix, where n is the dimension of the task space.

It was observed by Griffis and Duffy (1993) and Ciblak and Lipkin (1994) that the Cartesian stiffness matrix associated with a linear elastic coupling is, in general, asymmetric if the forces and moments exerted by the linear elastic coupling are not zero. Pigoski *et al.* (1992) and Ciblak and

J. Lenarčič and V. Parenti-Castelli (eds.), Recent Advances in Robot Kinematics, 119-128.
© 1996 *Kluwer Academic Publishers. Printed in the Netherlands.*

Lipkin (1994) also showed that the stiffness matrix in the body-fixed reference frame is the transpose of the stiffness matrix in the inertial reference frame. Howard *et al.* (1995) derived these results in differential geometric setting.

In this paper, we study the Cartesian stiffness matrix with the tools of differential geometry. The set of rigid body displacements in $\mathbb{R}^3$ is a Lie group $SE(3)$. We show that an affine connection must be defined on $SE(3)$ in order to compute the stiffness matrix. We define a $\binom{0}{2}$ stiffness tensor and demonstrate that the components of the stiffness matrix are obtained by evaluating this tensor on a pair of basis twists. This allows us to prove that the stiffness matrix is independent of the choice of parameterization of the manifold $SE(3)$. We demonstrate that by choosing any symmetric connection the stiffness matrix will be always symmetric. Finally, we show that the standard definition (Pigoski *et al.*, 1992; Ciblak and Lipkin, 1994; Howard, 1995) of the stiffness matrix uses a connection which is asymmetric. In general, this results in an asymmetric stiffness tensor and an asymmetric stiffness matrix.

2. Kinematics and differential geometry

Consider a rigid body moving in free space. Assume any inertial reference frame {F} fixed in space and a frame {M} fixed to the body at point O' as shown in Figure 1. At each instance, the configuration (position and orientation) of the rigid body can be described by a homogeneous transformation matrix corresponding to the displacement from frame {F} to frame {M}. These transformations form a Lie group $SE(3)$, the special Euclidean group in three-dimensions (Murray *et al.*, 1994).

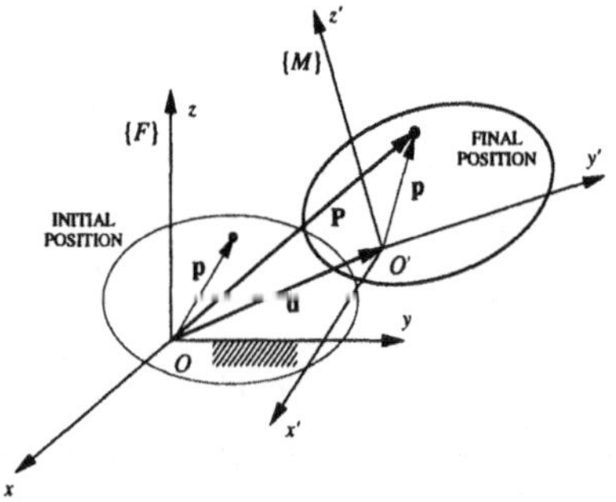

Figure 1. The inertial (fixed) frame and the moving frame attached to the rigid body.

On a Lie group, the tangent space at the group identity has the structure of a Lie algebra. The Lie algebra of $SE(3)$ is denoted by $se(3)$ and is given by:

$$se(3) = \left\{ \begin{bmatrix} \Omega & v \\ 0 & 0 \end{bmatrix}, \Omega \in \mathbb{R}^{3\times 3}, v \in \mathbb{R}^3, \Omega^T = -\Omega \right\}. \tag{1}$$

A 3×3 skew-symmetric matrix Ω can be uniquely identified with a vector $\omega \in \mathbb{R}^3$ so that for an arbitrary vector $x \in \mathbb{R}^3$, $\Omega x = \omega \times x$, where $\times$ is the cross product in $\mathbb{R}^3$. Each element $T \in se(3)$ can be thus identified with a vector pair $\{\omega, v\}$.

Since $se(3)$ is a vector space, any element can be expressed as a 6×1 vector of components corresponding to a chosen basis. The standard basis that will be used throughout the paper is:

$$L_1 = \begin{bmatrix} 0 & 0 & 0 & 0 \\ 0 & 0 & -1 & 0 \\ 0 & 1 & 0 & 0 \\ 0 & 0 & 0 & 0 \end{bmatrix} \quad L_2 = \begin{bmatrix} 0 & 0 & 1 & 0 \\ 0 & 0 & 0 & 0 \\ -1 & 0 & 0 & 0 \\ 0 & 0 & 0 & 0 \end{bmatrix} \quad L_3 = \begin{bmatrix} 0 & -1 & 0 & 0 \\ 1 & 0 & 0 & 0 \\ 0 & 0 & 0 & 0 \\ 0 & 0 & 0 & 0 \end{bmatrix}$$
$$L_4 = \begin{bmatrix} 0 & 0 & 0 & 1 \\ 0 & 0 & 0 & 0 \\ 0 & 0 & 0 & 0 \\ 0 & 0 & 0 & 0 \end{bmatrix} \quad L_5 = \begin{bmatrix} 0 & 0 & 0 & 0 \\ 0 & 0 & 0 & 1 \\ 0 & 0 & 0 & 0 \\ 0 & 0 & 0 & 0 \end{bmatrix} \quad L_6 = \begin{bmatrix} 0 & 0 & 0 & 0 \\ 0 & 0 & 0 & 0 \\ 0 & 0 & 0 & 1 \\ 0 & 0 & 0 & 0 \end{bmatrix} \tag{2}$$

This basis has the property that the components of an element $T \in se(3)$ are given precisely by the vector pair $\{\omega, v\}$ mentioned above.

The product operation on a Lie algebra is called a Lie bracket. The Lie bracket of two elements $T_1, T_2 \in se(3)$ is defined by:

$$[T_1, T_2] = T_1 T_2 - T_2 T_1. \tag{3}$$

The coefficients C_{ij}^k corresponding to the Lie brackets of the basis vectors:

$$[L_i, L_j] = \sum_k C_{ij}^k L_k, \tag{4}$$

are called *structure constants* of the Lie algebra (Sattinger and Weaver, 1986). The nonzero structure constants for the basis (2) are:

$$C_{12}^3 = C_{31}^2 = C_{23}^1 = C_{15}^6 = C_{26}^4 = C_{34}^5 = C_{42}^6 = C_{53}^4 = C_{61}^5 = 1 \tag{5}$$
$$C_{21}^3 = C_{13}^2 = C_{32}^1 = C_{51}^6 = C_{62}^4 = C_{43}^5 = C_{24}^6 = C_{35}^4 = C_{16}^5 = -1.$$

2.1. RIGHT INVARIANT VECTOR FIELDS

A differentiable vector field on a manifold is a smooth assignment of a tangent vector to each element of the manifold. At each point, a vector field defines a unique *integral curve* to which it is tangent (do Carmo, 1992). Formally, a vector field X is a derivation operator which, given a real-valued differentiable function f, returns its derivative along the integral curves of X.

On $SE(3)$, an example of a differentiable vector field, X, is obtained by setting:

$$X(A) = \hat{T}(A) = TA, \tag{6}$$

where $T \in se(3)$ and $A \in SE(3)$. Such a vector field is called *a right invariant vector field* and we use the notation $\hat{T}$ to indicate that the vector field was obtained by the right translation of the Lie algebra element T. By construction, the space of right invariant vector fields is isomorphic to the Lie algebra $se(3)$. In particular, $[\hat{L}_i, \hat{L}_j] = \widehat{[L_i, L_j]} = \sum_k C_{ij}^k \hat{L}_k$ (do Carmo, 1992).

Since the vectors $L_1, L_2, \ldots, L_6$ are a basis for the Lie algebra $se(3)$, the vectors $\hat{L}_1(A), \ldots, \hat{L}_6(A)$ form a basis of the tangent space at any point $A \in SE(3)$. Therefore, any vector field X can be expressed as $X = \sum_{i=1}^6 X^i \hat{L}_i$, where the coefficients X^i are real-valued functions. By defining $\omega = [X^1, X^2, X^3]^T$ and $v = [X^4, X^5, X^6]^T$, we can associate a vector pair $\{\omega, v\}$ to an arbitrary vector field X.

2.2. VELOCITY OF A RIGID BODY

Take a rigid body moving in space. The motion of the rigid body can be described by a curve $A(t) : \mathbb{R} \to SE(3)$. The tangent vector to this curve, $\frac{dA}{dt}$, is the velocity of the rigid body. The tangent vector $\frac{dA}{dt}$ can be mapped to an element T of the Lie algebra $se(3)$ by:

$$T = \dot{A}A^{-1} = \begin{bmatrix} \dot{R}R^T & -\dot{R}R^T d + \dot{d} \\ 0 & 0 \end{bmatrix}. \tag{7}$$

It can be shown that T does not depend on the choice of the body-fixed frame {M} and is called the right invariant representation of the velocity. If T is expressed in the basis (2), the vector of components $\{\omega, v\}$ corresponds to the twist describing the instantaneous velocity of the rigid body. More precisely, ω is the angular velocity of the rigid body while v is the linear velocity of the point on the rigid body that is instantaneously at the origin O of the frame {F}, both expressed in the inertial frame {F}. The Lie algebra $se(3)$ thus represents the space of twists. In this space, the basis vectors L_1, L_2 and L_3 correspond to instantaneous rotations about and L_4, L_5 and L_6 to instantaneous translations along the Cartesian axes x, y and z, respectively.

Equation (7) shows that:

$$\dot{A} = T\,A = \left(\sum_{i=1}^{6} T^i L_i\right) A = \sum_{i=1}^{6} T^i (L_i A) = \sum_{i=1}^{6} T^i \hat{L}_i. \tag{8}$$

We conclude that if the velocity of the rigid body is expressed in the basis $\hat{L}_1, \ldots, \hat{L}_6$, the components are equal to the components of the twist describing the instantaneous velocity. We thus call the basis vector fields $\hat{L}_1, \ldots, \hat{L}_6$ *the basis twists.*

3. Static analysis on $SE(3)$

3.1. TWISTS AND WRENCHES

If a wrench $W = \{\tau, F\}$ acts on the rigid body during a displacement $\Delta T = \{\omega, v\}\Delta t$, it produces work, $\Delta E = (F^T v + \tau^T \omega)\Delta t$, which is a scalar. Wrenches therefore belong to the dual of the space of twists, $W \in se^*(3)$. This implies that a force[1] acting on the rigid body that moves in space represents a one-form (Simo, 1992).

Given a basis for the vector fields, there exists a natural basis for one-forms, called *the dual basis*. If $\{\hat{L}_i\}$ is a basis for the vector fields, the dual basis for the one-forms, $\{\hat{\lambda}^i\}$, satisfies:

$$\ll \hat{\lambda}^i ; \hat{L}_j \gg = \delta^i_j, \tag{9}$$

where $\ll \hat{\lambda}^i ; \hat{L}_j \gg$ represents the action of a one-form $\hat{\lambda}^i$ on a vector field $\hat{L}_j$ and δ^i_j is the Kronecker δ. If a one-form $\mathcal{F} = \mathcal{F}_i \hat{\lambda}^i$ is expressed in the dual basis, it is easy to see that its action on a vector field $V = V^j \hat{L}_j$ is given by $\ll \mathcal{F} ; V \gg = \mathcal{F}_i V^i$. Here and in the rest of the chapter we use the Einstein summation convention.

It is not difficult to see that the components of the force one-form in the basis $\{\hat{\lambda}^i\}$ that is dual to the basis twists $\{\hat{L}_i\}$, are exactly the components of the corresponding wrench in $se^*(3)$.

3.2. FORCES IN A POTENTIAL FIELD

We are interested in the static analysis of conservative systems in which the associated potential field, Φ, is a function of position only. In $\mathbb{R}^3$, the force generated by a potential field ϕ is equal to the negative gradient of the potential field, $F = -\text{grad}(\phi)$. This can be generalized to an arbitrary manifold if we generalize the notion of gradient (Schutz, 1980). A gradient of a real-valued function f, denoted by df, is a one-form, whose action on an arbitrary vector field X is defined by:

$$\ll df ; X \gg = X(f). \tag{10}$$

Here we remind the reader that a vector field on a manifold must be regarded as a derivation (see Section 2.1).

The force one-form, $\mathcal{F}$, corresponding to a potential field Φ is therefore:

$$\mathcal{F} = -d\Phi. \tag{11}$$

[1]By force we mean the generalized force consisting of both, forces and torques acting on the rigid body.

To obtain the wrench that corresponds to the force one-form $\mathcal{F}$ at a point $A \in SE(3)$, the one form must be expressed in the basis dual to the basis twists. We can see from Eq. (9) that the components of the wrench are given by $W_i = \ll \mathcal{F}\,;\, \hat{L}_i \gg = -\hat{L}_i(\Phi)$ and that $\mathcal{F} = W_i\,\hat{\lambda}^i$.

3.3. COVARIANT DERIVATIVES

The Cartesian stiffness matrix describes how the components of a wrench change in the directions given by the basis twists. This suggests that the wrench has to be differentiated along the basis twists. But a wrench belongs to the space $se^*(3)$, the dual of the space of twists, so it can not be differentiated along a vector field, which is defined over the entire manifold. The quantity that can be differentiated with respect to a vector field is the force one-form $\mathcal{F}$.

Differentiation of vector fields and one forms on the manifold is not defined unless the manifold is endowed with *an affine connection.* Given a curve $A(t)$ on the manifold, the affine connection specifies how an element X of the tangent space at point $A(t_1)$ can be mapped to an element X' of the tangent space at some other point $A(t_2)$. The vector X' is called the parallel transport of X along $A(t)$. In this way, we can define a *covariant derivative* of a vector field $X(t)$ along a curve $A(t)$ by:

$$\left.\frac{DX}{dt}\right|_{t_0} = \lim_{h\to 0} \frac{X^{t_0}(t_0+h) - X(t_0)}{t}, \tag{12}$$

where $X^{t_0}(t_0+h)$ is the parallel transport of the vector $X(A(t_0+h))$ along $A(t)$ to the point $A(t_0)$. By taking covariant derivatives of a vector field X along the integral curves of another vector field Y, we obtain a covariant derivative, $\nabla_Y X$, of the vector field X with respect to the vector field Y:

$$\nabla_Y X|_{A_0} = \left.\frac{DX}{dt}\right|_{t_0}, \tag{13}$$

where $\frac{DX}{dt}$ is taken along the integral curve of Y passing through A_0 at $t = t_0$. The coefficients Γ^k_{ji} of the covariant derivative of a basis vector field along another basis vector field,

$$\nabla_{\hat{L}_i} \hat{L}_j = \Gamma^k_{ji} \hat{L}_k, \tag{14}$$

are called *Christoffel symbols*[2].

[2]Some texts (do Carmo, 1992) reserve the term only for the coordinate basis vectors. We follow the more general definition from (Schutz, 1980) in which the basis vectors can be arbitrary.

Remark 3.1 If for all vector fields X and Y, a connection satisfies:

$$\nabla_X Y - \nabla_Y X = [X, Y], \tag{15}$$

the connection is said to be *symmetric.*

A covariant derivative of a one-form can be defined through the covariant derivative of a vector field. Loosely speaking, to obtain a covariant derivative $\nabla_U \mathcal{F}$ of a one-form $\mathcal{F}$ along a vector field U, we use a "generalization" of the Leibniz' rule (Schutz, 1980). Let $\mathcal{F}$ be a one-form and U and V vector fields. We then have:

$$\nabla_U \ll \mathcal{F}\,;V \gg = \ll \nabla_U \mathcal{F}\,;V \gg + \ll \mathcal{F}\,;\nabla_U V \gg . \tag{16}$$

Since $\ll \mathcal{F}\,;V \gg$ is a real-valued function, $\nabla_U \ll \mathcal{F}\,;V \gg = U(\ll \mathcal{F}\,;V \gg)$. The rule for a covariant derivative of $\mathcal{F}$ along U is therefore:

$$\ll \nabla_U \mathcal{F}\,;V \gg = U(\ll \mathcal{F}\,;V \gg) - \ll \mathcal{F}\,;\nabla_U V \gg . \tag{17}$$

3.4. THE CARTESIAN STIFFNESS MATRIX AND STIFFNESS TENSOR

The Cartesian stiffness matrix is obtained by differentiating the force one-form in the directions of the basis twists. Since we are interested in changes of the wrench components in the directions of the basis twists, the resulting one-forms $\nabla_{\hat{L}_i} d\Phi$ must be expressed in the basis dual to the basis twists. This leads to the following definition for the coefficients of the stiffness matrix:

$$K_{ij} = \ll \nabla_{\hat{L}_j} d\Phi\,;\hat{L}_i \gg . \tag{18}$$

(The minus sign is omitted to conform to the usual definition of the stiffness matrix in the literature.) Equation (18) can be expanded using Eq. (17):

$$\begin{aligned} K_{ij} &= \ll \nabla_{\hat{L}_j} d\Phi\,;\hat{L}_i \gg & (19) \\ &= \hat{L}_j(\ll d\Phi\,;\hat{L}_i \gg) - \ll d\Phi\,;\nabla_{\hat{L}_j}\hat{L}_i \gg = (\hat{L}_j\hat{L}_i - \nabla_{\hat{L}_j}\hat{L}_i)(\Phi). \end{aligned}$$

A closer look at Eq. (18) reveals that the right-hand side is linear in both vector components:

1. $\ll \nabla_{f_j\hat{L}_j + f_k\hat{L}_k} d\Phi\,;\hat{L}_i \gg = f_j \ll \nabla_{\hat{L}_j} d\Phi\,;\hat{L}_i \gg + f_k \ll \nabla_{\hat{L}_k} d\Phi\,;\hat{L}_i \gg$ (linearity of the covariant derivative).
2. $\ll \nabla_{\hat{L}_j} d\Phi\,; f_i\hat{L}_i + f_k\hat{L}_k \gg = f_i \ll \nabla_{\hat{L}_j} d\Phi\,;\hat{L}_i \gg + f_k \ll \nabla_{\hat{L}_j} d\Phi\,;\hat{L}_k \gg$ (linearity of the one-form).

This suggests introducing a $\binom{0}{2}$ *stiffness tensor* $K = \nabla d\Phi$ defined for arbitrary vector fields X and Y by:

$$K(X, Y) = \ll \nabla_Y d\Phi\,;X \gg . \tag{20}$$

Equation (18) thus becomes:

$$K_{ij} = K(\hat{L}_i, \hat{L}_j). \tag{21}$$

The last equation leads to the following propositions.

Proposition 3.2 *The component K_{ij} of the Cartesian stiffness matrix is obtained by evaluating the stiffness tensor $K = \nabla d\Phi$ on the pair of basis twists $\hat{L}_i$ and $\hat{L}_j$.*

Proposition 3.3 *The Cartesian stiffness matrix does not depend on the choice of the coordinates for $SE(3)$ (parameterization of the space).*

Proof: Equation (21) suggests that the entries of the stiffness matrix depend on the $\binom{0}{2}$ tensor K and on the basis twists $\{\hat{L}_i\}$. By definition, the tensor K is independent of the choice of the coordinates (this makes it a tensor). Further, the basis twists are defined by the right translation of the canonical basis of the Lie algebra $se(3)$ and this operation is again coordinate invariant.

□

Remark 3.4 The independence of the Cartesian stiffness matrix from the parameterization of $SE(3)$ is also demonstrated by the fact that we never introduced the coordinates for $SE(3)$!

Proposition 3.5 *If the affine connection used in the definition of the stiffness tensor K is symmetric, then the stiffness tensor is symmetric.*

Proof: For a symmetric connection (Eq. 15), we obtain:

$$\begin{aligned} K(X,Y) &= (XY - \nabla_X Y)(\Phi) \quad\quad (22)\\ &= (XY - [X,Y] - \nabla_Y X)(\Phi) = (YX - \nabla_Y X)(\Phi) = K(Y,X). \end{aligned}$$

□

Corollary 3.6 *If the connection used to define the stiffness tensor in Eq. (20) is symmetric the Cartesian stiffness matrix is always symmetric.*

3.5. STANDARD DEFINITION OF THE CARTESIAN STIFFNESS MATRIX

In this section we compare our coordinate free definition of the Cartesian stiffness matrix based on the stiffness tensor with the definition that is usually found in the literature (Pigoski *et al.*, 1992; Ciblak and Lipkin, 1994; Howard, 1995). In these works, the coefficient of the stiffness matrix K_{ij} is computed by taking a small displacement ΔT^j in the direction of the basis twist $\hat{L}_j$, computing the corresponding change of the wrench component W_i and taking the limit of the quotient of the two as the displacement goes to 0:

$$K_{ij} = -\lim_{\Delta T^j \to 0} \frac{\Delta W_i}{\Delta T^j}. \tag{23}$$

We observe that this corresponds precisely to derivative of W_i along the integral curve of $\hat{L}_j$:

$$K_{ij} = -\hat{L}_j(W_i). \tag{24}$$

But $W_i = - \ll d\Phi\, ;\, \hat{L}_i \gg = -\hat{L}_i(\Phi)$, so the expression for the element K_{ij} becomes:

$$K_{ij} = \hat{L}_j \hat{L}_i(\Phi). \tag{25}$$

If we compare Eq. (25) with Eq. (20), it is immediately apparent that the term with the connection is missing in (25). Hence, we conclude that the standard definition of the stiffness matrix requires:

$$\nabla_{\hat{L}_i} \hat{L}_j = 0 \quad \forall i, j. \tag{26}$$

This equation completely specifies the connection ∇. If the vector fields are expressed with the basis twists $\{\hat{L}_i\}$, the Christoffel symbols for this connection all vanish:

$$\Gamma^k_{ji} = 0 \quad \forall i, j, k. \tag{27}$$

This result immediately implies that the connection (26) is not symmetric:

$$0 = \nabla_{\hat{L}_i} \hat{L}_j - \nabla_{\hat{L}_j} \hat{L}_i \neq [\hat{L}_i, \hat{L}_j]. \tag{28}$$

(See Eq. (6) for the values of the Lie brackets $[\hat{L}_i, \hat{L}_j]$.)

A consequence of this fact is stated in the following proposition (Ciblak and Lipkin, 1994; Howard *et al.*, 1995).

Proposition 3.7 *The Cartesian stiffness matrix, as defined in (*Pigoski *et al.*, 1992; Ciblak and Lipkin, 1994; Howard, 1995*), is in general asymmetric.*

Proof: This statement is the consequence of the asymmetry of the connection (see Proposition 3.5). From Eq. (25) we compute:

$$K_{ij} - K_{ji} = [\hat{L}_j, \hat{L}_i](\Phi). \tag{29}$$

Since the basis twists are not the coordinate basis, the Lie brackets do not vanish. It is also easy to see that at stationary points of the potential field Φ, the stiffness matrix becomes symmetric (at a stationary point, $X(\Phi) = 0$ for an arbitrary vector field X).

□

4. Conclusion

We presented a coordinate-free formulation of the Cartesian stiffness matrix for conservative mechanical systems in which the potential field Φ is a function of position only. We chose a basis for the vector fields so that in this basis the components of the velocity vector field are equal to the components of the twist associated with the motion of the rigid body. We argued that the force generated by the potential field Φ is a one-form. Further, in the basis dual to the basis chosen for the vector fields, the components of this one-form are equal to the components of the associated wrench. We demonstrated that an affine connection must be introduced to the manifold $SE(3)$ in order to compute the stiffness matrix and observed that the components of the stiffness matrix are obtained by evaluating a $\binom{0}{2}$ tensor on a pair of basis twists. This enabled us to prove that the stiffness matrix is independent of the choice of parameterization of the space (the manifold $SE(3)$). We demonstrated that by choosing a symmetric connection, the stiffness matrix is always symmetric. Finally, we showed that the standard definition of the Cartesian stiffness matrix assumes a connection which is asymmetric. In general, this results in an asymmetric stiffness tensor and an asymmetric stiffness matrix.

Acknowledgment

We acknowledge discussions with W. Stamps Howard. It was his work (Howard, 1995) that motivated our paper. Support for this research was provided by the NSF grants BCS 92-16691, MSS-91-57156, CISE/CDA 88-22719, ARPA Grant N00014-92-J-1647, NATO grant CRG 911041 and Army Grant DAAH04-96-1-0007. The first author was also supported by a fellowship from the Institute for Research in Cognitive Science at the University of Pennsylvania.

References

N. Ciblak and H. Lipkin. Asymmetric Cartesian stiffness for the modeling of compliant robotic systems. In *Proc. of the 23rd Biennial ASME Mechanisms Conf.*, Minneapolis, MN, 1994.

M. P. do Carmo. *Riemannian geometry.* Birkhauser, Boston, 1992.

M. Griffis and J. Duffy. Global stiffness modeling of a class of simple compliant couplings. *Mechanism and Machine Theory*, 28(2):207–224, 1993.

W. S. Howard, M. Žefran, and V. Kumar. On the 6x6 stiffness matrix for three dimensional motions. In *9th World Congress of the IFToMM*, Milano, Italy, 1995.

W. S. Howard. *Stability of Grasped Objects : beyond force closure.* PhD thesis, University of Pennsylvania, 1995.

R. M. Murray, Z. Li, and S. S. Sastry. *A Mathematical Introduction to Robotic Manipulation.* CRC Press, 1994.

T. Pigoski, M. Griffis, and J. Duffy. Stiffness mappings employing different frames of reference. In *Proc. 22nd Biennial ASME Mechanisms Conf.*, Scottsdale, AZ, 1992.

D.H. Sattinger and O.L. Weaver. *Lie groups and algebras with applications to physics.* Springer-Verlag, New York, 1986.

B. F. Schutz. *Geometrical Methods of Mathematical Physics.* Cambridge University Press, Cambridge, 1980.

J. C. Simo. The (symmetric) Hessian for geometrically nonlinear models in solid mechanics: intrinsic definition and geometric interpretation. *Computer Methods in Applied Mechanics and Engineering*, 96(2):189–200, 1992.

STABILITY ANALYSIS OF COMPLIANT MECHANISMS

D. MARSH
Department of Mathematics, Napier University, 219 Colinton Road, Edinburgh, UK
Tel: +44-131-455-4382 Fax: +44-131-455-4232 Email: dmarsh@maths.napier.ac.uk

R. HINES & J. DUFFY
Department of Mechanical Engineering, University of Florida, Gainesville, Florida, USA
Tel: +1-352-392-0814 Fax: +1-352-392-1071 Email: cimar@cimar.me.ufl.edu

Abstract. This paper describes how a stability analysis for compliant mechanisms may be performed using catastrophe theory. The philosophy is demonstrated by an analysis of the stability of the planar two-spring and spatial three-spring systems. Catastrophe locus plots are used to predict where a change in stability will occur. The approach presented yields a stability analysis which does not require an inverse analysis of the mechanism.

1. Introduction

Compliant mechanisms are well known for their applications in mounting and suspension systems. More recently, compliant mechanisms have been instrumental in the field of robotic force control [3, 6]. One of the earliest examples used to illustrate Catastrophe Theory was a mechanism consisting of a single rotating link and two springs called the Zeeman catastrophe machine [7]. In this paper we describe how catastrophe theory may be used to determine the conditions for a change in the physical stability of compliant mechanisms, and to classify the different ways in which these changes in stability occur. We show that the instability can be visualised using computer graphics thus providing an intuitive tool for design and/or analysis of compliant mechanisms. Further details of the two-spring system may be found in [5].

2. Elementary Catastrophes

A *catastrophe* occurs whenever a smooth change of parameters gives rise to a discontinuous change in behaviour. The reader may be familiar with Euler buckling where a continuous increase in the load on a beam causes a sudden buckling. When a force is applied to a spring system the position of the mechanism is governed by the fact that the potential energy in the system should be at a local minimum so that the mechanism is in a state of equilibrium. When the applied force is varied the position and lengths of the springs adjust to maintain the potential energy at a local minimum. A catastrophe occurs whenever a continuous change in the force causes the mechanism to jump to a new equilibrium position. The mechanism undergoes a drastic change in order to find a new position at which the potential energy is a local minimum. Catastrophe Theory provides classifications of catastrophes and presents a philosophy for understanding them. Suitable introductions to Catastrophe Theory are [7, 2].

J. Lenarčič and V. Parenti-Castelli (eds.), Recent Advances in Robot Kinematics, 129-138.
© 1996 *Kluwer Academic Publishers. Printed in the Netherlands.*

DEFINITIONS:

1. x = $(x_1,...,x_m)$ is said to be a *critical point* (*equilibrium* point) of a function f : $R^m \to R$ *whenever* $\frac{\partial f}{\partial x_i}(x) = 0$ *for* $1 \leq i,j \leq m$.

2. A critical point x of a function f is said to be *degenerate* whenever the determinant of the Hessian matrix $H = \left[\frac{\partial^2 f}{\partial x_i \partial x_j}\right]_{1 \leq i,j \leq m}$ evaluated at x is zero, and *non-degenerate* otherwise.

Let f_u : $R^m \to R$ be a *r*-parameter family of functions parametrized by u = $(u_1,...,u_r)$. The family may be expressed as a single function F : $R^m \times R^r \to R$ defined by $F(x_1,...,x_m, u_1,...,u_r) = f_u(x_1,...,x_m)$. For example, the family $f_{a,b}(t) = t^2 + at + b$ may be expressed as $F(t,a,b) = t^2 + at + b$.

The set of all critical points of *F*, $M_F = \{(x,u): \frac{\partial f_u}{\partial x_i} = 0, \ i = 1...m\}$ is called the *catastrophe set*. Let H_F denote the hessian matrix of *F* with respect to x_i (i = 1...m). Then we define the *singular set* : $S_F = \{(x,u) \in M_F : \det(H_F) = 0\}$, and the *bifurcation set* : $D_F = \{u : (x,u) \in S_F\}$.

A theorem of R. Thom [7] states that, in general, a *r*-parameter family of functions F : $R^m \times R^r \to R$ only contains catastrophes which are right-equivalent to (i.e. which in a neighbourhood of the critical point behaves identically to) one of the following families in a neighbourhood of the origin:

Fold:	$x_1^3 + u_1x_1 \pm x_2^2 \pm ... \pm x_m^2$ (r ≥ 1)
Cusp:	$x_1^4 + u_1x_1^2 + u_2x_1 \pm x_2^2 \pm ... \pm x_m^2$ (r ≥ 2)
Swallowtail:	$x_1^5 + u_1x_1^3 + u_2x_1^2 + u_3x_1 \pm x_2^2 \pm ... \pm x_m^2$ (r ≥ 3)
Hyperbolic Umbilic:	$x_1^2x_2 + x_2^3 + u_1x_1^2 + u_2x_1 + u_3x_2 \pm x_3^2 \pm ... \pm x_m^2$ (r ≥ 3)
Elliptic Umbilic:	$x_1^2x_2 - x_2^3 + u_1x_1^2 + u_2x_1 + u_3x_2 \pm x_3^2 \pm ... \pm x_m^2$ (r ≥ 3)
Butterfly:	$x_1^6 + u_1x_1^4 + u_2x_1^3 + u_3x_1^2 + u_4x_1 \pm x_2^2 \pm ... \pm x_m^2$ (r ≥ 4)
Parabolic Umbilic:	$x_1^2x_2 + x_2^4 + u_1x_1^2 + u_2x_2^2 + u_3x_1 + u_4x_2 \pm x_3^2 \pm ... \pm x_m^2$ (r ≥ 4)

Thus, in the families of potential functions arising from spring systems studied in this paper, one expects to find degenerate critical points described in the above list. Other types of critcial point might arise but they occur very rarely (in precise a mathematical sense).

Example 1 (Fold): $f_a(x) = x^3 + ax$ has a degenerate critical point when $f_a'(x) = 3x^2 + a = 0$ and $f''(x) = 6x = 0$. Thus $x = 0$ and $a = 0$ and hence only $f_0(x) = x^3$ has a degenerate critical point. The bifurcation set is a single point $a = 0$ on the a-axis. There are three (equivalence classes of) qualitatively differing functions in the family. When $a < 0$, $f_a(x)$ has two critical points, a local maximum and a local minimum; when $a > 0$, $f_a(x)$ has no critical points; and when $a = 0$, $f_a(x)$ has a degenerate critical point.

Example 2 (Cusp): $f_{a,b}(x) = x^4 + ax^2 + bx$ has degenerate critical points when $f_{a,b}'(x) = 4x^3 + 2ax + b = 0$, and $f_{a,b}''(x) = 12x^2 + 2a = 0$. Hence, $a = -6x^2$, $b = 8x^3$ which defines a *cuspidal cubic* curve, parametrized by x, in the (a,b)–plane as shown in figure 1. The curve is called the *discriminant* or *bifurcation* curve.

The bifurcation curve partitions the (a,b)–plane into the following regions: (i) the origin (0,0), (ii) the bifurcation curve with the origin removed, (iii) the area to the left of the curve, and (iv) the area to the right of of the curve. In region (iii) all points correspond to functions in the family with two minimums and one maximum. All points in the region (iv) correspond to functions with just one minimum. Points on the bifurcation curve correspond to functions with an inflexion (a degenerate critical point) and one minimum, and the origin corresponds to the function x^4 with a degenerate critical point. Each region may be represented by a single function illustrating the number and type of critical points as showed in Fig. 1. Similar analyses may be performed for each of the elementary catastrophes. The reader is refered to [7,2] for further details.

3. The Planar Two-Spring System

Consider the planar two-spring system which consists of two linear springs with one end of each connected at a common pivot P with coordinates (x,y), and the remaining ends having revolute joints at fixed base points, see Fig. 2.

Let the springs have elastic constants k_1 and k_2, free lengths l_{01} and l_{02}, and let the fixed ends be a distance d apart. When a force $F = F_x i + F_y j$ at point P is specified, the steady state positions of P correspond to the roots of a sextic polynomial in $t = \tan(\theta/2)$ [4].

In general, compliant systems such as the planar two-spring reside in static equilibrium corresponding to a potential minimum. Exceptionally, the system encounters a *catastrophe*. The effect is that the system becomes unstable and jumps to a new potential minimum. The set of all configurations for which the system experiences a change of stability is called the *catastrophe locus*. The catastrophe locus may be obtained in the position (x,y)– or (l_1,l_2)–plane (or in any space completely describing the system (θ_1, θ_2)–plane, etc). The set of forces (F_x,F_y) which gives rise to an unstable configuration is called the *bifurcation curve*. The stability of a physical system in equilibrium is related to the eigenvalues of the Hessian matrix of the potential energy of the system. The potential energy function of the two-spring system is

$$V = \frac{1}{2}k_1(l_1 - l_{01})^2 + \frac{1}{2}k_2(l_2 - l_{02})^2 - xF_x - yF_y. \tag{1}$$

Equilibrium is obtained when the potential energy is at a critical point, that is, the partial derivatives of the potential energy vanish:

$$k_1(l_1 - l_{01})q_{1x} + k_2(l_2 - l_{02})q_{2x} - F_x = 0, \tag{2}$$
$$k_1(l_1 - l_{01})q_{1y} + k_2(l_2 - l_{02})q_{2y} - F_y = 0, \tag{3}$$

where q_{ix} and q_{iy} are the projections of leg i onto the x- and y- axes. The Hessian matrix H, with respect to l_1 and l_2, for the two-spring system is:

$$H = \begin{bmatrix} \frac{\partial^2 V}{\partial l_1^2} & \frac{\partial^2 V}{\partial l_1 \partial l_2} \\ \frac{\partial^2 V}{\partial l_1 \partial l_2} & \frac{\partial^2 V}{\partial l_2^2} \end{bmatrix} = \begin{bmatrix} k_1(1-\rho_1 s_1^2) + k_2(1-\rho_2 s_2^2) & k_1\rho_1 c_1 s_1 + k_2\rho_2 c_2 s_2 \\ k_1\rho_1 c_1 s_1 + k_2\rho_2 c_2 s_2 & k_1(1-\rho_1 c_1^2) + k_2(1-\rho_2 c_2^2) \end{bmatrix} \tag{4}$$

where $\rho_i = l_{0i} / l_i$.

Suppose that, for a given applied force, one of the eigenvalues of the Hessian at a critical point is zero. Then the Hessian is singular and the corresponding critical point is degenerate. This occurs when

$$|H| = (k_1 + k_2)[k_1(1-\rho_1) + k_2(1-\rho_2)] + k_1 k_2 \rho_1 \rho_2 \sin^2(\theta_2 - \theta_1) = 0. \tag{5}$$

Applying the cosine rule, $\cos(\theta_2 - \theta_1) = (l_1^2 + l_2^2 - d^2) / 2l_1l_2$, using the fact that $\sin^2(\theta_2 - \theta_1) = 1 - \cos^2(\theta_2 - \theta_1)$ and substituting $\rho_i = l_{0i} / l_i$ ($i = 1,2$) we obtain a sextic polynomial in l_1 and l_2, defining the *catastrophe locus*:

$$(k_1 + k_2)[k_1k_2(l_1 - l_{01}) + k_2 l_1(l_2 - l_{02})]l_1^2 l_2^2 + k_1k_2 l_{01} l_{02}[l_1^2 l_2^2 - \frac{1}{4}(l_1^2 + l_2^2 - d^2)^2] = 0 \tag{6}$$

An example of a catastrophe locus is illustrated in Fig. 3. Note that not all points in the (l_1, l_2)–plane give rise to real configurations of the mechanism. Necessary and sufficient conditions for the two springs to connect to a single real point are: $|l_1| - |l_2| \leq d$, $|l_2| - |l_1| \leq d$, and $|l_1| + |l_2| \geq d$. For instance, the shaded regions in Fig. 3 correspond to regions that do not satisfy these restrictions and are not constructible configurations of the mechanism.

It is also possible to obtain a similar curve in the (x,y) πλανε showing the possible end points P for which there is a change of stability in the system. The curve can be determined from the one in the (l_1,l_2)–plane by the pair of mappings: $(x,y) \mapsto ((l_1^2 - l_2^2 + d^2)/2d, \pm\sqrt{\Phi}/2d)$ where $\Phi = (-l_1 + l_2 + d)(l_1 + l_2 - d)(l_1 - l_2 + d)(l_1 + l_2 + d)$. Thus each point in the (l_1, l_2)–plane corresponds to a pair of points in the (x,y) πλανε. In contrast, for a given point in the (x,y) πλανε, four possible values of l_1, l_2 can be obtained: $l_1 = \pm\sqrt{x^2 + y^2}, l_2 = \pm\sqrt{(x-d)^2 + y^2}$, each corresponding to a separate configuration. The catastrophe locus in the (l_1, l_2)–plane has four singular points: $(0, \pm d)$ and $(\pm d, 0)$ corresponding to singularities of the potential function. These points correspond in the (x,y) πλανε to the base points A(0,0) and B(d,0) of the mechanism. The union of the singular points and the catastrophe locus give a complete description of the instability.

3.1 Bifurcation Curve in the (F_x, F_y)–plane

The bifurcation set in the (F_x,F_y)–plane has a different interpretation to the catastrophe loci. Recall that for any given point in the (F_x,F_y)–plane there corresponds up to six real points in the (l_1,l_2)– and (x,y)–planes corresponding to the six stable equilibrium positions. Therefore, the entire system can be represented by a single point in the (F_x,F_y)–plane. The bifurcation curve in the (F_x,F_y)–plane can be obtained from the catastrophe locus in the (x,y) πλανε using equations (2) and (3). The four singular points, (±d, 0) and (0, ±d), correspond to four circles in the (F_x,F_y)–plane, centred on the F_x axis, with radii $k_1 l_{01}$ and $k_2 l_{02}$ [5].

The bifurcation set partitions the plane into distinct regions. Within each region the rank and signature of the Hessian remains the same. Thus each region may be designated by the number of real potential maximum, minimum, and saddles of the potential functions, or equivalently, by the number of stable and unstable configurations. In figure 4 each region partitioned by the bifurcation curve contains a label of the form (s, u) where s is the number of stable and u is the number of unstable configurations of the mechanism. So, for instance, a region labelled (4, 2) implies that all forces within that region give rise to potential functions with 4 stable and 2 unstable critcal points. Only real configurations are considered, so the sum of the stable and unstable configurations may change from region to region. From Fig. 4 it is clear that the degenerate critical points calculated directly from equation (6) correspond to a gain or loss in the number of real configurations of the mechanism, while the double points, which give rise to the circles in the (F_x,F_y)–plane, only correspond to a change in stability for a single configuration. It is possible to determine which paths come together as force crosses the bifurcation curve [5].

Example 3:

Let $k_1 = k_2 = l_{01} = l_{02} = d = 1$ and suppose force is incremented in the F_y-direction from F_y=0 to F_y=0.6. Figure 5 shows each increment of force as a '+' and its relation to the bifurcation curve. Equilibrium positions in the (x,y) πλανε at each increment are plotted on their bifurcation curve in the adjacent figure and numbered 1 to 6. As the force approaches the bifurcation curve near F_y=0.45, equilibrium paths 2 and 5 intersect on the bifurcation curve in the (x,y) πλανε. As the force is increased to F_y=0.6, paths 2 and 5 no longer exist in the real plane. Physically, path 2, which is stable, reaches a point where it changes to unstable at a point on the bifurcation curve. A catastrophe is encountered and the mechanisms jumps to configuration 1 where a potential minimum is found.

In addition to cusp catastrophes, for particular values of d, k_1, k_2, l_{01}, l_{02}, the two-spring exhibits an elliptic umbillic catastrophe. Further details, including coordinates of the cusps and the conditions for an elliptic umbillic, may be found in [5].

4. Analysis of the Spatial Three-Spring System

The three dimensional equivalent of the two-spring system is the compliant tetrahedron, Fig. 6. The tetrahedron features three linear springs with spring constants k_1, k_2, k_3, and natural spring lengths l_{01}, l_{02} and l_{03}, connected to ground on one end at points A (0, 0, 0), B $(x_2, 0, 0)$ and C $(x_3, y_3, 0)$ forming a triangular base. The other end of each spring is connected to a common pivot point $P(x, y, z)$. The inverse analysis has been performed by Zhang et al. [8] and Dietmaier [1]. Both papers find up to 22 equilibrium positions for a given force applied to point P of the tetrahedron.

The potential function of the three-spring is :

$$V = \frac{1}{2}k_1(l_1 - l_{01})^2 + \frac{1}{2}k_2(l_2 - l_{02})^2 + \frac{1}{2}k_3(l_3 - l_{03})^2 - xF_x - yF_y - zF_z. \qquad (7)$$

Instability of the system occurs when the determinant of the Hessian of V is zero. Space does not permit us to derive the resulting equation but it is of interest to note that the catastrophe locus is an algebraic surface defined by a twelfth order polynomial in l_1, l_2 and l_3.

Display of the catastrophe locus is more difficult than in the planar two-spring case since it is a surface with a complex geometry. The most straightforward way to visualize it is to plot a sequence of plane sections. Figure 7 shows sections in the l_1, l_2 plane for fixed values of l_3 ranging from -1.8 to 1.0 of the mechanism with $k_1 = k_2 = k_3 = 1$, $l_{01} = l_{02} = l_{03} = x_2 = 1$, $x_3 = 0.5, y_3 = \sqrt{3}/2$.

Using elementary trigonometry we may express x, y and z in terms of l_1, l_2 and l_3 giving a 1-2 correspondence between (l_1, l_2, l_3)– and (x, y, z)– space:

$$x = \frac{l_1^2 + x_2^2 - l_2^2}{2x_2}, \quad y = \frac{l_1^2 + x_3^2 + y_3^2 - l_3^2 - 2x_3 x}{2y_3}, \quad z = \pm\sqrt{l_1^2 - x^2 - y^2} \tag{8}$$

4.1 Bifurcation Set in (F_x, F_y, F_z)–space

The mechanism is in a state of static equilibrium when the partial derivatives of the potential functions are zero. These yield three equations defining a mapping from (l_1, l_2, l_3)– to (F_x, F_y, F_z)–space:

$$F_x = k_1(l_1 - l_{01})\frac{x}{l_1} + k_2(l_2 - l_{02})\frac{x - x_2}{l_2} + k_3(l_3 - l_{03})\frac{x}{l_3} \tag{9}$$

$$F_y = k_1(l_1 - l_{01})\frac{y}{l_1} + k_2(l_2 - l_{02})\frac{y}{l_2} + k_3(l_3 - l_{03})\frac{y - y_3}{l_3} \tag{10}$$

$$F_z = k_1(l_1 - l_{01})\frac{z}{l_1} + k_2(l_2 - l_{02})\frac{z}{l_2} + k_3(l_3 - l_{03})\frac{z}{l_3} \tag{11}$$

Note that the mappings are non-linear in l_i. Hence plane sections of the catastrophe locus map to space curves in the bifurcation set. To obtain a plane section of the bifurcation set, say F_z = constant, it is necessary to compute a surface-surface intersection in (l_1, l_2, l_3). This would be preferable to obtaining sections from an expression for the bifurcation set in (F_x, F_y, F_z) which would define an algebraic surface of high order.

In addition to the bifurcation set, one must include twelve spheres of points corresponding to the potential function being non-differentiable. This arises when one the springs compresses so that the pivot is one of the three base points (0, 0, 0), $(x_2, 0, 0)$, and $(x_3, y_3, 0)$. The spheres are analogous to the circles of points of the two-spring case where the potential function is non-differentiable. In addition to the bifurcation set, the spheres contribute a number of circles to the final instability plots. Thus the complete picture of instability of the three-spring is quite complex. Space does not permit us to describe the bifurcation set in any further detail, but additional results will be published in a future paper.

5. Conclusion

The paper describes how Catastrophe Theory may be used to obtain a stability analysis of compliant mechanisms. By way of example, the stability of the planar two-spring and spatial three-spring are considered. One of the strengths of the method described in the paper is that it can be applied without any knowledge of the inverse analysis of the mechanism. Future work will include a more complete analysis of the spatial three-spring system (including a classification of the catastrophes), and analyses of spring systems involving force and torque.

References

[1] P. Dietmaier. An Inverse Force Analysis of a Spatial Three-Spring System. In *ARK 4th International Workshop on Advances in Robot Kinematics and 1st International Workshop on Computation Geometry in Kinematics, Slovenia, July 1994*, 1994.

[2] R. Gilmore. *Catastrophe Theory for Scientists and Engineers*. Dover, 1981.

[3] M. Griffis. *A Novel Theory for Simultaneously Regulating Force and Displacement*. Ph.D. thesis, University of Florida, Gainesville, 1991.

[4] R. Hines. *Catastrophe Analysis of Compliant Mechanisms*. Master's thesis, University of Florida, Gainesville, 1995.

[5] R. Hines, D. Marsh, and J. Duffy. *Catastrophe Analysis of the Planar Two-Spring Mechanism*. In review.

[6] T. Pigoski, M. Griffis, and J. Duffy. Stiffness Mappings Employing Different Frames of Reference. In *Flexible Mechanisms Dynamics and Analysis, DE-Vol. 47, ASME Design Technical Conference, Scottsdale, Arizona*, 1992.

[7] T. Poston and I. Stewart. *Catastrophe Theory and its Applications*. Pitmann, 1978.

[8] Y. Zhang, C. Liang, J. Duffy, and E. J. F. Primrose. *A Reverse Force Analysis of a Spatial Three-Spring System*. In review, 1996.

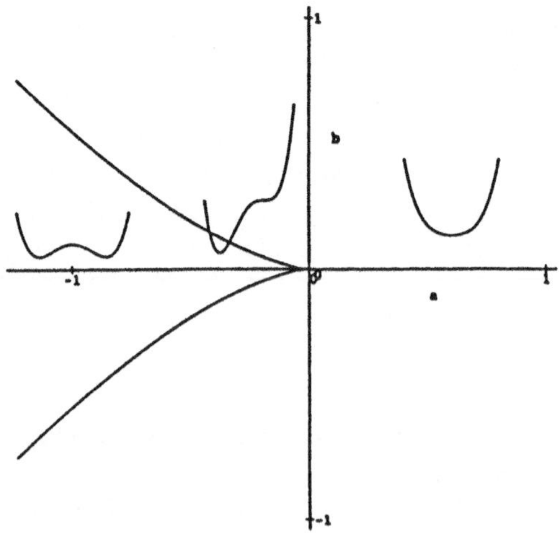

Figure 1: The bifurcation curve of the cusp catastrophe

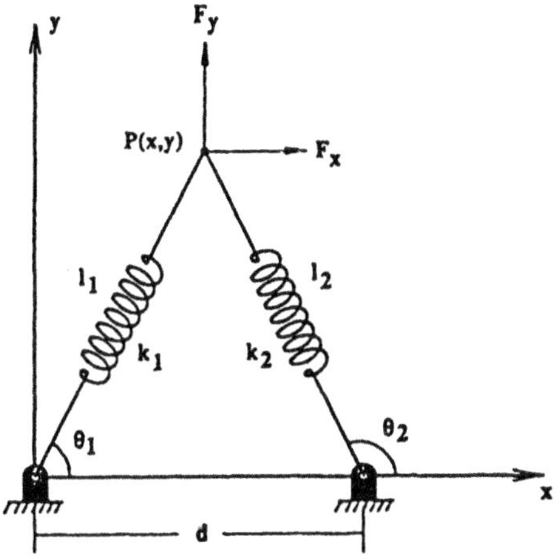

Figure 2: Planar two-spring mechanism

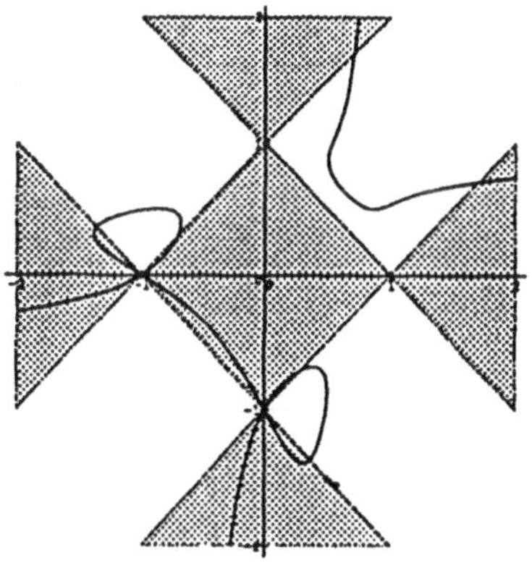

Figure 3

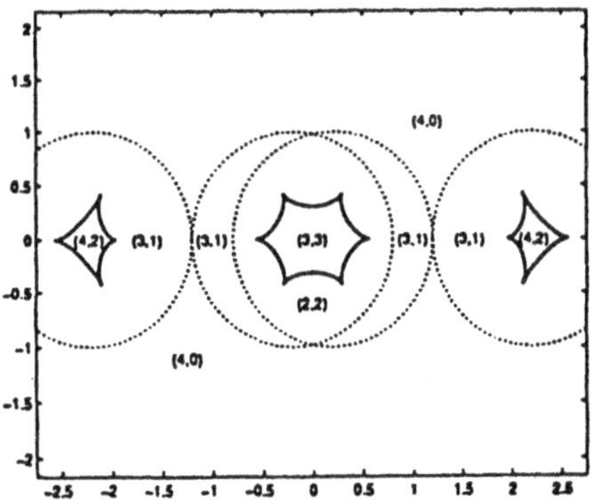

Figure 4

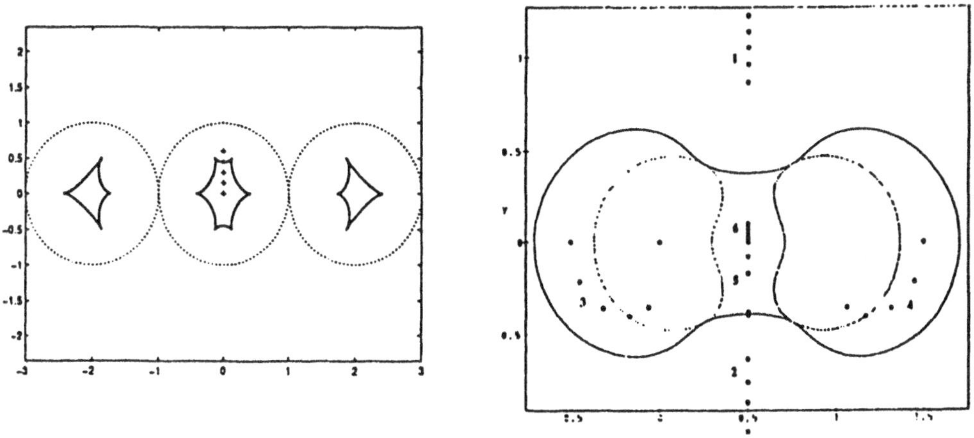

Figure 5: As the path of forces crosses the bifurcation curve two paths of equilibria coincide

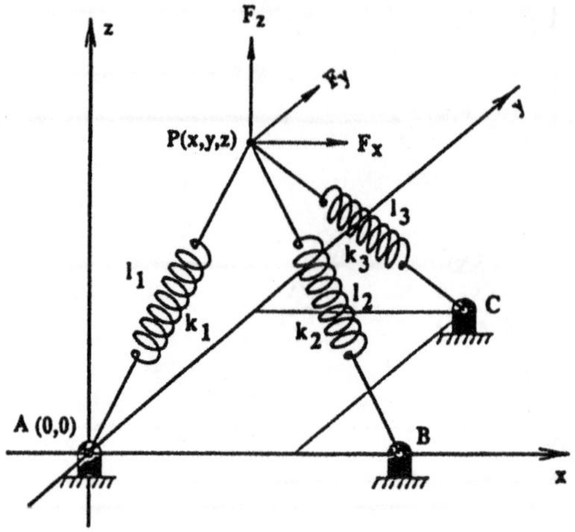

Figure 6: Three Spring Tetrahedron

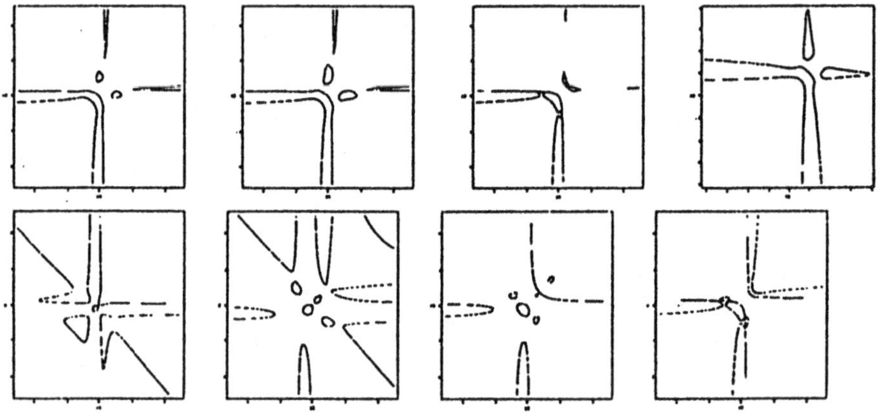

Figure 7: Sections of a Catastrophe Locus of a Three-Spring Mechanism

ACCELERATION ANALYSIS, VIA SCREW THEORY, AND CHARACTERIZATION OF SINGULARITIES OF CLOSED CHAINS

JOSÉ MARíA RICO M. AND JAIME GALLARDO A.
Departamento de Ingeniería Mecánica
Instituto Tecnológico de Celaya, 38000 Celaya, Gto. México

Abstract. It is shown that the acceleration analysis of closed chains provide a more accurate characterization of their singular positions, than that obtained by the velocity analysis only. Further, recent results have shown that the acceleration analysis of closed chains can be formulated very compactly via screw theory. Hence, it is possible to formulate a new methodology for the characterization of singular positions of closed chains. The authors believe that the approach expounded in this paper has some advantages over previously known methods.

1. Introduction

One of the most important topics in the analysis of linkages and manipulators is that of the determination and characterization of singularities. Unfortunately, the terminology, in this subject, is rather different depending whether the subject is analyzed from the mathematical (differentiable manifold theory), or the applied (analysis of linkages and manipulators) point of view. Moreover, the terminology changes from the analysis of closed chains (spatial linkages) to the analysis of open chains (manipulators).

Since the present contribution deals with single loop closed chains, the terminology adopted here stems from the work of Hunt (1978) and Sugimoto *et al* (1982), and Litvin and his coworkers (Litvin, 1980; Litvin *et al*, 1986a; Litvin *et al*, 1986b; Litvin and Tan, 1989; Kieffer, 1990; Kieffer, 1992; Kieffer, 1994; Kieffer and Litvin, 1991).

Consider a closed chain formed by $m-1$ rigid bodies joined by $m-1$ helical pairs, as shown in Figure 1, where bodies m and 1 have been rigidly joined to form a unique rigid body. Later, it will be shown that the velocity analysis of the closed chain is given by

J. Lenarčič and V. Parenti-Castelli (eds.), Recent Advances in Robot Kinematics, 139-148.
© 1996 *Kluwer Academic Publishers. Printed in the Netherlands.*

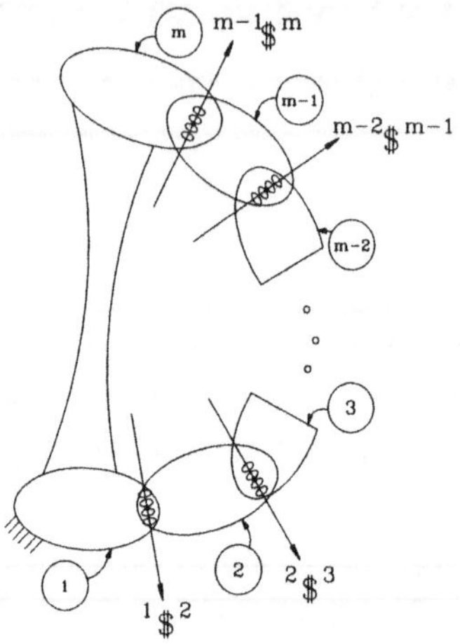

Figure 1. **A Closed Chain with m-1 Rigid Bodies.**

$$_1\omega_2{}^1\$^2 + {}_2\omega_3{}^2\$^3 + \cdots + {}_{m-2}\omega_{m-1}{}^{m-2}\$^{m-1} + {}_{m-1}\omega_1{}^{m-1}\$^1 = \vec{0} \quad (1)$$

If the solution of the velocity analysis requires that one or more, but not all, of the $_j\omega_{j+1} = 0$, for some j=1,2,...,m-1, the closed chain is in a singularity or uncertainty configuration depending whether $_j\omega_{j+1}$ is associated with an output or input link. Nevertheless, in this contribution the difference between input and output links is irrelevant. Hence, a configuration that satisfies the aforementioned condition will be called a singularity configuration without further qualification.

As far as the authors are aware, Hunt (1978) was the first to attempt a characterization of singular and uncertainty configurations of spatial linkages. A few years later, Sugimoto *et al* (1982) attempted, again, to characterize those configurations. The main tool behind these efforts was screw theory, more specifically the velocity equation of the closed loop associated with the linkage. Despite important progress in the characterization of configurations, Sugimoto *et al* (1982) recognized that the techniques developed by them were unable to distinguish between uncertainty configurations and structures; i.e. the transitory mobility predicted by the velocity equation did not agree, in some cases, with the degree of freedom of the linkage.

Litvin and his coworkers (Litvin, 1980; Litvin *et al*, 1986a; Litvin *et al*, 1986b; Litvin and Tan, 1989), in particular Kieffer (Kieffer, 1990; Kieffer, 1992; Kieffer, 1994; Kieffer and Litvin, 1991), approached the same problem by using the Taylor series expansion of the closure equations of the spatial linkage and following the path of the inverse kinematics problem. It should be noted that by employing the Taylor series expansion of

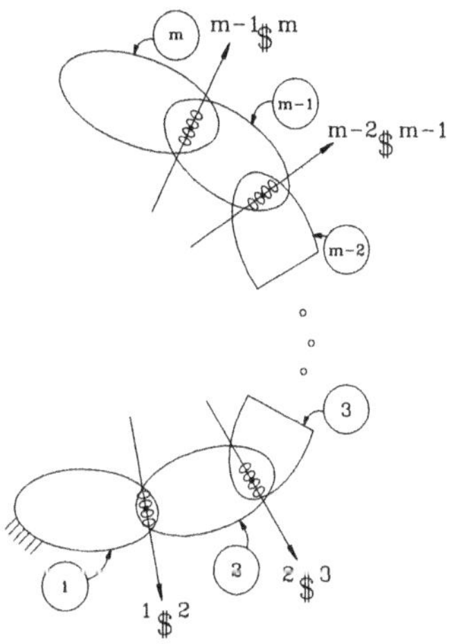

Figure 2. **An Open Serial Chain with m Rigid Bodies.**

the closure equations of a spatial linkage, it is possible to produce local models, or approximations, to the closure equations with different orders of approximation. First order or linear, second order or quadratic, third order or cubic, approximation. In that respect, Hunt's and Sugimoto's methods, which employ only the velocity equation of the spatial linkage, can be regarded as a linear approximation. Hence, it is not surprising that Hunt's and Sugimoto's approaches failed when applied to the kinematic analysis of spatial linkage, a highly non-linear problem.

An important disadvantage of Litvin's and Kieffer's approach is that the algorithm requires extensive numerical computations performed by a computer. Thus, it is impossible to obtain general results.

The approach presented in this paper is based in recently obtained results (Rico and Duffy, 1996) that allow to write the acceleration equation of spatial linkages in a very simple form involving the screws representing the kinematic pairs, of the linkage, and their Lie products. Those result show that screw theory can be expanded beyond a merely linear theory into a higher-order theory.

2. Fundamentals of the Method

Consider an open serial chain formed by m rigid bodies, where adjacent links are connected by helical pairs (see Figure 2). It is well known that revolute and prismatic pairs are particular cases of helical pairs and that spherical and cylindrical joints can be modelled by a series of revolute and prismatic pairs. Therefore, any serial chain can be modelled as indicated here, for more details see (Rico and Duffy, 1996).

The kinematic pair that connects the bodies j and j+1 is represented by

$$^{j}\$^{j+1} = (^{j}\vec{s}^{j+1}; {}^{j}\vec{s}_{O}^{j+1}), j = 1, 2, \ldots, m-1, \tag{2}$$

where

$^{j}\vec{s}^{j+1}$ = A unit vector in the direction of the kinematic pair.
$^{j}\vec{s}_{O}^{j+1} = {}_{j}h_{j+1}{}^{j}\vec{s}^{j+1} + {}^{j}\vec{s}^{j+1} \times {}^{j+1}\vec{q}^{m}$
where

$_{j}h_{j+1}$ = The pitch of the helical kinematic pair.

$^{j+1}\vec{q}^{m}$ = The position vector of an arbitrary point O, fixed in body 1, with respect to an arbitrary point along the axis of the kinematic pair.

The velocity state of body m, with respect to body 1 is defined by

$$^{1}\vec{V}^{m} = \begin{bmatrix} ^{1}\omega^{m} \\ ^{1}\vec{v}_{O}^{m} \end{bmatrix}, \tag{3}$$

where

$^{1}\omega^{m}$ = The angular velocity of body m with respect to body 1.
$^{1}\vec{v}_{O}^{m}$ = The velocity of that point O, which belongs to body m and it is instantaneously coincident with a point O, fixed in body 1.

It is well known that the velocity state $^{1}\vec{V}^{m}$ is given by

$$^{1}\vec{V}^{m} = {}_{1}\omega_{2}{}^{1}\$^{2} + {}_{2}\omega_{3}{}^{2}\$^{3} + \cdots + {}_{m-2}\omega_{m-1}{}^{m-2}\$^{m-1} + {}_{m-1}\omega_{m}{}^{m-1}\$^{m} \tag{4}$$

where $_{j}\omega_{j+1}$ represents the joint rate of the corresponding kinematic pair.

Further, it has been recently shown (Rico and Duffy, 1996) that if $^{1}\omega^{m} = \vec{0}$, the acceleration state of body m with respect to body 1, defined in this particular case by

$$^{1}\vec{A}^{m} = \begin{bmatrix} ^{1}\alpha^{m} \\ ^{1}\vec{a}_{O}^{m} \end{bmatrix}, \tag{5}$$

is given, in terms of the screws representing the kinematic pairs, by

$$\begin{aligned} ^{1}A^{m} = {} & {}_{1}\dot{\omega}_{2}{}^{1}\$^{2} + {}_{2}\dot{\omega}_{3}{}^{2}\$^{3} + \cdots + {}_{m-2}\dot{\omega}_{m-1}{}^{m-2}\$^{m-1} + {}_{m-1}\dot{\omega}_{m}{}^{m-1}\$^{m} \\ & + [{}_{1}\omega_{2}{}^{1}\$^{2} \quad {}_{2}\omega_{3}{}^{2}\$^{3} + \cdots + {}_{m-2}\omega_{m-1}{}^{m-2}\$^{m-1} + {}_{m-1}\omega_{m}{}^{m-1}\$^{m}] \\ & + [{}_{2}\omega_{3}{}^{2}\$^{3} \quad {}_{3}\omega_{4}{}^{3}\$^{4} + \cdots + {}_{m-2}\omega_{m-1}{}^{m-2}\$^{m-1} + {}_{m-1}\omega_{m}{}^{m-1}\$^{m}] \\ & + \cdots + [{}_{m-2}\omega_{m-1}{}^{m-2}\$^{m-1} \quad {}_{m-1}\omega_{m}{}^{m-1}\$^{m}], \end{aligned} \tag{6}$$

where $_{j}\dot{\omega}_{j+1}$ represents the time derivative of the joint rate of the corresponding kinematic pair, and the brackets represent Lie products.

Consider now the closed chain formed by rigidly joining body m with body 1, see Figure 1. Then, m = 1, and

1. The velocity state $^{1}\vec{V}^{m} = \vec{0}$, in particular $^{1}\omega^{m} = \vec{0}$. Hence, equation 5 becomes

$$ {}_1\omega_2\,{}^1\$^2 + {}_2\omega_3\,{}^2\$^3 + \cdots + {}_{m-2}\omega_{m-1}\,{}^{m-2}\$^{m-1} + {}_{m-1}\omega_m\,{}^{m-1}\$^m = \vec{0} \qquad (7) $$

Equation 1 represents the velocity analysis of the closed chain. It is easy to observe that the Jacobian matrix of the closed chain is

$$ J = \left[{}^1\$^2 \quad {}^2\$^3 \quad \cdots \quad {}^{m-2}\$^{m-1} \quad {}^{m-1}\$^1\right] \qquad (8) $$

2. The acceleration state ${}^1\vec{A}^m = \vec{0}$. Furthermore, since ${}^1\omega^m = \vec{0}$, equation 7 becomes

$$ \begin{aligned} &{}_1\dot{\omega}_2\,{}^1\$^2 + {}_2\dot{\omega}_3\,{}^2\$^3 + \cdots + {}_{m-2}\dot{\omega}_{m-1}\,{}^{m-2}\$^{m-1} + {}_{m-1}\dot{\omega}_m\,{}^{m-1}\$^m \\ &+ \left[{}_1\omega_2\,{}^1\$^2 \quad {}_2\omega_3\,{}^2\$^3 + \cdots + {}_{m-2}\omega_{m-1}\,{}^{m-2}\$^{m-1} + {}_{m-1}\omega_m\,{}^{m-1}\$^m\right] \\ &+ \left[{}_2\omega_3\,{}^2\$^3 \quad {}_3\omega_4\,{}^3\$^4 + \cdots + {}_{m-2}\omega_{m-1}\,{}^{m-2}\$^{m-1} + {}_{m-1}\omega_m\,{}^{m-1}\$^m\right] \\ &\qquad + \cdots + \left[{}_{m-2}\omega_{m-1}\,{}^{m-2}\$^{m-1} \quad {}_{m-1}\omega_m\,{}^{m-1}\$^m\right] = \vec{0}. \end{aligned} \qquad (9) $$

Equation 9 represents the acceleration analysis of the closed chain, in terms of the screws representing the kinematic pair of the closed chain.

So far, screw theory has been applied to the determination of transitory mobility and the characterization of singularities of closed chains by employing only the velocity equation of the closed chain (see equation 1). In fact, equation 1 represents a linear combination of m-1 screws equated to the zero screw. If the screw set $S = \{{}^1\$^2, {}^2\$^3, \cdots, {}^{m-2}\$^{m-1}, {}^{m-1}\$^1\}$ generates a n-dimensional subspace $1 < n < 6$ of the six-dimensional screw algebra (isomorphic to the Lie algebra e(3) of the Euclidean group) or, in the language of classical screw theory, if S form a n-screw system, then the dimension of the solution subspace of the homogeneus system, S_s, is given by

$$ dim\ S_s = (m-1) - n = m - (n+1). \qquad (10) $$

The dimension of S_s is also equal to the transitory mobility of the closed chain. However, it is not equal to the degree of freedom of the linkage, in general. The reason is quite simple, the degree of freedom is a concept related with the finite mobility of the linkage, sometimes called "full cycle mobility"; while the transitory mobility is an infinitesimal concept, related with the velocity analysis of the linkage.

The idea behind the new methodology proposed in this paper is equally simple. A necessary condition for a linkage to have finite or "full cycle" mobility is that not only the velocity analysis of the linkage but also the acceleration and higher order analyses of the linkage have a non-zero and meaningful solution.

3. Examples

1. Consider the RSRR spatial linkage shown in Figure 3 and proposed by Prof. Hunt, (1992). It should be noted that the lenghts a, b, and c are not independent, but they are related by the condition

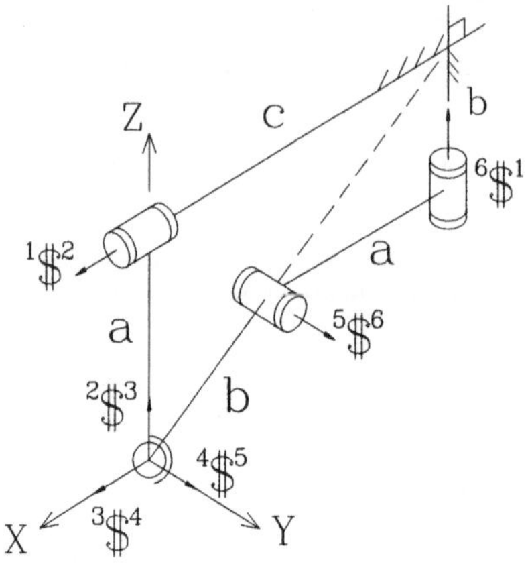

Figure 3. **A RSRR Spatial Linkage.**

$$c^2 = \left(\sqrt{a^2 + b^2} + b\right)^2 - a^2 \tag{11}$$

Choosing the center of the spherical pair, as the origin of the coordinate system, with the coordinate axes as indicated in Figure 3, the screw coordinates are given by

$$\begin{aligned} {}^1\$^2 &= (1, 0, 0; 0, a, 0), & {}^2\$^3 &= (0, 0, 1; 0, 0, 0), \\ {}^3\$^4 &= (1, 0, 0; 0, 0, 0), & {}^4\$^5 &= (0, 1, 0; 0, 0, 0), \\ {}^5\$^6 &= (0, 1, 0; 0, 0, -(c - a)), & {}^6\$^1 &= (0, 0, 1; 0, c, 0). \end{aligned}$$

The velocity analysis requires the solution of the equation

$${}_1\omega_2 {}^1\$^2 + {}_2\omega_3 {}^2\$^3 + {}_3\omega_4 {}^3\$^4 + {}_4\omega_5 {}^4\$^5 + {}_5\omega_6 {}^5\$^6 + {}_6\omega_1 {}^6\$^1 = 0. \tag{12}$$

The solution of the velocity analysis yields

$${}_1\omega_2 = -\frac{c}{a} {}_6\omega_1,\ {}_2\omega_3 = -{}_6\omega_1,\ {}_3\omega_4 = \frac{c}{a} {}_6\omega_1,\ {}_4\omega_5 = 0,\ {}_5\omega_6 = 0. \tag{13}$$

A possible interpretation of the results of the velocity analysis is that the RSRR linkage has transitory mobility 1, and in the position shown in Figure 3, the linkage has a double stationary configuration (${}_4\omega_5 = {}_5\omega_6 = 0$).

In what follows, it will be shown that the acceleration analysis of the linkage yields a more thorough characterization of the linkage mobility and of the singular configuration. After the substitution of ${}_4\omega_5 = {}_5\omega_6 = 0$, the equation of the acceleration analysis becomes

$$
\begin{aligned}
{}_1\dot{\omega}_2{}^1\$^2 + {}_2\dot{\omega}_3{}^2\$^3 + {}_3\dot{\omega}_4{}^3\$^4 + {}_4\dot{\omega}_5{}^4\$^5 + {}_5\dot{\omega}_6{}^5\$^6 + {}_6\dot{\omega}_1{}^6\$^1 \\
+ \left[{}_1\omega_2{}^1\$^2 \quad {}_2\omega_3{}^2\$^3 + {}_3\omega_4{}^3\$^4 + {}_6\omega_1{}^6\$^1\right] \\
+ \left[{}_2\omega_3{}^2\$^3 \quad {}_3\omega_4{}^3\$^4 + {}_6\omega_1{}^6\$^1\right] + \left[{}_3\omega_4{}^3\$^4 \quad {}_6\omega_1{}^6\$^1\right] &= \vec{0}, \quad (14)
\end{aligned}
$$

or, in the final form,

$$
{}_1\dot{\omega}_2{}^1\$^2 + {}_2\dot{\omega}_3{}^2\$^3 + {}_3\dot{\omega}_4{}^3\$^4 + {}_4\dot{\omega}_5{}^4\$^5 + {}_5\dot{\omega}_6{}^5\$^6 + {}_6\dot{\omega}_1{}^6\$^1 = \$_L. \quad (15)
$$

where

$$
\begin{aligned}
\$_L \equiv & -\left[{}_1\omega_2{}^1\$^2 \quad {}_2\omega_3{}^2\$^3 + {}_3\omega_4{}^3\$^4 + {}_6\omega_1{}^6\$^1\right] \\
& - \left[{}_2\omega_3{}^2\$^3 \quad {}_3\omega_4{}^3\$^4 + {}_6\omega_1{}^6\$^1\right] - \left[{}_3\omega_4{}^3\$^4 \quad {}_6\omega_1{}^6\$^1\right] \quad (16)
\end{aligned}
$$

The computation of the Lie products and vector additions yields

$$
\$_L = (0, 2\frac{c}{a}{}_6\omega_1{}^2, 0; c\,{}_6\omega_1{}^2, 0, -\frac{c}{a}{}_6\omega_1{}^2). \quad (17)
$$

Equation 16 represents a non-homogeneus linear system in the unknowns ${}_j\dot{\omega}_{j+1}$, for j = 1,2,3,4,5,6; where 7 is identified with 1. Obtaining the row reduced echelon form of the augmented matrix of the system, it follows that the system has a solution if, and only if,

$$
{}_6\omega_1{}^2\, c = 0. \quad (18)
$$

It should be noted that $c = 0$ implies that either $a = 0$, or $b = 0$. In any of these two cases, a closed chain with a least four intersecting revolute pairs is obtained. These linkages have finite mobility. However, the conditions $a = 0$, or $b = 0$ modify the geometry of the closed chain under consideration, and these possibilities must be eliminated. Therefore, the only remaining posibility is ${}_6\omega_1{}^2 = 0$. Hence, the closed chain with transitory mobility 1 is in reality a structure; i.e. an isolated singularity (Kieffer, 1990).

2. Consider the spatial linkage RCRCR, proposed by Sugimoto *et al* (1982), and shown in Figure 4. The linkage parameters are

$$
a_{12} = \frac{1}{\sqrt{3}}u.l.,\ a_{23} = 1u.l.,\ a_{34} = 2\sqrt{2}u.l.,\ a_{45} = 0u.l.,\ a_{45} = \frac{2}{\sqrt{3}}u.l.
$$

$$
S_{11} = (8 - 3\sqrt{3})u.l.,\ S_{33} = 2\sqrt{2}u.l.,\ S_{55} = 3u.l.
$$

$$
\alpha_{12} = 150^o,\ \alpha_{23} = 135^o,\ \alpha_{34} = 270^o,\ \alpha_{45} = 90^o,\ \alpha_{51} = 330^o,
$$

where u.l. stands for units of length. The position analysis yields

$$
\theta_1 = 0^o,\ \theta_2 = 0^o,\ \theta_3 = 90^o,\ \theta_4 = 105^o,\ \theta_5 = 270^o,
$$

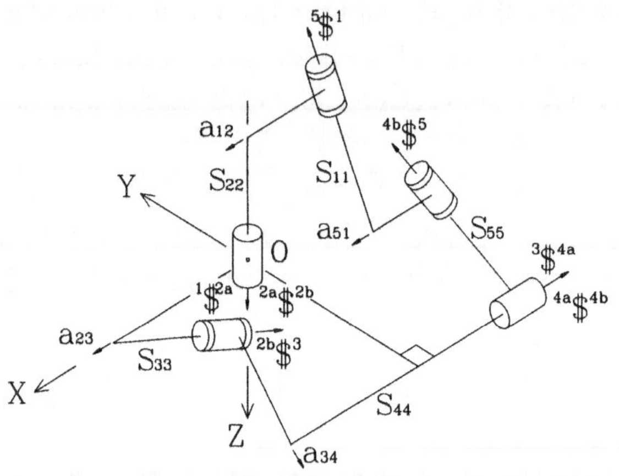

Figure 4. **A RCRCR Spatial Linkage.**

and

$$S_{22} = (4\sqrt{3} - 3)u.l., \, S_{44} = (1 + \sqrt{3})u.l.$$

The coordinates of the screws representing the kinematic pairs of the linkage, with respect to the coordinate system shown in the figure, are given by

$$^{5}\$^{1} = (0, \tfrac{1}{2}, -\tfrac{\sqrt{3}}{2}; 2\sqrt{3} - \tfrac{3}{2}, \tfrac{1}{2}, -\tfrac{1}{2\sqrt{3}}), \quad ^{1}\$^{2a} = (0,0,0;0,0,1),$$
$$^{2a}\$^{2b} = (0,0,1;0,0,0), \quad ^{2b}\$^{3} = (0, -\tfrac{1}{\sqrt{2}}, -\tfrac{1}{\sqrt{2}}; 0, \tfrac{1}{\sqrt{2}}, -\tfrac{1}{\sqrt{2}}),$$
$$^{3}\$^{4a} = (0,0,0;-1,0,0), \quad ^{4a}\$^{4b} = (-1,0,0;0,0,-4),$$
$$^{4b}\$^{5} = (0, \tfrac{\sqrt{3}}{2}, -\tfrac{1}{2}; 2, -\tfrac{\sqrt{3}}{2}, -\tfrac{3}{2}),$$

where the screws $^{1}\$^{2a}$ and $^{2a}\$^{2b}$ are associated with the cylindrical pair whose axis passes through the origin of the coordinate system. Similarly, $^{3}\$^{4a}$ and $^{4a}\$^{4b}$ are associated with the remaining cilyndrical pair.

The velocity analysis is given by

$${}_{5}\omega_{1}\,{}^{5}\$^{1} + {}_{1}\omega_{2a}\,{}^{1}\$^{2a} + {}_{2a}\omega_{2b}\,{}^{2a}\$^{2b} + {}_{2b}\omega_{3}\,{}^{2b}\$^{3} + {}_{3}\omega_{4a}\,{}^{3}\$^{4a} + {}_{4a}\omega_{4b}\,{}^{4a}\$^{4b} + {}_{4b}\omega_{5}\,{}^{4b}\$^{5} = \vec{0}. \tag{19}$$

However, the screw set $S = \{^{5}\$^{1}, ^{1}\$^{2a}, ^{2a}\$^{2b}, ^{2b}\$^{3}, ^{3}\$^{4a}, ^{4a}\$^{4b}, ^{4b}\$^{5}\}$ generates a five dimensional subspace of the Lie algebra, e(3), of the Euclidean space, or a five system. Hence, the linkage has, in this position, a transitory mobility of 2. Further, the solution of the velocity analysis is

$${}_{5}\omega_{1} = -\frac{\sqrt{3}}{2}\,{}_{4b}\omega_{5}, \; {}_{1}\omega_{2a} = {}_{4b}\omega_{5}, \; {}_{2a}\omega_{2b} = -{}_{4b}\omega_{5},$$

$${}_{2b}\omega_{3} = {}_{3}\omega_{4a} + \left(4 - \frac{3\sqrt{3}}{2}\right){}_{4b}\omega_{5}, \; {}_{4a}\omega_{4b} = 0. \tag{20}$$

The results seemingly indicate that the linkage is in a singular position associated with the second cilyndrical pair, ${}_{4a}\omega_{4b} = 0$.

The acceleration analysis requires the solution of the non-homogeneous linear system.

$$ {}_{5}\dot{\omega}_{1}\,{}^{5}\$^{1} + {}_{1}\dot{\omega}_{2a}\,{}^{1}\$^{2a} + {}_{2a}\dot{\omega}_{2b}\,{}^{2a}\$^{2b} + {}_{2b}\dot{\omega}_{3}\,{}^{2b}\$^{3} + {}_{3}\dot{\omega}_{4a}\,{}^{3}\$^{4a} + {}_{4a}\dot{\omega}_{4b}\,{}^{4a}\$^{4b} + {}_{4b}\dot{\omega}_{5}\,{}^{4b}\$^{5} = \$_L, $$

where

$$ \begin{aligned} \$_L \equiv -[&{}_{5}\omega_{1}\,{}^{5}\$^{1} \\ &{}_{1}\omega_{2a}\,{}^{1}\$^{2a} + {}_{2a}\omega_{2b}\,{}^{2a}\$^{2b} + {}_{2b}\omega_{3}\,{}^{2b}\$^{3} + {}_{3}\omega_{4a}\,{}^{3}\$^{4a} + {}_{4a}\omega_{4b}\,{}^{4a}\$^{4b} + {}_{4b}\omega_{5}\,{}^{4b}\$^{5}] \\ &- [{}_{1}\omega_{2a}\,{}^{1}\$^{2a} \quad {}_{2a}\omega_{2b}\,{}^{2a}\$^{2b} + {}_{2b}\omega_{3}\,{}^{2b}\$^{3} + {}_{3}\omega_{4a}\,{}^{3}\$^{4a} + {}_{4a}\omega_{4b}\,{}^{4a}\$^{4b} + {}_{4b}\omega_{5}\,{}^{4b}\$^{5}] \\ &- [{}_{2a}\omega_{2b}\,{}^{2a}\$^{2b} \quad {}_{2b}\omega_{3}\,{}^{2b}\$^{3} + {}_{3}\omega_{4a}\,{}^{3}\$^{4a} + {}_{4a}\omega_{4b}\,{}^{4a}\$^{4b} + {}_{4b}\omega_{5}\,{}^{4b}\$^{5}] \\ &- [{}_{2b}\omega_{3}\,{}^{2b}\$^{3} \quad {}_{3}\omega_{4a}\,{}^{3}\$^{4a} + {}_{4a}\omega_{4b}\,{}^{4a}\$^{4b} + {}_{4b}\omega_{5}\,{}^{4b}\$^{5}] \\ &- [{}_{3}\omega_{4a}\,{}^{3}\$^{4a} \quad {}_{4a}\omega_{4b}\,{}^{4a}\$^{4b} + {}_{4b}\omega_{5}\,{}^{4b}\$^{5}] - [{}_{4a}\omega_{4b}\,{}^{4a}\$^{4b} \quad {}_{4b}\omega_{5}\,{}^{4b}\$^{5}] \end{aligned} \tag{21} $$

The acceleration analysis has a solution if, and only if, $\$_L$ belongs to the subspace generated by the screw set S or, alternatively, if, and only if, the rank of the augmented matrix is equal to the rank of the coefficient matrix. It can be shown, after lengthy computations, that the acceleration analysis has a solution if, and only if,

$$ {}_{3}\omega_{4a}{}^{2} + \sqrt{3}\,{}_{4b}\omega_{5}{}^{2} - \sqrt{3}\,{}_{3}\omega_{4a}\,{}_{4b}\omega_{5} = 0. \tag{22} $$

Hence, solving ${}_{3}\omega_{4a}$ in terms of ${}_{4b}\omega_{5}$, the equation yields

$$ {}_{3}\omega_{4a} = {}_{4b}\omega_{5}\left(\frac{\sqrt{3}}{2} \pm \frac{\sqrt{3-4\sqrt{3}}}{2}\right). \tag{23} $$

Nevertheless, $3 - 4\sqrt{3} < 0$; therefore, the relationship is imaginary. Thus, a real and meaningful solution of the acceleration analysis requires that

$$ {}_{3}\omega_{4a} = {}_{4b}\omega_{5} = 0. \tag{24} $$

The final conclusion is that the linkage is not in a singular configuration but, instead, is an structure. Furthermore, the transitory mobility of the linkage, found via the velocity analysis exclusively, does not hold when examined into the light of the acceleration analysis.

4. Conclusions

This present paper has shown that the acceleration analysis of a closed chain, formulated via screw theory, provides a compact and powerful metod for a more accurate analysis of the transitory or infinitesimal mobility, and the characterization of singularities of closed chains. It was quite surprising to find that both test cases presented here were solved by looking into their acceleration analyses. It should be kept in mind that the conditions for finite mobility introduced in this paper are only necessary. Thus, it is possible to produce an structure that has solvable, in the real sense, velocity and acceleration analyses, and whose lack of finite mobility would have to be decided by looking into higher-order kinematic analyses.

5. Acknowledgement

The results reported in this paper were, in part, obtained by the second author in fulfillment of the partial requirements of a doctoral degree at the Dirección General de Institutos Tecnológicos (México). The initial part of the research was carried out, by the first author, during a summer visit at the Center for Intelligent Machines and Robotics (CIMAR), Mechanical Enginnering Department, University of Florida. The authors thank Prof. Joseph Duffy for his encouragement and fruitful discussions.

References

Gosselin, C., and Angeles, J. (1990), Singularity Analysis of Closed-Loop Kinematic Chains. *IEEE Transactions on Robotics and Automation*, Vol. 6, No. 3, pp. 281-290.

Hunt, K.H. (1978) *Kinematic Geometry of Mechanisms*, Oxford University Press, Oxford, U.K.

Hunt, K.H. (1992), Private communication with Prof. J. Duffy.

Kieffer, J. (1990), *Selected Topics in Mechanisms and Robotics: Singularities, Inverse Kinematics, and Collision Detection*, Ph. D. Dissertation, University of Illinois at Chicago.

Kieffer, J. (1992), Manipulator Inverse Kinematics For Untimed End-Effector Trajectories with Ordinary Singularities. *International Journal of Robotics Research*, Vol. 11, No. 3, pp. 225-237.

Kieffer, J. (1994), Differential Analysis of Bifurcations and Isolated Singularities for Robots and Mechanisms. *IEEE Transactions on Robotics and Automation*, Vol. 10, No. 1, pp. 1-10.

Kieffer, J. and Litvin, F.L. (1991), Local Parametric Representation of Displacement Functions for Linkages and Manipulators. *Mechanism and Machine Theory*, Vol. 26, No. 1, pp. 41-53.

Litvin, F.L. (1980), Application of Theorem of Implicit Function System Existence for Analysis and Synthesis of Linkages. *Mechanism and Machine Theory*, Vol. 15, pp. 115-125.

Litvin, F.L., Zhang, Y., Parenti Castelli, V. and Innocenti, C. (1986a), Singularities, Configurations, and Displacement Functions for Manipulators. *International Journal of Robotics Research*, Vol. 5 No. 2, pp. 52-65.

Litvin, F.L., Fanghella, P., Tan, J. and Zhang, Y. (1986b), Singularities in Motion and Displacement Functions of Spatial Linkages. *Journal of Mechanisms, Transmissions, and Automation in Design*, Vol. 108, pp 517-523.

Litvin, F.L. and Tan, J. (1989), Singularities in Motion and Displacement Functions of Constrained Mechanical Systems. *International Journal of Robotics Research*, Vol. 8, No. 2, pp. 30- 43.

Rico, J.M. and Duffy, J. (1996), An Application of Screw Algebra to the Acceleration Analysis of Serial Chains. *Mechanism and Machine Theory*, Vol. 31, No. 4, pp. 445-457.

Sugimoto, K., Duffy, J. and Hunt, K. H. (1982), Special Configurations of Spatial Mechanisms and Robot Arms. *Mechanism and Machine Theory*, Vol. 17, No. 2, pp. 119-132.

4. Workspace and trajectory analysis

M.G. Wagner, B. Ravani: *Computer aided design of robot trajectories using rational motions*

A. Doria: *Optimum design of closed kinematic chains in the presence of obstacles*

G.S. Chirikjian: *Synthesis of discretely actuated manipulator workspaces via harmonic analysis*

N.A. Aspragathos: *Optimal location of path following tasks in the workspace of a manipulator using genetic algorithms*

M. Ceccarelli: *Feasible workspace regions for a two-revolute manipulator design*

E.E. Peisach: *Determination of the workspace of 4R manipulator: basic case and particular cases*

COMPUTER AIDED DESIGN OF ROBOT TRAJECTORIES USING RATIONAL MOTIONS

MICHAEL G. WAGNER AND BAHRAM RAVANI
Department of Mechanical and Aeronautical Engineering
University of California at Davis

Abstract. This paper discusses some special applications of the theory of rational b-spline motions to the computer aided design of robot trajectories. The main advantage of this approach is the data conformity to state of the art CAD systems which allows straightforward data transfer from CAM to CAD systems. In particular the paper focusses on applications in which the motion of the end effector is implicitly defined by task level considerations. Such applications include for example arc welding, spray painting and the scanning of surfaces with robot equipment.

1. Introduction

Task space programming of robot manipulators usually requires the specification of the continuous path of a tool center point and the orientation of the tool or the end–effector of the robot. Early approaches involved geometrically simple point trajectories such as straight lines or a set of straight lines with smooth transitions (see e.g. [8]). In more complex applications such as arc welding or contouring operations, geometrically more complex point trajectories have to be used; however, most approaches provide very little utilization of the powerful techniques from the field of Computer Aided Geometric Design (CAGD) although they are standard in state of the art CAD systems.

On the other hand, some more recent methods for motion design based on so–called rational motions (see [4, 7]) would allow the application of a variety of Bézier and b-spline techniques to the path planning of robot manipulators. Using this approach the design of a Cartesian motion of an end–effector is usually done by specifying a set of key control configurations which are interpolated or approximated. As the main advantage, the

J. Lenarčič and V. Parenti-Castelli (eds.), Recent Advances in Robot Kinematics, 151-158.
© 1996 *Kluwer Academic Publishers. Printed in the Netherlands.*

resulting motion can be described in terms of the well known Bernstein or b-spline basis functions which provides a possibility for a very straightforward data exchange with industrial CAD systems. For example, surfaces generated through a milling process with a cylindrical or conical cutting tool are tensor pruduct NURBS surfaces (cf. [7]) and can therefore be directly handled by CAD systems for simulation purposes.

In industrial applications of robotics, the task level considerations usually determine the desired Cartesian space trajectory of the end–effector. Furthermore, in some tasks, it is not necessary to completely define a set of key control configurations for the end–effector and then determine its trajectory as the motion that interpolates these key control configurations. The desired trajectory of the end–effector often is implicitly described by or in part by the task level specifications. For example, in arc welding one would like to move the welding torch such that the relative position of the electrode with respect to the bead of the weld stays unchanged.

In this paper, we therefore consider a class of tasks for which the trajectory of the end–effector is described by the path of the tool center point and the direction of the z-axis of the hand coordinate frame. These types of robot trajectories are usually encountered in arc welding, surface scanning, spray painting, and in general in any type of complex contouring operations for which the geometry of the working profile is known, sensed, or is part of the task specification. Our main goal is to design the end–effector motion as a rational motion given in NURBS representation. We shall see, however, that this can only be achieved, if we restrict ourselves to a rational tool center point path and a rational unit vector field that describes the direction of the moving z-axis. Nevertheless, since it is always possible to approximate a nonrational curve by a rational curve with any wanted numerical accuracy, these restrictions are only of theoretical importance.

We are currently implementing the resulting algorithms in a general purpose programming system. The system can be used in off-line robot programming in many industrial applications requiring continuous robot path planning and programming.

2. Rational Motions

To simplify notations we will describe points with help of homogeneous coordinates collected in homogeneous coordinate vectors $\mathbf{P} = (p_0, p_1, p_2, p_3)^T$ and denote $\mathbf{p} = (p_1, p_2, p_3)^T$. For points not at infinity, i.e. $p_0 \neq 0$, the corresponding inhomogeneous Cartesian coordinates are $\underline{\mathbf{p}} = (x_p, y_p, z_p)^T = \frac{1}{p_0}\,\mathbf{p}$. by Analogously we are going to use homogeneous plane coordinates $\mathbf{P}^*$ for the description of planes. A point $\mathbf{P}$ lies in the plane $\mathbf{P}^*$ if $\langle \mathbf{P}, \mathbf{P}^* \rangle = 0$. Here and in the sequel $\langle , \rangle$ shall denote the canonical scalar product in R^3

or R^4.

In terms of homogeneous coordinates the position of the end effector of a robot with respect to the fixed coordinate system is given by a real 4×4 matrix of type

$$\mathrm{M} = \left(\begin{array}{c|ccc} v_0 & 0 & 0 & 0 \\ \hline v_1 & r_{1.1} & r_{1.2} & r_{1.3} \\ v_2 & r_{2.1} & r_{2.2} & r_{2.3} \\ v_3 & r_{3.1} & r_{3.2} & r_{3.3} \end{array} \right) = \left(\begin{array}{c|ccc} v_0 & 0 & 0 & 0 \\ \hline v_1 & & & \\ v_2 & & \mathrm{R} & \\ v_3 & & & \end{array} \right), \quad v_0 \neq 0, \qquad (1)$$

where the 3×3 submatrix R satisfies $\mathrm{R} \cdot \mathrm{R}^T = v_0^2 \, \mathrm{I}_3$. I_3 denotes the identity matrix. The matrix M is often called *position matrix* or *hand matrix*. For any point $\widehat{\mathbf{P}} = (\widehat{p}_0, \widehat{p}_1, \widehat{p}_2, \widehat{p}_3)^T$ of the end effector its position results from the linear transformation

$$\mathbf{P} = \mathrm{M} \cdot \widehat{\mathbf{P}}. \qquad (2)$$

A continuous one parameter set of positions is called a one-parameter motion $\mathrm{M} = \mathrm{M}(t)$. Its parameter t is assumed to be the time. If all element functions of M are (piecewise) polynomials of degree n, the motion M is said to be a *(piecewise) rational motion* of degree n. In this case all point trajectories $\mathbf{P}(t) = \mathrm{M}(t) \cdot \widehat{\mathbf{P}}$ are rational curves and can be written in NURBS form

$$\mathbf{P}(t) = \mathrm{M}(t) \cdot \widehat{\mathbf{P}} = \sum_{i=0}^{m} N_{i.\mathbf{t}}(t)\, \mathrm{A}_i \cdot \widehat{\mathbf{P}} = \sum_{i=0}^{m} N_{i.\mathbf{t}}(t)\, \mathbf{P}_i \qquad (3)$$

with so-called *control points* $\mathbf{P}_i$. The b-spline basis functions $N_{i.\mathbf{t}}(t)$ (see e.g. [5]) of degree n are supposed to be defined over an appropriate knot vector $\mathbf{t}$. In (3) we use the NURBS representation

$$\mathrm{M}(t) = \sum_{i=0}^{m} N_{i.\mathbf{t}}(t)\, \mathrm{A}_i \qquad (4)$$

of the motion M as introduced in [7]. The matrices A_i are not of type (1) and therefore do not describe Euclidean positions of the end effector. However, it is possible to visualize these so-called control matrices of M with help of affine, hence distorted, copies of the moving object.

For the construction of rational motions we need a special representation formula which has been discussed in [6]. It can be obtained by applying dual quaternion calculus (see e.g. [1]). It can be shown that a motion is rational if and only if its dual quaternion representation is a rational curve in dual quaternion space. Thus every rational motion of degree n can be written as follows

$$\mathrm{M} = \left(\begin{array}{c|ccc} \overline{v}_0 \, \|\mathbf{D}\|^2 & 0 & 0 & 0 \\ \hline v_1 & & & \\ v_2 & & \overline{v}_0 \, \mathrm{D} & \\ v_3 & & & \end{array} \right) \tag{5}$$

with $\mathbf{D} = (d_0, d_1, d_2, d_3)^T$ and

$$\mathrm{D} = \begin{pmatrix} d_0^2 + d_1^2 - d_2^2 - d_3^2 & 2(d_1 d_2 - d_0 d_3) & 2(d_1 d_3 + d_0 d_2) \\ 2(d_1 d_2 + d_0 d_3) & d_0^2 - d_1^2 + d_2^2 - d_3^2 & 2(d_2 d_3 - d_0 d_1) \\ 2(d_1 d_3 - d_0 d_2) & 2(d_2 d_3 + d_0 d_1) & d_0^2 - d_1^2 - d_2^2 + d_3^2 \end{pmatrix}. \tag{6}$$

Here $\overline{v}_0$, $\mathbf{v} = (v_1, v_2, v_3)^T$, and $\mathbf{D}$ are polynomials of degree $n - 2k$, n, and k, respectively, $0 \leq 2k \leq n$. The four parameters $d_0, \ldots, d_3$ collected in the real quaternion $\mathbf{D}$ are well known Euler's parameters. They describe the real part of the dual quaternion representation of M.

The design of rational motions is done most efficiently by computing polynomials $\overline{v}_0$, $\mathbf{v}$, $\mathbf{D}$ that match certain prescribed constraints. Once these polynomials have been found, the b-spline representation of M can be obtained by inserting them into (5). For a detailed description of the corresponding algorithms the reader is referred to [7].

3. Rational robot trajectrories for a special class of tasks

In the following we want to derive an algorithm for the automated design of rational motions that describe the Cartesian space trajectory of a robot's end effector. We assume that the motion is implicitly described by the task level considerations, such that the trajectory of the tool center point is a given curve $\underline{\mathbf{c}}(t)$ and the direction of the z-axis of the hand coordinate system is described by a vector field $\mathbf{n}(t)$. Furthermore we assume that $\mathbf{n}$ is normalized, i.e. $\|\mathbf{n}(t)\| = 1 \; \forall t$. In arc welding, for example, $\underline{\mathbf{c}}(t)$ would be the seam and $\mathbf{n}(t)$ a vector field related to the relative positions of the electrode with respect to the bead of the weld. In applications that deal with the scanning of surfaces such as in aircraft inspection on the other hand, $\underline{\mathbf{c}}(t)$ would be a known curve on the surface that has to be scanned and $\mathbf{n}(t)$ the corresponding unit normal vector to the surface

First we observe that $\underline{\mathbf{c}}(t)$ has to be a rational curve if we want to achieve a rational robot trajectory. For most applications addressed here, this is in general not the case. It is therefore necessary to approximate the tool center point path by a rational curve $\underline{\mathbf{c}}(t)$, for example by a cubic NURBS curve. Such approximation techniques are often used in CAD systems, which are mainly based on rational representations of curves and surfaces. We therefore just refer to the standard textbooks [5, 3] and assume that $\underline{\mathbf{c}}(t)$ is rational.

For a rational motion, the unit vector $\mathbf{z}(t)$ of the moving z-axis has to be a rational vector valued function, which describes a curve on the unit sphere S^3. Analogously to the tool center point path, the optimal direction $\mathbf{n}(t)$ of the end effector in general is not rational. Thus we have to approximate $\mathbf{n}(t)$ by a rational curve $\mathbf{z}(t)$ on S^3 and construct a motion, which follows the approximated direction $\mathbf{z}(t)$.

Methods for rational curve design on the sphere have been discussed in [2]. This approach is based on a so-called *generalized stereographic projection* which maps a rational curve $\mathbf{D_n} = (d_0, \mathbf{d})$ in threedimensional projective space onto a rational curve

$$\mathrm{N} = (n_0, n_0\mathbf{n})^T = \begin{pmatrix} d_0^2 + d_1^2 + d_2^2 + d_3^2 \\ 2(d_0 d_2 + d_1 d_3) \\ 2(d_2 d_3 - d_0 d_1) \\ d_0^2 - d_1^2 - d_2^2 + d_3^2 \end{pmatrix} \tag{7}$$

on the unit sphere. For details the reader is referred to [2]. Using the generalized stereographic projection, approximation is simple. We just have to compute a possible preimage $\mathbf{D_n}(t)$ of the optimal curve $\mathbf{n}(t)$ and approximate it with a rational curve $\mathbf{D}(t)$.

Finally $\mathbf{z}(t)$ can be obtained as the image of $\mathbf{D}$ under the generalized stereographic projection. Note that $\mathbf{D_n}(t)$ is not unique and thus the actual approximation depends on the choice of $\mathbf{D_n}(t)$. According to [2] the set of all possible curves $\mathbf{D_n}(t)$ forms a ruled surface. By solving (7) for the unkowns d_i and using Plücker's line coordinates (see e.g. [1]), this surface reads

$$\Phi = \left(n_x, n_y, n_z + \sqrt{n_x^2 + n_y^2 + n_z^2}, n_x, n_y, n_z - \sqrt{n_x^2 + n_y^2 + n_z^2}\right)^T, \tag{8}$$

$\mathbf{n} = (n_x, n_y, n_z)^T$. Since $\mathbf{n}(t)$ is normalized, (8) reduces to

$$\Phi = (n_x, n_y, n_z + 1, n_x, n_y, n_z - 1). \tag{9}$$

We can now obtain a very straightforward solution for $\mathbf{D_n}(t)$ by intersecting Φ with the plane $\mathbf{O}^* = (1, 0, 0, 0)^T$. This yields

$$\mathbf{D_n}(t) = (0, n_x, n_y, n_z + 1). \tag{10}$$

Furthermore, if we choose (10) as the preimage of $\mathbf{n}(t)$, we reduce the above approximation problem to a two-dimensional problem.

By comparing (7) to (6) we immetiately recognize that the curve $\mathbf{D}$ which has been used for the approximation of $\mathbf{n}$ can directly be interpreted as the kinematic image of the spherical part of M. Together with the tool center point path

$$(v_1, v_2, v_3)^T = (d_0^2 + d_1^2 + d_2^2 + d_3^2)\, \underline{\mathbf{c}} \tag{11}$$

equation (5) yields a rational robot trajectory which satisfies the task specific constraints. Summing up, we get the following algorithm

1. Approximate the tool center point path by a rational curve $\underline{\mathbf{c}}$.
2. Compute $\mathbf{D_n}(t)$ according to (10) and approximate it by a rational curve $\mathbf{D}$.
3. Compute the motion $\mathrm{M}(t)$ of the end effector according to (11) and (5).

As an example of the application of the above algorithm consider welding a lap joint. The orientation of the welding torch with respect to a Frenet–Serret coordinate frame of the seam $\underline{\mathbf{c}}$ is described by angles α and β as shown in Figure 1. We want to generate a continuous rational motion of the end–effector such that α and β have nearly fixed values for the entire trajectory. Thus we have to compute a motion M which interpolates $\underline{\mathbf{c}}$ and approximates the direction $\mathbf{n}$ of the welding torch. The resulting motion is shown in Figure 2.

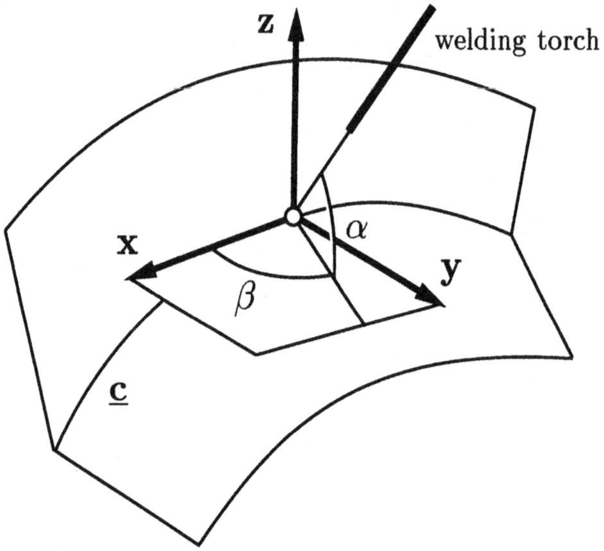

Figure 1. Welding torch orientation.

4. Additional constraints for a second axis

As mentioned in the previous section, the motion M which is obtained by applying the above algorithm is certainly not the only rational motion that satisfies the task specific constraints. The kinematic images of the spherical parts of all rational solutions describe a rational ruled surface in the kinematic image space, which reads

$$(z_x, z_y, z_z + 1, z_x, z_y, z_z - 1)^T, \quad \mathbf{z} = (z_x, z_y, z_z)^T, \quad \|\mathbf{z}(t)\| = 1 \ \forall t. \quad (12)$$

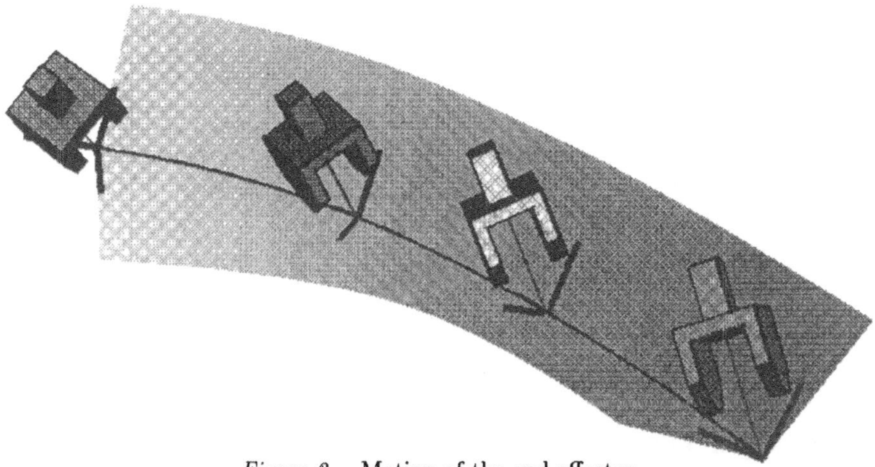

Figure 2. Motion of the end effector.

Obviously, any curve on (12) can be used for the construction of M. In particular one could try to introduce additional constraints for a second axis of the hand coordinate system. In the following we want to outline an algorithm which provides a robot trajactory such that the x-axis of the hand coordinate system stays horizontal during the whole motion.

According to (2) the direction of the x-axis at time t with respect to the fixed coordinate system is given by $\mathrm{D}(t) \cdot (1,0,0)^T$ with D from (6). In order to guarantee that the direction of the x-axis of the hand coordinate system is perpendicular to the z-axis of the fixed coordinate system, the motion has to satisfy

$$(0,0,1) \cdot \mathrm{D} \cdot (1,0,0)^T = 0 \ \forall t \tag{13}$$

which is quadratic in the coordinates of $\mathbf{D}$. The curve $\mathbf{D}(t)$ therefore has to be the intersection of the quadric (13) and the rational ruled surface (12) and can be obtained with a standard surface–surface intersection algorithm. Note that the resulting motion in general is only algebraic; however, it can be approximated by a rational motion either by approximating the intersection curve $\mathbf{D}(t)$ or by applying the technique used in [9] for planar algebraic motions.

More general, a robot trajectory satisfying that a given vector $\widehat{\mathbf{h}}$ in the hand coordinate system stays perpendicular to a given vector $\mathbf{f}$ measured in the fixed coordinate system throughout the whole motion can be obtained by intersecting (12) with the quadric

$$\mathbf{f}^T \cdot \mathrm{D} \cdot \widehat{\mathbf{h}} = 0 \tag{14}$$

instead of (13).

5. Conclusions

In this paper we have studied some applications of the theory of rational b-spline motions the design of robot trajectories and outlined the advantages of this approach. We have considered certain task specific constraints which are encountered, for instance, in arc welding, spray painting, or the scanning of surfaces with robot equipment. Finally, we presented a simple algorithm for the construction of rational robot trajectories which are implicitly defined by these task level considerations.

Acknowledgements

The first author gratefully acknowledges the financial support by the Austrian Science Foundation through FWF project J01126–MAT. This work was also supported by California Department of Transportation through the AHMCT research program.

References

1. Bottema, O. and Roth, B. (1979) *Theoretical Kinematics*, North–Holland Publishing Company, Amsterdam, New York, Oxford.
2. Dietz, R., Hoschek, J. and Jüttler, B. (1994) An algebraic approach to curves and surfaces on the sphere and on other quadrics, *Computer Aided Geometric Design*, **Vol. no. 10**, pp. 211–229.
3. Farin, G. (1992) *Curves and Surfaces for Computer Aided Geometric Design*, 3rd edition, Academic Press.
4. Ge, Q.J. and Ravani, B. (1994) Computer Aided Geometric Design of Motion Interpolants, *ASME Journal of Mechanical Design*, **Vol. no. 116**, pp. 756–762.
5. Hoschek, J. and Lasser, D. (1993) *Fundamentals of Computer Aided Geometric Design*, Peters, Wellesley.
6. Jüttler, B. (1993) Über zwangläufige rationale Bewegungsvorgänge, *Sitzungsberichte der Österreichischen Akademie der Wissenschaften*, **Vol. no. 202**, pp. 117–132.
7. Jüttler, B. and Wagner, M.G. (1996) Computer Aided Design with Spatial Rational B-spline Motions, *ASME Journal of Mechanical Design*, **Vol. no. 118**.
8. Paul, R.P. (1981) *Robot Manipulators. Mathematics, Programming, and Control.* MIT Press, p. 279.
9. Ravani, B. and Roth, B. (1983) Motion Synthesis Using Kinematic Mappings, *Journal of Mechanisms, Transmission, and Automation in Design*, **Vol. no. 105**, pp. 460–467.
10. West, T., Velinsky, S. and Ravani, B. (1995) Advanced Highway Maintainance and Construction Technology Applications, *Transportation Research News*, **Vol. no. 176**, pp. 17–23.

OPTIMUM DESIGN OF CLOSED KINEMATIC CHAINS IN THE PRESENCE OF OBSTACLES

A. DORIA
Department of Mechanical Engineering
University of Padova
Via Venezia 1, 35131 Padova, Italy

Abstract. The problem of optimum design of closed kinematic chains with one or more degrees of freedom and obstacles in the workspace is considered. An optimum design method with a penalty function embedding the collision control algorithm is developed. Several numerical examples which show the effectiveness of the proposed method are presented.

1. Introduction

Closed-loop mechanisms are often used in automatic machines and in robots to generate complex paths, to guide rigid bodies or to realise a functional relation between the motion of the motor-link and the follower-link. In many industrial application the workspace of automatic machines and robots is limited by the presence of obstacles, which may be other machines, fixtures, or parts of the same machine (auto collision). Therefore the designer often has to cope with the problem of dimensional synthesis of mechanisms in the presence of obstacles.

In the last years dimensional synthesis of mechanisms based on optimisation techniques has developed into one of the most powerful synthesis techniques. It is based on the automatic search of a set of design parameters which minimise a penalty function, which represents the error between the actual behaviour of the mechanism and the specified one. The minimisation is usually constrained, since the design parameters have to satisfy constraints in order to guarantee mechanism assembly, to prevent branching or to limit pressure angle. Some authors have embedded the constraints in the penalty function with successful results.

The optimum synthesis approach is well suited to solve the problem of the design of mechanisms in the presence of obstacles because it is powerful and the requirement of collision free motion can be included in the penalty function. The collision control algorithm included in the penalty function is run at every step of optimisation, therefore it has to be simple and precautionary to avoid a long time of calculation. The method for optimum synthesis of mechanisms in the presence of obstacles presented in this paper includes a collision control algorithm based on embedding of links and obstacles into circles or polygonals.

J. Lenarčič and V. Parenti-Castelli (eds.), Recent Advances in Robot Kinematics, 159-168.
© 1996 *Kluwer Academic Publishers. Printed in the Netherlands.*

The core of the optimisation procedure is that developed at Department of Mechanical Engineering of Padova University for optimum synthesis without obstacles (Cossalter *et al* (1992),(1993)); it allows the synthesis of mechanisms which can be decomposed into Assur groups and uses a non-derivative algorithm.

2. Collision detection

2.1 COLLISION DETECTION APPROACHES

Collision detection procedures have been studied in the last decades owing to the development of robots. They can be divided into collision detection procedures in the Cartesian space and collision detection procedures in the joint space. In the first approach the shapes of the links of the robot and of the obstacles are described in the workspace (the Cartesian space) and collisions are checked looking for the intersections between the volumes which represent the links and the volumes which represent the obstacles. Since the shape of links and obstacles may be rather complex, usually they are embedded into simpler convex volumes whose parameters are determined on the basis of the required accuracy (Mitsi and Bouzakis (1993)).

In the second approach first the obstacles are mapped into the joint space (it is the space where robot configuration is described by the vector of joint variables); then the collisions are detected by comparing the trajectory of the manipulator in the joint space with the obstacles described in the same space (Red *et al* (1987), Shiller and Gwo (1993)). This approach is very promising if new collision-free paths have to be planned quickly; but in the field of mechanism design, where the motion is planned at the design stage, does not show particular advantages over the method working in the Cartesian space. For this reason a collision detection algorithm in the Cartesian space is developed in the paper.

2.2 COLLISION DETECTION ALGORITHM

Since a plane problem is dealt with the links of the mechanisms are embedded into a series of circles, whereas the obstacles are embedded into circles or polygonals.

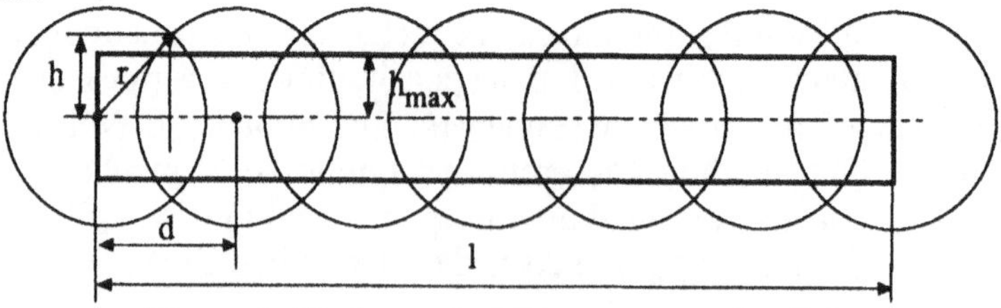

Figure 1. Circles embedding a rectangular link.

The parameters of the circles which enclose the links (figure 1) are determined in the following way. The axis of the link is considered and the maximum distance of the points of the actual boundary of the link from the axis is calculated (h_{max}). The minimum distance ($h>h_{max}$) of the embedding contour from the link axis and the maximum distance (the radius r of the circle) of the embedding contour from the link axis are then fixed. The parameters h and r allow the determination of the distance (d) of circles centres along the link axis:

$$d = 2h\sqrt{\left(\frac{r}{h}\right)^2 - 1} \qquad (1)$$

The number of circles which embed the link is then:

$$n = \frac{l}{d} + 1 \qquad (2)$$

where l is link length and n is rounded up to the closest integer number. As r/h decreases the accuracy of the approximation increases, but the number of circles increases. If the obstacle is embedded into a circle, the collision of a link with this obstacle is checked by calculating the distances s_i of the centres of the circles (that embed the link) from the centre of the circle which embeds the obstacle and comparing these distances with the sum of the radii. The collision criterion is:

$$\frac{s_i}{r+r_o} = \frac{\sqrt{(x_i - x_o)^2 + (y_i - y_o)^2}}{r+r_o} \leq 1 \qquad (3)$$

where x_i, y_i are the coordinates of the ith circle which embeds the link and x_o, y_o are the coordinates of the circle which embeds the obstacle, the radius of this circle is r_o. The reciprocal of ratio (3) gives an indication of collision gravity.
If the obstacle is represented by a polygonal, the interference of the circles which embed the link with each side of the polygonal is checked. The collision between the ith circle, whose equation is:

$$(x - x_i)^2 + (y - y_i)^2 = r^2 \qquad (4)$$

and a side of the polygonal which belongs to the line:

$$ax + by + c = 0 \qquad (5)$$

and is limited by points (x_1,y_1) and (x_2,y_2) is checked in the following way. First the distance s_i of the centre of the ith circle from the line is calculated and it is compared with the radius of the circle; if:

$$\frac{s_i}{r} = \frac{\dfrac{|ax_i + by_i + c|}{\sqrt{a^2 + b^2}}}{r} \leq 1 \qquad (6)$$

there are one or two intersections between the circle and the line. These intersections are actual collisions if they belong to the side of the polygonal. Therefore the coordinates of the intersection points are found by solving the system:

$$\begin{aligned} &(x - x_i)^2 + (y - y_i)^2 = r^2 \\ &ax + by + c = 0 \end{aligned} \qquad (7)$$

If the distance of an intersection point from both points (x_1,y_1) and (x_2,y_2) is less than the length of the side, the intersection is an actual collision. In this case the reciprocal of ratio (6) gives an indication of collision gravity.

When the obstacle is enclosed into a circle if condition (3) is not verified the link does not intersect the obstacle and it is outside the obstacle. On the other hand equations (6) and (7) allow to check the collisions of the link with the sides of the polygonal but do not allow to verify if the link is inside or outside the obstacle. The serial nature of open and closed kinematic chains simplifies the solution of this problem. A mechanism or a robot can move inside a closed area without collisions with the boundaries only if all the links and the frame are inside the closed area, otherwise at least one link collides with the boundary. Therefore, in order to avoid the automatic synthesis of a mechanism moving inside a polygonal obstacle, it is sufficient to verify the position of the fixed pivots. A fixed pivot is inside a polygonal if an half-line starting from the pivot has an odd number of intersections with the boundary.

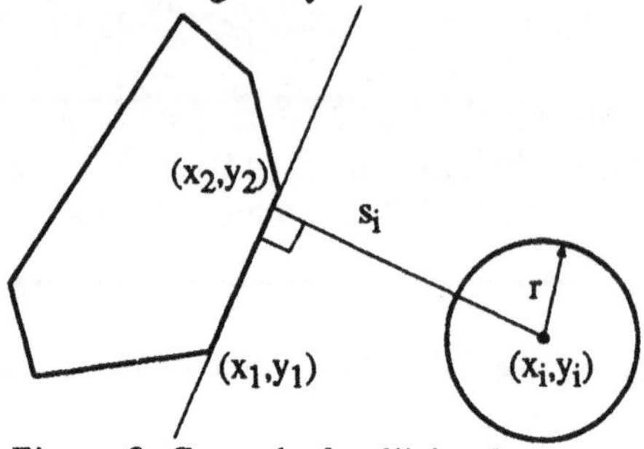

Figure 2. Control of collision between a circle and a polygonal obstacle.

3. Optimum synthesis

3.1 INPUT DATA

The input data are:

-The kind of planar mechanism and the Assur groups which compose it; the mechanism may have one or more closed chains and one or more degrees of freedom.

-The design parameters which are link lengths and angles; their maximum number is 50.

-The first guess mechanism.

-The function, path or rigid body motion that the mechanism has to generate, it is described with a maximum of 50 design points.

-The minimum distance (h) of the embedding contour from link axis and the radius (r) of the circles.

-The geometry of obstacles.

Different distances (h) and radii (r) can be specified for the links. In this way the link which is closer to the obstacles and provably will collide during the motion can be precisely approximated with a small r/h ratio and therefore with a large number of circles, whereas the other links which are farther from the obstacles can be approximated in a coarse way. The adopted minimisation method gives the design parameters a wide range of variability, the first guess mechanism needs not satisfy any particular requirement, it cannot be assembled over the full range of motion or it can collide with obstacles. Often a good solution can be reached after some attempts starting from different first guess mechanisms;

therefore the possibility of random generation of the first guess mechanism was included in the optimisation code.

3.2 PENALTY FUNCTION

The penalty function is the sum of four terms:

$$F = F_s + F_a + F_c + F_{fp} \tag{8}$$

The first term F_s is the structural term (Cossalter *et al* (1992)) and takes into account the error between the desired motion and the motion obtained with the current vector of design parameters; it is defined as follows:

$$F_s = w_x^2 f_x + w_y^2 f_y + w_\theta^2 f_\theta \tag{9}$$

f_x f_y and f_θ are the sums of the squares of the errors in the design points on x, y and on angle θ respectively; weights w_x, w_y, w_θ allow the desired type of synthesis to be performed:

if $w_x = w_y = 0$ and $w_\theta \neq 0$, function generation is obtained, in this case θ is the follower-link rotation correlated to motor-link rotation;

if $w_x \neq 0, w_y \neq 0$ and $w_\theta = 0$, path generation is obtained;

if $w_x \neq 0, w_y \neq 0$ and $w_\theta \neq 0$, rigid body guide is obtained, and in this case θ is rigid body rotation.

The second term F_a is related to the non-assembly condition (Cossalter *et al* (1992)) and it is added only if the mechanism cannot be assembled in all positions; it is defined as follows:

$$F_a = F_s^* + C_1^{\exp} \tag{10}$$

where F^*_s is the structural penalty function calculated in the previous iteration, C_1 and *exp* are two constants.

The third term F_c is related to collisions and is added only if collisions are found; it is defined as follows:

$$F_c = \left(\sum_{j=1}^{nc} r_j / s_j \right) C_2 + F_s \tag{11}$$

where r_j/s_j is the gravity of each collision between the circles embedding the links and the obstacles and the summatory extends over the number of collisions (nc). Constant C_2, which multiplies the summatory, and the structural penalty function F_s, which is added to the summatory, are used to increase this term in order to shift the search of optimum design parameters far from collision conditions.

The last term F_{fp} *is* added only if the fixed pivots are inside the obstacle and is:

$$F_{fp} = C_3 F_s \tag{12}$$

where C_3 is a constant.

Minimisation of penalty function is carried out by using the BCPOL subroutine of the IMSL library (1987), based on a direct search Complex algorithm; no derivative information is taken into account.

4. Numerical examples

4.1 EXAMPLE 1

A four-bar linkage is synthesised with a point belonging to the coupler plane able to generate for a crank rotation of 40° a rectilinear path (from x=2.93 to x=7.08 units). This path is divided into 19 equidistant design points. The eight design parameters, which are named *hi*, are represented in figure 3. The distance between the fixed pivots is constant and equal to 10 units. First the synthesis was carried out without obstacles in the workspace and the mechanism of figure 3 was obtained.

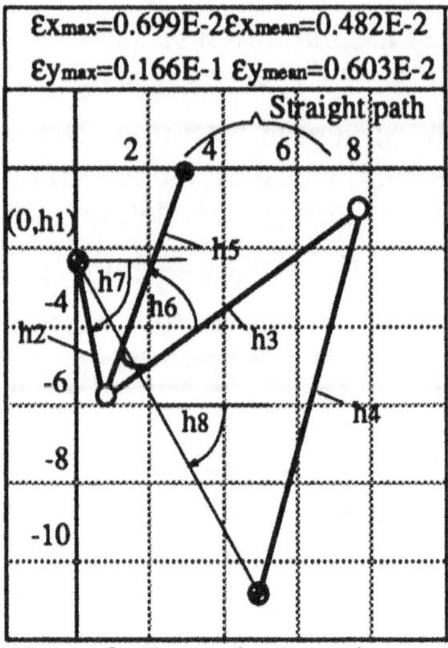

Parameters	Range of design parameters		First guess	Optimum
h1	-10	10.0	-2.50	-2.23
h2	1.00	20.0	3.50	3.65
h3	1.00	20.0	9.00	8.58
h4	1.00	20.0	10.0	10.4
h5	1.00	20.0	6.50	6.22
h6	-3.1	+3.1	1.00	0.59
h7	-3.1	+3.1	-1.35	-1.38
h8	-3.1	+3.1	-1.00	-1.05

Figure 3. Example 1; optimum synthesis of a four-bar linkage without obstacles in the workspace (only the skeleton of the mechanism is shown).

The tables of figure 3 show the range of design parameters, the parameters of the guess mechanism and the parameters of the optimum mechanism. ε_{xmax} and ε_{ymax} are the maximum errors between the desired path and the path generated by the optimised mechanism in direction x and y respectively ε_{xmean} and ε_{ymean} are the mean errors between the desired path and the path generated by the optimised mechanism in direction x and y respectively (the signs of errors are not taken into account).

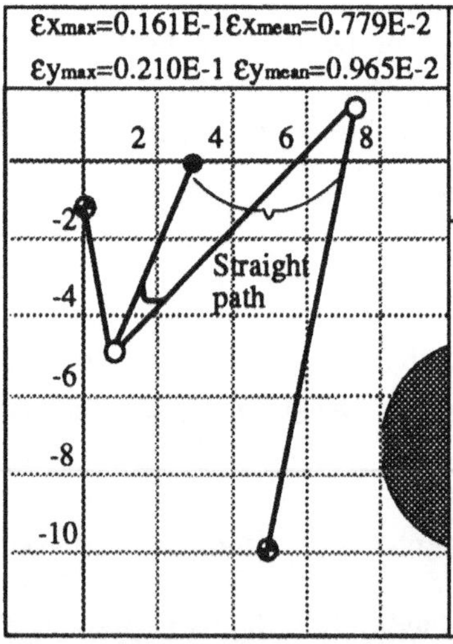

Parameters	Range of design parameters		First guess	Optimum
h1	-3.0	-1.0	-1.23	-1.20
h2	1.82	5.48	2.15	3.88
h3	4.29	12.9	6.02	9.45
h4	5.20	15.6	7.60	11.8
h5	3.11	9.33	7.00	5.50
h6	-3.1	+3.1	2.63	0.36
h7	-3.1	+3.1	1.90	-1.40
h8	-1.1	+0.9	-0.90	-1.05

Figure 4. Example 1; optimum synthesis of a four-bar linkage with a circular obstacle in the workspace.

Then a circular obstacle with the centre in the point (10.5,-7.5) and a radius of 2.5 units was introduced in the workspace. The axis of the driven link of the mechanism of figure 3 collides with this obstacle, therefore a new optimisation was carried out taking into account the obstacle. All the links were embedded in circles, the minimum distance of the embedding contour from the link axis was 0.2 units, whereas the radii of the circles were 0.26 units.

Several attempts were carried out both with a first guess mechanism equal to the optimum mechanism of figure 3 and with random first guess mechanisms. The best results were obtained in the second case.

Figure 4 shows the optimum mechanism with the optimisation parameters and the errors. The fixed pivots of the optimum mechanism which avoids collisions are shifted in direction y, the link lengths and the shape of the coupler are different from those of the mechanism synthesised without obstacle. The maximum and mean errors of the mechanism of figure 4 are about twice the errors of the mechanism of figure 3.

4.2 EXAMPLE 2

In example 1 the motor link has an oscillating motion and the return stroke is symmetric to the advance stroke. In many applications the motor-link is driven by an electric motor and has a continuous rotary motion, therefore the return stroke is different from the advance stroke and the trace point line is a closed curve. In example 2 the four-bar linkage has to generate the same path of example 1 in the advance stroke (40°), but it has a different return stroke. The only constraint on the return stroke is that a collision free motion is required. The obstacle and the circles which embed all the links are the same of example 1. Optimisation was carried out starting from random generated first guess mechanisms.

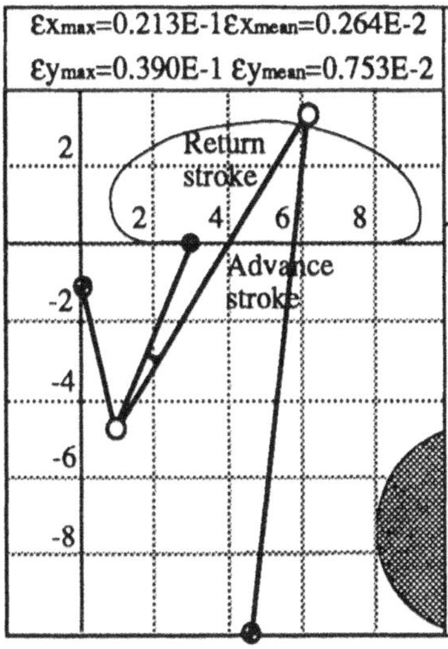

Parameters	Range of design parameters		First guess	Optimum
h1	-3.0	0.00	-1.14	-1.06
h2	1.80	5.40	4.10	3.85
h3	4.25	12.4	9.66	9.76
h4	5.50	16.0	7.66	13.5
h5	3.15	9.45	5.63	5.17
h6	-3.1	+3.1	1.36	0.16
h7	-3.1	+3.1	-0.47	-1.34
h8	-1.5	-0.5	-1.29	-1.09

Figure 5. Example 2; optimum synthesis of a four-bar linkage with a circular obstacle in the workspace, different advance and return strokes are considered.

Figure 5 shows the optimum mechanism with the optimisation parameters; the errors are similar to those of example 1; but there is a larger difference between mean and maximum errors.

4.3 EXAMPLE 3

The third example deals with optimum design of a two degree of freedom five-bar linkage. The axis of the central pivot of the mechanism, which can be the axis of the wrist of an end-effector, has to generate an elliptical path for a complete rotation of the two motor links. The centre of the ellipse is in the point (15,15), the major axis is 8 units long, the minor axis is 5 units long; 50 design points are considered. The fixed pivots of the mechanism have to belong to a square area with a side of 30 units.

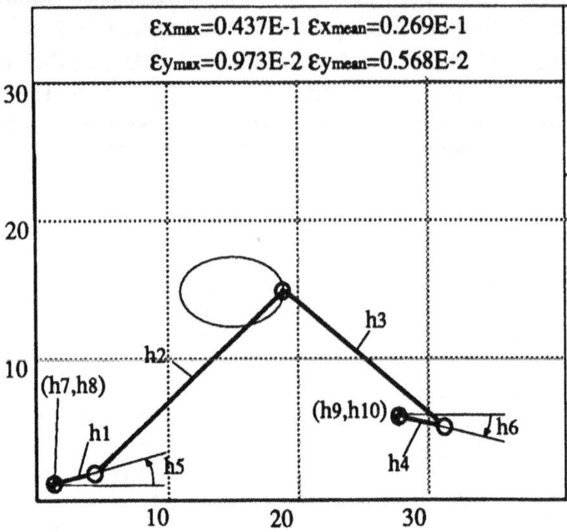

Parameters	Range of design parameters		First guess	Optimum
h1	1.00	6.00	1.13	3.34
h2	10.0	25.0	18.8	19.7
h3	10.0	30.0	20.5	15.4
h4	1.00	6.00	1.03	3.55
h5	-3.1	+3.1	-1.8	0.23
h6	-3.1	+3.1	0.15	-0.2
h7	0.00	30.0	29.5	1.05
h8	0.00	30.0	19.9	1.16
h9	0.00	30.0	8.34	27.5
h10	0.00	30.0	23.8	5.98

Figure 6. Example 3; optimum synthesis of a five-bar linkage without obstacles in the workspace.

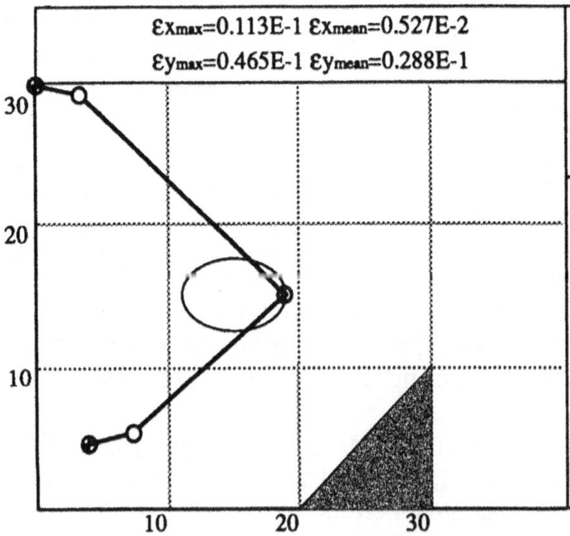

Parameters	Range of design parameters		First guess	Optimum
h1	1.00	6.00	4.71	3.33
h2	10.0	25.0	11.0	21.2
h3	10.0	30.0	15.2	15.2
h4	1.00	6.00	1.09	3.38
h5	-3.1	+3.1	-0.45	-0.23
h6	-3.1	+3.1	-2.13	0.22
h7	0.00	30.0	24.0	0.11
h8	0.00	30.0	28.3	30.0
h9	0.00	30.0	14.3	3.96
h10	0.00	30.0	19.8	4.64

Figure 7. Example 3; optimum synthesis of a five-bar linkage with a triangular obstacle in the workspace.

First the synthesis was carried out without obstacles in the workspace. Figure 6 shows design parameters, the optimised mechanism and the errors.

The triangular obstacle shown in figure 7 was then introduced in the workspace, since it is located in the lower left corner of the workspace it prevents optimum solutions like that found without obstacles in the workspace. A new synthesis was carried out starting from random generated first guess mechanisms; all the links of the mechanism were embedded into circles, the parameters *h* and *r* were 0.2 and 0.26 units respectively. The optimum mechanism and its errors are reported in figure 7. The link lengths of this mechanism are similar to those of the mechanism optimised without obstacles in the workspace, whereas the coordinates of the fixed pivots are very different. The errors of the mechanisms optimised with and without obstacles in the workspace are comparable; the mechanism of figure 6 shows smaller errors in direction y, whereas the mechanism of figure 7 shows smaller errors in direction x.

4.4 EXAMPLE 4

The last example deals with the same problem of example 3, but two obstacles, a circle and a polygonal, were introduced in the workspace. Figure 8 shows the optimum mechanism, optimisation parameters and the errors. The optimum mechanism is rather different from those of example 3, but its errors are less than those of the five-bar linkage synthesised without obstacles in the workspace.

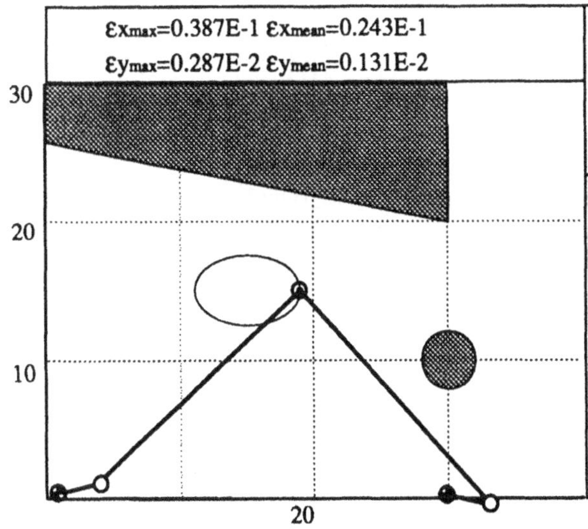

Parameters	Range of design parameters		First guess	Optimum
h1	1.00	6.00	3.50	3.33
h2	10.0	25.0	21.5	20.6
h3	10.0	30.0	24.6	21.0
h4	1.00	6.00	2.39	3.35
h5	-3.1	+3.1	-0.71	0.23
h6	-3.1	+3.1	1.37	-0.22
h7	0.00	30.0	27.8	0.54
h8	0.00	30.0	16.5	0.42
h9	0.00	30.0	13.5	30.0
h10	0.00	30.0	6.91	0.32

Figure 8. Example 4; optimum synthesis of a five-bar linkage with two obstacles in the workspace.

5. Conclusions

Numerical examples show that the proposed method allows optimum design of plane mechanism with one or two degrees of freedom in the presence of circular or polygonal obstacles.

The errors of the paths of the mechanisms optimised in the presence of obstacles are like those of the mechanisms optimised without obstacles in the workspace.

Since the links of the mechanisms can move in parallel planes auto collisions were not considered, they could be easy included in a further development of the method.

References

V. Cossalter, A. Doria, M. Pasini, C. Scattolo, A simple numerical approach for optimum synthesis of a class of planar mechanisms, *Mech. Mach. Theory, Vol 27, No. 3, pp.357-366, 1992.*

V. Cossalter, M. Da Lio, A. Doria, Optimum synthesis of spatial function generator mechanisms, *Meccanica 28, pp 263-268, 1993.*

S. Mitsi, K.D. Bouzakis, Simulation of redundant manipulator for collision avoidance in manufacturing and assembly environments, *Mech. Mach. Theory, Vol 28, No. 1, pp. 13-21, 1993.*

W.E. Red, H.V. Troung-Cao, K.H. Kim, Robot Path Planning in Three-Dimensions Using the Direct Subspace, *Journal of Dynamic systems, Measurement, and Control, Vol. 109, pp 238-244, 1987.*

Z. Shiller, Y.R. Gwo, Collision-Free Path Planning of Articulated Manipulators, *Journal of Mechanical Design, Journal of Mechanical Design, Vol. 115, pp 901-908, 1993.*

IMSL Manual, Math/Library, Vol.8, Version 1.0, *IMSL Inc., Houston, 1987.*

Acknowledgements

The author is grateful to Professor V. Cossalter for his advice and encouragement during the preparation of the paper and acknowledges the useful work done by the graduating student A. Camelos, with his thesis.

The author expresses his appreciation to the Italian Department of University and Scientific Research for the financial support provided (40% grant).

SYNTHESIS OF DISCRETELY ACTUATED MANIPULATOR WORKSPACES VIA HARMONIC ANALYSIS

GREGORY S. CHIRIKJIAN
Dept. of Mechanical Engineering
Johns Hopkins University
Baltimore, Maryland, 21218 USA

Abstract.

Discrete actuators have a finite number of states, e.g. stepper motors and bistable pneumatic cylinders. Given that the number of distinct configurations attainable by discretely actuated manipulators grows exponentially in the number of actuated degrees of freedom, i.e. K^n for n actuators or actuated modules each with K states, brute force representation of discrete manipulator workspaces is not feasible in the highly actuated case. Approximating the workspaces of segments of discrete manipulators as density functions on the Euclidean group (which describes workspace positions and orientations) the whole workspace can be approximated as an n-fold convolution of these functions. A generalization of the Fourier transform (and the convolution theorem) is used in this paper to efficiently compute convolutions and to solve the inverse problem of synthesizing discrete manipulators which have prescribed workspace properties.

1. Introduction

Mechanisms and robotic manipulators with discrete actuators can be used reliably without feedback, and thus have the potential to be much cheaper than manipulators with standard continuous actuation. In fact, the "binary" manipulator constructed from 15 pneumatic cylinders shown in Figure 1 was built together with a simple computer interface for approximately \$ 1000 in parts and materials - an order of magnitude cheaper than if continuous actuation had been used. Each actuator is driven to its stops, and thus has two states (one bit). For a detailed look at the kinematics of this manipulator, see [1].

J. Lenarčič and V. Parenti-Castelli (eds.), Recent Advances in Robot Kinematics, 169-178.
© 1996 *Kluwer Academic Publishers. Printed in the Netherlands.*

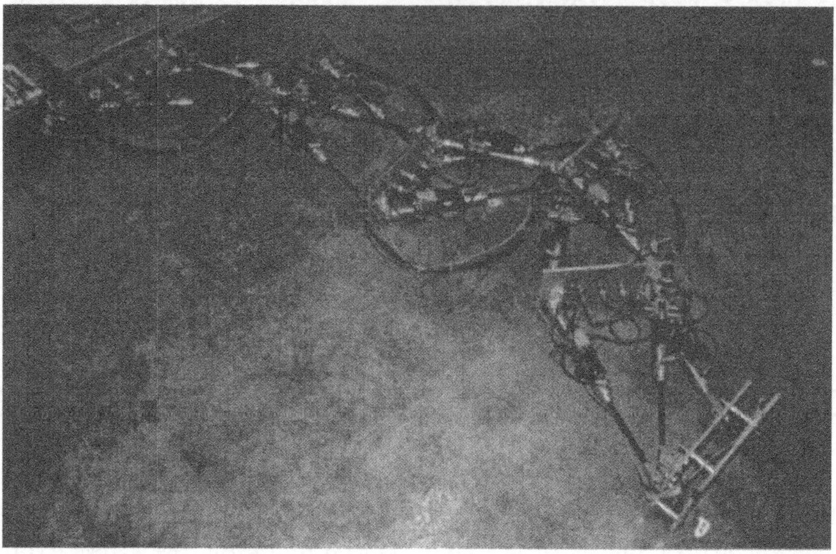

Figure 1: Hardware Implementation of a 15-bit Variable-Geometry-Truss

The bottleneck in the use of binary manipulators has been the planning of relatively smooth motion, and overcoming the problems associated with the combinatorially explosive number of configurations that result from numerous actuators. In fact, this is a major reason why open-loop discrete actuation is not more widely used in robotics. The mathematical methods used in this paper for dealing with the kinematics of these devices is one step toward realizing the cost benefits of discrete actuation.

The determination of manipulator workspaces (the set of positions *and* orientations attainable by the end-effector) is a central factor in the design of any robotic manipulator. In the context of standard continuous range-of-motion manipulators, dexterity measures (scalar functions defined over the workspace) are also important [8]. In the context of manipulators with open-loop discrete actuation such as stepper motors or bistable pneumatic cylinders, the *density* of reachable frames (number of frames per unit workspace volume) replaces dexterity measures as a scalar function of primary importance defined over the workspace. For example, workspace density plays a very important role in inverse kinematics [4] of discretely actuated manipulators. Computing discretely actuated manipulator workspace densities using brute force requires K^n evaluations of the forward kinematic equations, where K is the resolution of each of the n actuators or actuated modules. This makes the synthesis of discrete manipulators with a prescribed distribution of reachable frames intractable using brute force when n is large.

This paper discusses methods based on the concept of convolution which can approximate the same densities in computational time proportional to n. To this end, the concept of a convolution product of two scalar-valued functions on subsets of the Special Euclidean Group (which describes motion in Euclidean space) is defined and used to generate manipulator workspaces and workspace densities of discretely actuated manipulators.

In the following sections we review mathematical concepts dealing with convolution of functions and harmonic analysis on groups. This provides the definitions and tools needed to use the concept of convolution applicable to the current problem.

2. Mathematical Background

A review of terminology, notation, and significant results pertaining to the current problem are presented in this section.

Definition: A *group* is a set G together with an operation $\circ$ which takes two elements of the set and returns an element of the set. All elements $g, g_1, g_2, g_3 \in G$ must satisfy the following properties:

- $g_1 \circ (g_2 \circ g_3) = (g_1 \circ g_2) \circ g_3$.
- There exists an element $e \in G$ such that $e \circ g = g \circ e = g$.
- For every element $g \in G$ there is an element $g^{-1} \in G$ such that $g^{-1} \circ g = g \circ g^{-1} = e$.

In order to distinguish a group from the set, the former is denoted $(G, \circ)$, unless the operation is understood from the context, in which case G refers to both the set and the group.

Definition: The Special Euclidean Group, $SE(N)$, (also called Euclidean Motion group) is the group which describes rigid motions (rotation and translation) in N dimensional Euclidean space, [6].

We can represent any element of $SE(N)$ with an $(N+1) \times (N+1)$ homogeneous transform matrix of the form

$$g = \begin{pmatrix} R & \overline{b} \\ \overline{0}^T & 1 \end{pmatrix},$$

where $\overline{b} \in \mathbb{R}^N$, and $R \in SO(N)$, i.e. R is an $N \times N$ special orthogonal matrix ($\det(R) = 1$ and $RR^T = I$, where I is the identity matrix). The group operation corresponding to this representation is matrix multiplication of two homogeneous transform matrices.

Given two real-valued functions with group-valued argument, one defines the convolution of these functions as:

$$(f_1 * f_2)(g_x) = \int_G f_1(g_\xi) f_2(g_\xi^{-1} \circ g_x) d\mu(g_\xi), \tag{1}$$

where the integral is taken over the group (which in this case is $G = SE(N)$) and $d\mu(\cdot)$ is the volume element for $SE(N)$.

Just as in the case of convolution of functions on the real line, the physical meaning in the generalizations above is that the function $f_2(\cdot)$ is shifted, multiplied by $f_1(\cdot)$, and the product is integrated. In the case when $G = SE(N)$ this "shift" is a translation and rotation.

The concept of convolution of functions on $SE(N)$ is useful for discretely actuated manipulator workspace generation because each function $f_i(g)$ can be taken to be a continuous density function which approximates the finite but large set of discrete frames generated by all configurations of the ith module in a discretely actuated manipulator. The convolution of the density functions of adjacent modules results in an approximation of the density of frames reachable by the structure composed of both modules. Furthermore, for a manipulator composed of n discretely actuated modules, the workspace is given by the convolution of each module density function:

$$f_W(g) = (f_1 * \ldots * f_n)(g).$$

It must be emphasized that this approach becomes more accurate as the number of discrete states increases. For example, the manipulator in Figure 1 can be viewed as a cascade of five modules each with three binary actuators. Because each module only has eight reachable frames, fitting a continuous density function to the frames reachable by each module will not yield accurate results. Furthermore, since there are relatively few (2^{15}) states in this case, the workspace can be generated using brute force. However, if each actuator in this same kinematic structure had three bits, then the whole structure would have 2^{45} states - and it would not be possible to generate the workspace using brute force. In this case, the continuous density function associated with each module would be constructed based on 2^9 points, and a continuous approximation will be relatively accurate. Thus, the approach of this paper is an asymptotic approach which becomes better with more states - which is precisely when brute force computation becomes intractable.

In the two dimensional case, the homogeneous transforms g_x and g_ξ can be parametrized as $g_x = g(x_1, x_2, \theta)$ and $g_\xi = g(\xi_1, \xi_2, \alpha)$ where

$$g(a,b,c) = \begin{pmatrix} \cos c & -\sin c & a \\ \sin c & \cos c & b \\ 0 & 0 & 1 \end{pmatrix}$$

and a volume element with which to integrate is:

$$d\mu(g_\xi) = C d\xi_1 d\xi_2 d\alpha.$$

The constant C is chosen such that

$$\int_{SE(N)} f_i(g)d\mu(g) = K_i,$$

where K_i is the number of frames the top of module i can reach with respect to its base. By appropriately scaling $f_i(g)$, we choose $C = 1$ without loss of generality. This means that the convolution product of functions on $SE(2)$ can be expressed in the form:

$$(f_1*f_2)(x_1, x_2, \theta) = \int_{-\infty}^{\infty}\int_{-\infty}^{\infty}\int_{-\pi}^{\pi} f_1(\xi_1, \xi_2, \alpha)\times f_2\left((x_1 - \xi_1)c\alpha + (x_2 - \xi_2)s\alpha,\right.$$

$$\left.-(x_1 - \xi_1)s\alpha + (x_2 - \xi_2)c\alpha, (\theta - \alpha)\mathrm{mod}\ 2\pi\right) \quad d\xi_1 d\xi_2 d\alpha,$$

where $c\alpha = \cos\alpha$ and $s\alpha = \sin\alpha$. In the current context we use mod 2π to indicate that 2π is subtracted from (or added to) the argument as many times as is required for the result to be in the range $[-\pi, \pi]$.

In the above equation, the notational abbreviation $f(g(x_1, x_2, \theta)) = f(x_1, x_2, \theta)$ has been made since the meaning is clear from the context.

3. Unitary Representations and the Fourier Transform

Recall that one of the most powerful properties of the Fourier transform of functions on the real line is that the Fourier transform of the convolution of two functions is the product of the Fourier transforms of the functions. This useful property extends to the concept of a Fourier transform for functions on $SE(N)$.

In the case of the Fourier transform on the real line, the function $u(x, a) = e^{iax}$ plays a very important role. An analogous function is needed for the group $SE(N)$ so that the Fourier transform pair can be defined for this case. However, since $SE(N)$ is not commutative, a scalar function $u(g, a)$ is not sufficient to reflect the noncommutative nature of $SE(N)$. Instead of a unitary scalar function, we must look at a unitary matrix $U(g, a)$ with elements $u_{mn}(g, a)$. A number of works including [9] and [7] have shown that an appropriate unitary matrix $U(g, a)$ such that

$$U(g_1 \circ g_2, a) = U(g_1, a)U(g_2, a)$$

for all $g_1, g_2 \in SE(N)$ can be defined in terms of the standard functions of mathematical physics. A matrix with this property is called a *representation* of the group. In particular, for the case of $SE(2)$ the elements of the infinite dimensional matrix $U(g, a)$ are

$$u_{mn}(g(r, \phi, \theta), a) = i^{n-m}e^{-i[n\theta+(m-n)\phi]}J_{n-m}(ar) \tag{2}$$

where the group elements are expressed in polar coordinates:

$$g = g(r, \phi, \theta) = \begin{pmatrix} \cos\theta & -\sin\theta & r\cos\phi \\ \sin\theta & \cos\theta & r\sin\phi \\ 0 & 0 & 1 \end{pmatrix}$$

and $J_\nu(x)$ is the νth order Bessel function.

Given this background we are ready for the following definition:

Definition [9],[7] The Fourier transform of a rapidly decreasing square integrable function [1] f, and the inverse transform are defined as:

$$\mathcal{F}(f) = \hat{f}(a) = \int_{SE(2)} f(g) U(g^{-1}, a) d\mu(g)$$

and

$$f(g) = \mathcal{F}^{-1}(\hat{f}) = \int_0^\infty \text{trace}(\hat{f}(a) U(g, a)) a da.$$

Recall that the trace of a matrix $A \in I\!R^{n \times n}$ is the sum of the diagonal elements: $\text{trace}(A) = \sum_{i=1}^n a_{ii}$, and the trace is invariant under transformations of the form QAQ^T where $Q \in SO(n)$. That is, $\text{trace}(A) = \text{trace}(QAQ^T)$. Furthermore $\text{trace}(AB) = \text{trace}(BA)$.

The matrix elements of the transform can be calculated using the matrix elements of $U(g, a)$ defined in Equation 2 as:

$$\hat{f}_{mn}(a) = \int_{SE(2)} f(g) u_{mn}(g^{-1}, a) d\mu(g).$$

Likewise, the inverse transform can be written in terms of elements as:

$$f(g) = \int_0^\infty \hat{f}_{mn}(a) u_{nm}(g, a) a da$$

where summation notation is used for $n, m \in [-\infty, \infty]$.

One of the most powerful properties of the Fourier transform of functions on $I\!R^N$ is that the Fourier transform of the convolution of two functions is the product of the Fourier transforms of the functions. This property extends to the concept of a Fourier transform for functions on $SE(2)$. Namely, we have that for functions $f_1(g)$ and $f_2(g)$ which decay rapidly to zero outside of a compact subset of $SE(2)$ that

$$\mathcal{F}(f_1 * f_2) = \mathcal{F}(f_2)\mathcal{F}(f_1),$$

[1] $f(\cdot)$ is rapidly decreasing if $\lim_{r\to\infty} r^n f(g(r, \phi, \alpha)) = 0$ for all $n \in \mathbf{Z}^+$. Examples include functions $f(\cdot)$ which is zero outside of a compact subset of $SE(2)$ (i.e., if $f(\cdot)$ has compact support), and $f(g) = e^{-a^2 r^2} F(\phi, \alpha)$ where $f(\cdot)$ is bounded.

where the convolution is defined as in Equation 1. The order of the product of these Fourier transform matrices is important since $SE(2)$ is not commutative.

In the next section, we will use this property to localize the workspace synthesis problem for discretely actuated manipulators to the design of a single module.

4. Applications to Discretely Actuated Manipulators

Given the developments of the previous sections, the discretely actuated manipulator workspace problem is formulated as follows.

It is assumed that a manipulator is composed of n identical modules which are cascaded to form a "macroscopically serial" structure. Each of the modules has K states and the distribution of frames reachable by one module is approximated using a continuous density function $f(g)$ such that

$$\int_{SE(N)} f(g) d\mu(g) = K.$$

The whole workspace will then have a density function $f_W(g)$ which is the n-fold convolution of the function $f(g) > 0$. That is,

$$f_W(g) = (\underbrace{f * \ldots * f}_{n\ times})(g).$$

Furthermore,

$$\int_{SE(N)} f_W(g) d\mu(g) = K^n.$$

We seek to solve the following problem: Given $f_W(g)$ (the desired density of frames in the workspace) find the module density function $f(g)$ (and thus localize the problem of module design) which achieves this goal as closely as possible. Taking the Fourier transform of both sides one finds

$$\hat{f}_W(a) = [\hat{f}(a)]^n \quad \text{or} \quad f_W(g) = \mathcal{F}^{-1}([\hat{f}(a)]^n).$$

Likewise, the inverse problem becomes

$$\hat{f}(a) = [\hat{f}_W(a)]^{\frac{1}{n}} \quad \text{or} \quad f(g) = \mathcal{F}^{-1}([\hat{f}_W(a)]^{\frac{1}{n}}).$$

While in general this involves numerical computations and the selection of a root, we focus on a closed form example which is examined here to illustrate the procedure.

Consider the case when we are given a desired workspace density function

$$f_W(g) = F_1(r)(C + 2A\cos\alpha) + 2BF_2(r)\cos\phi \tag{3}$$

where $A, B, C \geq 0$ and $C^2 \geq 4(A^2 + B^2)$ are given real numbers which ensure that $f_W(g) > 0$ for all $g \in SE(2)$. This allows us to consider workspaces where the density is highest when the end-effector orientation angle is closest to zero, and tapers to a lower value for high orientation angle. It also allows us to consider workspaces which vary with the angle ϕ in the plane. Therefore, while this is a restrictive example (in order that closed form solutions can be generated), it illustrates some of the properties of the more general solution, and serves as a benchmark against which future numerical codes can be compared.

The function $f_W(g)$ can be rewritten in the complex form:

$$f_W(g) = F_1(r)\left(C + A(e^{i\alpha} + e^{-i\alpha})\right) + BF_2(r)(e^{i\phi} + e^{-i\phi}).$$

The only nonzero part of the Fourier transform matrix is then the block

$$\hat{f}_W(a) = \int_{SE(2)} f_W(g)U(g^{-1}, a)d\mu(g) = \begin{pmatrix} \gamma(a) & 0 & 0 \\ \beta(a) & \kappa(a) & \beta(a) \\ 0 & 0 & \gamma(a) \end{pmatrix} \quad (4)$$

where

$$\gamma(a) = A\int_0^\infty F_1(r)J_0(ar)rdr \equiv A\epsilon_1(a),$$

$$\kappa(a) = C\int_0^\infty F_1(r)J_0(ar)rdr \equiv C\epsilon_1(a),$$

and

$$\beta(a) = -iB\int_0^\infty F_2(r)J_1(ar)rdr \equiv -iB\epsilon_2(a).$$

Orthogonality causes all other elements in the Fourier transform matrix to be zero. More generally, if $f_W(g)$ is any bandlimited function on $SE(2)$, all but a finite number of elements of $\hat{f}_W(a)$ will be zero.

For rapidly decreasing functions F_1 and F_2 these integrals are approximated efficiently, particular when these functions have compact support. In the following subsections, the Fourier transform matrix in Equation 4 will be raised to various powers, and the inverse transform will be computed by multiplying the unitary matrix representation of $SE(2)$, taking the trace of this product, and integrating the result.

The part of the unitary matrix $U(g, a)$ that will multiply the nonzero block of the Fourier transform matrix and contribute to the inverse transform is

$$U(g, a) = \begin{pmatrix} u_{-1,-1}(g, a) & u_{-1,0}(g, a) & u_{-1,1}(g, a) \\ u_{0,-1}(g, a) & u_{0,0}(g, a) & u_{0,1}(g, a) \\ u_{1,-1}(g, a) & u_{1,0}(g, a) & u_{1,1}(g, a) \end{pmatrix}.$$

All the powers of the matrix $\hat{f}_W$ that we will deal with have nonzero blocks of the form:

$$\hat{f}_W^k(a) = \begin{pmatrix} x(a) & 0 & 0 \\ z(a) & y(a) & z(a) \\ 0 & 0 & x(a) \end{pmatrix}.$$

In particular, $x(a)$, $y(a)$, and $z(a)$ are defined below for the case when $k = \frac{1}{2^m}$ for some positive integer m. One then calculates

$$\mathrm{tr}(\hat{f}_W^k(a)U_a(g)) =$$

$$(u_{-1,-1}(g,a) + u_{1,1}(g,a))x(a) + y(a)u_{0,0}(g,a)(u_{-1,0}(g,a) + u_{1,0}(g,a))z(a).$$

From the definition of the elements of the unitary matrix $U(g,a)$ given in Equation 2, we get

$$\mathrm{tr}(\hat{f}_W^k(a)U_a(g)) = 2\cos\alpha J_0(ar)x(a) + J_0(ar)y(a) + 2i\cos\phi J_1(ar)z(a).$$

We will now illustrate the workspace synthesis problem. That is, given $f_W(g)$, find $f(g)$ such that

$$\underbrace{(f * ... * f)}_{2^m\ times}(g) = f_W(g).$$

This is solved by taking the Fourier transform of both sides, and observing that for fractional powers of this matrix with $k = 1/n$ where $n = 2^m$ we get:

$$\begin{pmatrix} \gamma(a) & 0 & 0 \\ \beta & \kappa(a) & \beta(a) \\ 0 & 0 & \gamma(a) \end{pmatrix}^{\frac{1}{2^m}} =$$

$$\begin{pmatrix} \gamma^{1/2^m}(a) & 0 & 0 \\ \frac{\beta(a)}{\epsilon_1^{(1-\frac{1}{2^m})}(a)\prod_{j=1}^m (A^{1/2^j}+C^{1/2^j})} & \kappa^{1/2^m}(a) & \frac{\beta(a)}{\epsilon_1^{(1-\frac{1}{2^m})}(a)\prod_{j=1}^m (A^{1/2^j}+C^{1/2^j})} \\ 0 & 0 & \gamma^{1/2^m}(a) \end{pmatrix}.$$

Therefore,

$$f(g) = \int_0^\infty \mathrm{tr}([\hat{f}_W(a)]^{\frac{1}{2^m}} U_a(g))ada$$

$$= (2A^{\frac{1}{2^m}}\cos\alpha + C^{\frac{1}{2^m}}) \int_0^\infty J_0(ar)\epsilon_1^{\frac{1}{2^m}}(a)ada +$$

$$\frac{2B\cos\phi}{\prod_{j=1}^{m}(A^{1/2^j}+C^{1/2^j})}\int_0^\infty J_1(ar)\frac{\epsilon_2(a)}{\epsilon_1^{(1-\frac{1}{2^m})}(a)}ada.$$

From this example it is clear that the Fourier transform is a natural tool to localize the module design problem for a prescribed workspace. By determining the module density function $f(g)$ for given workspace density $f_W(g)$, the relatively few states of a single module can be altered using techniques presented in [2], and the combinatorially explosive design problem of treating the manipulator as one unit is avoided.

5. Conclusion

In this work, it is shown how the workspace shape and density of frames reachable by a discretely actuated manipulator can be synthesized using the concepts of the convolution product and Fourier transform of scalar functions on the Euclidean Motion Group. While motivational examples were presented which yield closed form solutions, much work remains in developing general numerical codes.

Acknowledgements

This work was made possible by a Presidential Faculty Fellow Award. Thanks go to Joshua Houck for his development of binary mechanical hardware (Figure 1), and Derek Seabury for developing a parallel port interface to run the prototype from a PC.

References

1. Chirikjian, G.S.,"A Binary Paradigm for Robotic Manipulators," *Proceedings of the 1994 IEEE International Conference on Robotics and Automation*, San Diego, CA, May 1994.
2. Chirikjian, G.S., "Kinematic Synthesis of Mechanisms and Robotic Manipulators with Binary Actuators," *ASME Journal of Mechanical Design*, Vol. 117, No. 4., pp 573-580, Dec. 1995
3. Ebert-Uphoff, I., Chirikjian, G.S.,"Efficient Workspace Generation for Binary Manipulators with Many Actuators," Journal of Robotic Systems, June, 1995, pp 383-400.
4. Ebert-Uphoff, I., Chirikjian, G.S., "Inverse Kinematics of Discretely Actuated Hyper-Redundant Manipulators Using Workspace Densities," *Proceedings of the 1996 IEEE International Conference on Robotics and Automation*, Minneapolis, MN, April 1996.
5. Gosselin, C.M., Guillot, M., "The Synthesis of Manipulators with Prescribed Workspace," *Proc. of the 1990 Mechanisms Conference*, DE-Vol. 25, pp. 315-320.
6. Murray, R. M., Li, Z., Sastry, S.S., *A Mathematical Introduction to Robotic Manipulation*, CRC Press, Ann Arbor MI, 1994.
7. Orihara, A., "Bessel Functions and the Euclidean Motion Group," Tohoku Mathematical Journal, Vol. 13, 1961, pp. 66-71.
8. Park, F.C., Brockett, R.W., "Kinematic Dexterity of Robotic Mechanisms," *The International Journal of Robotics Research*, Vol. 13, No. 1, February 1994, pp 1-15.
9. Sugiura, M., *Unitary Representations and Harmonic Analysis*, 2nd edition, North-Holland, Amsterdam, 1990.

OPTIMAL LOCATION OF PATH FOLLOWING TASKS IN THE WORKSPACE OF A MANIPULATOR USING GENETIC ALGORITHMS

N. A. ASPRAGATHOS
Mechanical Engineering Department
University of Patras, Patras, Greece

Abstract. In this paper a method is presented, for estimating the best location, from dexterity point of view, of a path following task into the workspace of an industrial manipulator. The best location of the path is defined as the location with the higher mean value of the manipulability along the path. The optimisation problem is formulated in a procedural way and solved using Genetic Algorithms (GA).
The efficiency of the method is demonstrated through experiments carried out on a planar manipulator and on a PUMA robot in a simulation environment.

1. Introduction

In the design of a robot workcell the locations of various tasks into the workspace are determined for most of the cases, by intuition gained by experience. Current off-line programming and simulation packages do not provide tools for best locating a task in the workspace. The designer of a workcell has to consider criteria such as: ergonomic effectiveness, best exploitation of the robot and auxiliary equipment, flexibility to accomplish the greatest variety of tasks, least cycle time, best dexterity, and avoidance of singularity points in order to find the best layout of a workcell, which is a very complicated task, taken into account both the lack of tools and of a relevant theory.

The problem of avoiding singularities or developing algorithms for least time or energy consumption has received considerable attention by Yoshikawa(1984), and Shin, Mckay(1986). However, these papers consider a given location of the task and then algorithms have been invented to avoid singularities or to travel on a given path in near-minimum time.

The dexterity of a manipulator is related to the ability of the manipulator to move its end-effector in all directions. The farther a manipulator is from a singular configuration the better its dexterity. The manipulability index (Yoshikawa (1990) given by

$$w = \sqrt{\det(J(\underline{q})\ J^{T}(\underline{q})}$$

J. Lenarčič and V. Parenti-Castelli (eds.), Recent Advances in Robot Kinematics, 179-188.
© 1996 *Kluwer Academic Publishers. Printed in the Netherlands.*

has been proposed as a measure of the dexterity of a manipulator, where J is the Jacobian matrix and q the vector of joint co-ordinates.
A little of work has been devoted to the automatic identification of the best location of a task. An algorithm has been proposed by Nelson and Donath (1990) for optimising the location of an assembly task in a manipulator's workspace. In this algorithm, a gradient function of manipulability has been developed based on an explicit determination of the manipulability function in joint space. The formulated problem of optimisation has been solved by a modified steepest descent procedure to locate automatically the assembly tasks away from singularities. This method requires the explicit calculation of det(J(q)) and $\partial\, |\det(J(q))| / \partial q$ for a non redundant manipulator. Therefore this method of solution cannot be used when the determinant of the Jacobian is non analytic or is non continuous.
In this paper a method based on Genetic Algorithms (see Goldberg (1989)) is proposed to locate a continuous path in the location with the higher manipulability. An optimisation problem is formulated where the mean value of the manipulability in a number of intermediate points estimated using an algorithm introduced by Taylor (1979) for a path described by a straight line segment or by a generalised Taylor's algorithm introduced by Aspragathos (1992) for a path represented by a 3D curve is calculated. The optimisation problem has been formulated in a procedural way and the objective function is non-linear and non continuous. The solution of the optimisation problem is based on a simple Genetic Algorithm.
In the following of the paper the formulation of the optimisation problem is analysed, then the genetic algorithm is presented and finally simulated experiments are carried out to demonstrate the efficiency of the introduced method.

2. Optimisation problem statement

The problem of estimating an optimal location of a desired path of manipulator's hand is defined as follows: Determine the location of the desired path represented by a curve in 3D space so that the mean of the manipulability index along the path be maximum. The method will be presented using a straight line segment as the desired path for simplicity and then will be extended to any 3D curve.

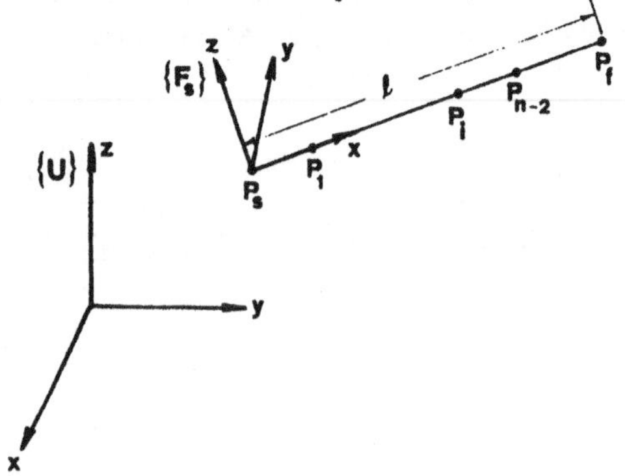

Figure 1. The definition of the universe and the local frame at the starting point of the path.

The position and orientation of the straight line segment is defined by a local Cartesian co-ordinate frame $\{F_S\}$ and the length of the segment. As it is illustrated in Fig. 1, the starting frame $\{F_S\}$ coincides with the staring point of the path while the x axis is along the path and the z axis is perpendicular to the straight line segment.
The frame $\{F_S\}$ is defined in a parameter form by $\underline{x}_s = [x\ y\ z\ \vartheta\ \phi\ \psi]^T$, where x, y, z are the Cartesian co-ordinates of the starting point of the segment defined in the universal frame {U} and ϑ, φ, ψ are the Euler angles of the frame $\{F_S\}$ referenced to frame {U}. Alternatively the location of the frame $\{F_S\}$ can be defined through the transformation

$$^{U}T_{F_s}(\underline{q}_s) \quad \text{where} \quad \underline{q}_s = [q_1 \ldots . q_k]^T \quad \text{and} \quad \underline{q}_s^L \leq \underline{q}_s \leq \underline{q}_s^U \tag{1}$$

is the vector of joint co-ordinates of the manipulator when the frame of the tool coincides with the frame $\{F_S\}$. In Eq. (1) k is the number of degrees of freedom of the manipulator. It is supposed for simplicity that the tool is perpendicular to the segment. The Cartesian or the joint co-ordinates defining the local frame $\{F_S\}$ are the variables of the objective function representing the mean value of the manipulability index along the path. The objective function will be approximated in a procedural way based on Taylor's bounded deviation method for planning and execution of straight line manipulator trajectories. It is well known, that given two configuration frames of the hand; the starting frame $\{F_S\}$ and the frame $\{F_f\}$ corresponding to the final point of the path, the Taylor's algorithm calculates the corresponding joint co-ordinates $\underline{q}_s$ and $\underline{q}_f$. Subsequently the Cartesian co-ordinates of the joint midpoint is calculated and is compared to the Cartesian configuration corresponding to the midpoint of the straight line segment. If the two configurations "deviate" more than a allowed amount, the final point configuration is replaced by the Cartesian midpoint configuration and the algorithm is applied recursively to this straight line segment. This generation of midpoints terminates when the deviation is smaller than the specified amount. The arm motion is linearly interpolated between successive midpoints. Taylor proposed the Euclidean metric as a position deviation function. This technique reduces substantially the real time computation.
Let suppose that the Taylor's algorithm determines P_i (i=1, n-2) intermediate knot points, then an index of the path manipulability is given by

$$w_p = \left(\sum_{i=1}^{i=n-2} w_i + w_s + w_f\right) / n \tag{2}$$

where w_i, w_s, w_f, are the manipulability indices at an intermediate point, the starting point and the final point of the segment respectively. The calculated path manipulability is an approximation of the average manipulability of the path which depends upon the permitted maximum position deviation used in Taylor's algorithm. This dependence is complicated enough so it is difficult to estimate the error. However, using the proposed approximation, a considerable amount of computational time is saved.

Every intermediate knot position/orientation of the tool on the straight line segment is given as a function of the parameters defining the frame at the starting position/orientation by

$$^{U}T_{F_i} = {}^{U}T_{F_s}(\underline{q}_s)\ [L_i\ 0\ 0]^T \qquad (3)$$

The joint co-ordinates $\underline{q}_S$ can be replaced by the Cartesian co-ordinates $\underline{x}_S$ but, in the following, only the joint co-ordinates will be used. Solving the inverse kinematic problem at every intermediate and the final position\orientation of the tool, the joint co-ordinates of the manipulator are determined and then the corresponding Jacobian matrix

$$J(\underline{q}_i({}^{U}T_{F_i}(\underline{q}_s)) \qquad (4)$$

For genetic algorithm search it is not necessary to obtain an analytic formula for the path manipulability so the inverse kinematic problem can be solved numerically if it is impossible to obtain an analytic solution,. Using Eq. (4) the manipulability index at every intermediate knot point and at the two end points of the segment are calculated as functions of $\underline{q}_S$, and finally the path manipulability is calculated as formulated in Eq. (2). So at the end of the analysis the path manipulability is determined as a function of the joint co-ordinates of the frame $\{F_S\}$

$$w_p(\underline{q}_S) = \left(\sum_{i=1}^{i=n-2} w_i(\underline{q}_S) + w_S(\underline{q}_S) + w_f(\underline{q}_S)\right)/n \qquad (5)$$

The number and the position of the intermediate knot points depends upon the location of the path into the workspace so that the objective function is non continuous.

In a previous work by Aspragathos (1992), the Taylor's method has been generalised to be applicable to a Cartesian trajectory represented by any 3D curve. The algorithm again calculates enough intermediate points in order to generate a bounded position deviation trajectory of the tip of the hand of the manipulator. Therefore the proposed formulation of the path manipulability can be applied to a 3D curve using the generalised Taylor's method.

By maximising the path manipulability subject to the constraints of optimisation variables given in Eq. (1), one can find the location of the path where the manipulability of the robot is the higher. This optimisation problem can not be solved using a classical gradient method (see Cooper and Steinberg (1970)) because the objective function has a procedural nature and in the most general case is non analytic and non continuous. The selection of the Genetic Algorithms to solve this problem is based on the advantages of the GAs in search of very complex and unstructured environments.

3. The genetic algorithm solution

A solution to the formulated optimisation problem based on a simple Genetic Algorithm is introduced. There are at least three basic reasons that make GAs attractive in solving such a problem: First, they are proven to provide robust search in complex spaces finding nearly global optima. In general the bisection recursive method used to define the knot points under bounded position deviation provides a non continuous and very complex non linear search space. Therefore classical optimisation methods depending upon restrictive requirements of continuity and derivative existence, and due to their inherent local scope of search, are unsuitable for the solution of this problem. Second, GAs do not require an analytic form of the objective function and of the constraints. Third, the GAs allow additional constraints to be specified easily, so the introduced method can be extended to work in an environment cluttered with obstacles.

A critical aspect in designing a GA is the basic mechanism that links the GA to the real problem which has to be solved. This mechanism is twofold: firstly an evaluation of function (fitness) that returns a measure of the expected goodness of the solution, and secondly a way of encoding solutions to the real problem on artificial chromosomes.

GAs are essentially unconstrained search procedures within the given representation space. Goldberg (1989) proposed the penalty function concept to convert the constrained problem to an unconstrained one. Another formulation is to design and apply special recombination operators that maintain feasibility. In the present GA solution a different approach introduced Nearchou and Aspragathos (1995) has been adopted, so when a candidate solution violates the constraints, the fitness function takes a value equal to zero and has almost zero possibility to survive and no hope to get copies in the next generations.

Therefore the fitness function related to the objective function given by Eq. (5) and the joint coordinate constraints (see Eq. (1)) is given by

$$\text{fitness} = \begin{cases} w_p(\underline{q}_s), & \text{if } \underline{q}_s^L \leq \underline{q}_s \leq \underline{q}_s^U \\ 0, & \text{if } \underline{q}_s \leq \underline{q}_s^L \text{ or } \underline{q}_s^U \leq \underline{q}_s \end{cases} \tag{6}$$

The chromosome selected in this work is an m-bit string with the following syntax: the (m/k)-bits correspond to the first joint of the robot, the next (m/k)-bits correspond to the second joint, etc. Therefore the jth joint co-ordinate can be calculated from the values of the corresponding (m/k)-bit using the following mapping:

$$q_i = q_i^L + [\underline{\text{Bitvalue}} / 2^{m/k}]\, q_i^U \tag{7}$$

where Bitvalue is the decimal number corresponding to string of the jth joint.

4. Discussion of the results

The efficiency of the proposed method for estimating the location of a straight line path with the higher path manipulability has been demonstrated through simulated experiments using a planar 3-DOF manipulator and a PUMA robot.
It is assumed that the tip of the hand of the planar 3-DOF manipulator shown in Fig. 2 has to move along the straight line segment AB, with the axis of its end-effector perpendicular to the segment. It is a prerequisite to locate the segment AB in the location with the higher path manipulability.

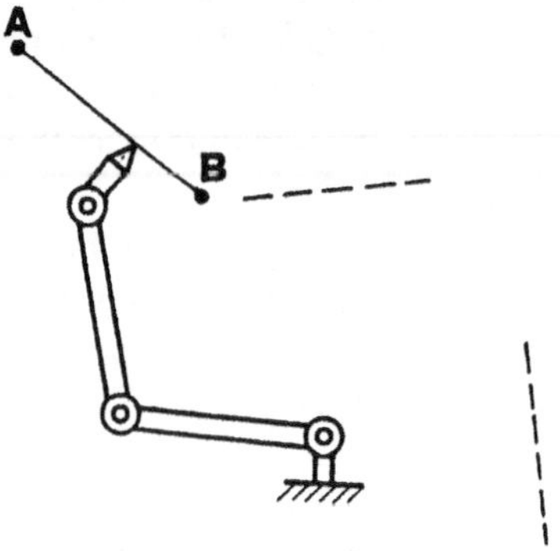

Figure 2. The planar manipulator and the location of segments with near-maximum path manipulability.

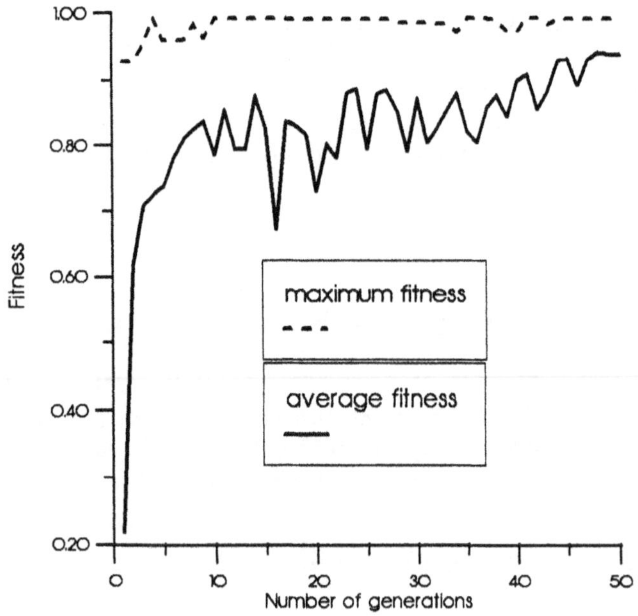

Figure 3. The convergence of the genetic algorithm.

The proposed algorithm has been tested in a number of experiments with straight line segments of various lengths. Fig. 3 illustrates the convergence of the GA with the following control parameters:Population size = 50, Maximum number of generations = 50, Probability of crossover = 0.4, Probability of mutation = 0.015. The best-of-generation value and the generation average are presented for one run of the GA solution. As it is shown in this figure the convergence criterion proposed by De Jong (1975) according to which the (average fitness/best fitness) ratio has to be greater than 95% is satisfied in this case.The locations of the estimated near-maximum manipulability in some of the executed runs are shown in Fig. 2 with dashed line. The algorithm estimated the near-maximum path manipulability at the location of the straight line path where the starting point has the following co-ordinates (3.7, 2.3) and the angle between the segment and the x axis is -1.43 rad. This path was followed by the tip of the hand using the Resolved Motion Rate Control (RMRC) introduced by Whitney (1972). The manipulability and the normalised Manipulator Velocity Ratio (MVR) introduced by Dubey and Luh (1986) for each step of RMRC algorithm are plotted in Fig. 4 in order to show the effectiveness of the algorithm in locating a path as far as possible from singularities and obtaining the lower joint velocity for a desired Cartesian velocity. The graph shows that there is a point where the manipulability is very low. So the proposed method does not guarantee the avoidance of singularity points when the desired maximum deviation in Taylor's algorithm is too high, since the lower the number of the intermediate points the higher the possibility to locate one or more points of the path close to singular

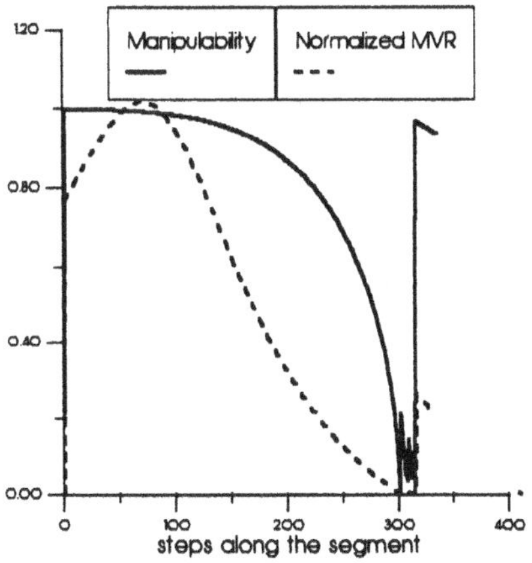

Figure 4. The manipulability and the MVR along the path located by the proposed method into the workspace of the planar manipulator.

points. Since Taylor's algorithm is very fast the user can choose a small maximum deviation to secure a location with maximum path manipulability and as far as possible from singular points. In the present paper, only the effect of the manipulability on the MVR has been considered. However it is well known that the ability of the robot to achieve high velocity of its end-effector in some direction depends not only on the manipulability but on the shape and orientation of the manipulability ellipsoid too, which are not considered in this paper.

A considerable number of runs of the introduced algorithm has been done using a PUMA robot with a variation of GA control parameters for paths of different lengths. In every knot point the algorithm selects the inverse kinematic solution with the highest manipulability, since multiple solutions are obtained in solving the inverse kinematics of PUMA (see Craig (1986)). Under the same GA control parameters and for the same length of the segment, 20 runs have been executed and the mean value of the near-maximum path manipulability and the standard deviation of them are presented in Table 3.1.

TABLE 3.1

	12 inches		20 inches		36 inches	
Case	Mean value	Standard deviation	Mean value	Standard deviation	Mean value	Standard deviation
A	7433.25	114.25	7256.1	230.4	6879.8	368.2
B	7468.7	46.92	7346.45	119.8	6965	330.3
C	7491.4	38.9	7454.2	40.4	7014	193.6

where the control parameters for each case are:

Case	Population	Max number of generations	Crossover Probability	Mutation Probability
A	50	50	0.6	0.033
B	50	100	0.6	0.033
C	100	100	0.4	0.04

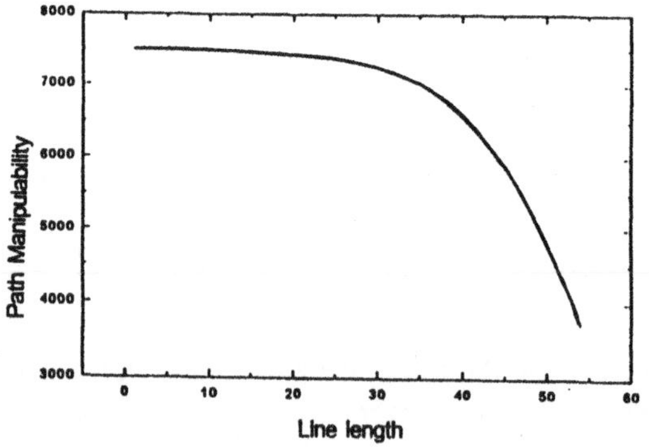

Figure 5. The near-maximum path manipulability versus the length of the straight line segment.

As can be concluded from the Table 3.1, the higher path manipulability and the lower standard deviation are achieved by using the control parameters of GA chosen in case C. The mean value of the maximum path manipulability from 25 runs has been calculated for various lengths of the path segment using the GA control parameters of case C.

The near-maximum path manipulabilty versus line length has been plotted in Fig. 5. From this diagram the limits of the length of the segment to be followed efficiently by the end-effector of the PUMA manipulator is about 25-30 inches, since for longer straight paths the rate of reduction of the path manipulability is quite high. For segments longer than 50 inches the algorithm failed to locate the path into the workspace of the PUMA manipulator. The path manipulability for a segment of one inch found equal to 7519.5 which is very close to the higher manipulability of PUMA equal to 7575, which has been estimated by exhaustive search of the workspace. This is a strong indication that the proposed algorithm works correctly. In the case that a low maximum deviation has been chosen, the proposed method finds a location of the path where every point of the segment has a high manipulability, as it is shown in Fig. 6.

Of course the proposed method estimates a location of the path with near-maximum average manipulability since the number and the position of the knot points depends upon the desired bound of the position deviation. However satisfactory position deviation does not implies also an acceptable approximation of the average manipulability along the path. The relation between the position deviation and the error in average manipulability calculation has to be farther investigated.

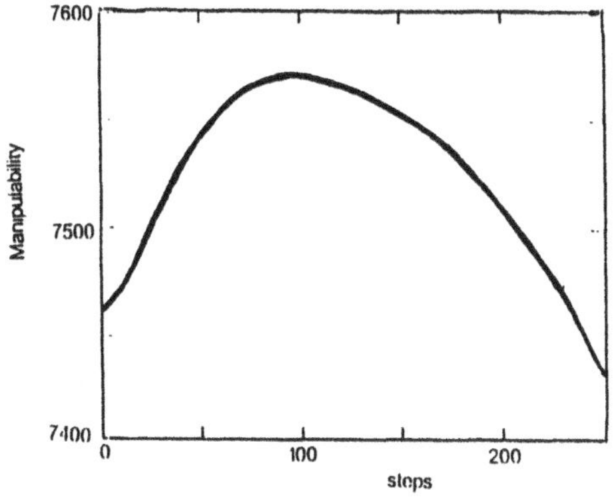

Figure 6. The manipulability along a segment of 12 inches located by the proposed method into the workspace of the PUMA robot.

Since the GA is a heuristic procedure of searching complex spaces, an experimental verification of their efficiency is necessary. Unfortunately the page limits of this paper do not permit a more extended presentation of experimental results.

5. Conclusions

In this paper an algorithmic procedure has been introduced for locating a path of a robot in a location with near-maximum average manipulability along the path. By

farther elaborating this method, tools can be developed for designing more efficient layouts of robot workcells.
The method formulates an index called path manipulability by averaging the manipulability at knot points of the path calculated by a recursive method providing a bounded position deviation. The location with near-maximum path manipulability has been estimated using a simple genetic algorithm.
The ability of the GA to search complex and unstructured spaces has been proved experimentally in the introduced problem. The formulation of the problem and the advantages of the GAs promises a successful extension of the introduced method in more complex spaces such as environments cluttered with obstacles.

References

Yoshikawa T. (1984) Analysis and Control of Robot Manipulators with Redundancy, in Robotics Research: The First International Symposium, ed. M. Brady and R. Paul, MIT Press, pp. 735-747.
Shin K. G., Mckay N. D. (1986) Selection of Near-Minimum Time Geometric Paths for Robotic Manipulators, IEEE Trans. on Automatic Control, Vol. AC-31, No 6.
Yoshikawa 3. T. (1990) Foundations of Robotics: Analysis and Control", The MIT Press, Cambridge, Massachusetts.
Nelson B., Donath M. (1990) Optimizing the Location of Assembly Tasks in a Manipulator's Workspace, Journal of Robotic Systems 7(6), pp 791-811.
Goldberg D. E. (1989) Genetic Algorithms in Search, Optimization and Machine Learning, Addison-Wesley Publishing,.
Taylor R. H. (1979) Planning and Execution of Straight Line Manipulator Trajectories, IBM Journal of Research and Development 23.
Aspragathos, N. A. (1992) Cartesian Path Generation by Robots under Maximum Allowed Deviation, 3rd International Workshop on Advances in Robot Kinematics, Ferrara, Italy.
Cooper L., Steinberg D., (1970) Introduction to Methods of Optimization, W. B. Saunders Company.
Nearchou A. C., Aspragathos N. A. (1995), Obstacle Avoidance Control of Redundant Robots using Genetic Algorithms, 3rd IEEE Mediterranean Symposium on New Directions in Control and Automation, Limassol, Cyprus.
De Jong K. A. (1975), An Analysis of the Behaviour of Genetic Adaptive Systems, Doctoral Dissertation, University of Michigan.
Whitney D. E. (1972), The Mathematics of Co-ordinated Control of Prosthetic Arms and Manipulators, Journal of Dynamic Systems Measurement, and Control, pp 303-309.
Dubey R., J. Y. S. Luh, (1986) Performance Measures and their Improvement for Redundant Robots, Robotics Theory and Application, ASME Winter Annual Meeting.
13. J. J. Craig, Introduction to Robotics: Mechanics and Control, Addison Wesley, 1986.

Acknowledgement: The results of this paper are obtained by simulated experiments done by my student N. Kaloumenos.

FEASIBLE WORKSPACE REGIONS FOR A TWO-REVOLUTE MANIPULATOR DESIGN

M. CECCARELLI
Dipartimento di Ingegneria Industriale, Universita di Cassino
Via Di Biasio 43, 03043 CASINO (FR), Italy
e-mail: cecca@ing.unicas.it

Abstract - In this paper the concept of feasible workspace regions for prescribing workspace in a two-revolute manipulator design has been proposed to determine the limits of arbitrarily given workspace points. A design formulation has been used to deduce analytical procedures for individuating the feasible regions as regions within which feasible solutions for the manipulator design can be obtained. Particularly, the feasible workspace regions have been geometrically characterised and specific topologies have been shown for prescribing hole existence in the workspace in order to demonstrate the feasibility of the procedure with respect to additional design constraints.

1. Introduction

Prescribing a suitable workspace for manipulators design can be considered as a basic datum, since the workspace characteristics are recognized as fundamental for a proper application and evaluation of a robot manipulator.
A first problem in a manipulator design can be stated as how to prescribe a suitable workspace so that a design algorithm may give feasible solutions?
In this paper, the problem has been approached for two-revolute open chain manipulators, since they are widely used and they have the basic workspace geometry which is believed as a basis for future extension of the herein proposed ideas and formulation.
The geometry of the toroidal workspace of two-revolute open chains has been described and geometrically analyzed by Fichter E. F. and Hunt K. H., [1], with the aim to deduce theorems useful both in analysis and synthesis procedures. They have also demonstrated by means of geometric argues the existence of four cognate two-revolute dyads which may trace the same toroidal surface. However, the design problem has been approached in details only by Roth B., [2], who developed a design algorithm based on the

J. Lenarčič and V. Parenti-Castelli (eds.), Recent Advances in Robot Kinematics, 189-198.
© 1996 *Kluwer Academic Publishers. Printed in the Netherlands.*

analytical expression of a torus. Although the expression of a torus has been used, the design algorithm is numerical in nature.

In a previous paper, [3], the concept of feasible regions for prescribing workspace of two-revolute manipulators has been introduced as a result of an attempt to deduce a closed-form solution for the design problem. Particularly, suitable design equations have been formulated by using the torus equation and a discussion on the form of the design formulation has given the expression of general feasible workspace regions for designing two-revolute manipulators with a given base location and no more constraints.

In this paper, we have further analyzed and characterised the peculiar geometry of the general feasible regions and we have gone further into the design problem by including additional constraints to improve the design formulation and to investigate their effect on the form and the extension of the feasible workspace regions.

2. A formulation for workspace design of two-revolute manipulators

The design parameters of a general two-revolute manipulator are, Fig. 1:

a_1, a_2 which are the link lengths giving the common normal distance between the two joint axes Z_1 and Z_2, and the distance of a reference point H from Z_2, respectively;

α_1 which is a twist angle between Z_1 and Z_2, measured positive clockwise about X_1;

d_2 which is the link offset giving the distance between the joint centers along Z_2.

The angle θ_1 and θ_2 are the revolution angles of the joints about the joint axes measured between X axis and a_1 direction, a_1 and a_2 directions, respectively. The link offset $d_1 = 0$ is meaningless from design viewpoint since it may only shift up or down the workspace.

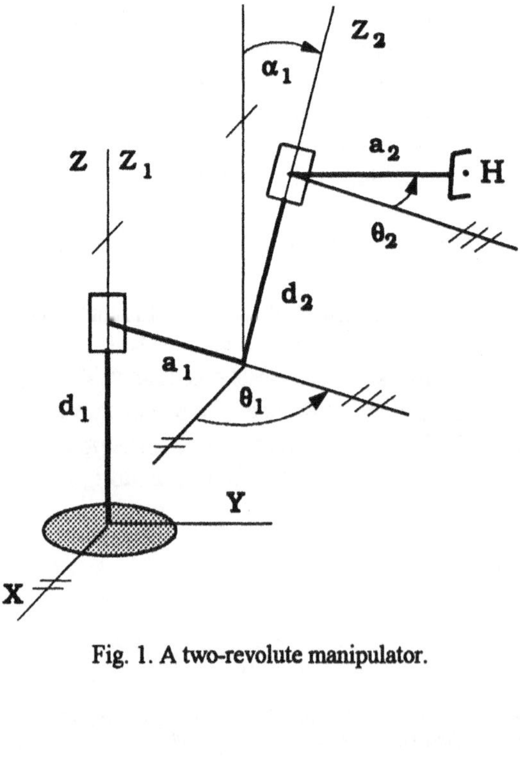

Fig. 1. A two-revolute manipulator.

A design formulation can be obtained by using the algebraic expression of the toroidal workspace of a reference point H on the two-revolute manipulator with respect to a base frame XYZ and with the hypotheses $a_1 \neq 0$, $a_2 \neq 0$, $\alpha_1 \neq 0$, in the form

$$\left(r^2 + z^2 - A\right)^2 + (Cz + D)^2 + B = 0 \tag{1}$$

in which the so-called structural parameters A, B, C, D are given by

$$A = a_1^2 + a_2^2 + d_2^2 \ , \ B = -\ 4a_1^2 a_2^2 \ , \ C = 2a_1/\text{sen}\,\alpha_1 \ , \ D = -\ 2a_1 d_2 \cos\alpha_1/\text{sen}\,\alpha_1 \tag{2}$$

and r and z are the radial and axial reaches, respectively.
In particular, a design procedure can be proposed, either numerically, [2], or algebraically, [3], by using an inversion of Eqs.(1) and (2), separately.
Thus, by using Eq. (1) for a solvable design problem we may write a linear system

$$U\,x = w \tag{3}$$

where the unknowns A, C^2, and CD are the components of a design vector x. The given data in the form of prescribed workspace reaches r_i and z_i (i=1,...,4) can be used to compute

$$U = \begin{bmatrix} -\ 2(\rho_2 - \rho_1) & \left(z_2^2 - z_1^2\right) & 2(z_2 - z_1) \\ -\ 2(\rho_3 - \rho_1) & \left(z_3^2 - z_1^2\right) & 2(z_3 - z_1) \\ -\ 2(\rho_4 - \rho_1) & \left(z_4^2 - z_1^2\right) & 2(z_4 - z_1) \end{bmatrix} \quad w = \begin{bmatrix} -\left(\rho_2^2 - \rho_1^2\right) \\ -\left(\rho_3^2 - \rho_1^2\right) \\ -\left(\rho_4^2 - \rho_1^2\right) \end{bmatrix} \tag{4}$$

in which $\rho_i = r_i^2 + z_i^2$ (i=1,...,4) are the square total reaches of the prescribed points.
A design problem for prescribed workspace through four reachable points can be formulated in a closed-form algorithm by inverting algebraically the design equations (2) and (3) in a step by step procedure.
Thus, we can obtain the solution of the structural coefficients in the form, [3],

$$A = \frac{\rho_1^2 k_1 - z_1^2 k_2 + z_1 k_3 - k_4}{8K_U} \qquad C^2 = \frac{\rho_1^2 k_5 - \rho_1 k_2 + z_1 k_6 - k_7}{4\,K_U}$$

$$CD = -\ \frac{z_1^2 k_6 + \rho_1^2 k_8 - \rho_1 k_3 - k_9}{8K_U} \qquad B = -\ (\rho_1 - A)^2 - (Cz_1 + D)^2 \tag{5}$$

with

$$K_U = z_1^2\,k_{10} - z_1\,k_3 + k_{11} - r_1^2\,k_1$$

where

$$k_1 = z_3 z_4 \Delta\, z_{43} - z_2 z_4 \Delta\, z_{42} + z_2 z_3 \Delta\, z_{32} \ ,$$

$$k_2 = \rho_2^2 \Delta\, z_{43} - \rho_3^2 \Delta\, z_{42} + \rho_4^2 \Delta\, z_{32} \ ,$$

$$k_3 = \rho_2^2 \Delta\, z_{43}^2 - \rho_3^2 \Delta\, z_{42}^2 + \rho_4^2 \Delta\, z_{32}^2 \ ,$$

$$k_4 = \rho_2^2 z_3 z_4 \Delta\, z_{43} - \rho_3^2 z_2 z_4 \Delta\, z_{42} + \rho_4^2 z_2 z_3 \Delta\, z_{32} \ , \tag{6}$$

$$k_5 = \rho_2 \Delta z_{43} - \rho_3 \Delta z_{42} + \rho_4 \Delta z_{32} \ , \qquad (6)$$

$$k_6 = \rho_3 \rho_4 \Delta \rho_{43} - \rho_2 \rho_4 \Delta \rho_{42} + \rho_2 \rho_3 \Delta \rho_{32} \ ,$$

$$k_7 = z_2 \rho_3 \rho_4 \Delta \rho_{43} - z_3 \rho_2 \rho_4 \Delta \rho_{42} + z_4 \rho_2 \rho_3 \Delta \rho_{32} \ ,$$

$$k_8 = \rho_2 \Delta z_{43}^2 - \rho_3 \Delta z_{42}^2 + \rho_4 \Delta z_{32}^2 \ ,$$

$$k_9 = z_2^2 \rho_3 \rho_4 \Delta \rho_{43} - z_3^2 \rho_2 \rho_4 \Delta \rho_{42} + z_4^2 \rho_2 \rho_3 \Delta \rho_{32} \ ,$$

$$k_{10} = r_2^2 \Delta z_{43} - r_3^2 \Delta z_{42} + r_4^2 \Delta z_{32} \ ,$$

$$k_9 = r_2^2 \rho_3 \rho_4 \Delta \rho_{43} - r_3^2 \rho_2 \rho_4 \Delta \rho_{42} + r_4^2 \rho_2 \rho_3 \Delta \rho_{32}$$

in which

$$\Delta \rho_{ij} = \rho_i - \rho_j \ , \quad \Delta z_{ij} = z_i - z_j \ , \quad \Delta z_{ij}^2 = z_i^2 - z_j^2 \qquad i = 2, \ldots, 4$$

Then the design parameters can be computed as

$$a_2^2 = -\frac{B}{4a_1^2} \ , \quad d_2^2 = -\frac{D^2}{4a_1^2 - C^2} \ , \quad \alpha_1 = \mathrm{atan2}\left(\frac{2a_1}{C} \ , \ -\frac{D}{d_2 C}\right) \qquad (7)$$

(atan2 is the atan operator for oriented angles) when the cubic equation

$$\left(a_1^2\right)^3 - \frac{4A + C^2}{4}\left(a_1^2\right)^2 + \frac{AC^2 - D^2 - B}{4}\left(a_1^2\right) + \frac{BC^2}{16} = 0 \qquad (8)$$

has been solved to give the a_1 length. It has been found that the cubic has two feasible roots or none. Each one can be used to synthesize two two-revolute manipulators because of α_1, [3].

3. Constraints on prescribing workspace points

The design equations (1) have four unknown structural parameters so that a solvable design formulation can be deduced by using four distinct given workspace points.
However, the given workspace points cannot be prescribed arbitrary.
Infact, the workspace formulation itself gives some constraints for designing two-revolute manipulators.
Thus, the solvability of Eqs.(3) requires the determinant of U not to be zero, giving the condition that any three given workspace points cannot lay on a horizontal line. No constraints exist for laying on a vertical line since the "banana shape" of a general torus may allow this occurrence, [3].
However, since the torus equation is of fourth order, a fourth given point is needed to individuate a specific torus. Thus most of the constraints for a solvable design formulation are restricted to the fourth given point when the other three can be given arbitrary with the only condition of not being on a horizontal line. The fourth point will be given so that a torus will across over all the four points. This condition has been thought useful to introduce the concept of feasible workspace regions as regions within which the fourth point can be located to give a torus and thus a feasible design.

4. Feasible workspace regions

Feasible workspace regions can be yet determined by using the torus formulation whose design conditions and constraints will determine the regions within which the fourth point, let's say P_1 may be located to give a solvable design problem.
Particularly, when three points P_2, P_3, P_4 have been properly assigned further workspace design analytical constraints for the fourth given point P_1 can be obtained by using the design equations (5) for the basic constraints on the design unknowns A, C^2 and CD.
When P_2, P_3, P_4 are given feasible workspace regions for P_1 can be determined by the following constraints:

- the structural parameter A must be positive so that a feasible design region R_A can be individuated as
$$R_A(P_1) = (r_1, z_1) : A > 0 \tag{9}$$
- the structural parameter C must be real so that a feasible design region R_C can be individuated as
$$R_C(P_1) = (r_1, z_1) : C^2 \geq 0 \tag{10}$$
- the determinant of U, which has been expressed as K_U in Eqs.(5), cannot be zero so that a feasible design region R_U can be individuated as
$$R_U(P_1) = (r_1, z_1) : K_U \neq 0 \tag{11}$$

A numerical determination of the abovementioned design regions can be obtained by using Eqs.(5) to individuate the boundaries of the regions through the conditions $A = 0$, $C^2 = 0$, $K_U = 0$. Then a feasible workspace region $R_W(P_1)$ can be determined as the intersection of the feasible design regions, i.e.

$$R_W(P_1) = R_A \cap R_U \cup R_C' \tag{12}$$

with $$R_C'(P_1) = R_C - R_C \cap (R_A \cap R_U) \tag{13}$$

where $\cap$ is the intersection operator and $\cup$ is the union operator.
$(R_C - R_C')$ is an unfeasible design region in a sense that no possible solutions can be obtained for the two-revolute manipulator design when P_1 is selected within it. It is to note that this region determines a hole in the feasible workspace region R_W with the peculiarities that it is characterised by the four points which connect the feasible regions. Three of these points are always defined by the prescribed three points P_2, P_3, P_4 and the fourth P* cannot a prescribed one, since all the designed tori will cross over them. If P_1 is selected coincident with P* the torus is undetermined and all the tori of R_W can be considered as a design feasible result. Point P* can be easy determined as the intersection point between the curves $A = 0$ and $C^2 = 0$ and distinct from P_2, P_3, P_4. Its coordinates can determined by means of Eq.(1) and from the analytical z expression of the intersection which can be expressed as

$$z^2 k_2 - z k_3 + k_4 =$$
$$\left[z^2 + \frac{z^2 k_2 k_{10} + z(k_1 k_6 - k_3 k_4 - k_3 k_{10}) + k_4 k_{10} + (k_4 - k_7) k_1}{k_1 k_2} \right]^2 k_1 \tag{14}$$

This is a fourth order equation, whose three solutions are known as P_2, P_3, P_4. Therefore, we may use the properties of algebraic equations, which gives the sum of the roots as the ratio between the coefficients of the terms of order four and three to obtain, after some algebraic manipulations

$$k_S = 2\,\frac{r_2^2\,\Delta z_{43}^2 - r_3^2\,\Delta z_{42}^2 + r_4^2\,\Delta z_{32}^2}{k_5} \tag{15}$$

Consequently, the axial reach z^* of P^* can be computed as

$$z^* = k_S - z_2 - z_3 - z_4 \tag{16}$$

Fig.2 is an illustrative example of a general topology of $R_W(P_1)$.

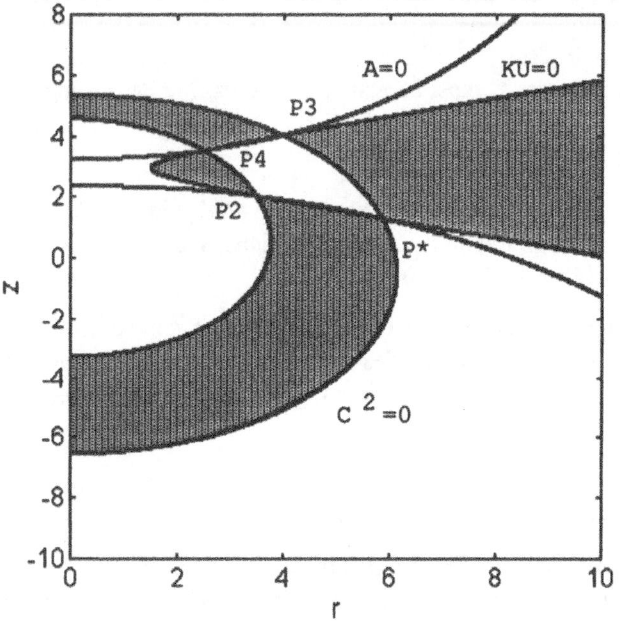

Fig.2 Characteristic topology of a feasible workspace region $R_W(P_1)$.

An important attribute of manipulator workspaces is the existence or the absence of a hole. It is to note that $R_W(P_1)$ has been obtained with no conditions about hole existence so that P_1 selection within it may give or may not give a toroidal workspace with hole.

The proposed formulation can be extended to include a design constraint so that if the fourth point P_1 belongs to a particular region R_h, the resulting workspace will have a hole. The boundary of this region can be obtained noting that a two-revolute manipulator workspace will have a hole when it intersects the Z axis in the points whose axial reaches are solutions of the following equation

$$z^4 + \left(C^2 - 2A\right)z^2 + (2CD)z + \left(A^2 + D^2 + B\right) = 0 \tag{17}$$

Therefore, the workspace of a two-revolute manipulator has a hole when Eq.(17) does not give real solutions. Moreover, since the third order term in Eq.(17) is vanished the sum of the roots is null so that they are a couple of complex and conjugate solutions with the real part equal and opposite in sign. This conditions can be expressed analytically and after some algebraic manipulations the discriminant of Eq.(17) can be expressed as

$$K_D = 16\left[-2A + C^2\right]^4\left[A^2 + D^2 + B\right] - 4\left[-2A + C^2\right]^3\left[2CD\right]^2 - 128\left[-2A + C^2\right]^2\left[A^2 + D^2 + B\right]^2 + 144\left[-2A + C^2\right]\left[2CD\right]^2\left[A^2 + D^2 + B\right] + 256\left[A^2 + D^2 + B\right]^3 - 27\left[2CD\right]^4 \quad (18)$$

Thus, a feasible design region R_h, prescribing a hole in the toroidal workspace, can be defined as

$$R_h(P_1) = (r_1, z_1) : K_D > 0 \text{ and } K_h < 0 \quad (19)$$

in which

$$K_h = \left(C^2 - 2A\right)^2 - 4\left(A^2 + D^2 + B\right) \quad (20)$$

The boundaries can be computed with the limit conditions for the inequalities in Eq.(19).

Thus a feasible region R_{WH}, prescribing hole existence in a designed torus, can be determined as the intersection of the general R_W and R_h, i.e.

$$R_{WH}(P_1) = R_W \cap R_h \quad (21)$$

Fig.3 illustrates the condition of Eq.(21) for the case of Fig.2.

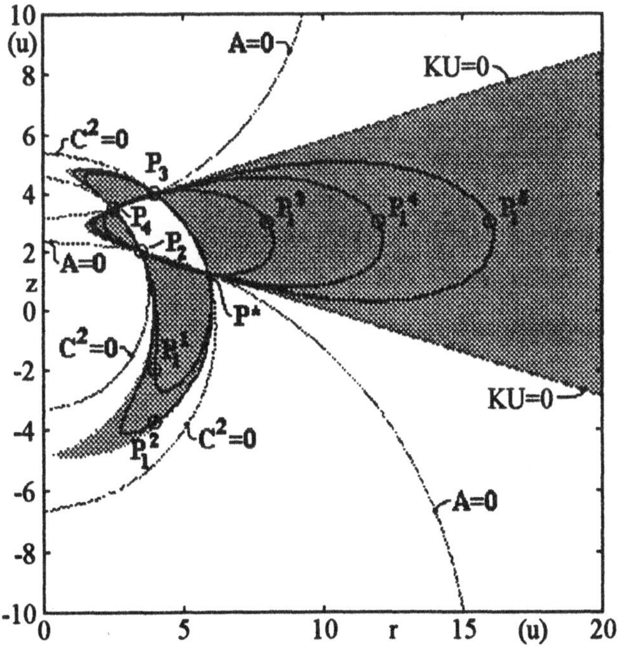

Fig.3 Characteristic topology of a feasible workspace region $R_{WH}(P_1)$ prescribing a hole.

Particularly, Figs. 2 and 3 show a general topology of feasible regions $R_W(P_1)$ and $R_{WH}(P_1)$, which are both composed of four different design regions, which are jointed at the four characteristic points. Each region has a typical shape although its size may vary considerably depending on the relative position of the given points P_2, P_3, P_4. It can be observed that the curve $C^2 = 0$ may determine or not a nozzle region depending on the position of point P_2 with respect to P_3 and P_4. The examples reported in the paper illustrate the basic case: if P_2 is at near left of P_3 but outside of P_4 - P_3 radial range, a nozzle will appear on the top side of $R_W(P_1)$ and $R_{WH}(P_1)$, Fig.4; if P_2 is at near right of P_3 but outside of P_4 - P_3 radial range, a nozzle will appear on the bottom side of $R_W(P_1)$ and $R_{WH}(P_1)$, Fig.5; finally, if P_2 is inside P_4 - P_3 radial range, a nozzle will not appear and the curve $C^2 = 0$ shows two separate branches intersecting both the Z axis, Fig.3.

Some other particular topologies may arise looking at the variation on the curves $A = 0$ and $K_U = 0$. Specifically, the curve $K_U = 0$ generally shows a parabolic horizontal shape in the r-z plane determining a small nozzle region on the left of P_4 , Figs.2-5. This region may collapse because of a variation of $K_U = 0$, but also because the contemporaneously degeneration of the curve $A = 0$, which is usually composed of a two hyperbolic branches intersecting separately the Z axis. However, in the case of degeneration the curve $A = 0$ may change to a parabolic shape, Fig.5, which may then collapse on the curve $K_U = 0$ itself, giving very particular feasible workspace regions as the case of Fig.6.

Finally, Fig 3 shows how different solutions of P_1 location may give different tori both in size and shape depending on the region of selection. Infact P_1 is alternatively selected in the regions with $K_U>0$ or $C^2>0$ and a torus may have a cross-section as an egg-shaped

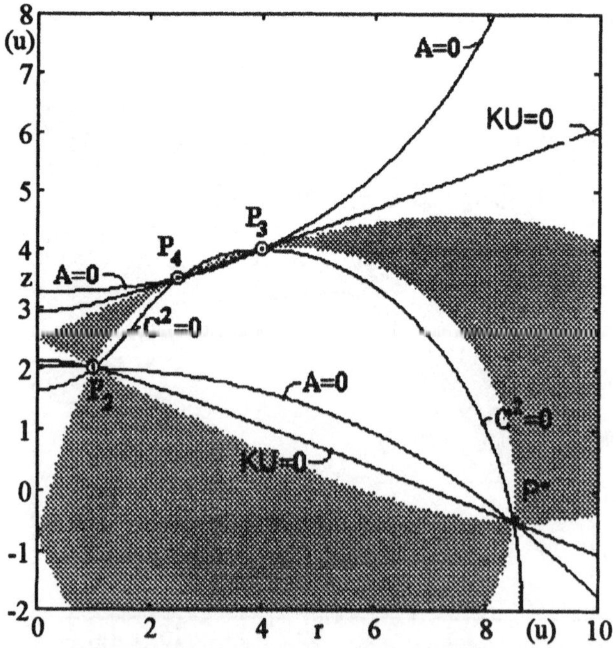

Fig.4 Feasible workspace region $R_{WH}(P_1)$ with reduced nozzle area between P_3 and P_4.

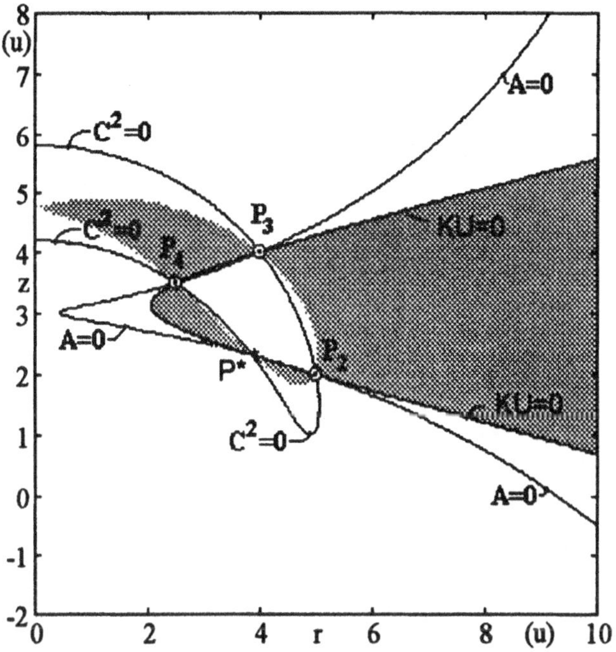

Fig.5 Feasible workspace region $R_{WH}(P_1)$ with nozzle area on the bottom side near P_2.

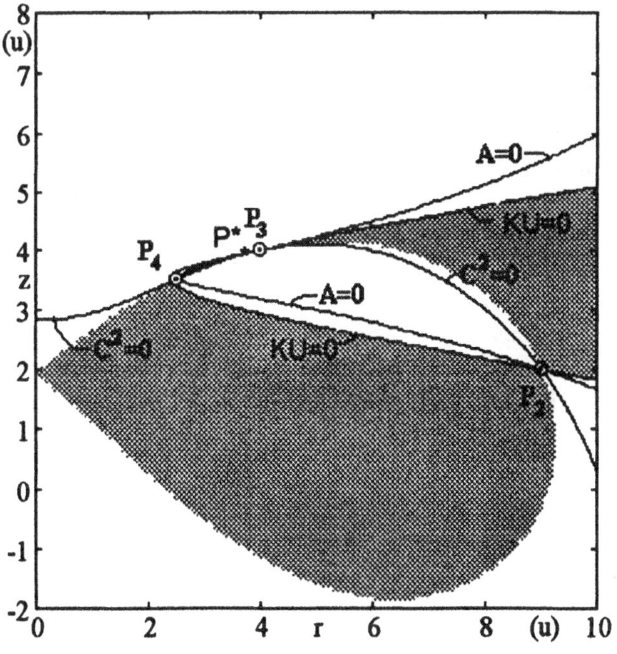

Fig.6 Feasible workspace region $R_{WH}(P_1)$ for a degenerated case.

one (named as flattened section in [1]) or a banana shaped one (named as concavo-convex section in [1]), respectively.
It is worth to note that any further condition or constraint, which can be expressed as a function of structural parameters because of Eqs.(5), can be included easily from a theoretical viewpoint as a further intersection between the feasible workspace region R_W and the resulting feasible design region, as the case of hole existence has illustrated.

5. Conclusion

In this paper a new concept has been developed in the form of feasible workspace regions to determine regions within which a toroidal workspace can be prescribed for a solvable design problem of two-revolute manipulators.
In particular, the constraints on prescribing four workspace points have been restricted to only one point whose location must be selected within the feasible workspace regions, which have been computed by using three arbitrary given workspace points and design constraints from a suitable design formulation.
The procedure of determining a feasible workspace region has been successfully extended by adding the further constraint for hole existence in the workspace with the aim to obtain novel topologies and design charts and to show that the proposed concepts and formulation can be easy generalised, although some computational complexity may arise by using the torus formula for manipulators, [4].

References

1.Fichter E.F., Hunt K.H., "The Fecund Torus, its Bitangent-Circles and Derived Linkages", Mechanism and Machine Theory, Vol.10, 1975, pp.167-176.
2.Roth B., "Analytical Design of Two-Revolute Open Chains", Preprints of the Sixth CISM-IFToMM Symposium on Theory and Practice of Robots and Manipulators, Cracow, 1986, pp. 180-187.
3.Ceccarelli M., Scaramuzza G., "Analytical Constraints for a Workspace Design of 2R Manipulators", Computational Kinematics '95, Merlet J.P and Ravani B. (Eds), Kluwer, Dordrecht, 1995, pp.173-182.
4.Ceccarelli M., "Determination of the Workspace Boundary of a General N-Revolute Manipulator", in: Advances in Robot Kinematics and Computational Geometry, Lenarcic J. and Ravani B. (Editors), Kluwer, Dordrecht, 1994, pp.39-48.

DETERMINATION OF THE WORKSPACE OF *4R* MANIPULATOR: BASIC CASE AND PARTICULAR CASES

E.E. PEISACH
St-Petersburg State University of Technology and Design
B. Morskaya 18, St-Petersburg, Russia

Abstract. The paper treats the workspace of a *4R* manipulator. The emphasis is given to the existence domain of the basic case, the classification of the particular cases, the classification of the envelope branches and geometry of the workspace.

1. Introduction

To determine of the manipulator workspace in the form of an open kinematic chain with single revolute pairs, different numerical and analytical methods are utilised (Gupta and Roth, 1982, Rastegar, 1998; Ceccarelli, 1989: Ceccarelli and Vinciguerra, 1998; and others). The problem of determination of the workspace for *3R* manipulator can be considered completely solved, since the algorithms for any possible values of constant parameters, including all particular cases, has been in principle deduced (Peisach, 1993). However, the problem of determining of the workspace for a *4R* manipulator requires further study.

The purpose of the present work is to investigate the workspace of a *4R* manipulator on a basis of the envelope theory (Peisach, 1995). The emphasis is on the existence domain of the basic case, the classification of the particular cases in the considered problem, the classification of the envelope branches, as well as on the geometry of the workspaces.

2. Problem statement

Let us denote (see Fig. 1): *0* is the fixed link; *1,2,3,4* are the movable links of the *4R* manipulator; *M* is the hand reference point; $O_iX_iY_iZ_i$ is a frame of reference invariable associated with the link *i* ($i = 0,1,2,3,4$); h_1, θ_1, l_2, h_2, θ_2, l_3, h_3, θ_3,

J. Lenarčič and V. Parenti-Castelli (eds.), Recent Advances in Robot Kinematics, 199-204.
© 1996 *Kluwer Academic Publishers. Printed in the Netherlands.*

l_4, h_4 are the constant parameters of the manipulator; φ_1, φ_2, φ_3, φ_4 are the variable parameters of the manipulator (the twist angles in revolute joints); x_i, y_i, z_i are the coordinates of the point M in the reference frame $O_iX_iY_iZ_i$; r_i is the distance from the point M to the axis z_i.

It is required to find the workspace of the *4R* manipulator, that is the field of the space which is the geometric locus of all possible positions of the point M in the fixed reference frame $O_0X_0Y_0Z_0$ as a result of variations of the angles φ_1, φ_2, φ_3, φ_4 from 0 to 2π. The workspace has the form of a rotation body with the rotation axis Z_0. Denote as Ω the cross section of the workspace by the half-plane going through the axis Z_0 and bounded by this axis.

Let us introduce the coordinate plane Orz in which the planar figure Ω is placed. Let us consider that the point O coincides with the point O_0, and the axis Oz with the axis Z_0. We refer to the point M' with the coordinates $r = r_0$, $z = z_0$ as the mapping of the point M onto the plane Orz. The position of the point M' in the plane Ω is defined by the values of the angles φ_2, φ_3, φ_4.

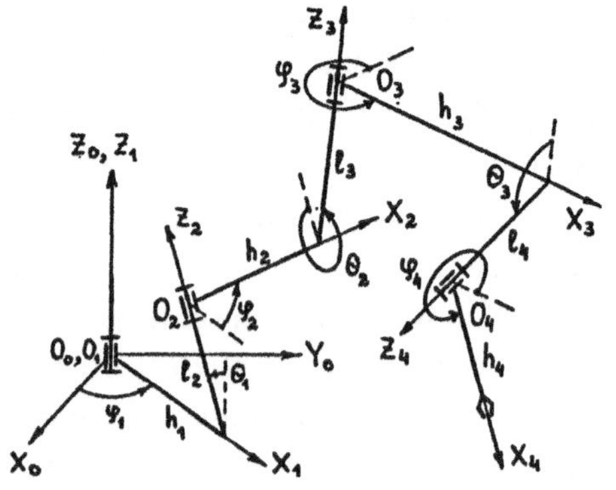

Fig. 1: Kinematic diagram of *4R* manipulator

The problem of determination of the workspace turns into the determination of the planar figure Ω or, more precisely, of its boundary contour in the plane Orz. For the determination of the boundary contour of the region Ω it is preliminary required to find the envelope of the family of curves, described by point M' as functions of the values of angles φ_2, φ_3, φ_4 between 0 to 2π.

In choosing the constant parameters of the *4R* manipulator the following *common restrictions* should be considered:

$$h_1,\ l_2,\ h_2,\ l_3,\ h_3,\ l_4 \geq 0,\ h_4 > 0,\ 0 \leq \theta_1, \theta_2, \theta_3 \leq 2\pi \ ; \quad (1)$$

$$h_i^2 + sin^2\,\theta_i \neq 0\ (i = 1,2,3)\ ; \quad (2)$$

$$sin^2\,\theta_1 + sin^2\,\theta_2 + sin^2\,\theta_3 \neq 0,\ h_1^2 + l_2^2 + h_2^2 + l_3^2 + h_3^2 \neq 0\ . \quad (3)$$

The conditions (1) are obvious. If at least one of conditions (2) is not satisfied, the *4R* manipulator will degenerate into a manipulator with less degrees of freedom than *4*. If one of inequalities (3) is not satisfied, the workspace will collapse onto a surface.

3. Basic case and particular cases

The relation of the coordinates r and z of the point M' of the field Ω from the variable parameters φ_2, φ_3, φ_4 may be represented in the form of the two following equations:

$$z = B_2 + V_2 p_1\ ;\ r^2 = A_2 + 2h_1 U_2 - z^2\ , \quad (4)$$

where

$$A_2 = h_1^2 + r_2^2 + (z_2 + l_2)^2\ ,\ B_2 = (z_2 + l_2)q_1\ ;$$
$$U_2 = x_2 c_2 - y_2 s_2\ ,\ V_2 = x_2 s_2 + y_2 c_2\ ;$$
$$x_2 = U_3 + h_2\ ,\ y_2 = V_3 q_2 - (z_3 + l_3)p_2\ ;$$
$$z_2 = V_3 p_2 + (z_3 + l_3)q_2\ ,\ r_2^2 = x_2^2 + y_2^2\ ;$$
$$U_3 = x_3 c_3 - y_3 s_3\ ,\ V_3 = x_3 s_3 + y_3 c_3\ ;$$
$$x_3 = h_4 c_4 + h_3\ ,\ y_3 = h_4 q_3 s_4 - l_4 p_3\ ;$$
$$z_3 = h_4 p_3 s_4 + l_4 q_3\ ,\ r_3^2 = x_3^2 + y_3^2\ ;$$
$$p_i = sin(\theta_i)\ ,\ q_i = cos(\theta_i)\ ,\ (i = 1,2,3)\ ;$$
$$s_i = sin(\varphi_i)\ ,\ c_i = cos(\varphi_i)\ ,\ (i = 1,2,3,4)\ .$$

In the determination of the envelope of the family of curves, described by the point M', it isn't necessary to obtain the uniform algorithm suitable for all range of the constant parameters of the *4R* manipulator satisfying the common restrictions (1-3). The mentioned range is divided into two subranges, one corresponds to *the basic case*, and the other to *the particular cases.*

The basic case is characterised by the following indications: 1) the corresponding subrange of constant parameters (i.e. the existence domain for the basic case) are reflected by the conditions only in the form of inequalities; 2) there is the uniform algorithm for the determination of the envelope suitable for all points of the mentioned subrange.

Each of the particular cases is characterised by the following indications: 1) among the conditions reflected the corresponding subrange of constant parameters (i.e. the existence domain for a particular case) there is at least one condition in the form of equality; 2) at least one of the branches of the envelope is determined by the use of a specific algorithm suitable only for the given particular case.

In Peisach (1995) we presented the analytical algorithm of determination of the envelope of the curve family (4) for the basic case. This algorithm has been named as *the basic algorithm*.

It has been established that the envelope of the family of curves (4) consists of two groups of branches. All branches of the first group have been referred to as *the sub-envelope E1*, and all branches of the second group as *the sub-envelope E2*. It has been shown that the sub-envelope *E2* represents one or several segments placed along the axis *Oz*.

Note that in Ceccarelli and Vinciguerra (1995), only the sub-envelope *E1* has been obtained. Neglecting the sub-envelope *E2* may result in inaccuracies of the shape of the workspace in a number of numerical examples.

The analysis of the basic algorithm shows that it can be used in fulfil the following conditions in the form of inequalities:

$$h_1 > 0\,,\ p_1 \neq 0\,,\ q_1 \neq 0\,,\ h_2 > 0\,,\ p_2 \neq 0\,,\ q_2 \neq 0\,,\ h_3 > 0\,,\ p_3 \neq 0\,,\ h_4 > 0\ ; \quad (5)$$
$$\{h_3 > h_4\} \text{ or } \{h_3 \leq h_4 \text{ and } (h_4^2 - h_3^2)q_3^2 \neq l_4^2 p_3^2\}\ .$$

Thus, *the existence domain of the basic case* is defined by conditions (5).

4. Classification of particular cases

If from the region defined by inequalities (1-3), which reflect the common restrictions, we exclude the existence domain (5) of the basic case, the existence domain of all particular cases will be obtained.

We propose to classify all particular cases into nine groups. Several particular cases are contained in each group. The common existence domain of all particular cases entering in a group is expressed by one condition in the form of equality and several conditions in the form of inequalities. These conditions are as follows:

1st group: $h_1 = 0;\ p_1 \neq 0,\ q_1 \neq 0,\ h_2^2 + p_2^2 \neq 0,\ h_3^2 + p_3^2 \neq 0,$
$h_2^2 + h_3^2 + l_2^3 + l_3^3 \neq 0,\ h_4 > 0;$

2nd group: $p_1 = 0;\ h_1 > 0,\ h_2^2 + p_2^2 \neq 0,\ h_3^2 + p_3^2 \neq 0,\ p_2^2 + p_3^2 \neq 0,\ h_4 > 0;$

3rd group: $q_1 = 0;\ h_1 > 0,\ h_2^2 + p_2^2 \neq 0,\ h_3^2 + p_3^2 \neq 0,\ h_4 > 0;$

4th group: $h_2 = 0;\ h_1 > 0,\ p_1 \neq 0,\ q_1 \neq 0,\ p_2 \neq 0,\ h_3^2 + p_3^2 \neq 0,\ h_4 > 0;$

5th group: $p_2 = 0;\ h_1 > 0,\ p_1 \neq 0,\ q_1 \neq 0,\ h_2 > 0,\ h_3^2 + p_3^2 \neq 0,\ h_4 > 0;$

6th group: $q_2 = 0;\ h_1 > 0,\ p_1 \neq 0,\ q_1 \neq 0,\ h_2 > 0,\ h_3^2 + p_3^2 \neq 0,\ h_4 > 0;$

7th group: $h_3 = 0;\ h_1 > 0,\ p_1 \neq 0,\ q_1 \neq 0,\ h_2 > 0,\ p_2 \neq 0,\ q_2 \neq 0,\ p_3 \neq 0,\ h_4 > 0;$

8th group: $p_3 = 0;\ h_1 > 0,\ p_1 \neq 0,\ q_1 \neq 0,\ h_2 > 0,\ p_2 \neq 0,\ q_2 \neq 0,\ h_3 > 0,\ h_4 > 0;$

9th group: $(h_4^2 - h_3^2)q_3^2 = l_4^2 p_3^2;\ h_1 > 0,\ p_1 \neq 0,\ q_1 \neq 0,\ h_2 > 0,\ p_2 \neq 0,\ q_2 \neq 0,$
$h_3 > 0,\ p_3 \neq 0,\ h_4 > h_3.$

5. Classification of envelope branches

All envelope branches can be divided into two types: *the branches of the first type* and *the branches of the second type*. The first ones are characterised by the following property: to a continuos movement of point M' along any of those branches corresponds a continuos change of the variable parameters φ_2, φ_3, and φ_4. All other points are reached with one and same value of any of one or two variable parameters φ_2, φ_3, φ_4.

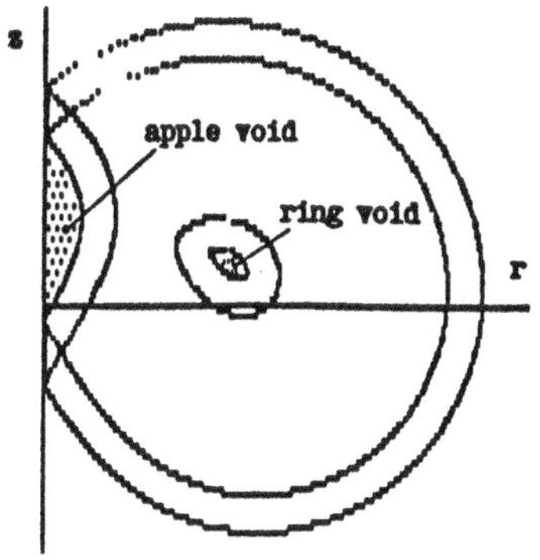

Fig. 2: The workspace cross section

All envelope branches found for the basic case fall into the branches of the first type. At the same time, some envelope branches for the particular cases meet with the branches of the second type. In many particular cases, the envelope consists of branches of the first and the second type. There are also particular cases in which the envelope consists only of the branches of the second type.

In particular cases, the branches are defined by the basic algorithm for all values of the independent variable parameter φ_4, except in very few (as a rule, one or two) singular values. The search of the envelope branches for the singular values of parameter φ_4 is provided by the special algorithm that is suitable only for the given particular case. For the particular cases relating to the 1st, 2nd, and 3rd groups none of the envelope branches may be found by means of the basic algorithm.

The envelope branches (closed curves) and the axis *Oz* separate the coordinate plane *Orz* in some enclosed regions. Among them there are regions that form the cross section Ω of workspace, and regions that represent the voids inside of workspace.

Voids of two kinds are met with: *ring void* and *apple void* (Ceccarelli, 1989) or *cavity* and *lacuna* (in terms of Peisach, 1993). Our algorithm includes the procedure of automatic *identification* of the enclosed regions.Fig. 2 shows the envelope of the family of curves (4) for the *4R* manipulator with the following values of parameters: $h_1 = 10$, $\theta_1 = 60^\circ$, $l_2 = 2$, $h_2 = 8$, $\theta_2 = 50^\circ$, $l_3 = 2$, $h_3 = 6$, $\theta_3 = 40^\circ$, $l_4 = 0$, $h_4 = 1$. The sub-envelope *E1* consists of four branches. The sub-envelope *E2* involves two segments placed along the axis *Oz*. The sub-envelopes *E1*, *E2* and the axis *Oz* produce *8* enclosed regions. Six of them form the cross section Ω of the workspace. Two regions are voids: one is cavity, the other is lacuna.

6. Conclusions

The domain of admissible values of the constant parameters of 4R manipulator is divided into two subranges, one of which corresponds to the basic case, and the second to the particular cases. For any particular case at least one of the boundary branches of the workspace is determined by the specific algorithm. It is proposed to classify all particular cases into nine groups. The common existence domain of all particular cases entering in some group is expressed by one condition in form of equality and several conditions in form of inequalities. All boundary branches can be divided into two types.

References

Gupta, KC., Roth, B. (1982). Design considerations for manipulator workspace, *Journal of Mech. Design*, Vol. 104, pp. 704-711.

Rastegar, J. (1988). Workspace analysis of 4R manipulators with various degrees of dexterity, *Journal of Mech. Transmi. Auto. In Design*, Vol. 110, pp. 42-47.

Ceccarelli, M. (1989). On the workspace of 3R robot arms, *Proc. 5th SIROM IFToMM*, Bucharest, Vol. II-1, pp. 37-46.

Peisach, E.E. (1993). An algorithm of determination of 3R manipulator workspace, *Proc. 6th SIROM IFToMM*, Bucharest, pp. 199-208.

Peisach, E.E. (1995). Algorithm of determination of 4R manipulator workspace on the base of the envelopes theory, *Problems of Mech. Engin. And Machine Reliability*, Moscow, No. 2, pp. 90-95.

Ceccarelli, M., Vinciguerra, A. (1995). On the workspace of general 4R manipulator, *Int. Journal Robotics Research*, Vol. 14, No. 2, pp. 152-160.

5. Modelling and computation

E. Ceresole, P. Fanghella, C. Galletti: *A discussion on metric relations in spatial kinematics: Implicit and explicit formulas and functional dependence*

V. Milenkovic, P.H. Milenkovic: *Tongue model for characterizing vocal tract kinematics*

G. Gogu, A. Barraco, P. Coiffet: *A new and efficient method for symbolical calculation of the Jacobian matrix*

K. Dobrovodsky : *Quaternion based force-torque analysis of open serial kinematic structures*

M.L. Keler: *Dual vectors of the kinematic and the dynamic kind in matrices*

K. Hao: *Dual numbers, Lie algebra and 6R inverse kinematics*

J. Lenarčič: *Backbone curves to approximate minimum joint torque configurations of planar multiple-link manipulators in presence of obstacles*

A DISCUSSION ON METRIC RELATIONS IN SPATIAL KINEMATICS: IMPLICIT AND EXPLICIT FORMULAS AND FUNCTIONAL DEPENDENCE.

E. CERESOLE, P. FANGHELLA, and C. GALLETTI
Istituto di Meccanica Applicata alle Macchine
University of Genoa, Via Opera Pia 15-A, 16145 Genova, Italy

Abstract. This paper deals with the functional dependence of closure equations on unknown pair variables. It is shown that the problem can be studied by considering a small set of metric relations obtained from displacement group theory. Symbolic expressions for such relations are provided and their systematic analysis is performed in order to determine the algebraic and geometric conditions of the functional dependence. Obtained results are also interpreted as relations among invariant geometric properties of displacement groups.

1. Introduction

Research on symbolic robot modelling has mainly been focused on single-loop open-chain mechanisms. Minor attention has been devoted to single-loop spatial closed chains. In recent years, the importance of efficient automatic approaches to multiloop spatial mechanisms has been recognized, and there has been a growing interest in automatic generation of symbolic mathematical models consisting of a minimum number of constraint equations. These topics are dealt with, for instance, in [1-5] and in the references cited therein.

Current approaches to multibody kinematics based on symbolic manipulation and on minimization of the number of equations are acknowledged to provide mathematical models that perform better in various circumstances. A discussion on their advantages is presented in [6]. In particular, it is stressed that these approaches allow closed-form solutions to kinematic closure equations for many mechanisms (even very complex ones) of technical interest, for which they furnish all real solutions to closure equations. Moreover, they provide recursive solutions (a triangular system of polynomial equations with only one unknown at a time) for a large number of mechanisms, and do not require the numerically defined, initial, closed positions of all links, nor the numerical assembly of a mechanism.

It has also been noticed that commercially available symbolic manipulation programs are not adequate to solving multiloop mechanisms, due to the complexity of the expressions to be developed; more practical ways can be followed by using topological and structural information properly [7-12]. For complete, fully automatic achievements of the aforesaid goals, there exist efficient, systematic procedures that allow one to

J. Lenarčič and V. Parenti-Castelli (eds.), Recent Advances in Robot Kinematics, 207-216.
© 1996 *Kluwer Academic Publishers. Printed in the Netherlands.*

"reduce" the mathematical models of many multiloop mechanisms to a sequence of single-loop models.

Reference [8] discusses a formal approach (based on the invariant properties of displacement groups) to accomplishing the following tasks:

1) A constraint analysis for choosing implicit metric relations, i.e., suitable scalar representations of the constraints in a kinematic chain to express the closure equations with a minimum coupling. These relations can have alternative expressions, but, to obtain correct results, they must represent the actual constraints of a mechanism unambiguously.

2) A procedure: i): to state if implicit metric relations are dependent on one (or more) unknown variable(s) of the kinematic chain, and ii) to render such dependence explicit through pair and link transformations; this procedure can also give information on the numerical conditioning of the explicit closure equations.

Some preliminary remarks on the applicability of such a formal approach have been presented in [6]. The main purpose of the present paper is to furnish a general discussion on the dependence of the metric relations on the variables of the pairs in a kinematic chain. This topic is dealt with by a systematic approach, and mechanisms with fully spatial motion are considered.

2. Summary of Metric Relations

There are various well-known ways of deriving loop closure equations for a single-loop chain. The efficiency of special-purpose, local-kinematics-based, structure-oriented methods is evident. The basic idea is that, in a kinematic chain, it is possible to identify particular geometric entities and define reference-independent invariant relations among them that are not affected by the transformations on the kinematic pairs of certain subpaths of the chain. Therefore, if we regard a given kinematic chain (Fig. 1) as a closed sequence of four subchains, p_1, p_2, p_3, and p_4, and if we can find particular geometric entities, IP_1 and IP_2 (at the ends of subchains p_3 and p_4) that are unaffected by the transformations through paths p_1 and p_2, respectively, then we can state that the invariant relations between IP_1 and IP_2 depend only on the variables of the pairs in p_3 and p_4. Closure equations for the single loop considered can be derived by equating the two set of expressions obtained by writing the relations between the IPs as functions of the geometric parameters and of the pair variables in the two sub-chains, p_3 and p_4, respectively. In this way, it is possible to generate directly a set of closure equations with fewer unknowns than the total number of pairs in the chain.

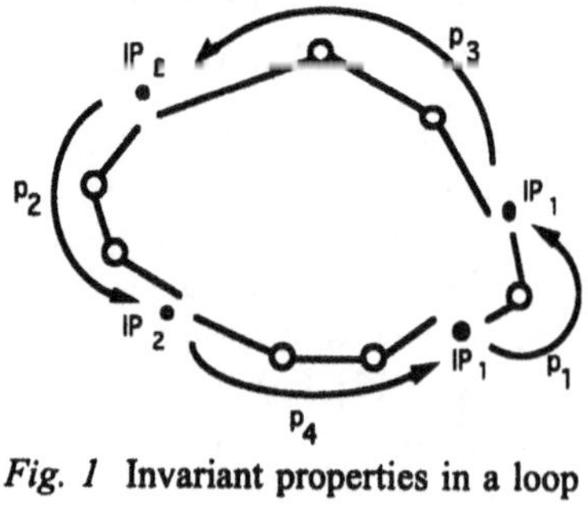

Fig. 1 Invariant properties in a loop

Displacement group theory offers powerful tools for deriving the invariant properties and metric relations suited to representing kinematic constraints. Here, the set of

Invariant Properties, IP, is the set of geometric properties that do not change under all transformations of a group; IP strictly depends on the kind of transformations of a group. In order to write metric relations, IP can be uniquely defined by means of a small number of representative geometric elements (points and unit vectors), as shown in Tab. 1.

Note that the geometric entities used to represent IP can coincide with IP (as in the case of group S); they can be a subset of IP (as in the cases of group R); they can differ from IP and do not need to be invariant for a displacement in the corresponding group (as in the case of the point K for the group F). A complete theory of metric relations is presented in [10], where kinematic constraints (KCs) are defined to deal also with displacements that are subsets of displacement groups. It has been shown that when two displacements in different groups G_1 and G_2 are composed in order to achieve a displacement in a group $G_\cup$, a restriction on the composed displacement into a proper subset of $G_\cup$ can be obtained (by definition, $G_\cup = \min G \supseteq G_1 \cup G_2$). Such a restriction can be expressed through relations between the invariant geometric entities of the groups G_1 and G_2 with respect to $G_\cup$. In the following, such relations will be called "metric relations" (MRs). Since they relate only group invariant properties, they depend on the groups G_1, G_2 and $G_\cup$, and are independent of the actual displacements in G_1 and G_2. We can express them as: $MR(G_1, G_2 \mid G_\cup)$.

TABLE 1 - Invariant properties of displacement groups

Group	Invariant properties	Invariant properties representation in Metric Relations
R	the axis of rotation; all points of the axis; every plane normal to the axis; the direction of rotation	a point (K) of the axis; the rotation axis unit vector ($\mathbf{v}^1$)
P	every direction; lines parallel to the translation axis; every plane containing transl. axis	a point (K) the translation axis unit vector ($\mathbf{v}^1$) a unit vector ($\mathbf{v}^2$) normal to unit vect. ($\mathbf{v}^1$)
C	the axis of displacement; the direction of rotation	a point (K) of the axis; the rotation axis unit vector ($\mathbf{v}^1$)
P2	the planes of displacement; every direction	a point (K); unit vector ($\mathbf{v}^1$) normal to displ. planes; a unit vector ($\mathbf{v}^2$) normal to unit vect. ($\mathbf{v}^1$)
S	the centre-point of rotation	the centre-point (K)
F	all planes of displacement	a point (K); unit vector ($\mathbf{v}^1$) normal to displ. planes
P3	every direction	not used
RP3	the direction of rotation	unit vector ($\mathbf{v}^1$) of the rotation direction

Table 2 presents the set of MRs and their expressions in terms of IP representation. The numbers in the first row of the Tab. 2 specifies the types of metric relations to be applied: expressions for all considered MRs in terms of IP representations are given in the following list, where subscripts 1 and 2 make reference to IPs of G_1 and G_2, respectively:

1: $\mathbf{v}^1_2 \bullet \mathbf{v}^1_1$ 2: $(K_2-K_1) \bullet (\mathbf{v}^1_2 \times \mathbf{v}^1_1)$; 3: $(K_2-K_1) \bullet \mathbf{v}^1_1$;

4: $(K_1-K_2) \bullet \mathbf{v}^1_2$ 5: $(K_2-K_1) \bullet \mathbf{v}^2_1$; 6: $(K_2-K_1) \bullet (\mathbf{v}^1_1 \times \mathbf{v}^2_1)$

7: $[(K_2-K_1) \times \mathbf{v}^1_1] \bullet [(K_2-K_1) \times \mathbf{v}^1_1]$ 8: $[(K_2-K_1) \times \mathbf{v}^1_2] \bullet (\mathbf{v}^1_1 \times \mathbf{v}^1_2)$

9: $(K_2-K_1) \bullet (K_2-K_1)$.

Non-empty entries in Tab. 1 state the types of metric relations to be used for all binary combinations of groups. For examples, two non-parallel revolutes (G_1 = R, G_2 = R, $G_\cup$ = D) imply 4 metric relations of types from 1 to 4; two parallel revolutes (G_1 = R, G_2 = R, $G_\cup$ = F) state the existence of 1 metric relation of type 7; two intersecting revolutes (G_1 = R, G_2 = R, $G_\cup$ = S) state the existence of 1 metric relation of type 1. The meaning of the numbers in each entry is discussed in the following. In few cases, different MRs must be specified according to particular geometric relations between IP_1 and IP_2; the corresponding information is provided in the last column "rel" of the table.

TABLE 2 - Metric relations between two displacement groups

G_1-G_2	$G_\cup$	1	2	3	4	5	6	7	8	9	rel
R-R	D	2	2	2	2						
	F							2			
	S	2									
R-P	RP3		2						2/4		
	F		2								
R-C	D	2	2						2/4		
	RP3							2			
R-P2	RP3				2						
R-S	D			2						2	
R-F	D	2			2						
R-RP3	D	2									
P-C	RP3		2								
P-S	D					2	2				
C-C	D	2	2								
	RP3							2			
C-P2	RP3				2						e
C-F	D	2									-e
	D	2			2						e
C-S	D							4			-g
C-RP3	D	2									
P2-S	D			2							
S-S	D									2	
S-F	D				2						
F-F	D	2									
F-RP3	D	2									
RP3-RP3	D	2									

Legenda: e = displacement axis of C is parallel to the displacement planes of F or P2
g = centre-point of S belongs to the displacement axis of C

A geometric interpretation can be assigned to each MR:
1: angle between two directions; 2: distance between two lines; 3, 4, 5, 6: distance between a point and a plane; 7: distance between a point and a line; 8: distance between a point and a line perpendicular to two lines; 9: distance between two points.
For instance, in R-S case, relation (3) defines the distance of the point K_2 (centre-point of S) to a plane passing through a point K_1 (on the axis of R) and normal to the R axis. Relations (3)-(6) have analogous meanings; therefore only the first one is discussed in the paper.
An insight into previous expressions, useful for further discussion, can be gained by rendering explicit all MRs according to a standard Hartenberg-Denavit notation (Fig. 2)

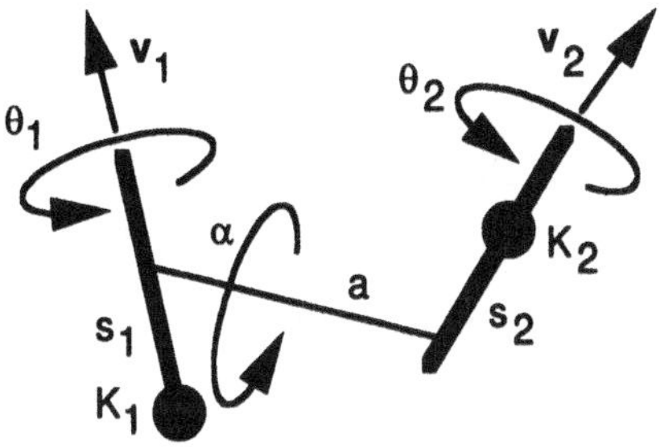

Fig. 2 Notation for explicit expressions of metric relations

1': $\cos(\alpha)$; 2': $a \sin(\alpha)$; 3': $s_2 \cos(\alpha)-s_1$; 7': $(s_2)^2 \sin^2(\alpha)+a^2$;
8': $(s_1)^2 \sin^2(\alpha)$ 9': $(s_1)^2+(a)^2+(s_2)^2-2s_1 s_2 \cos(\alpha)$

These explicit expressions confirm that MRs are independent of the variables of the groups G_1 and G_2. Every time a term θ or s appears in an expression, the expression is applied to a group for which the term is not variable (it is a geometric parameter). For instance, relation (3') contains the terms s_1 and s_2 that are fixed because the relation is applied to R-R, R-S and P2-S binary combinations in Tab. 2.

In order to obtain useful explicit expressions for closure equations from MRs, their values must depend explicitly on the values of the unknown variables of the pairs in chains p_3 and/or p_4 of Fig. 1 (functional dependence). Such a dependence can be established in two ways:
- by writing symbolically the complete explicit expressions for closure equations in the phase of modelling a specific mechanism, taking into account the actual structure of the chain, and by examining if these expressions contain the unknown pair variables;
- in a more efficient way, the dependence can be stated once, and for all by a systematic analysis of the functional dependence on unknown pair variables of all metric relations. This second option, obtained by a generalization of expressions (1')-(9'), is presented in the paper and we give the explicit expressions for all metric relations in terms of unknown variables of both revolute and prismatic pairs. Such expressions allow us to discuss the geometric conditions that assure their functional dependence on the unknowns. The case of a unique unknown pair in each subchain p_3 (or p_4) is considered to derive explicit equations, but this allow us to obtain general results.

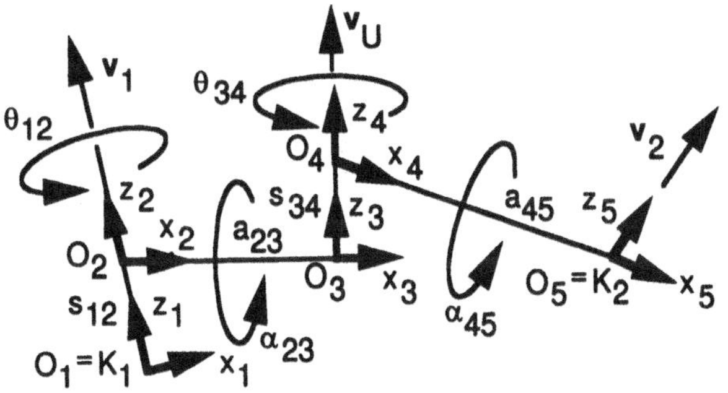

Fig. 3 Two groups with an unknown pair

3. Explicit expressions for MRs for a subchain with one unknown pair (R, P, C)

The desired expressions can be obtained by symbolic evaluation of all MRs through a chain containing: a constraint KC_1 in a group G_1, an unknown pair (P or R), and a constraint KC_2 in a group G_2. In Fig. 3 a Hartenberg-Denavit representation is used to relate the IPs of two generic constraints KC_1 and KC_2 connected by an intermediate pair J^u, whose pair variable is s_{34} if the pair is prismatic, or θ_{34} if the pair is revolute.
Explicit expressions for MRs are:

MR:1 $-\sin\alpha_{23}\sin\alpha_{45}\cos\theta_{34}+\cos\alpha_{23}\cos\alpha_{45}$

MR:2 $a_{45}\cos\alpha_{23}\sin\alpha_{45}+a_{23}\sin\alpha_{23}\cos\alpha_{45}+a_{45}\sin\alpha_{23}\cos\alpha_{45}\cos\theta_{34}-$
$s_{34}\sin\alpha_{23}\sin\alpha_{45}\sin\theta_{34}+a_{23}\cos\alpha_{23}\sin\alpha_{45}\cos\theta_{34}$

MR:3 $a_{45}\sin\alpha_{23}\sin\theta_{34}+s_{34}\cos\alpha_{23}+s_{12}$

MR:7 $a_{45}^2\cos^2\theta_{34}+2\,a_{23}a_{45}\cos\theta_{34}+a_{45}^2\cos^2\alpha_{23}\sin^2\theta_{34}-$
$2\,a_{45}\sin\alpha_{23}\cos\alpha_{23}s_{34}\sin\theta_{34}+s_{34}^2\sin^2\alpha_{23}+a_{23}^2$

MR:8 $a_{45}\sin\alpha_{23}\sin\theta_{34}+s_{34}\cos\alpha_{23}\sin^2\alpha_{45}+s_{34}\sin\alpha_{23}\sin\alpha_{45}\cos\alpha_{45}\cos\theta_{34}+$
$a_{23}\sin\alpha_{23}\sin^2\alpha_{45}\cos\theta_{34}\sin\theta_{34}-a_{23}\cos\alpha_{23}\cos\alpha_{45}\sin\alpha_{45}\sin\theta_{34}+$
$s_{12}\sin^2\alpha_{45}\sin^2\theta_{34}+s_{12}\cos^2\alpha_{23}\sin^2\alpha_{45}\cos^2\theta_{34}+$
$2\,s_{12}\sin\alpha_{23}\cos\alpha_{23}\sin\alpha_{45}\cos\alpha_{45}\cos\theta_{34}+s_{12}\sin^2\alpha_{23}\cos^2\alpha_{45}$

MR:9 $a_{45}^2+2\,a_{23}a_{45}\cos\theta_{34}+s_{34}^2+a_{23}^2+2\,s_{12}s_{34}\cos\alpha_{23}+s_{12}^2+2\,s_{12}a_{45}\sin\alpha_{23}\sin\theta_{34}$

From these expressions, it is possible to derive the degree of the corresponding polynomials with the unknown s_{34} for J^u =P or with the unknown $\tan(\theta_{34}/2)$ for J^u =R. In the remaining part of the paper, only the case J^u = R is considered, for which s_{34} is a constant parameter; the case J^u = P can be derived in an analogous way.
The expressions for MR:2, MR:3, and MR:9 always give polynomials of fixed degree two. This is shown in Tab. 2 by the number "2" written in each non-empty entry. The polynomial derived from MR:7 has a maximum degree 4 due to the presence of the terms $\sin^2\theta_{34}$ and $\cos^2\theta_{34}$: the value of the actual degree depends on G_1 and G_2, as shown in Tab. 2. The polynomial derived from MR:8 has a maximum degree 4 due to quadratic and mixed sin-cos terms: in this case, the value of the actual degree can be 2, depending on geometric relations between IP_1 and IP_2, as shown in Tab. 2. This problem is addressed in the following section.
If more unknown pairs are inserted between G_1 and G_2, the structures of the previous expressions do not change, but the parameters a_{ij} and α_{ij} are not constant and depend on the variable of the inserted pairs.

4. Functional dependence by explicit expressions

A sufficient condition to make the dependence of MRs on the unique unknown pair variable θ_{34} vanish consists in imposing that all coefficients of θ_{34} be zero. This section presents an exhaustive search for these conditions.

<u>*MR:1*</u> $\sin\alpha_{23}\sin\alpha_{45}\cos\theta_{34}$
Term to be equal to zero: 1.1) $\sin\alpha_{23}\sin\alpha_{45}$
Two conditions guarantee thet this term will vanish, i.e.:
$\alpha_{23}=0$ or π; $\alpha_{45}=0$ or π
whose geometrical interpretation (Fig. 3) is:
i) Axis of KC_1 parallel to the axis of J^u; ii) Axis of KC_2 parallel to the axis of J^u

<u>*MR:2*</u> $a_{45}\sin\alpha_{23}\cos\alpha_{45}\cos\theta_{34}-s_{34}\sin\alpha_{23}\sin\alpha_{45}\sin\theta_{34}+a_{23}\cos\alpha_{23}\sin\alpha_{45}\cos\theta_{34}$
Terms to be equal to zero: 2.1) $a_{45}\sin\alpha_{23}\cos\alpha_{45}$; 2.2) $s_{34}\sin\alpha_{23}\sin\alpha_{45}$;

2.3) $a_{23} \cos \alpha_{23} \sin \alpha_{45}$

The seven conditions (A-E") according to which terms 2.1 to 2.3 are equal to zero are shown in the following table: for instance, the condition D' states that, for the values $\alpha_{23} = 0$ or $\alpha_{23} = \pi$, terms 1 and 2 of MR:2 with an R unknown vanish.

condition	A	B	C	D'	D"	E'	E"
parameter	a_{45}	s_{34}	a_{23}	α_{23}	α_{23}	α_{45}	α_{45}
value	0	0	0	$0, \pi$	$\pi/2, 3\pi/2$	$0,\pi$	$\pi/2, 3\pi/2$
terms = 0	1	2	3	1,2	3	2,3	1

By combining the conditions from A to E", seven different sets of values of the parameters that make all terms 2.1-2.3 vanish are shown in the following table.

set	conditions	parameters
i	A+E'	$a_{45} = 0$, $\alpha_{45} = 0,\pi$
ii	C+D'	$a_{23} = 0$, $\alpha_{23} = 0,\pi$
iii	D'+E'	$\alpha_{23} = 0,\pi$, $\alpha_{45} = 0,\pi$
iv	A+B+C	$a_{45} = 0$, $s_{34} = 0$, $a_{23} = 0$
v	A+B+D"	$a_{45} = 0$, $s_{34} = 0$, $\alpha_{23} = \pi/2,3\pi/2$
vi	B+C+E"	$s_{34} = 0$, $a_{23} = 0$, $\alpha_{45} = \pi/2,3\pi/2$
vii	B+D"+E"	$s_{34} = 0$, $\alpha_{23} = \pi/2,3\pi/2$, $\alpha_{45} = \pi/2,3\pi/2$

Using the geometric interpretations of the metric relations given above and making reference to Fig. 3, the following geometrical conditions are deduced from the last table: they can be used to verify when the dependence of MR:2 on the unique unknown pair θ_{34} vanishes:

i) axes of KC_2 and J^u coincident; ii) axes of KC_1 and J^u coincident; iii)axes of KC_1, KC_2 and J^u parallel; iv) axes of KC_1, KC_2 and J^u intersecting; v) axes of KC_1 and J^u perpendicular; axes of KC_2 and J^u intersecting; vi) axes of KC_2 and J^u perpendicular; axes of KC_1 and J^u intersecting; vii) axes of KC_1 and J^u perpendicular; axes of KC_2 and J^u perpendicular; axes of KC_1 and KC_2 coplanar.

_MR:3_ $a_{45} \sin \alpha_{23} \sin \theta_{34}$

Term to be equal to zero: 3.1) $a_{45} \sin \alpha_{23}$

Conditions for zeroes: i) $a_{45} = 0$; ii) $\alpha_{45} = 0, \pi$

Geometrical interpretation:

i) Point K_2 belongs to the axis of J^u; ii) Axis of J^u normal to the plane with normal v_1

_MR:7_ $a_{45}^2 \cos^2\theta_{34} + 2\, a_{23} a_{45} \cos\theta_{34} + a_{45}^2 \cos^2\alpha_{23} \sin^2\theta_{34} + 2\, a_{45}\; s_{34} \sin\alpha_{23} \cos\alpha_{23} \sin\theta_{34}$

Terms to be equal to zero

7.1) a_{45}; 7.2) $a_{23} a_{45}$; 7.3) $a_{45}^2 \cos^2 \alpha_{23}$; 7.4) $a_{45}\, s_{34} \sin \alpha_{23} \cos \alpha_{23}$

or 7.2) $a_{23}\, a45$; 7.4) $a_{45}\, s_{34} \sin \alpha_{23} \cos \alpha 23$; 7.5) $\cos \alpha_{23} \pm 1$

Note that condition 7.5) makes the unknown disappear from the sum ($a_{45}^2 \cos^2 \theta_{34}$ + $a_{45}^2 \cos^2 \alpha_{23} \sin^2 \theta_{34}$). In this case, the quadratic terms in MR:7 disappear, then the degree of the corresponding polynomial with $\tan(\theta_{34}/2)$ is 2 instead of 4, as in the general case of $\cos \alpha_{23} \neq \pm 1$. This conclusion corresponds to the geometrical meanings of the problem. From Tab. 2, MR:7 is used to relate G_1 and G_2 in four situations:

a) two revolutes with parallel axes (G_1=R, G_2=R, $G_\cup$=F);

b) a revolute and a cylindrical pair with parallel axes (G_1=R, G_2=C, $G_\cup$=RP3);
c) two cylindrical pairs with parallel axes (G_1=R, G_2=C, $G_\cup$=RP3);
d) a cylindrical pair and a spherical pair (G_1=C, G_2=S, $G_\cup$=D).
Cases a) to c) are those for which a specific angle α_{23} between axes can be defined, and, in order to have permanent parallel axes between G_1, and G_2, also the axis of J^u must be parallel to them, so that $\cos\alpha_{23} = \pm 1$. For those cases, a quadratic polynomial expression for a closure equation is correct. In case d), an arbitrary angle α_{23} can be assumed, depending on the choice of the references in Fig. 3; then condition $\cos\alpha_{23} = \pm 1$ has no meaning and the most general case of a quartic polynomial closure equation still stands.
Conditions for zeroes:

condition	A	B	C	D'	D"
parameter	a_{23}	s_{34}	a_{45}	α_{23}	α_{23}
value	0	0	0	0, π	$\pi/2, 3\pi/2$
terms = 0	2	4	1,2,3,4	4,5	3,4

Sets of values of the parameters that make all terms vanish:

set	conditions	parameters
i	C	$a_{45} = 0$
ii	A+D'	$a_{23} = 0, \alpha_{23} = 0,\pi$

Geometrical conditions to verify for dependence to vanish.
i) Point K_2 belongs to the axis of J^u; ii) Axes of KC_1 and J^u coincident

MR:8 $a_{45}\sin\alpha_{23}\sin\theta_{34} + s_{34}\sin\alpha_{23}\sin\alpha_{45}\cos\alpha_{45}\cos\theta_{34} + s_{12}\sin^2\alpha_{45}\sin^2\theta_{34} + a_{23}\sin\alpha_{23}\sin^2\alpha_{45}\cos\theta_{34}\sin\theta_{34} - a_{23}\cos\alpha_{23}\cos\alpha_{45}\sin\alpha_{45}\sin\theta_{34} + s_{12}\cos^2\alpha_{23}\sin^2\alpha_{45}\cos^2\theta_{34} + s_{12}\sin\alpha_{23}\cos\alpha_{23}\sin\alpha_{45}\cos\alpha_{45}\cos\theta_{34}$

Terms to be equal to zero
8.1) $a_{45}\sin\alpha_{23}$; 8.2) $s_{34}\sin\alpha_{23}\sin\alpha_{45}\cos\alpha_{45}$; 8.3) $a_{23}\sin\alpha_{23}\sin^2\alpha_{45}$
8.4) $a_{23}\cos\alpha_{23}\cos\alpha_{45}\sin\alpha_{45}$; 8.5) $s_{12}\sin^2\alpha_{45}$; 8.6) $s_{12}\cos^2\alpha_{23}\sin^2\alpha_{45}$;
8.7) $s_{12}\sin\alpha_{23}\cos\alpha_{23}\sin\alpha_{45}\cos\alpha_{45}$
Conditions for zeroes:

condition	A	B	C	D	E'	E"	F'	F"
parameter	s_{12}	a_{23}	s_{34}	a_{45}	α_{23}	α_{23}	α_{45}	α_{45}
value	0	0	0	0, π	0, π	$\pi/2, 3\pi/2$	0,π	$\pi/2, 3\pi/2$
terms = 0	5,6,7	3,4	2	1	1,2,3,7	4,6,7	2,3,4,5,6,7	2,4,7

Sets of values of the parameters that make all terms vanish:

set	conditions	parameters
i	D+F'	$a_{45} = 0, \alpha_{45} = 0,\pi$
ii	E'+F'	$\alpha_{23} = 0,\pi, \alpha_{45} = 0,\pi$
iii	(A)+B+E'	$(s_{12} = 0), a_{23} = 0, \alpha_{23} = 0,\pi$
iv	(A)+E'+F"	$(s_{12} = 0), \alpha_{23} = 0,\pi, \alpha_{45} = \pi/2,3\pi/2$
v	A+B+C+D	$s_{12} = 0, a_{23} = 0, s_{34} = 0, a_{45} = 0$
vi	A+B+D+F"	$s_{12} = 0, a_{23} = 0, a_{45} = 0, \alpha_{45} = \pi/2,3\pi/2$

Note that an inspection of MR:8 shows that sets iii) and iv) do not require $s_{12} = 0$. For instance, the set of values iii) furnishes, for $a_{23} = 0$, and $\alpha_{23} = 0$, the expression for MR:8, $s_{12} \sin^2 \alpha_{45} \sin^2 \theta_{34} + + s_{12} \sin^2 \alpha_{45} \cos^2 \theta_{34} = s_{12} \sin^2 \alpha_{45}$, independent of the variable θ_{34} also for $s_{12} \neq 0$.
When $\alpha_{23} = 0, \pi$ or $\alpha_{45} = 0, \pi$, all quadratic and mixed sin-cos terms in MR:8 vanish. Therefore, in this case, the degree of the corresponding polynomial with $\tan(\theta_{34}/2)$ is 2. This fact is shown by two numbers in the corresponding entries of Tab. 2.
Geometrical conditions to verify for dependence to vanish.
i) axes of J^u and KC_2 coincident; ii) axes of KC_1, J^u and KC_2 parallel (not meaningful); iii) axes of KC_1 and J^u coincident; iv) axes of KC_1 and J^u parallel; axes of KC_2 and J^u perpendicular (not meaningful); v) all three axes intersecting at a common point; vi) axes of KC_1 and J^u intersecting, and axes of KC_2 and J^u intersecting and perpendicular.
Set ii) is not meaningful because MR:8 can be applied to groups R-P and R-C that have no parallel axes (see $G_\cup$ in Tab. 2); for analogous reasons, also set iv) is not meaningful.

MR:9 $\quad 2\, a_{23}\, a_{45} \cos \theta_{34} + 2\, s_{12}\, a_{45} \sin \alpha_{23} \sin \theta_{34}$
Terms to be equal to zero; 9.1) $a_{23}\, a_{45}$; 9.2) $s_{12}\, a_{45} \sin \alpha_{23}$
Conditions for zeroes:

condition	A	B	C	D
parameter	s_{12}	a_{23}	a_{45}	α_{23}
value	0	0	0	0, π
terms = 0	2	1	1,2	2

Sets of values of the parameters that make all terms vanish:

set	conditions	parameters
i	C	$a_{45} = 0$
ii	A+B	$s_{12} = 0$, $a_{23} = 0$
iii	B+D	$a_{23} = 0$, $\alpha_{23} = 0, \pi$

Geometrical conditions to verify for dependence to vanish:
i) K_2 belongs to the axis of J^u; ii) and iii) K_1 belongs to the axis of J^u

5. Concluding remarks

From the previous results (and from similar results obtained for a prismatic pair), it is easy to observe that the dependence of a metric relation on the variable of a pair vanishes in two different situations. The first one is related to group theory and can be summarized in the following general statement.
If G^u is the group of the unknown pair J^u (or, more generally, of a generic constraint KC^u), the dependence vanishes if the invariant properties IP^u of G^u contain the invariant properties IP^1 of G_1 or IP^2 of G_2 that are used in the metric relation MR. Here, the invariant properties are to be intended as the geometric entities listed in column 2 of Tab. 1, not as their representations (column 3 of Tab. 1).
The second situation of independence cannot be expressed in general form. Such a situation cannot be foreseen by group theory and must be deduced from algebraic manipulations of metric relations, as shown in the paper.
For instance, from the seven geometrical conditions that make the dependence of MR:2

on the unknown θ_{34} vanish, we observe that all conditions except v), and its symmetric counterpart vi), correspond to the first situation (group-related conditions). On the contrary, conditions v) and vi) correspond to the second situation.
In a recent work [6], the authors have studied this particular case deeply, and have verified that the existence of an overconstrained mobile mechanism can be established. A systematic investigation on this subject is still in progress.

References

1 W. Schiehlen (ed.) 1990 "Multibody Systems Handbook," Springer-Verlag.

2 S. Stifter and J. Lenarcic (eds) 1990 "Proc. 2nd Int. Workshop on Advances in Robot Kinematics," Springer-Verlag.

3 W. Schiehlen (ed.) 1993 "Advanced Multibody System Dynamics. Simulation and Software Tools," Kluwer Academic Publ.

4 J. Angeles, G. Hommel and P. Kovács (eds) 1993 "Computational Kinematics," Kluwer Academic Publ.

5 P. Nikravesh. and G. Gwanghum 1993 "Systematic Construction of the Equations of Motion for Multibody Systems Containing Closed Kinematics Loops," ASME J. Mechanical Design, pp. 143-149.

6 P. Fanghella and C. Galletti 1995 " Functional Dependence of Metric Relations in Symbolic Kinematics of Robots and Mechanisms," Proc. 9th World Congress on Theory of Machines and Mechanisms, Vol. 2, pp. 803-807, Milano (I).

7 A. Kecskeméthy "On Closed Form Solutions of Multiple-Loop Mechanisms," in [4] pp 263-274

8 P. Fanghella and C. Galletti "A Modular Method for Computational Kinematics," in [4] pp 275-284.

9 M. Hiller and A. Kecskeméthy 1993 "Dynamics of Multibody Systems with Minimal Coordinates," NATO Advanced Study Institute on "Computer Aided Analysis of Rigid and Flexible Mechanical Systems," June 27 - July 9, Troia (P).

10 P. Fanghella and C. Galletti 1994 "Metric Relations and Displacement Groups in Mechanism and Robot Kinematics," ASME, J. Mechanical Design, Vol. 117, pp. 470- 478.

11 G. Kramer 1992 "Solving Geometric Constraint Systems," MIT Press.

12 J. Hervé 1978, "Analyse Structurelle des Mécanismes par Groupe des Déplacements," Mechanism and Machine Theory, Vol. 13, pp. 437-450.

Acknowledgement
This work was supported by a grant of Italian MURST.

TONGUE MODEL FOR CHARACTERIZING VOCAL TRACT KINEMATICS

VELJKO MILENKOVIC
1222 Highway 42, Sturgeon Bay, WI 54235 USA

AND

PAUL H. MILENKOVIC
Department of Electrical and Computer Engineering
University of Wisconsin-Madison
1415 Engineering Dr., Madison, WI 53706 USA

Abstract. Tongue position, which shapes the vocal airway, is a key link in the chain between the intention to speak and production of an acoustic wave. Scientific study of how the brain controls tongue shape to produce speech has motivated the collection of point-parameterized measurements of the tongue with a computer-controlled x-ray tracking system. The sparse nature of these data, necessitated by exposure limitation, makes interpretation challenging. Curve fitting methods developed from work in robot motion planning can 1) visualize tongue shape, 2) reduce the number of kinematic parameters representing tongue shape, and 3) map tongue position relative to the roof of the mouth in an acoustically relevant manner. These new tools for linguistic data visualization and data reduction are applicable to the scientific description of a large (n=60) database of speakers and the synthesis of speech from an articulatory model. (Work supported by USPHS grant DC-00820.)

1. Introduction

Robots simulate motions and tasks performed by biological systems, and even their nomenclature reflects biology: arm, shoulder, elbow, hand, and wrist. The theory of robots and human prosthetic devices is interrelated [1]. In the present work we study the tongue: a biological system. While

J. Lenarčič and V. Parenti-Castelli (eds.), Recent Advances in Robot Kinematics, 217-224.
© 1996 *Kluwer Academic Publishers. Printed in the Netherlands.*

a machine counterpart to the tongue does not exist on the factory floor, the tongue is primarily mechanical in function, having many attributes of a robotic end effector.

The source-filter theory of speech production states that while pitch, stress in speaking and melody in singing, originates in oscillation of the vocal folds in the larynx, the shape of the supraglottal vocal conduit produces the sound colorations that distinguish the vowels [2]. Furthermore, occlusions in the vocal conduit interrupt sound to produce the consonants. A major determinant of the vocal conduit is the rapidly changing shape and positioning of the tongue relative to the fixed shape of the human palate. Methods for representing the resulting geometrical differences between speech sounds are the primary focus of this paper.

This work is motivated by the University of Wisconsin X-ray Microbeam Speech Production Database [3]. Point-parameterized measurements of the tongue, jaw, and lips were made with a computer-controlled x-ray tracking system. The Database includes measurements for 60 subjects performing a phonetically rich set of speaking tasks. These data are relevant to the comparison of how different speakers say the same thing, of critical importance in machine recognition of speech. These data are also relevant to the synthesis of speech from an articulatory description, of importance in improving the naturalness of machine synthesized speech.

The x-ray system measures the tongue in a midsagittal x-y plane at the locations of four radiopaque pellets. The x-axis denotes tongue thrust to the teeth and the y-axis denotes tongue elevation to the roof of the mouth. The z-axis, normal to the midsagittal plane, measures lateral displacement of the tongue, of lesser importance in speech movement. The sparse sampling of the tongue greatly reduces x-ray exposure. The roof of the mouth is measured from a plaster dental cast, and the back of the throat is estimated from an x-ray scan. These surfaces are represented by many more points as the pellet tracking system doesn't come into play. The expression of the tongue pellets in terms of "head space", a coordinate system defined by the bite surface of the upper teeth taken from the dental cast, is obtained from "reference" pellets afixed to the nose and upper front tooth.

We present solutions to three problems: 1) fitting a curve representing the tongue surface through the given four points, 2) reducing the tongue parameters from 8 (4 x-y points) to 5 (x-y of tongue tip, slope of tongue tip, initial value and rate of change of linearly-varying tongue curvature), and 3) mapping the proximity of points on the tongue to points on the palate in a manner consistent with vocal tract acoustics. Curve fitting the tongue surface has been incorporated into a computer animation of articulatory movement with synchonized audio playback. The moving picture is similar to the higher dosage x-ray movies once used for the visualization of speech movement. Reducing the number of parameters may aid the comparisons of speech sounds made by different speakers. Mapping tongue proximity to the palate may help distinguish tongue movements having an acoustic effect from tongue movements that represent redundant degrees of freedom, aiding understanding of the complex motions between speech sounds.

The curve fitting, parameter reduction, and tongue mapping algorithms are encoded in a set of Pascal language subroutines contained in the text file `tinterp.pas`. This 315 line file gives the complete algorithmic and formulaic expression of our methods, and it is available via E-mail to `milenkovic@engr.wisc.edu`. A discussion of the assumptions and methodologies underlying these routines follows.

2. Model for Tongue Surface

A sizable amount of earlier work in robotics had addressed curve fitting between specified points for purposes of end-effector path control ([4], [5], [6], and [7]). We treat fitting the tongue surface to the measured points as a path control problem.

While many fitting techniques exist, each with different results, the decision to measure the tongue at four points is fortuitous when it comes to choosing a method. Four points, sequentially arranged along a non-reentrant path, uniquely define a curve in which local curvature varies linearly with curvilinear distance. This is the most "regular" curve in the 4-point case, in the same sense that a curve with constant curvature (a circular arc), uniquely specified by three points, is the most regular curve for the 3-point case.

The "Curvic" interpolation, developed specifically for modelling the tongue, approximates the linearly varying curvature by three cubic curve segments between the four points. These cubic curves differ from the well-known cubic spline method in having end conditions better suited to the tongue. The cubic curves are used to interpolate 4 points between each measured pair of points: 12 interpolated points plus 4 measured gives 16 points for purposes of the computer animation and other data reductions.

The Curvic algorithm sequences through the segments joining pairs of measured points – see Figure 1. In each segment, the coordinate system is locally rotated to pass the x-axis through the segment endpoints. The interpolated points are computed from a cubic function $y = f(x)$ where y and x are in the rotated coordinate system. When slope $f'(x)$ is small, curvature is approximated by $f''(x)$. For a shallow curve, distance along the x-axis is nearly the same as arc length. Thus, the cubic, with f'' linear in x, approximates linearly varying curvature with arc length. That the tongue does not "bulge out" significantly between adjacent pairs of marker pellets supports these assumptions.

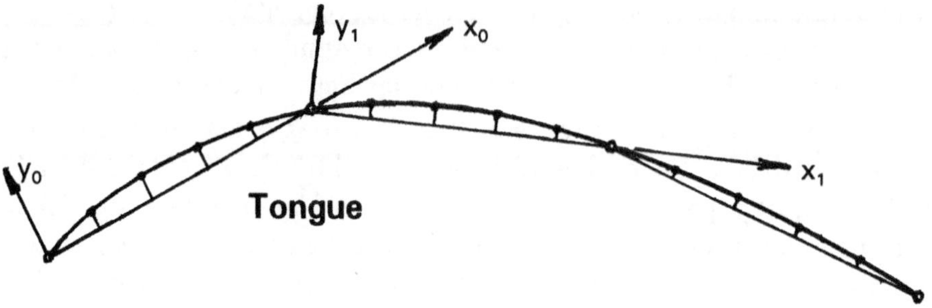

Figure 1. Cubic curve tongue surface.

We place the four interpolated points for a segment on a cubic curve, without explicitly evaluating a cubic. The points are given equal increment of local x; their heights y are chosen to have second differences forming an arithmetic progression, as they should for a cubic curve. Values for these second differences are based on "parameter" computation of the two end curvatures and on requisite continuity conditions in the two places where the three segments join.

Curvic is incorporated into a computer animation of tongue shape with simultaneous audio playback. Trigonometric evaluations are not required, permitting display of 70 frames per second on a PC with a 486 DX2-66 MHz processor.

3. Five-Parameter Tongue Shape

Of the four pellet markers on the tongue, only the tongue tip marker needs to be uniquely specified. For the set of four points that define a particular constant-rate curve, relocating the three points past the tongue tip on that curve will define that same curve. This follows from the fact that the constant-rate curve through any four non-collocated points is unique. Since each of the three points has a degree of freedom, movement along the defined curve, we can reduce 8 parameters (4 x-y points) to 5.

The distance between pellet markers in fact changes during speaking, as observed in computer animation of the x-ray tracking data. This stretching or compression of the tongue surface may be significant for mechanical modelling of tongue musculature, but it is of no acoustic consequence if the tongue surface follows the same curve. Since only the outward surface geometry is relevant for our purposes, five parameters representing the tongue surface are

1, 2 x-y coordinates of tongue tip – define translation,

3 initial slope (angle) of tongue curve – defines orientation,

4, 5 initial curvature, rate of change of curvature with distance – define shape.

Parameters 1 and 2 are in the given data. Computation of the remaining three parameters uses the length of three lines that connect consecutive input points and the sines of two angles formed by these lines. These are readily computed from the pellet marker x-y coordinates without prior need for interpolation. In fact, parameters 4 and 5 help specify the endpoint curvatures used in the Curvic interpolation.

4. Conformal Grid for Vocal Conduit

Calculation of the acoustic properties requires applying a suitable grid system to the vocal conduit. Prior efforts ([8], [9]) have employed an ad-hoc grid system to compute the area function: cross-sectional area as a function of a curvilinear axis starting at the lips and ending at the larynx. The 2-d midsagittal conduit is thus reduced to a 1-d acoustic transmission line, simplifying calculation of acoustic modes.

We proposed replacing the ad-hoc methods with a conformal grid. A 2-d conformal grid follows the rules governing irrotational potential flow, which is the limiting case for the acoustic flow field when wavelength is larger than crosswise dimensions of the conduit (i.e. no "cross modes"). Two perpendicularly intersecting sets of curves constitute the grid: longitudinal streamlines follow the direction of flow or wave propagation, and transverse equipotential lines mark the wave fronts. The density of streamlines gives the fluid momentum (acoustic "L"), and the area of a region bounded by equipotential lines gives the bulk fluid compressibility (acoustic "C"). The conformal grid gives equipotential lines the same local spacing as the streamlines; where the vocal conduit is narrow the grid cells are nearly square.

Closed-form expressions for conformal mapping only exist for boundaries having a simple mathematical description, and mapping is accomplished with the aid of analytic functions of complex variables [10]. To deal with empirical, somewhat irregular boundaries, we have developed rules of geometric constuction to approximate the conformal map. Avoiding singularities or indeterminacies in the construction requires smooth boundaries. The Curvic interpolation of the tongue furnishes a smooth lower-bounding streamline. The palate surface, measured from a dental cast, exhibits surface roughness and corners (notably palate with teeth and palate with back of throat). Solely for the grid constuction, we replace the given palate points by a second-order smoothing of palate points, furnishing the upper-bounding streamline.

Grid computation starts with construction of individual equipotential curves as illustrated in Figure 2. Selecting a lower terminal point T on the tongue, we must find the corresponding upper terminal point P on the palate. A computer search locates an initial P between two given palate points and estimates where on that segment P falls. We assign a "local curvature" k_p to this location for P and run a circular arc of radius $1/k_p$ through the two given palate points (concave downward as illustrated, if $k_p > 0$). The value k_p is computed to run the arc through the next given palate point to the left if P located to the left, through the next palate point to the right if P is on the right, and a weighted compromise for P in the middle.

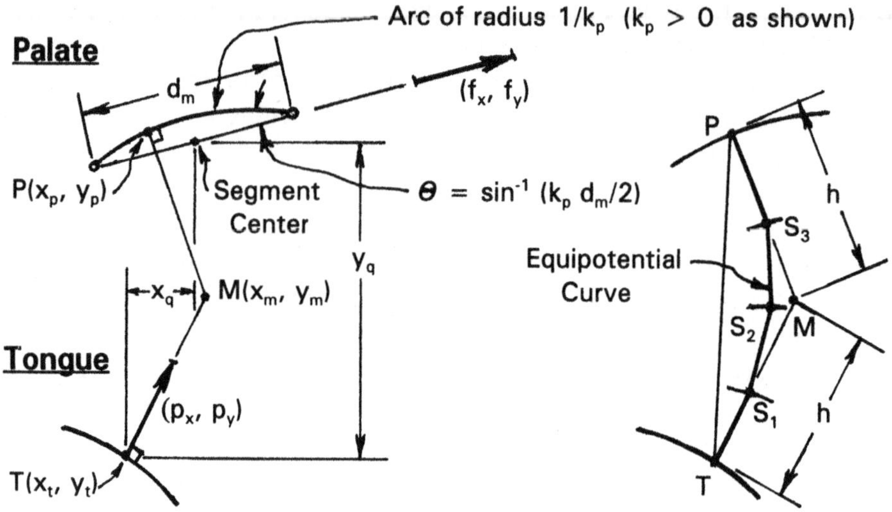

Figure 2. Construction of conformal grid equipotential curve.

The final estimate of P is located so that the radial line normal to the arc at P, and the normal to the tongue at T, intersect at a point M equidistant from P and T (see Figure 2). Points P and M satisfying this condition are computed

$$
\begin{aligned}
h &= \frac{(f_x y_q - f_y x_q)\cos\theta - (x_q^2 + y_q^2 - d_m^2/4)k_p/2}{1 + (f_x p_y - f_y p_x)\cos\theta - (p_x x_q + p_y y_q)k_p}, \\
x_m &= x_t + p_x h, \quad y_m = y_t + p_y h, \\
x_p &= x_t + (p_x - f_y \cos\theta - x_q k_p)h/(1 - hk_p), \\
y_p &= y_t + (p_y + f_x \cos\theta - y_q k_p)h/(1 - hk_p),
\end{aligned}
$$

(symbols defined in Figure 2). This formulation is significant for its use of curvature (reciprocal radius); curvature can range from concave up ($k_p <$

0) to flat ($k_p = 0$) to convave down ($k_p > 0$) without encountering the singularity of infinite radius.

The next step is to locate the intersections of the equipotential line with each of three streamlines. The intersection points S_1, S_2, and S_3 (see Figure 2) depend on local palate curvature k_p, local tongue curvature k_t, and equipotential semilength h. Because the conformal grid spacing is equal in the streamline and equipotential directions, the equipotential spacing so determined locates the next point T on the tongue surface from which to "shoot" another equipotential line. With the initial T on the tongue tip, the grid algorithm works its way along the tongue surface towards the throat.

5. Results

Tongue gestures (Figure 3) were extracted from the "Grandfather Passage" recited by subject 18 (male) from the Database. This script about a hypothetical grandfather typifies fluent, connected speech. The vowel /u/ has a central constriction of the tongue and a rounded narrowing of the lips (not shown in the plot). The vowel /i/ has a rather narrow front constriction of the tongue. The liquid /r/ has a more complex tongue shape, that varies considerably between speakers. Note how local convexity of the tongue surface reduces the spacing between tongue intersection points T and brings the streamlines closer to the tongue surface while concavity does the opposite. Palate curvature k_p has a similar effect on spacing of intersection points P and distance between streamlines.

These results are preliminary inasmusch as they do not completely map the vocal conduit. Information about the lips and front teeth can be determined from the other x-ray markers; information about the throat (pharynx) cavity may have to be gleaned from magnetic resonance (MR) imaging results [9] or extrapolated through the use of acoustic constraints [11]. On the other hand, the tongue model by itself can describe acoustically significant aspects of speech movement.

References

1. D. E. Witney, "The Mathematics of Coordinated Controls of Prosthestic Arms and Manipulators", J. Dynamic Systems, Measurement and Control, (Trans. ASME), 303-309 (1972).
2. G. Fant, Acoustic Theory of Speech Production, Mouton, The Hague (1960).
3. J. Westbury, X-ray Microbeam Speech Production Database User's Handbook, Madison, WI, USA (1994).
4. V. Milenkovic, "Computer Synthesis of Continuous Path Robot Motion", Proc. 5th World Cong. Theory of Machines and Mechanisms, (ASME), 1332-1335, Montreal, Canada (1979).
5. V. Milenkovic, "Coordinates Suitable for Angular Motion Synthesis in Robots", Robots VI Conf. Proc., (SME), 407-420, Detroit, MI, USA (1982).

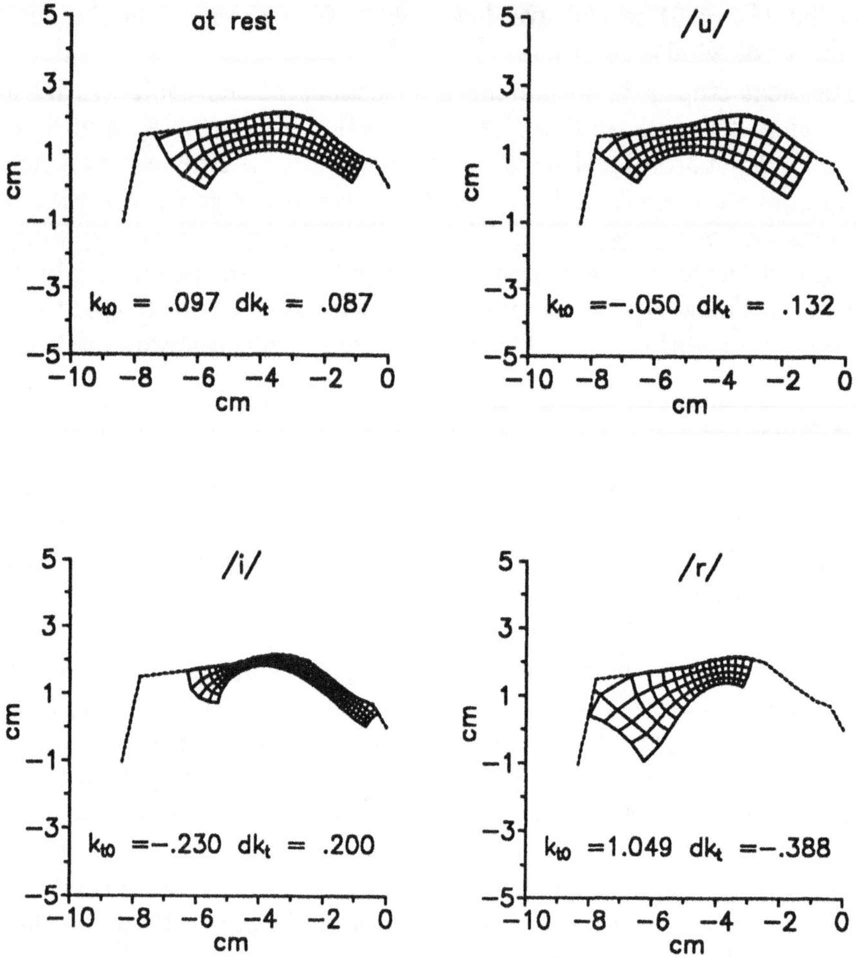

Figure 3. Interpolated tongue shape and conformal grid. Back of the throat is on the left, front teeth on the right. k_{t0} is initial tongue curvature, dk_t rate of change of curvature.

6. M. Loo, and V. Milenkovic, "Multicircular Curvilinear Robot Path Generation", Robots 11 Conf. Proc. and 17th Internat. Sympos. Industrial Robots (SME), 18-19 to 18-27, Chicago, Il, USA (1987).
7. M. Loo, Y. A. Hamidieh, and V. Milenkovic, "Generic Path Control for Robot Applications", Robots 14 Conf. Proc., (SME), 10/49-10/64, Detroit, MI, USA (1990).
8. P. Mermelstein, "Articulatory Model for Study of Speech Production," J. Acoust. Soc. Am. 53, 1070-1082 (1973).
9. T. Baer, J. C. Gore, L. C. Gracco, and P. W. Nye, "Analysis of Vocal Tract Shape and Dimensions using Magnetic Resonance Imaging: Vowels," J. Acoust. Soc. Am. 90, 799-828 (1991).
10. V. L. Streeter, Fluid Dynamics, McGraw-Hill, New York (1948).
11. P. Milenkovic, and J. Dembowski, "Efficient Computation of One-to-Many Mapping between Formant Frequencies and Vocal Tract Shape," ASHA, New Orleans, USA (1994).

A NEW AND EFFICIENT METHOD FOR SYMBOLICAL CALCULATION OF THE JACOBIAN MATRIX

G. GOGU*, A. BARRACO*, P. COIFFET**
Ecole Nationale Supérieure d'Arts et Métiers Paris
151 bld de l'Hôpital, 75013 Paris, France
***Laboratoire de Robotique de Paris*
10-12 Avenue de l'Europe, 78140 Vélizy, France

Abstract. The paper presents a new and efficient method for symbolical computation of the Jacobian matrix available for any general redundant or non-redundant *n* degrees-of-mobility manipulator. Compared with the existing methods the number of required mathematical operations is considerably smaller, making the method remarkably more efficient than the previous methods.

1. Introduction

In industrial robots the Jacobian matrix relates the joints coordinates to the operational coordinates. At present, several methods have been presented for this purpose (Paul, 1981; Vukobratovic and Potkonjak, 1979; Renaud, 1980; Waldron, 1982; Lenarcic, 1983; Orin and Schrader, 1983; Chang-Jin Li *et al.* 1991; Coiffet, 1992; Barraco, 1995). A comparison of the different existing methods can be found in Orin and Schrader (1983). The main characteristics of the new method presented in this paper are:
*The method uses the algorithm presented by one of the authors for geometric modelling of industrial robots (Gogu, 1991). The homogeneous operators of rotation, translation, rotation-translation and translation-rotation are used (Duditza *et al*, 1987; Gogu and Coiffet, 1996; Gogu *et al*, 1996) to simplify both the calculation and the notations specific to Denavit-Hartenberg representation.
*The Jacobian matrix can be calculated with reference to any coordinate system *e*, located at any point on the kinematic chain of a *n* degrees-of-mobility non-redundant or redundant robot.
*A symbolic expression of the Jacobian matrix is obtained, computed using the elements of the 4x4 matrices used for geometric modelling by homogeneous operators.
*Compared with the existing methods the number of the required mathematical operations is considerably smaller, making the method remarkably more efficient than the previous methods.
* The method allows to determine the coordinate system *e* in which the expression of the Jacobian matrix is the simplist one. This coordinate system is called optimal coordinate system for the symbolic calculation of the Jacobian matrix.

J. Lenarčič and V. Parenti-Castelli (eds.), Recent Advances in Robot Kinematics, 225-234.
© 1996 *Kluwer Academic Publishers. Printed in the Netherlands.*

2. Symbolical calculation of the Jacobian matrix

In robot kinematics, the following equations can be written between infinitesimal and finite displacement operators:

$$dA_{f-1,e} = A_f' A_{f-1,e} = A_{f-1,e} A_e' \tag{1}$$

if $f \leq e$ $(f=1, 2,...,n)$ and

$$dA_{e,f} = A_e' A_{e,f} = A_{e,f} A_f' \tag{2}$$

if $f > e$ $(f=1, 2,...,n)$.

From the equations (1) and (2) we have

$$A_e' = A_{f-1,e}^{-1} A_f' A_{f-1,e} \tag{3}$$

if $f \leq e$ $(f=1, 2,...,n)$ and

$$A_e' = A_{e,f} A_f' A_{e,f}^{-1} \tag{4}$$

if $f > e$ $(f=1, 2,...,n)$. The operators from the equations (1)-(4) have the following meaning:

$A_{f-1,e}$ - homogeneous operator giving the position and orientation of the coordinate system e with reference to the coordinate system f-1,

$A_{e,f}$ - homogeneous operator giving the position and the orientation of the coordinate system f with reference to the coordinate system e,

$dA_{f-1,e}$ - operator of the infinitesimal variation of position and orientation of the coordinate system e,

$dA_{e,f}$ - operator of the infinitesimal variation of position and orientation of the coordinate system f,

A_e' - infinitesimal displacement operator with reference to the coordinate system e,

A_f' operator of the infinitesimal displacement in the joint $(f\text{-}1,f)$, $f=1,2,...,n$.

The expressions of the operators A_f' are defined for each translational and rotational joints as it is presented in the Table 1. The infinitesimal operators A_e' can be calculated using the equations (3) and (4) if we know the matrices of the homogeneous operators $A_{f-1,e}$ and $A_{e,f}$ as well as the infinitesimal operators A_f' for each robot joint $(f=1,2,...,n)$. The hexadimensional vector of the infinitesimal displacement corresponding to A_e', calculated for each joint $(f=1,2,...,n)$ with unit values for the infinitesimal displacements $(d_{xf} = 1,\ d_{yf} = 1,\ d_{zf} = 1,\ \delta_{xf} = 1,\ \delta_{yf} = 1,\ \delta_{zf} = 1)$, represents the column vector ${}^eJ_f$ of the Jacobian matrix ${}^eJ_e$:

$$\left[{}^eJ_e\right] = \left[{}^eJ_1 \quad {}^eJ_2 \quad . \quad . \quad . \quad {}^eJ_n\right] \tag{5}$$

where

$$\left[{}^eJ_f\right] = \left[d_{xe} \quad d_{ye} \quad d_{ze} \quad \delta_{xe} \quad \delta_{ye} \quad \delta_{ze}\right]^T, f = 1,2,\ldots,n. \tag{6}$$

The kinematic equation between the linear velocity ${}^ev$ and the angular velocity ${}^e\omega$ and the joint velocity $\dot{q}$ can be written, using the Jacobian matrix ${}^eJ_e$, as:

$$\begin{bmatrix} {}^{e}v \\ {}^{e}\omega \end{bmatrix}_{e} = \begin{bmatrix} {}^{e}J_{e} \end{bmatrix} \begin{bmatrix} \dot{q} \end{bmatrix}, \; e \geq 0. \tag{7}$$

The operator $A_{f-1,e}$ $(f \leq e)$ can be a translation, rotation or rotation-translation homogeneous operator (Gogu and Coiffet, 1996):

$$\left[A_{f-1,e}\right] = \begin{bmatrix} 1 & 0 & 0 & 0 \\ {}^{f-1}r_e^x & {}^{f-1}i_e^x & {}^{f-1}j_e^x & {}^{f-1}k_e^x \\ {}^{f-1}r_e^y & {}^{f-1}i_e^y & {}^{f-1}j_e^y & {}^{f-1}k_e^y \\ {}^{f-1}r_e^z & {}^{f-1}i_e^z & {}^{f-1}j_e^z & {}^{f-1}k_e^z \end{bmatrix} = \begin{bmatrix} 1 & 0 & 0 & 0 \\ {}^{f-1}\bar{r}_e & {}^{f-1}\bar{i}_e & {}^{f-1}\bar{j}_e & {}^{f-1}\bar{k}_e \end{bmatrix}, \tag{8}$$

The vector ${}^{f-1}\bar{r}_e$ gives the position of the origin of the coordinate system e in the coordinate system $f\text{-}1$ (${}^{f-1}\bar{r}_e = \left[O_{f-1}O_e\right]_{f-1}$). The unit vectors ${}^{f-1}\bar{i}_e$, ${}^{f-1}\bar{j}_e$ and ${}^{f-1}\bar{k}_e$ give the orientation of the axes $O_e x_e$, $O_e y_e$ and $O_e z_e$ in the coordinate system $f\text{-}1$. From equation (3), A_e' and implicitely ${}^{e}J_f$ for $f \leq e$ $(f=1, 2,...,e)$ can be expressed, as it is presented in Table 2.

The operator $A_{e,f}$ $(e<f)$ can be a translation, rotation or translation-rotation homogeneous operator (Gogu and Coiffet, 1996):

$$\left[A_{e,f}\right] = \begin{bmatrix} 1 & 0 & 0 & 0 \\ {}^{e}r_f^x & {}^{e}i_f^x & {}^{e}j_f^x & {}^{e}k_f^x \\ {}^{e}r_f^y & {}^{e}i_f^y & {}^{e}j_f^y & {}^{e}k_f^y \\ {}^{e}r_f^z & {}^{e}i_f^z & {}^{e}j_f^z & {}^{e}k_f^z \end{bmatrix} = \begin{bmatrix} 1 & 0 & 0 & 0 \\ {}^{e}\bar{r}_f & {}^{e}\bar{i}_f & {}^{e}\bar{j}_f & {}^{e}\bar{k}_f \end{bmatrix}, \tag{9}$$

The vector ${}^{e}\bar{r}_f$ gives the position of the origin of the coordinate system f in the coordinate system e (${}^{e}\bar{r}_f = \left[O_e O_f\right]_e$). The unit vectors ${}^{e}\bar{i}_f$, ${}^{e}\bar{j}_f$ and ${}^{e}\bar{k}_f$ give the orientation of the axes $O_f x_f$, $O_f y_f$ and $O_f z_f$ in the coordinate system e. From equation (4), A_e' and implicitely ${}^{e}J_f$ for $e<f$ $(f=e+1,e+2,...,n)$ can be expressed, as it is presented in Table 3.

3. Example

The expression of the Jacobian matrix ${}^{3}J_3$ is presented, as an example, in the Table 4, for the kinematic chain in Fig.1. The matrices $A_{f-1,e}$ and $A_{e,f}$ $(f=1,2,...,6$ and $e=3)$ used in the calculation of the Jacobian matrix by the method presented in the paper are also given in the Table 4.

TABLE 1. Expressions of the infinitesimal operators A_f' corresponding to translational and rotational joints *(f-1,f)*

Translational joint	Rotational joint
$\left[A_f'\right]=\left[T_f'(x,d_{xf})\right]=\begin{bmatrix}0&0&0&0\\1&0&0&0\\0&0&0&0\\0&0&0&0\end{bmatrix}d_{xf}$	$\left[A_f'\right]=\left[R_f'(x,d_{xf})\right]=\begin{bmatrix}0&0&0&0\\0&0&0&0\\0&0&0&-1\\0&0&1&0\end{bmatrix}d_{xf}$
if the joint *(f-1,f)* has as axes $O_f x_f \equiv O_{f-1}x_{f-1}$	
$\left[A_f'\right]=\left[T_f'(y,d_{yf})\right]=\begin{bmatrix}0&0&0&0\\0&0&0&0\\1&0&0&0\\0&0&0&0\end{bmatrix}d_{yf}$	$\left[A_f'\right]=\left[R_f'(y,\delta_{yf})\right]=\begin{bmatrix}0&0&0&0\\0&0&0&1\\0&0&0&0\\0&-1&0&0\end{bmatrix}\delta_{yf}$
if the joint *(f-1,f)* has as axes $O_f y_f \equiv O_{f-1}y_{f-1}$	
$\left[A_f'\right]=\left[T_f'(z,d_{zf})\right]=\begin{bmatrix}0&0&0&0\\0&0&0&0\\0&0&0&0\\1&0&0&0\end{bmatrix}d_{zf}$	$\left[A_f'\right]=\left[R_f'(z,\delta_{zf})\right]=\begin{bmatrix}0&0&0&0\\0&0&-1&0\\0&1&0&0\\0&0&0&0\end{bmatrix}\delta_{zf}$
if the joint *(f-1,f)* has as axes $O_f z_f \equiv O_{f-1}z_{f-1}$	

TABLE 2. Expressions of the infinitesimal operators A_e' and of the column vector ${}^eJ_f$, $f=1,2,\dots,e$ *(f≤e)* corresponding to the infinitesimal displacement in the joint *(f-1,f)*

Translational joint

$$\left[A_e'\right]=\begin{bmatrix}0&0&0&0\\ {}^{f-1}i_e^x&0&0&0\\ {}^{f-1}j_e^x&0&0&0\\ {}^{f-1}k_e^x&0&0&0\end{bmatrix} \quad\Rightarrow\quad \left[{}^eJ_f\right]=\left[{}^{f-1}i_e^x\quad {}^{f-1}j_e^x\quad {}^{f-1}k_e^x\quad 0\quad 0\quad 0\right]^T$$

if the translational joint *(f-1,f)* has as axes $O_f x_f \equiv O_{f-1}x_{f-1}$.

$$\left[A_e'\right]=\begin{bmatrix}0&0&0&0\\ {}^{f-1}i_e^y&0&0&0\\ {}^{f-1}j_e^y&0&0&0\\ {}^{f-1}k_e^y&0&0&0\end{bmatrix} \quad\Rightarrow\quad \left[{}^eJ_f\right]=\left[{}^{f-1}i_e^y\quad {}^{f-1}j_e^y\quad {}^{f-1}k_e^y\quad 0\quad 0\quad 0\right]^T$$

if the translational joint *(f-1,f)* has as axes $O_f y_f \equiv O_{f-1}y_{f-1}$.

TABLE 2. (Continuation)

$$[A_e'] = \begin{bmatrix} 0 & 0 & 0 & 0 \\ {}^{f-1}i_e^z & 0 & 0 & 0 \\ {}^{f-1}j_e^z & 0 & 0 & 0 \\ {}^{f-1}k_e^z & 0 & 0 & 0 \end{bmatrix} \quad \Rightarrow \quad [{}^eJ_f] = \left[{}^{f-1}i_e^z \quad {}^{f-1}j_e^z \quad {}^{f-1}k_e^z \quad 0 \quad 0 \quad 0\right]^T$$

if the translational joint *(f-1,f)* has as axes $O_f z_f \equiv O_{f-1} z_{f-1}$.

Rotational joint

$$A_e' = \begin{bmatrix} 0 & 0 & 0 & 0 \\ {}^{f-1}r_e^y\,{}^{f-1}i_e^z - {}^{f-1}r_e^z\,{}^{f-1}i_e^y & 0 & -{}^{f-1}k_e^x & {}^{f-1}j_e^x \\ {}^{f-1}r_e^y\,{}^{f-1}j_e^z - {}^{f-1}r_e^z\,{}^{f-1}j_e^y & {}^{f-1}k_e^x & 0 & -{}^{f-1}i_e^x \\ {}^{f-1}r_e^y\,{}^{f-1}k_e^z - {}^{f-1}r_e^z\,{}^{f-1}k_e^y & -{}^{f-1}j_e^x & {}^{f-1}i_e^x & 0 \end{bmatrix} \quad \Rightarrow$$

$$[{}^eJ_f] =$$

$$\left[{}^{f-1}r_e^y\,{}^{f-1}i_e^z - {}^{f-1}r_e^z\,{}^{f-1}i_e^y \,\middle|\, {}^{f-1}r_e^y\,{}^{f-1}j_e^z - {}^{f-1}r_e^z\,{}^{f-1}j_e^y \,\middle|\, {}^{f-1}r_e^y\,{}^{f-1}k_e^z - {}^{f-1}r_e^z\,{}^{f-1}k_e^y \,\middle|\, {}^{f-1}i_e^x \,\middle|\, {}^{f-1}j_e^x \,\middle|\, {}^{f-1}k_e^x\right]^T$$

if the rotational joint *(f-1,f)* has as axes $O_f x_f \equiv O_{f-1} x_{f-1}$.

$$A_e' = \begin{bmatrix} 0 & 0 & 0 & 0 \\ {}^{f-1}r_e^z\,{}^{f-1}i_e^x - {}^{f-1}r_e^x\,{}^{f-1}i_e^z & 0 & -{}^{f-1}k_e^y & {}^{f-1}j_e^y \\ {}^{f-1}r_e^z\,{}^{f-1}j_e^x - {}^{f-1}r_e^x\,{}^{f-1}j_e^z & {}^{f-1}k_e^y & 0 & -{}^{f-1}i_e^y \\ {}^{f-1}r_e^z\,{}^{f-1}k_e^x - {}^{f-1}r_e^x\,{}^{f-1}k_e^z & -{}^{f-1}j_e^y & {}^{f-1}i_e^y & 0 \end{bmatrix} \quad \Rightarrow$$

$$[{}^eJ_f] =$$

$$\left[{}^{f-1}r_e^z\,{}^{f-1}i_e^x - {}^{f-1}r_e^x\,{}^{f-1}i_e^z \,\middle|\, {}^{f-1}r_e^z\,{}^{f-1}j_e^x - {}^{f-1}r_e^x\,{}^{f-1}j_e^z \,\middle|\, {}^{f-1}r_e^z\,{}^{f-1}k_e^x - {}^{f-1}r_e^x\,{}^{f-1}k_e^z \,\middle|\, {}^{f-1}i_e^y \,\middle|\, {}^{f-1}j_e^y \,\middle|\, {}^{f-1}k_e^y\right]^T$$

if the rotational joint *(f-1,f)* has as axes $O_f y_f \equiv O_{f-1} y_{f-1}$.

$$A_e' = \begin{bmatrix} 0 & 0 & 0 & 0 \\ {}^{f-1}r_e^x\,{}^{f-1}i_e^y - {}^{f-1}r_e^y\,{}^{f-1}i_e^x & 0 & -{}^{f-1}k_e^z & {}^{f-1}j_e^z \\ {}^{f-1}r_e^x\,{}^{f-1}j_e^y - {}^{f-1}r_e^y\,{}^{f-1}j_e^x & {}^{f-1}k_e^z & 0 & -{}^{f-1}i_e^z \\ {}^{f-1}r_e^x\,{}^{f-1}k_e^y - {}^{f-1}r_e^y\,{}^{f-1}k_e^x & -{}^{f-1}j_e^z & {}^{f-1}i_e^z & 0 \end{bmatrix} \quad \Rightarrow$$

$$[{}^eJ_f] = \left[{}^{f-1}r_e^x\,{}^{f-1}i_e^y - {}^{f-1}r_e^y\,{}^{f-1}i_e^x \,\middle|\, {}^{f-1}r_e^x\,{}^{f-1}j_e^y - {}^{f-1}r_e^y\,{}^{f-1}j_e^x \,\middle|\, {}^{f-1}r_e^x\,{}^{f-1}k_e^y - {}^{f-1}r_e^y\,{}^{f-1}k_e^x \,\middle|\, {}^{f-1}i_e^z \,\middle|\, {}^{f-1}j_e^z \,\middle|\, {}^{f-1}k_e^z\right]^T$$

if the rotational joint *(f-1,f)* has as axes $O_f z_f \equiv O_{f-1} z_{f-1}$.

TABLE 3. Expressions of the infinitesimal operators $A_e^{'}$ and of the column vector ${}^eJ_f$, $f=e+1,e+2,...,n$ $(e<f)$ corresponding to the infinitesimal displacement in joint $(f\text{-}1,f)$

Translational joint

$$\left[A_e^{'}\right]=\begin{bmatrix} 0 & 0 & 0 & 0 \\ {}^e i_f^x & 0 & 0 & 0 \\ {}^e i_f^y & 0 & 0 & 0 \\ {}^e i_f^z & 0 & 0 & 0 \end{bmatrix} \qquad \Rightarrow \qquad \left[{}^eJ_f\right]=\left[{}^e i_f^x \quad {}^e i_f^y \quad {}^e i_f^z \quad 0 \quad 0 \quad 0\right]^T$$

if the translational joint $(f\text{-}1,f)$ has as axes $O_f x_f \equiv O_{f-1} x_{f-1}$.

$$\left[A_e^{'}\right]=\begin{bmatrix} 0 & 0 & 0 & 0 \\ {}^e j_f^x & 0 & 0 & 0 \\ {}^e j_f^y & 0 & 0 & 0 \\ {}^e j_f^z & 0 & 0 & 0 \end{bmatrix} \qquad \Rightarrow \qquad \left[{}^eJ_f\right]=\left[{}^e j_f^x \quad {}^e j_f^y \quad {}^e j_f^z \quad 0 \quad 0 \quad 0\right]^T$$

if the translational joint $(f\text{-}1,f)$ has as axes $O_f y_f \equiv O_{f-1} y_{f-1}$.

$$\left[A_e^{'}\right]=\begin{bmatrix} 0 & 0 & 0 & 0 \\ {}^e k_f^x & 0 & 0 & 0 \\ {}^e k_f^y & 0 & 0 & 0 \\ {}^e k_f^z & 0 & 0 & 0 \end{bmatrix} \qquad \Rightarrow \qquad \left[{}^eJ_f\right]=\left[{}^e k_f^x \quad {}^e k_f^y \quad {}^e k_f^z \quad 0 \quad 0 \quad 0\right]^T$$

if the translational joint $(f\text{-}1,f)$ has as axes $O_f z_f \equiv O_{f-1} z_{f-1}$.

Rotational joint

$$A_e^{'}=\begin{bmatrix} 0 & 0 & 0 & 0 \\ {}^e r_f^y\,{}^e i_f^z - {}^e r_f^z\,{}^e i_f^y & 0 & -{}^e i_f^z & {}^e i_f^y \\ {}^e r_f^z\,{}^e i_f^x - {}^e r_f^x\,{}^e i_f^z & {}^e i_f^z & 0 & -{}^e i_f^x \\ {}^e r_f^x\,{}^e i_f^y - {}^e r_f^y\,{}^e i_f^x & -{}^e i_f^y & {}^e i_f^x & 0 \end{bmatrix} \qquad \Rightarrow$$

$$\left[{}^eJ_f\right]=\left[{}^e r_f^y\,{}^e i_f^z - {}^e r_f^z\,{}^e i_f^y \,\middle|\, {}^e r_f^z\,{}^e i_f^x - {}^e r_f^x\,{}^e i_f^z \,\middle|\, {}^e r_f^x\,{}^e i_f^y - {}^e r_f^y\,{}^e i_f^x \,\middle|\, {}^e i_f^x \,\middle|\, {}^e i_f^y \,\middle|\, {}^e i_f^z\right]$$

if the rotational joint $(f\text{-}1,f)$ has as axes $O_f x_f \equiv O_{f-1} x_{f-1}$.

$$A_e^{'}=\begin{bmatrix} 0 & 0 & 0 & 0 \\ {}^e r_f^y\,{}^e j_f^z - {}^e r_f^z\,{}^e j_f^y & 0 & -{}^e j_f^z & {}^e j_f^y \\ {}^e r_f^z\,{}^e j_f^x - {}^e r_f^x\,{}^e j_f^z & {}^e j_f^z & 0 & -{}^e j_f^x \\ {}^e r_f^x\,{}^e j_f^y - {}^e r_f^y\,{}^e j_f^x & -{}^e j_f^y & {}^e j_f^x & 0 \end{bmatrix} \qquad \Rightarrow$$

$$\left[{}^eJ_f\right]=\left[{}^e r_f^y\,{}^e j_f^z - {}^e r_f^z\,{}^e j_f^y \,\middle|\, {}^e r_f^z\,{}^e j_f^x - {}^e r_f^x\,{}^e j_f^z \,\middle|\, {}^e r_f^x\,{}^e j_f^y - {}^e r_f^y\,{}^e j_f^x \,\middle|\, {}^e j_f^x \,\middle|\, {}^e j_f^y \,\middle|\, {}^e j_f^z\right]$$

if the rotational joint $(f\text{-}1,f)$ has as axes $O_f y_f \equiv O_{f-1} y_{f-1}$.

TABLE 3. (Continuation)

$$A_e' = \begin{bmatrix} 0 & 0 & 0 & 0 \\ {}^e r_f^y\,{}^e k_f^z - {}^e r_f^z\,{}^e k_f^y & 0 & -{}^e k_f^z & {}^e k_f^y \\ {}^e r_f^z\,{}^e k_f^x - {}^e r_f^x\,{}^e k_f^z & {}^e k_f^z & 0 & -{}^e k_f^x \\ {}^e r_f^x\,{}^e k_f^y - {}^e r_f^y\,{}^e k_f^x & -{}^e k_f^y & {}^e k_f^x & 0 \end{bmatrix} \Rightarrow$$

$$\left[{}^e J_f\right] = \left[{}^e r_f^y\,{}^e k_f^z - {}^e r_f^z\,{}^e k_f^y \middle| {}^e r_f^z\,{}^e k_f^x - {}^e r_f^x\,{}^e k_f^z \middle| {}^e r_f^x\,{}^e k_f^y - {}^e r_f^y\,{}^e k_f^x \middle| {}^e k_f^x \middle| {}^e k_f^y \middle| {}^e k_f^z\right]$$

if the rotational joint *(f-1,f)* has as axes $O_f z_f \equiv O_{f-1} z_{f-1}$.

TABLE 4. Homogeneous operators $A_{f-1,f}$, $A_{f-1,e}$ and $A_{e,f}$ (*f=1,2,...,6* and *e=3*) used for the calculation of the Jacobian matrix ${}^3J_3$ by the method proposed by the authors

$$A_{01} = K_{01'}^z(\varphi_{10}) = R_{01}^z(\varphi_{10})T_{11'}^y(b_1)$$

$$A_{12} = K_{12'}^x(\varphi_{21}) = R_{1'2}^x(\varphi_{21})T_{22''}^x(a_2)T_{2''2'}^z(b_2)$$

$$A_{23} = R_{2'3}^x(\varphi_{32})$$

$$A_{34} = Q_{34}^z(\varphi_{43}) = T_{33''}^y(a_3)T_{3''3'}^x(b_3)R_{3'4}^z(\varphi_{43})$$

$$A_{45} = Q_{45}^z(\varphi_{54}) = T_{44''}^z(a_4)T_{4''4'}^x(b_4)R_{4'5}^z(\varphi_{54})$$

$$A_{56} = Q_{56}^x(\varphi_{65}) = T_{55'}^y(b_5)R_{5'6}^x(\varphi_{65})$$

$$A_{13} = A_{12}\,A_{23}$$

$$A_{03} = A_{01}\,A_{13} = A_{01}\,A_{12}A_{23}$$

$$A_{35} = A_{34}\,A_{45}$$

$$A_{36} = A_{35}\,A_{56} = A_{34}\,A_{45}A_{56}$$

$$[A_{01}] = \begin{bmatrix} 1 & 0 & 0 & 0 \\ -b_1S10 & C10 & -S10 & 0 \\ b_1C10 & S10 & C10 & 0 \\ 0 & 0 & 0 & 1 \end{bmatrix} \quad [A_{23}] = \begin{bmatrix} 1 & 0 & 0 & 0 \\ 0 & C32 & 0 & S32 \\ 0 & 0 & 1 & 0 \\ 0 & -S32 & 0 & C32 \end{bmatrix} \quad [A_{45}] = \begin{bmatrix} 1 & 0 & 0 & 0 \\ b_4 & C54 & -S54 & 0 \\ 0 & S54 & C54 & 0 \\ a_4 & 0 & 0 & 1 \end{bmatrix}$$

$$[A_{12}] = \begin{bmatrix} 1 & 0 & 0 & 0 \\ a_2 & 1 & 0 & 0 \\ -b2S21 & 0 & C21 & -S21 \\ b2C21 & 0 & S21 & C21 \end{bmatrix} \quad [A_{34}] = \begin{bmatrix} 1 & 0 & 0 & 0 \\ b_3 & C43 & -S43 & 0 \\ a_3 & S43 & C43 & 0 \\ 0 & 0 & 0 & 1 \end{bmatrix} \quad [A_{56}] = \begin{bmatrix} 1 & 0 & 0 & 0 \\ 0 & 1 & 0 & 0 \\ b_5 & 0 & C65 & -S65 \\ 0 & 0 & S65 & C65 \end{bmatrix}$$

$$[A_{13}] = \begin{bmatrix} 1 & 0 & 0 & 0 \\ a_2 & C32 & 0 & S32 \\ -b_2S21 & S21S32 & C21 & -S21C32 \\ b_2C21 & -C21S32 & S21 & C21C32 \end{bmatrix} \quad [A_{35}] = \begin{bmatrix} 1 & 0 & 0 & 0 \\ b_3+b_4C43 & C543 & -S543 & 0 \\ a_3+b_4S43 & S543 & C543 & 0 \\ a_4 & 0 & 0 & 1 \end{bmatrix}$$

TABLE 4. (Continuation)

$$[A_{03}]=\begin{bmatrix} 1 & 0 & 0 & 0 \\ -b_1S10+a_2C10+b_2S10S21 & C10C32-S10S21S32 & -S10C21 & C10S32+S10S21C32 \\ b_1C10+a_2S10-b_2C10S21 & S10C32+C10S21S32 & C10C21 & S10S32-C10S21C32 \\ b_2C21 & -C21S32 & S21 & C21C32 \end{bmatrix}$$

$$[A_{36}]=\begin{bmatrix} 1 & 0 & 0 & 0 \\ b_3+b_4C43-b_5S543 & C543 & -S543C65 & S543S65 \\ a_3+b_4S43+b_5C543 & S543 & C543C65 & -C543S65 \\ a4 & 0 & S65 & C65 \end{bmatrix}$$

$$[{}^3J_3]=$$

$$\begin{bmatrix} b_2S21C32-b_1C32+a_2S21S32 & 0 & 0 & a_3 & a_3+b_4S43 & -a_4S543 \\ a_2C21 & -b_2 & 0 & -b_3 & -b_3-b_4C43 & a_4C543 \\ -a_2S21C32-b_1S32+b_2S21S32 & 0 & 0 & 0 & 0 & b_3S543-b_5-a_3C543+b_4S54 \\ -C21S32 & C32 & 0 & 0 & 0 & C543 \\ S21 & 0 & 1 & 0 & 0 & S543 \\ C21C32 & S32 & 0 & 1 & 1 & 0 \end{bmatrix}$$

TABLE 5. Number of mathematical operations necessary for the calculation of the Jacobian matrix by different methods

Method	Jacobian matrix	Number of mathematical operations		
		Multiplications	Additions	Cosines/sines
Paul (1981)	${}^EJ_E$	162	69	12
Vukobratovic-Potkonjak (1979)	${}^EJ_E$	279	76	12
Renaud (1980)	${}^eJ_e$	93	24	10
Waldron (1982)	${}^0J_0$	125	52	10
Olson-Ribble	${}^0J_E$	169	88	12
Orin-Schrader(1983)	${}^EJ_E$	155	65	12
Li-Hemami-Sankar (1991)	${}^eJ_e$	46	31	10
Method proposed by the authors	${}^3J_3$	9	24	8

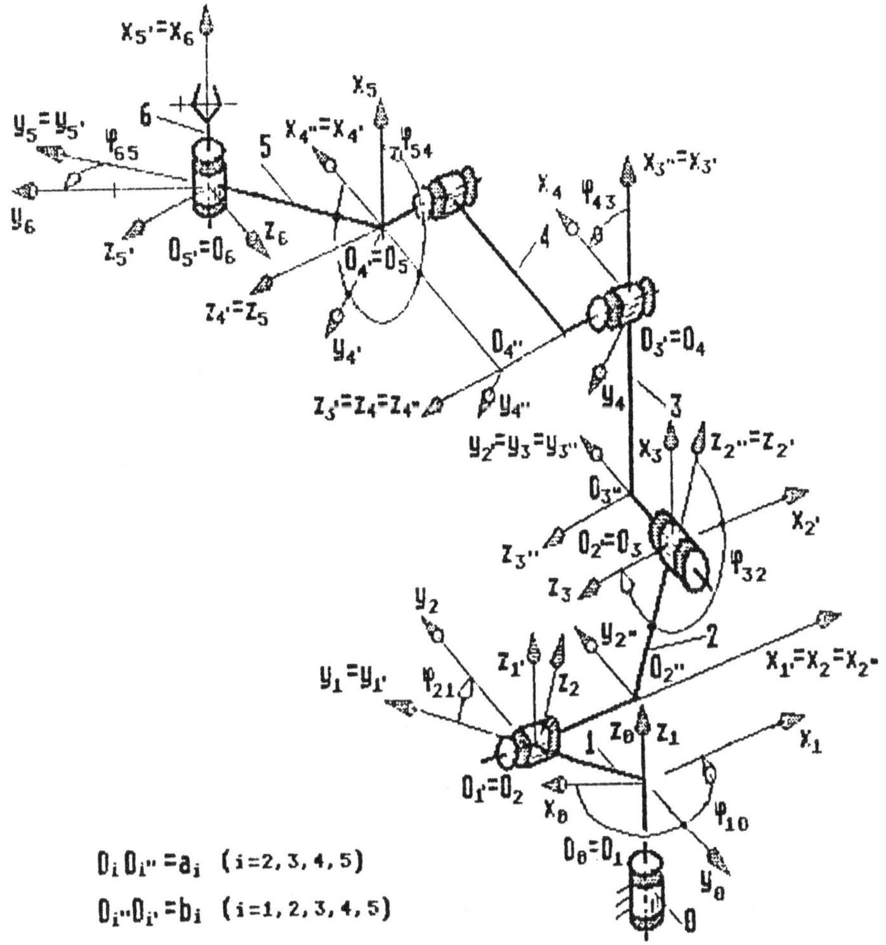

Figure 1. The coordinate systems used in the calculation of the Jacobian matrix $^{e}J_{e}$

4. Computational comparison

The Table 5 indicates the number of mathematical operations and functions called for several methods for the robot presented in Fig.1.

5. Conclusions

This paper has presented a new and efficient method for symbolical computation of the Jacobian matrix. As compared with other existing methods, the one proposed in this paper is the fastest. The comparison is based on the required number of mathematical operations for computing ${}^{e}J_{e}$ for the 6 degrees-of-mobility robot presented in Fig.1. The method allows to determine the coordinate system e in which the expression of the Jacobian matrix is the simplist one. This coordinate system is called optimal coordinate system for the symbolic calculation of the Jacobian matrix and its determination will be presented by the authors , for different kinematic chains in a future paper.

References

Paul, R.P. (1981) *Robot Manipulators: Mathematics, Programming, and control* MIT Press.

Vukobratovic, M., Potkonjak, V.(1979) Contribution of the forming of computer methods for automatic modelling of spatial mechanisms motions. *Mech. Mach. Theory*, Vol 14, , pp. 179-200.

Renaud, M. (1980) Calcul de la matrice Jacobienne nécessaire à la commande coordonnée d'un manipulateur. *Mech. Mach. Theory*, Vol. 5, No. 2, pp. 81-92.

Waldron, K. J. (1982) Geometrically based manipulator rate control algorithms. *Mech. Mach. Theory*, Vol. 17, No.6,pp 379-385.

Lenarcic, J. (1983) A new method for calculating the Jacobian for a robot manipulator. *Robotica,* Vol. 1, part 4, pp. 205-209.

Orin, D.E., Schrader, W.W. (1983) Efficient Jacobian determination for robot manipulators. *Proc. 6th World Congr. on Theory of Machines and Mechanisms*, New Delhi Vol. 2, pp. 994-997.

Chang-Jin L., Hemami, A., Sankar, T.S. (1991) An efficient computational method of Jacobian for robot manipulators, *Advances in Robot Kinematics (Stifter, S., Lenarcic J.,Eds.),* Springer-Verlag, Wien, pp. 63-70

Coiffet. P.(1992) *La Robotique: Principes et Applications*. 3ème Edition Hermes, Paris

Barraco, A.(1995) *Dynamique des systèmes*, ENSAM, DEA Robotique, Paris.

Duditza, Fl., Gogu, Gr., Diaconescu, D. (1987) Kinematics of robot guidance mechanisms based on homogeneous operators. *Proc. 7th World Congr. on Theory of Machines and Mechanisms,* Sevilla, Vol. 2, pp. 1183-1186.

Gogu, Gr., Coiffet, Ph. (1996) *Mathématiques pour la Robotique. Tome 1. Représentation des mouvements des corps*. Hermes, Paris.

Gogu, Gr., Coiffet, Ph., Barraco, A. *Mathématiques pour la Robotique. Tome 2.Modélisation géométrique et cinématique*, Hermes, Paris, 1996 (in press).

QUATERNION BASED FORCE–TORQUE ANALYSIS OF OPEN SERIAL KINEMATIC STRUCTURES

K. DOBROVODSKÝ
The Institute of Control Theory and Robotics,
Slovak Academy of Sciences
Dúbravská cesta 9, 842 37 Bratislava, Slovakia

Abstract. Force–torque analysis of open serial kinematic structures is considered in this work. The close relation between vector and pure quaternion representation of force and torque balance equalities is presented together with the quaternion position and orientation representation of links. These equalities embody the common starting point of the relevant problems concerned with the static analysis. One task is to establish prismatic joint forces and revolute joint torques caused by an external endpoint force, torque, and the gravity of links in the given endpoint position and orientation. Since the inverse kinematics is involved in this problem, redundant structures admit additional restrictions in the form of the cost function depending on joint variables. Another task is to establish the external force and torque, having measured forces in prismatic and torques in revolute joints in the given position of joints. The aim of the paper is to present the common base of the vector force–torque balance equalities and the quaternion representation of positions and orientations of any open multiple link redundant or nonredundant serial kinematic structure.

1. Introduction

The first goal of the analysis is to establish forces along axes of prismatic joints and torques about axes of revolute joints of the frictionless mechanism under the assumption that the force and the torque is applied to the endpoint of any open serial kinematic structure. If the weight of the links is neglected, the transpose of the Jacobian matrix is frequently used to express the relation between the joint force–torque n-tuple and the endpoint force

J. Lenarčič and V. Parenti-Castelli (eds.), Recent Advances in Robot Kinematics, 235-244.
© 1996 *Kluwer Academic Publishers. Printed in the Netherlands.*

and torque. Since not only the forces along and torques about the joint axes may be critical and the gravity of the links usually may not be neglected, we will study the general equalities that hold in the static position of a manipulator arm. We will express both, the position-orientation equalities and the vector force–torque balance equalities in terms of quaternions. Finally, we will show how the balance equalities and the position-orientation equalities provide the two problem formulations depending on what is apriori given and what are the unknowns.

In the case of a given position of the end effector of a nonredundant 6 DOF manipulator we have a finite (up to 16) number of positions of the links. For the given external force and/or torque applied to the endpoint of the structure, forces and torques between links for each set of link positions are unambiguous. Redundant kinematic structures admit optimization of these forces and/or torques by some kind of the cost function.

Vice versa, in the case of a given (measured) forces and torques in joints between the links, the external force and/or torque is the unknown. This problem formulation leads to a set of linear equations which matrix depends on the values of joint variables.

2. Quaternion position and orientation representation

An alternative way to vectors and tensors how to express positions and orientations of rigid bodies in the 3–D space are quaternion pairs introduced in Dobrovodský (1994). Recall that a quaternion is defined as

$$q = q_0 + q_1 i + q_2 j + q_3 k \tag{1}$$

where q_0, q_1, q_2, q_3 are real coordinates and unit quaternions i, j, k are quaternion generators. They correspond with the three unit vectors of an orthonormal basis. A quaternion, which has only the vector part (i.e. $q_0 = 0$), can represent a shift in a 3–dimensional space. We call it a pure (translating) quaternion. Thus a pure quaternion and a vector are synonymous.

The product of quaternion generators i, j, k is defined as

$$i^2 = j^2 = k^2 = -1, \quad ij = k, \quad jk = i, \quad ki = j \tag{2}$$

If i, j, k are generators having the property (2) and

$$i' = q^{-1} i q, \quad j' = q^{-1} j q, \quad k' = q^{-1} k q \tag{3}$$

where q is any nonzero quaternion, then i', j', k' are generators too. Generators i, j, k and generators i', j', k' are mutually rotated. A nonzero quaternion may express this rotation. Since we can replace a nonzero quaternion by the corresponding unit quaternion, the unit quaternion is used to represent an orientation of i, j, k in respect to i', j', k'. Thus we can write $\bar{q}$ instead of q^{-1}. Partially, when i, j, k is rotated an angle φ in respect to i', j', k' around the axis $i = i'$, $q = \cos(\varphi/2) + i\sin(\varphi/2)$.

We will represent a position and orientation of a body by a pair of quaternions (p, q), where p is a pure quaternion (corresponding with a position vector) and q is a unit quaternion (corresponding with an orientation tensor).

Any quaternion (force, torque, position, orientation) may be expressed by coordinates in respect to different bases

$$r = r_0 + r_1 i + r_2 j + r_3 k = r'_0 + r'_1 i' + r'_2 j' + r'_3 k' \tag{4}$$

We can express the change of coordinates $R = (r_0, r_1, r_2, r_3)$, $R' = (r'_0, r'_1, r'_2, r'_3)$ in the form

$$R = \bar{Q}\, R'\, Q \quad \text{or} \quad R' = Q\, R\, \bar{Q} \tag{5}$$

where $Q = (q_0, q_1, q_2, q_3)$ are the same coordinates of q from (3) in respect to both $1, i, j, k$ and $1, i', j', k'$.

For a serial open kinematic structure consisting of n elements we can express a position p of the origin of the last element in respect to the origin of the base and an orientation q of the basis i_n, j_n, k_n of the last element in respect to the basis i_0, j_0, k_0 of the base in the form

$$p = \sum_{m=1}^{n} p_m, \qquad q = \sum_{m=n}^{1} q_m \tag{6}$$

where p_m, q_m is a position-orientation pair of the element m in respect to the element $m-1$. Note that the product of quaternions q_m is not commutative and

$$i_{m-1} = \bar{q}_m i_m q_m \qquad j_{m-1} = \bar{q}_m j_m q_m \qquad k_{m-1} = \bar{q}_m k_m q_m \tag{7}$$

Equalities (7) allow us to reduce all quaternion generators to the basis $1, i_0, j_0, k_0$ according to (4), (5). Thus we have for coordinates P, Q of p, q from (6) in respect to the quaternion basis $1, i_0, j_0, k_0$:

$$P = P_1 + Q_1 P_2 \bar{Q}_1 + \ldots + Q_1 \ldots Q_{n-1} P_n \bar{Q}_{n-1} \ldots \bar{Q}_1 \tag{8}$$

$$Q = Q_1 Q_2 \ldots Q_n \tag{9}$$

Note that p_m, q_m from (6) is a position and orientation of the element m in respect to the element $m-1$. They may be expressed by coordinates in respect to any basis according to (4). P_m, Q_m are the coordinates of p_m, q_m

from (6) in respect to the quaternion basis $1, i_{m-1}, j_{m-1}, k_{m-1}$. Thus we have quaternion quadruples (not quaternions) of relative coordinates in (8), (9).

3. Kinematic structure notation

Suppose that we have a fixed coordinate system and a moving coordinate system, and let both be identical at the instant. We introduce the following symbols for constant and variable translations along and rotations about orthogonal basis vectors i, j, k, respectively: a, b, c for constant translations, A, B, C for constant rotations, x, y, z for variable translations and X, Y, Z for variable rotations. Constants correspond with link parameters and variables correspond with joint variables. Under this notation any finite sequence of letters from the alphabet $\{a, b, c, A, B, C, x, y, z, X, Y, Z\}$ defines an open serial kinematic chain, (Dobrovodský, 1994a). A subsequence from $\{a, b, c, A, B, C\}$ defines a link and any letter from $\{x, y, z, X, Y, Z\}$ defines a joint. We will call single letters elements of the kinematic structure. We attache a coordinate system to each element (a letter) of the structure. The next letter specifies the axis along or about which is the next coordinate system shifted or rotated. Origins of these coordinate systems will be the points of the force–torque investigation. This approach allows us to specify forces and torques even in critical points inside the links consisting of more letters from $\{a, b, c, A, B, C\}$, as we will see later.

The well known Denavit-Hartenberg notation yields a substring $ZcaA$ for revolute joint Z and a substring $CzaA$ for prismatic joint z. For example $\Omega = ZcXazZcXcZc$ uniquely defines an open kinematic chain with joints Z, X, z, Z, X, Z and links c, a, c, c, c. For this example of the kinematic structure we have the following correspondence of position and orientation quaternions

$$
\begin{array}{lll}
Z & p_1 = 0 & q_1 = c_1 + k_0 s_1 \\
c & p_2 = k_1 \delta_2 & q_2 = 1 \\
X & p_3 = 0 & q_3 = c_3 + i_2 s_3 \\
a & p_4 = i_3 \delta_4 & q_4 = 1 \\
z & p_5 = k_4 \delta_5 & q_5 = 1 \\
Z & p_6 = 0 & q_6 = c_6 + k_5 s_6 \\
c & p_7 = k_6 \delta_7 & q_7 = 1 \\
X & p_8 = 0 & q_8 = c_8 + i_7 s_8 \\
c & p_9 = k_8 \delta_9 & q_9 = 1 \\
Z & p_{10} = 0 & q_{10} = c_{10} + k_9 s_{10} \\
c & p_{11} = k_{10} \delta_{11} & q_{11} = 1
\end{array}
\tag{10}
$$

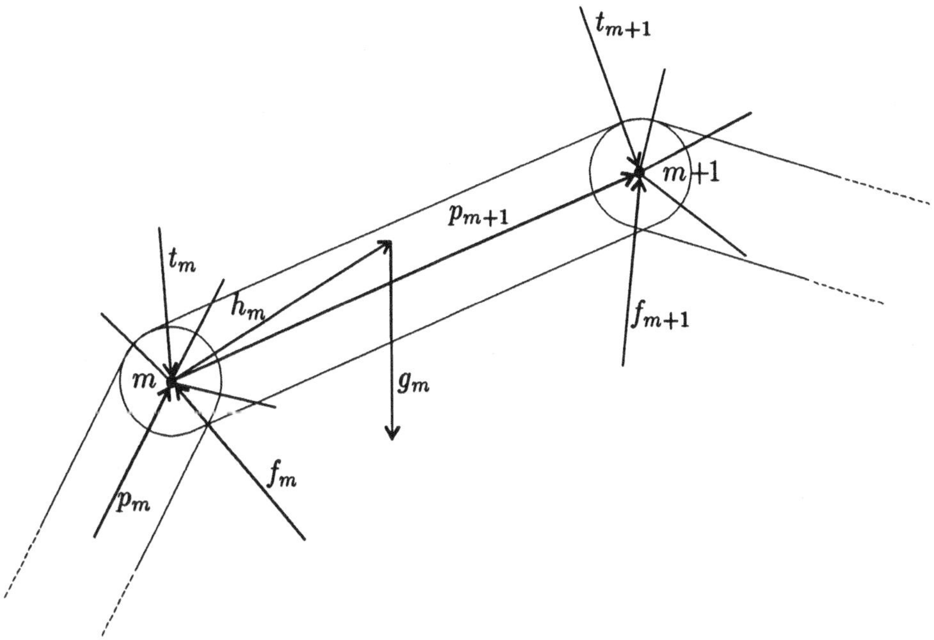

Figure 1. Decomposition of Forces and Torques

where parameters $\tau = (\delta_2, \delta_4, \delta_7, \delta_9, \delta_{11})$ are the link parameters, parameters $\vartheta = (\delta_1, \delta_3, \delta_5, \delta_6, \delta_8, \delta_{10})$ are the joint variables and $c_m = \cos(\delta_m/2)$, $s_m = \sin(\delta_m/2)$.

4. Force–torque balance equalities

We begin by considering the free rigid body of an individual link incorporated in an open kinematic chain, (Asada and Slotine, 1986). Let f_m is the linear force exerted by the link $m-1$ and acts upon the link m. The force applied to link m by link $m+1$ is then given by $-f_{m+1}$ since f_{m+1} is the force exerted by the link m and acts upon the link $m+1$, (Fig. 1). The gravity force acting at the centroid of the link m is denoted by g_m and is proportional to the scalar mass and the acceleration vector of gravity. The balance of linear forces acting at the link m is then given by

$$f_m - f_{m+1} + g_m = 0 \qquad m = 1, \ldots, n \tag{11}$$

Note that "link 0" is the base and "link $n+1$" is the external environment. External force acting on the last link n is $-f_{n+1}$.

Next we derive the balance of torques about selected points of links. We denote torques t_m the same way as forces. The linear forces also cause torques about the origins of elements. The balance of torques with respect to the origin of the first coordinate system moving together with the following link is given by

$$t_m - t_{m+1} + h_m \times g_m - p_{m+1} \times f_{m+1} = 0 \qquad m = 1, \ldots, n \tag{12}$$

where p_m is the position vector of the origin of the coordinate system of m-th element in respect to the origin of the coordinate system of (m-1)-th element and h_m is the position vector of the center of gravity of m-th element in respect to the origin of the coordinate system of m-th element. Note that all vectors are defined "point to point", thus no choice was made for coordinates of these vectors according to (4).

Since pure quaternions form a 3–D linear space of vectors, we will switch from 3–D vectors to corresponding pure quaternions. If r, s are pure quaternions, than the following equality holds

$$rs = r \times s - r \cdot s \tag{13}$$

where "$\times$" is a vector product and "$\cdot$" is a scalar product. Equality (13) implies

$$r \cdot s = -(rs + sr)/2 \tag{14}$$

$$r \times s = (rs - sr)/2 \tag{15}$$

We can replace vector products in (12) by quaternion expressions according to (15):

$$2t_m - 2t_{m+1} + h_m g_m - g_m h_m - p_{m+1} f_{m+1} + f_{m+1} p_{m+1} = 0 \quad m = 1, \ldots, n \tag{16}$$

Equalities (6), (7), (11), (12) create the common starting point for several problem formulations. In these equalities the concrete kinematic structure is entering through the relations like (10).

5. Forces and torques in joints

In the first force–torque problem we have given the force $-f_{n+1}$ and the torque $-t_{n+1}$ acting upon the last element of the structure, the gravity forces g_m, and the position vectors h_m of centers of gravity of elements in respect to origins of coordinate frames attached to the elements. The unknowns are the internal forces f_m and the torques t_m. The pure quaternion (vector) form of the solution is

$$f_m = f_{n+1} - \sum_{l=m}^{n} g_l \quad (17)$$

$$t_m = t_{n+1} - \sum_{l=m}^{n} (h_l g_l - g_l h_l)/2 + \sum_{l=m}^{n} (p_{l+1} f_{l+1} - f_{l+1} p_{l+1})/2 \quad (18)$$

where $m = n, n-1, \ldots, 1$. Note that the existence of the solution does not depend on the values of joint variables. We have given the coordinates of gravity forces g_l in respect to the basis i_0, j_0, k_0 and the coordinates of positions h_l in respect to the basis i_l, j_l, k_l, respectively. The orthonormal bases should be expressed by the basis i_0, j_0, k_0. Finally, coordinates of forces f_m and torques t_m should be transformed into coordinates in the basis i_m, j_m, k_m of the corresponding frame. The unknown joint force or torque is given by the single coordinate related with the joint axis in accordance with the kinematic structure specification string. For example, if $\Omega = ZcXazZcXcZc$, then the tuple of forces in prismatic and torques in revolute joints is

$$t_{1,3}, t_{3,1}, f_{5,3}, t_{6,3}, t_{8,1}, t_{10,3} \quad (19)$$

where the first index is the element number and the second index is the relative coordinate number 1 for x or X, 2 for y or Y and 3 for z or Z.

In the case of nonredundant kinematic structure, we will solve the inverse kinematics first, and, having expressed the relations between orthonormal bases, we can evaluate coordinates of forces and torques according to expressions (17), (18) in the basis i_0, j_0, k_0. Finally, we have to transform the coordinates into the appropriate relative coordinates in the basis i_m, j_m, k_m, respectively. This procedure is straightforward, except the inverse kinematics, where up to 16 solutions may occur.

In the case of redundant kinematic structure, inverse kinematics can not be solved first and separately. We define the cost function with values depending on joint variables. Since not only forces along and torques about the joint axes may be critical, the recommended cost function is the sum of square magnitudes of all forces and torques (or at least of forces and torques in joints):

$$R = \sum_{m=1}^{n} f_m \cdot f_m + \sum_{m=1}^{n} t_m \cdot t_m \quad (20)$$

Expressing forces and torques by quadruples of quaternion coordinates in respect to the basis $1, i_0, j_0, k_0$ we have

$$R = \sum_{m=1}^{n} F_m \bar{F}_m + \sum_{m=1}^{n} T_m \bar{T}_m \tag{21}$$

where F_m, T_m are the coordinates of solutions (17), (18) in respect to the basis $1, i_0, j_0, k_0$. As we can see in (17), if the the force f_{n+1} is constant in respect to the base, the coordinates F_m in respect to the basis $1, i_0, j_0, k_0$ do not depend on joint variables. In this case the first term may be omitted. Nevertheless, they may depend on joint variables in general.

Maxima and minima of a real function R of r variables ϑ subject to differentiable accessory conditions (constraints) (7), (8) may, in principle, be found after 6 of the r variables ϑ have been eliminated with the aid of the relations (7), (8). If it is impossible or impractical to eliminate 6 variables directly, the method of Lagrange multipliers may be used. The Lagrange function defined on quaternions is

$$L = R + \lambda_p \cdot (p - \sum_{m=1}^{n} p_m) + \lambda_q \cdot (\bar{q} \prod_{m=n}^{1} q_m - \prod_{m=n}^{1} q_m \bar{q}) \tag{22}$$

where R is from (20), pure quaternions λ_p, λ_q are Lagrange multipliers and "$\cdot$" is the scalar product of pure quaternions, (14). Since for any quaternions r, s the expression $(rs - sr)$ is a pure quaternion (vector), the last expression in (22) is a pure quaternion. This is the way how to incorporate orientation accessory condition into the Lagrange function.

The corresponding form of the Lagrange function based on quadruples of quaternion coordinates is

$$L = R + \bar{\Lambda}_p(P - \sum_{m=1}^{n}((\prod_{l=1}^{m-1} Q_l)P_m(\prod_{l=m-1}^{1} \bar{Q}_l))) + \bar{\Lambda}_q(\bar{Q} \prod_{m=1}^{n} Q_m - \prod_{m=1}^{n} Q_m \bar{Q}) \tag{23}$$

where R is from (21). The necessary condition for a maximum or minimum of R subject to the constraints (7), (8) is $\partial L / \partial \vartheta = 0$.

Only very special conditions lead to effective procedures for the solution. For example in Lenarčič (1994) the planar kinematic structure is $\Omega = ZaZa \ldots Za$ is investigated, where $\delta_2 = \delta_4 = \ldots = d$. According to this kinematic structure we have the following correspondence of position and orientation quaternions

Z	$p_m = 0$	$q_m = c_m + k_{m-1} s_m$	$m = 1, 3, \ldots$
a	$p_m = i_{m-1} d$	$q_m = 1$	$m = 2, 4, \ldots$

When orientation is not predesired, the structure $\Omega = ZaZa$ is nonredundant and we can solve the inverse kinematics first and separately. Otherwise, when $\Omega = ZaZaZa\ldots$, the structure is redundant and we can optimize the torques (18) by adding the cost function, (Lenarčič, 1994).

6. Endpoint force and torque

In the second force–torque problem we have given (measured) forces along and/or torques about joint axes of redundant or nonredundant kinematic structure in a given joint position. We are to obtain the endpoint force $-f_{n+1}$ and the torque $-t_{n+1}$ acting upon the last element of the structure from the equalities (11), (12). In this case r $(r \leq n)$ joint variables are known and we have 6n scalar equations with $6(n+1)-r$ unknown coordinates of forces f_m and torques t_m, $m = 1,\ldots,n+1$. Since single relative coordinates are given, we cannot solve these equations in a compact vector form like (17), (18). The matrix of the resulting system of linear equations and thus even the existence of the solution depends on the values of joint variables. If we have a nonredundant full position-orientation structure $(r = 6)$, the number of equations is equal to the number of the unknowns. If the kinematic structure is redundant, the system of equations becomes to be over constrained and the measured forces along and/or torques about joint axes should be considered to be erroneous. Pseudo inverse of the matrix of the system of equations may be used to find a pseudo solution.

7. Conclusions

We studied the static properties of a general open serial kinematic structure with the desired position and orientation and with the force and torque applied to the endpoint of the structure. Both, the position, orientation and static forces and torques, we expressed by quaternions. We have strictly distinguished between quaternions and coordinates of quaternions and showed how quaternions may be used to express force–torque balance vector equalities. We defined the cost function in order to minimize the sum of magnitudes of forces and torques in joints and even in any selected internal points of the structure. Finally, we discussed the case of the measured forces and torques with the goal to specify the force and the torque acting upon the endpoint of the kinematic structure.

References

Asada, H., Slotine, J.-J.E. (1986) *Robot Analysis and Control*, John Wiley & Sons, New York, Chichester, Brisbane, Toronto, Singapore, pp. 74–75

Dobrovodský, K. (1994) *Quaternion Position Representation in Robot Kinematic Structures*, IEE Int. Conf. on Control '94, pp. 561–564

Dobrovodský, K. (1994a) *On the Reduction of Parameters in Kinematic Equations*, Advances in Robot Kinematics and Computational, Kluwer Academic Publishers, Dordrecht/Boston/London, pp. 421–426

Lenarčič, J. (1994) *Minimum Joint Torque Configurations of Planar Multiple-Link Manipulator*, Advances in Robot Kinematics and Computational Geometry, Kluwer Academic Publishers, Dordrecht/Boston/London, pp. 281–288

DUAL VECTORS OF THE KINEMATIC AND THE DYNAMIC KIND IN MATRICES

M. L. KELER
Fachhochschule München
Lothstr. 34, 80335 München, Germany

Abstract - Dual vectors are a very efficient mathematical tool for the treating of line vectors and screws. Their great advantage is the consequent analogy to the algebra with the usual vectors. The real parts of dual vectors are in total accordance with the usual vectors. As there does not yet exist commonly a software using the dual unit we apply either subroutines or matrices. Matrices are into the bargain very useful in mechanics. Therefore we decide in this paper on the application of matrices and transfer the dual arithmetic and the dual vector algebra to matrix operations. The transition from kinematics to dynamics and vice versa is performed by conversion matrices for instance of inertia or of elastic compliance which convert real parts into dual parts and dual parts into real parts with changed physical dimensions (correlation of screws in ray coordinates with screws in axis coordinates). The dual vectors for the kinematic state of a rigid body with respect to another are represented in a coordinate system fixed in the considered body instantaneously coinciding in the base system. The absolute kinematic dual quantities in the body fixed system have to be multiplied after transformations with the inertia matrix of the body and also regarded the absolute turn of the body. The results are matrix formulas which yield kinetic wrenches as dual vectors verifiable by other methods [11], [13], [14], [17].

1. Introduction

A vector with three real components as real part means in this theory one of the ∞^1 line bound vectors with the action line through the origin. As each of these line bound vectors has the same effect on a rigid body we call this multitude just a line vector.
The dual part is multiplied with the **dual unit ε with the property $\varepsilon^2 = 0$**. A vector with real components multiplied with ε converts from a line vector into a free vector which is equivalent to anyone of the ∞^4 possible couples of line vectors.
A second multiplication of such a free vector with ε converts the couple of the line vectors into a pair of opposite free vectors which erase each other.
The sum of a line vector at the origin and of a free vector **perpendicular** to it can be taken for the unequivocal parallelly shifted line vector in the space as one line vector of the equivalent couple to the free vector while the other as the opposite line vector at the origin vanishes together with the original line vector.
The sum of a line vector at the origin and of a free vector **not perpendicular** to it is equivalent to a screw with the action line or axis in the space.

J. Lenarčič and V. Parenti-Castelli (eds.), Recent Advances in Robot Kinematics, 245-254.
© 1996 *Kluwer Academic Publishers. Printed in the Netherlands.*

2. Several kinds of screws represented by dual vectors

The dual unit vector $\hat{\vec{p}}^{ij} = \vec{p}_0^{ij} + \varepsilon\, \vec{p}_1^{ij}$ for an axis consists of the direction unit vector $\vec{p}_0^{ij}$ as real part and the moment vector $\vec{p}_1^{ij} = \vec{r} \times \vec{p}_0^{ij}$ as dual part with the position vector $\vec{r}$ from the origin to the axis. In the same way we get the dual unit vector $\hat{\vec{q}} = \vec{q}_0 + \varepsilon\, \vec{q}_1$ for an action line.

We know following **kinematic dual vectors** for the displacement of a rigid body i with respect to another rigid body or base j and their derivatives with resp. to *t* represented in a coordinate system fixed in the body i instantaneously coinciding in the system j:

Motion screw $\hat{\vec{B}}^{ij} = (\beta^{ij} + \varepsilon\, b^{ij})(\vec{p}_0^{ij} + \varepsilon\, \vec{p}_1^{ij}) = \vec{\beta}_O^{ij} + \varepsilon\, \vec{b}_O^{ij}$ with

β^{ij} the scalar arc of the angular displacement with the dimension rad = 1,

$\vec{\beta}_O^{ij} = \beta^{ij}\, \vec{p}_0^{ij}$ the angular displacement vector at the origin O in direction of the axis and

b^{ij} the linear displacement in direction of the axis,

$\vec{b}_O^{ij} = b^{ij}\, \vec{p}_0^{ij} + \beta^{ij}\, \vec{p}_1^{ij}$ the moment of the motion screw with respect to the origin with the dimension of a length, for instance m.

The motion screw is approximately derived from the exact motor after [3] by *Dimentberg* until to the accuracy of the IInd order in [6].

Velocity screw $\hat{\vec{V}}^{ij} = \dot{\hat{\vec{B}}} = (\omega^{ij} + \varepsilon\, v^{ij})(\vec{p}_0^{ij} + \varepsilon\, \vec{p}_1^{ij}) = \vec{\omega}_O^{ij} + \varepsilon\, \vec{v}_O^{ij}$

Acceleration screw $\hat{\vec{A}}^{ij} = \dot{\hat{\vec{V}}}^{ij} = \left(\alpha^{ij} + \varepsilon\, a^{ij}\right)\left(\vec{p}_0^{ij} + \varepsilon\, \vec{p}_1^{ij}\right) = \vec{\alpha}_O^{ij} + \varepsilon\, \vec{a}_O^{ij}$

The velocity of the instantaneously coinciding point O_i on the body i with respect to j is equal to the dual part

$$\vec{v}_{Oij} = \vec{v}_O^{ij} \qquad (1)$$

whereas the acceleration of this point is obtained, because of the turning system i, by

$$\vec{a}_{Oij} = \vec{a}_O^{ij} + \omega_O^{ij} \times v_O^{ij}. \qquad (2)$$

We have following **dynamic dual vectors**

Momentum screw $\hat{\vec{p}} = (p + \varepsilon\, L)(\vec{q}_0 + \varepsilon \vec{q}_1) = \vec{p}_O + \varepsilon\, \vec{L}_O$

Wrench (force screw) $\hat{\vec{F}} = (F + \varepsilon\, M)(\vec{q}_0 + \varepsilon\, \vec{q}_1) = \vec{F}_O + \varepsilon\, \vec{M}_O$ (3)

3. The representation of dual vectors by matrices and their algebra with matrices

The real parts are primarily the first rows of a column matrix.

Basic arithmetical dual operations

Equality, addition and subtraction follow the same rules for the elements of dual scalars, dual vectors and dual matrices in comparison with those elements of real matrices.

The product of two dual scalars $(a + \varepsilon b)(c + \varepsilon\, d) = ac + \varepsilon\, (bc + ad)$

can be written as a product of a matrix with a vector $\begin{pmatrix} a & 0 \\ b & a \end{pmatrix}\begin{pmatrix} c \\ d \end{pmatrix} = \begin{pmatrix} ac \\ bc + ad \end{pmatrix}$. (4)

The quotient of two dual scalars

if $a \neq 0$ $\quad \dfrac{c + \varepsilon d}{a + \varepsilon b} = \dfrac{c}{a} + \varepsilon \dfrac{-bc + ad}{a^2}$

becomes by the inverse in the matrix product

$$\begin{pmatrix} x_0 \\ x_1 \end{pmatrix} = \begin{pmatrix} a & 0 \\ b & a \end{pmatrix}^{-1} \begin{pmatrix} c \\ d \end{pmatrix} = \frac{1}{a^2} \begin{pmatrix} a & 0 \\ -b & a \end{pmatrix} \begin{pmatrix} c \\ d \end{pmatrix} = \begin{pmatrix} \frac{c}{a} \\ \frac{-bc+ad}{a^2} \end{pmatrix}, \tag{5}$$

if $a = 0$ and $c = 0$ and $b \neq 0$ $\dfrac{0+\varepsilon d}{0+\varepsilon b} = \dfrac{d}{b} + \varepsilon UN$ with UN = undetermined real scalar

is the result of the matrix equation $\begin{pmatrix} 0 & 0 \\ b & 0 \end{pmatrix} \begin{pmatrix} x_0 \\ x_1 \end{pmatrix} = \begin{pmatrix} 0 \\ d \end{pmatrix}$, $\begin{pmatrix} x_0 \\ x_1 \end{pmatrix} = \begin{pmatrix} \frac{d}{b} \\ UN \end{pmatrix}$. (6)

The *product of a dual scalar with a dual vector*

$$\hat{\vec{V}} = \hat{V}\hat{\vec{p}} = (\omega + \varepsilon v)(\vec{p}_0 + \varepsilon \vec{p}_1) = \omega \vec{p}_0 + \varepsilon (v\vec{p}_0 + \omega \vec{p}_1)$$

is dealt with the matrix product $\begin{pmatrix} \boldsymbol{w} & \boldsymbol{O} \\ \boldsymbol{v} & \boldsymbol{w} \end{pmatrix} \begin{pmatrix} \boldsymbol{p_0} \\ \boldsymbol{p_1} \end{pmatrix} = \begin{pmatrix} \omega \boldsymbol{p_0} \\ v\boldsymbol{p_0} + \omega \boldsymbol{p_1} \end{pmatrix}$ (7)

with the 3×3 scalar submatrices $\boldsymbol{w}$, $\boldsymbol{v}$ with the scalars ω and v and the null matrix $\boldsymbol{O}$ and the 3×1 column submatrices $\boldsymbol{p}_0$ and $\boldsymbol{p}_1$

The *dual scalar product of two dual vectors*

$$\hat{\vec{q}}\hat{\vec{p}} = (\vec{q}_0 + \varepsilon \vec{q}_1)(\vec{p}_0 + \varepsilon \vec{p}_1) = \vec{q}_0\vec{p}_0 + \varepsilon (\vec{q}_1\vec{p}_0 + \vec{q}_0\vec{p}_1)$$

is written in matrix form $\begin{pmatrix} \boldsymbol{q}_0^T & \underline{O}^T \\ \boldsymbol{q}_1^T & \boldsymbol{q}_0^T \end{pmatrix} \begin{pmatrix} \boldsymbol{p}_0 \\ \boldsymbol{p}_1 \end{pmatrix} = \begin{pmatrix} \boldsymbol{q}_0^T \boldsymbol{p}_0 \\ \boldsymbol{q}_1^T \boldsymbol{p}_0 + \boldsymbol{q}_0^T \boldsymbol{p}_1 \end{pmatrix}$, with $\underline{O}^T = (0 \quad 0 \quad 0)$.(8)

The factors must not be dual unit vectors.

The *cross product of two dual vectors*

$$\hat{\vec{V}}^{34} \times \hat{\vec{V}}^{23} = \left(\vec{\omega}_O^{34} + \varepsilon \vec{v}_O^{34}\right) \times \left(\vec{\omega}_O^{23} + \varepsilon \vec{v}_O^{23}\right) = \vec{\omega}_O^{34} \times \vec{\omega}_O^{23} + \varepsilon \left(\vec{v}_O^{34} \times \vec{\omega}_O^{23} + \vec{\omega}_O^{34} \times \vec{v}_O^{23}\right)$$

is dealt with the matrix product $\begin{pmatrix} \underline{\tilde{\omega}}_O^{34} & \boldsymbol{O} \\ \underline{\tilde{v}}_O^{34} & \underline{\tilde{\omega}}_O^{34} \end{pmatrix} \begin{pmatrix} \underline{\omega}_O^{23} \\ \underline{v}_O^{23} \end{pmatrix} = \begin{pmatrix} \underline{\tilde{\omega}}_O^{34} \underline{\omega}_O^{23} \\ \underline{\tilde{v}}_O^{34} \underline{\omega}_O^{23} + \underline{\tilde{\omega}}_O^{34} \underline{v}_O^{23} \end{pmatrix}$ (9)

with the skew symmetric cross product submatrices $\underline{\tilde{\omega}}_O^{34}$ and $\underline{\tilde{v}}_O^{34}$

The *transformation of a dual vector*
from a system with origin O into a system with origin C
can be done by a dual 3×3 matrix:

$$\hat{\boldsymbol{D}}^{CO} = \boldsymbol{D}_0^{CO} + \varepsilon \boldsymbol{D}_1^{CO} \tag{10}$$

The columns of the real part matrix consist of the unit vectors representing the right-handed rectangular coordinate system of the axes with the origin O in the system with the origin C:

$$\boldsymbol{D}_0^{CO} = (\boldsymbol{e}_1 \quad \boldsymbol{e}_2 \quad \boldsymbol{e}_3)^{CO} \text{ with } \boldsymbol{e}_i\boldsymbol{e}_j = 1 \text{ for } i = j \quad \boldsymbol{e}_i\boldsymbol{e}_j = 0 \text{ for } i \neq j. \tag{11}$$

The columns of the dual part matrix consist of
the cross products of the position vector from C to O in system C $\vec{r}_{CO}^{\,C}$
with the unit vectors of the axes. We obtain this matrix as the matrix product of the position vector's cross product matrix with the matrix of the unit vectors of the axes:

$$\boldsymbol{D}_1^{CO} = \tilde{\boldsymbol{r}}_{CO}^{C} \boldsymbol{D}_0^{CO} . \tag{12}$$

The transformation for instance of a velocity screw from system O into system C is written in dual form:

$$\vec{\omega}^C + \varepsilon \vec{v}^C = \left(\boldsymbol{D}_0^{CO} + \varepsilon \boldsymbol{D}_1^{CO}\right)\left(\vec{\omega}^O + \varepsilon \vec{v}^O\right) = \boldsymbol{D}_0^{CO}\vec{\omega}^O + \varepsilon \left(\boldsymbol{D}_1^{CO}\vec{\omega}^O + \boldsymbol{D}_0^{CO}\vec{v}^O\right)$$

and in matrix form with the partitioned 6×6 transformation matrix:

$$\begin{pmatrix} \underline{\omega} \\ \underline{v} \end{pmatrix}^C = \begin{pmatrix} D_0 & O \\ D_1 & D_0 \end{pmatrix}^{CO} \begin{pmatrix} \underline{\omega} \\ \underline{v} \end{pmatrix}^O = \begin{pmatrix} D_0^{CO}\underline{\omega}^O \\ D_1^{CO}\underline{\omega}^O + D_0^{CO}\underline{v}^O \end{pmatrix} \tag{13}$$

4. The conversion of kinematic dual vectors into dynamic dual vectors and vice versa

If the free vector for the linear velocity ***v*** of a point with the mass *m* is multiplied by the mass *m* then it results the line vector of the linear momentum, shortly momentum,

$$\vec{p} = m\vec{v}$$

on the line parallel to ***v*** through the mass point.

The product of the line vector of the angular velocity ω with the moment of inertia *J* of a principal axis parallel to ω yields the free vector of the angular momentum or moment of momentum

$$\vec{L} = J\vec{\omega}$$

The differentiation with respect to the time delivers correspondingly the line vector of the kinetic force ***F*** from the free vector of the linear acceleration ***a***

$$\vec{F} = m\vec{a}$$

on the line parallel to ***a*** through the mass point

and the free vector of the kinetic moment ***M*** from the line vector of the angular acceleration α parallel to the principal axis

$$\vec{M} = J\vec{\alpha}\,.$$

The line vector of the normal force F_n for instance on the central axis of a straight cantilever beam with a length *l*, the modulus *E* of elasticity and the cross-sectional area *A* produces the free vector of the translational displacement of the cross-section at the end of the beam

$$\vec{h}_n = \frac{l}{EA}\vec{F}_n\,.$$

The free vector of the torsional moment M_t at a beam is converted by the product with its torsional compliance into the line vector of the angular displacement of the beam end

$$\vec{\chi}_t = \frac{l}{GJ_t}\vec{M}_t$$

on the line parallel to the central axis through the shear center of the cross-sectional area.

We chose the arrangement of the kinematic dual vector components in a column matrix with the real part in the three upper elements and the dual part in the three lower elements (ray coordinates). For the dynamic dual vectors it is advisable to arrange the dual part in the upper elements and the real part in the lower elements of the partitioned column matrix (axis coordinates), for instance:

Force screw **Velocity screw**

$$\hat{F}^C = \begin{pmatrix} \underline{M} \\ \underline{F} \end{pmatrix}^C \begin{matrix} \text{dual part} \\ \text{real part} \end{matrix} \qquad \hat{V}^C = \begin{pmatrix} \underline{\omega} \\ \underline{v} \end{pmatrix}^C \begin{matrix} \text{real part} \\ \text{dual part} \end{matrix} \tag{14)(15}$$

This arrangement has following advantages:

- The conversion matrices between kinematic and dynamic screws are symmetrical, for instance:

$$\begin{pmatrix} \underline{L} \\ \underline{p} \end{pmatrix}^C = \begin{pmatrix} J & O \\ O & m \end{pmatrix}^C \begin{pmatrix} \underline{\omega} \\ \underline{v} \end{pmatrix}^C \tag{16}$$

- The matrix product as *simple scalar product of dual vectors*

$$\left(F^C\right)^T \hat{V}^C = \left[\begin{pmatrix} \underline{M} \\ \underline{F} \end{pmatrix}^C\right]^T \begin{pmatrix} \underline{\omega} \\ \underline{v} \end{pmatrix}^C = \begin{pmatrix} \underline{M}^T & \underline{F}^T \end{pmatrix}^C \begin{pmatrix} \underline{\omega} \\ \underline{v} \end{pmatrix}^C = \left(\underline{M}^C\right)^T \underline{\omega}^C + \left(\underline{F}^C\right)^T \underline{v} \tag{17}$$

delivers immediately *the dual part of the dual scalar product of dual vectors* for a kinematic and a dynamic screw which is in this example physical power. With a sufficient small motion screw instead of a velocity screw the result is physical work.
- The upper three and the lower three elements of each column matrix could also change places without loosing the above mentioned advantages. But in the proposed arrangement the adjacent elements 3, 4, 5 can easily be extracted for planar problems.

5. The transformation matrices for kinematic and dynamic screws and some rules

The *transformation of a kinematic screw* from system O into C, eq. (13) and (12),

$$\boxed{\hat{V}^C = T_{ki}^{CO} \hat{V}^O = \begin{pmatrix} \underline{\omega} \\ \underline{v} \end{pmatrix}^C = \begin{pmatrix} D_0^{CO} & O \\ \tilde{r}_{CO}^C D_0^{CO} & D_0^{CO} \end{pmatrix} \begin{pmatrix} \underline{\omega} \\ \underline{v} \end{pmatrix}^O}, \text{ with } r_{CO} = CO \tag{18}$$

transforms as well the factors as the result of a cross product of kinematic screws from system O in dual vector form and in matrix form, eq. (9),

$$\hat{\underline{V}}_1^O \times \hat{\underline{V}}_2^O = \hat{\tilde{V}}_1^O \hat{V}_2^O \qquad \text{with } \hat{\tilde{V}}_1^O = \begin{pmatrix} \underline{\tilde{\omega}}_1 & O \\ \underline{\tilde{v}}_1 & \underline{\tilde{\omega}}_1 \end{pmatrix}^O \text{ and } \hat{V}_2^O = \begin{pmatrix} \underline{\omega}_2 \\ \underline{v}_2 \end{pmatrix}^O$$

into system C in dual vector form and in matrix form:

$$\left(T_{ki}^{CO} \hat{V}_1^O\right) \times \left(T_{ki}^{CO} \hat{V}_2^O\right) = T_{ki}^{CO} \hat{\tilde{V}}_1^O \hat{V}_2^O \Rightarrow \hat{\tilde{V}}_1^C T_{ki}^{CO} \hat{V}_2^O = T_{ki}^{CO} \hat{\tilde{V}}_1^O \hat{V}_2^O$$

From this equation it can be concluded the formula

$$\boxed{\hat{\tilde{V}}_1^C T_{ki}^{CO} = T_{ki}^{CO} \hat{\tilde{V}}_1^O}. \tag{19}$$

With the partitioned matrices the real part yields for every cross product matrix, for instance $\underline{\tilde{\omega}}$ or $\tilde{r}_{CO}$ $\boxed{\underline{\tilde{\omega}}^C D_0^{CO} = D_0^{CO} \underline{\tilde{\omega}}^O \text{ or } \tilde{r}_{CO}^C D_0^{CO} = D_0^{CO} \tilde{r}_{CO}^O}$. (20)

The *transformation of a dynamic screw* from system C to system O is done by the matrix

$$\boxed{T_{dy}^{OC} = \begin{pmatrix} D_0^{OC} & \tilde{r}_{OC}^O D_0^{OC} \\ O & D_0^{OC} \end{pmatrix}}. \tag{21}$$

Its reverse transformation is obtained by the inverse matrix

$$\boxed{T_{dy}^{CO} = \left(T_{dy}^{OC}\right)^{-1} = \begin{pmatrix} \left(D_0^{OC}\right)^T & \left(\tilde{r}_{OC}^O D_0^{OC}\right)^T \\ O & \left(D_0^{OC}\right)^T \end{pmatrix}} \text{ and with} \tag{22}$$

$$\left(D_0^{OC}\right)^T = D_0^{CO} \text{ and } \left(\tilde{r}_{OC}^O D_0^{OC}\right)^T = \left(D_0^{OC}\right)^T \left(\tilde{r}_{OC}^O\right)^T = D_0^{CO}\left(-\tilde{r}_{OC}^O\right) = -\tilde{r}_{OC}^C D_0^{CO} = \tilde{r}_{CO}^C D_0^{CO}$$

we get
$$\boxed{T_{dy}^{CO} = \begin{pmatrix} D_0^{CO} & \tilde{r}_{CO}^{C} D_0^{CO} \\ O & D_0^{CO} \end{pmatrix}}. \tag{23}$$

Following formula shows that kinematic and dynamic dual vectors transform contragrediently, so that dual and simple scalar products of them are invariant:
$$\boxed{T_{ki}^{CO} = \left(T_{dy}^{OC}\right)^T = \begin{pmatrix} \left(D_0^{OC}\right)^T & O \\ \left(\tilde{r}_{OC}^{O} D_0^{OC}\right)^T & \left(D_0^{OC}\right)^T \end{pmatrix}} \tag{24}$$

The formula (19) with the right-handed multiplication by T_{ki}^{OC}

$\hat{\tilde{V}}_1^C T_{ki}^{CO} T_{ki}^{OC} = T_{ki}^{CO} \hat{\tilde{V}}_1^O T_{ki}^{OC}$ yields the similarity transformation:
$$\boxed{\hat{\tilde{V}}_1^C = T_{ki}^{CO} \hat{\tilde{V}}_1^O T_{ki}^{OC}}, \tag{25}$$

and those formulas (20) for the real parts also

$\underline{\tilde{\omega}}^C D_0^{CO} D_0^{OC} = D_0^{CO} \underline{\tilde{\omega}}^O D_0^{OC}$ or $\tilde{r}_{CO}^C D_0^{CO} D_0^{OC} = D_0^{CO} \tilde{r}_{CO}^O D_0^{OC}$
$$\boxed{\underline{\tilde{\omega}}^C = D_0^{CO} \underline{\tilde{\omega}}^O D_0^{OC} \quad \text{or} \quad \tilde{r}_{CO}^C = D_0^{CO} \tilde{r}_{CO}^O D_0^{OC}}. \tag{26}$$

6. The momentum screw of a rigid body

A rigid body can be considered as an infinite multitude of points X with constant distances from each other and with the infinitesimal mass *dm* for each point. If we know the velocity screw in a coordinate system C with the origin at the centroid C, we transform the velocity screw into a coordinate system with parallel axes and the origin in one point X of the body with the position vector r_{CX}^C = CX in system C.
The real part of the dual transformation matrix is the 3×3 unit matrix $D_0 = I$.
But in the dual part of the transformation matrix we need the cross product matrix of the negative position vector - r_{CX}^C that is from the considered point X to the centroid C:
$$D_1 = -\tilde{r}_{CX}^C :$$

$$\begin{pmatrix} \underline{\omega} \\ \underline{v} \end{pmatrix}^X = \begin{pmatrix} I & O \\ -\tilde{r}_{CX}^C & I \end{pmatrix}^{XC} \begin{pmatrix} \underline{\omega} \\ \underline{v} \end{pmatrix}^C . \tag{27}$$

It does not exist a moment of inertia for a mass point. Therefore the conversion matrix is $\begin{pmatrix} O & O \\ O & dm \end{pmatrix} = \begin{pmatrix} O & O \\ O & Idm \end{pmatrix}$

and the infinitesimal momentum
$$d\begin{pmatrix} \underline{L} \\ \underline{p} \end{pmatrix}^X = \begin{pmatrix} O & O \\ O & dm \end{pmatrix} \begin{pmatrix} \underline{\omega} \\ \underline{v} \end{pmatrix}^X . \tag{28}$$

This momentum has to be transformed contragrediently into the parallel system C:
$$d\begin{pmatrix} \underline{L} \\ \underline{p} \end{pmatrix}^C = \begin{pmatrix} I & \tilde{r}^C \\ O & I \end{pmatrix}^{CX} d\begin{pmatrix} \underline{L} \\ \underline{p} \end{pmatrix}^X \quad \text{with } \tilde{r}^C = \tilde{r}_{CX}^C = \left(-\tilde{r}_{CX}^C\right)^T . \tag{29}$$

The substitution of the intermediate terms yields
$$d\begin{pmatrix} \underline{L} \\ \underline{p} \end{pmatrix}^C = \begin{pmatrix} I & \tilde{r}^C \\ O & I \end{pmatrix}^{CX} \begin{pmatrix} O & O \\ O & dm \end{pmatrix} \begin{pmatrix} I & O \\ -\tilde{r}^C & I \end{pmatrix}^{XC} \begin{pmatrix} \underline{\omega} \\ \underline{v} \end{pmatrix}^C . \tag{30}$$

With the product of the three partitioned 6×6 matrices we obtain

$$d\begin{pmatrix}\underline{L}\\ \underline{p}\end{pmatrix}^C=\begin{pmatrix}-\tilde{r}^C\tilde{r}^C dm & \tilde{r}^C dm\\ -\tilde{r}^C dm & dm\end{pmatrix}\begin{pmatrix}\underline{\omega}\\ \underline{v}\end{pmatrix}^C \tag{31}$$

The integration over m yields

$$\begin{pmatrix}\underline{L}\\ \underline{p}\end{pmatrix}^C=\int_m d\begin{pmatrix}\underline{L}\\ \underline{p}\end{pmatrix}^C=\begin{pmatrix}-\int_m \tilde{r}^C\tilde{r}^C dm & \int_m \tilde{r}^C dm\\ -\int_m \tilde{r}^C dm & \int_m I dm\end{pmatrix}\begin{pmatrix}\underline{\omega}\\ \underline{v}\end{pmatrix}^C \tag{32}$$

For the centroid C vanishes the integral $\int_m \tilde{r}^C dm=0$; besides we obtain $\int_m I dm=m$.

The first integral delivers the matrix of the moments of inertia with respect to the axes x, y, z of the coordinate system C with the centroid C as origin (in general not the principal axes):

$$-\int_m \tilde{r}^C\tilde{r}^C dm=\begin{pmatrix}J_{xx} & -J_{xy} & -J_{xz}\\ -J_{xy} & J_{yy} & -J_{yz}\\ -J_{xz} & -J_{yz} & J_{zz}\end{pmatrix}^C=J^C\ . \tag{33}$$

Thus we obtain the conversion equation from a velocity screw to a momentum screw at the centroid C of a rigid body by its matrix of inertia as conversion matrix:

$$\begin{pmatrix}\underline{L}\\ \underline{p}\end{pmatrix}^C=\begin{pmatrix}J & O\\ O & m\end{pmatrix}^C\begin{pmatrix}\underline{\omega}\\ \underline{v}\end{pmatrix}^C\ . \tag{34}$$

7. The first derivative with respect to the time of the momentum screw

The differentiated momentum screw at the centroid of the body as the origin of the system C with respect to the time yields the kinetic wrench (force screw) in system C acting upon the body. It depends on the absolute kinematic state of the body which we got for each body by the kinematic analysis of the mechanism or robot for instance by the use of dual vectors:

$$\boxed{\begin{pmatrix}\underline{M}\\ \underline{F}\end{pmatrix}^C=\frac{d}{dt}\begin{pmatrix}\underline{L}\\ \underline{p}\end{pmatrix}^C=\begin{pmatrix}J & O\\ O & m\end{pmatrix}^C\begin{pmatrix}\underline{\alpha}\\ \underline{a}\end{pmatrix}^C+\begin{pmatrix}\tilde{\underline{\omega}} & O\\ O & \tilde{\underline{\omega}}\end{pmatrix}^C\begin{pmatrix}J & O\\ O & m\end{pmatrix}^C\begin{pmatrix}\underline{\omega}\\ \underline{v}\end{pmatrix}^C}\ . \tag{35}$$

The 1st term on the right-handed side of the equation depends on the acceleration screw, that is the change of the velocity screw with the angular acceleration $\underline{\alpha}^C$ as the real part and the moment $\underline{a}^C$ of the acceleration screw as the dual part of the kinematic dual vector. The acceleration s-crew is represented in a body fixed system C.
The 2nd term regards with the big cross product matrix of $\underline{\omega}^C$ the turning of the body with its fixed vectors of the moment of momentum $J^C\ \underline{\omega}^C$ and of the velocity $\underline{v}^C$ of the point C multiplied by the scalar mass m. The 2nd term depends on the velocity screw.

The sum of the upper parts of both the terms yields the dual part of the dynamic dual vector as a free vector of the kinetic moment $\underline{M}^C$
The sum of the lower parts of both the terms yields the kinetic force $\underline{F}^C$ as line vector at the centroid C for the acceleration of the point C with the concentrated mass m of the

body as the real part of the dynamic dual vector. So the acceleration of the point C in the instantaneously coinciding base system consists of the two parts: the moment $\underline{a}^C$ of the acceleration screw as dual part of the kinematic dual vector and the **not to the kinematic dual vector belonging cross product** $\underline{\omega}^C \times \underline{v}^C$.
To condense the equation above we write

$$\boxed{\begin{pmatrix} \underline{M} \\ \underline{F} \end{pmatrix}^C = \boldsymbol{M}_A^C \begin{pmatrix} \underline{\alpha} \\ \underline{a} \end{pmatrix}^C + \boldsymbol{M}_{\omega V}^C \begin{pmatrix} \underline{\omega} \\ \underline{v} \end{pmatrix}^C} \tag{36}$$

with the 1st and the 2nd 6×6 conversion matrices of inertia in system C are

$$\boldsymbol{M}_A^C = \begin{pmatrix} \boldsymbol{J} & \boldsymbol{O} \\ \boldsymbol{O} & \boldsymbol{m} \end{pmatrix}^C \qquad \boldsymbol{M}_{\omega V}^C = \begin{pmatrix} \underline{\tilde{\omega}}\boldsymbol{J} & \boldsymbol{O} \\ \boldsymbol{O} & \underline{\tilde{\omega}}\boldsymbol{m} \end{pmatrix}^C$$

8. The kinetic wrench in other coordinate systems than that one of the centroid

In general the computation of mechanical systems using dual vectors is performed in the base coordinate system for instance with origin O. Here we obtained the absolute kinematic results for each body in its body fixed coordinate system instantaneously coinciding in the base,

see [6]: $\qquad \hat{\boldsymbol{V}}^O = \begin{pmatrix} \underline{\omega} \\ \underline{v} \end{pmatrix}^O, \quad \hat{\boldsymbol{A}}^O = \begin{pmatrix} \underline{\alpha} \\ \underline{a} \end{pmatrix}^O$.

For the considered body are given:

- the matrix $\boldsymbol{J}^C$ of the moments of inertia with resp. to the axes through the centroid C and the matrix of the scalar mass *m* together in the 1st conversion matrix of inertia
  $$\boldsymbol{M}_A^C = \begin{pmatrix} \boldsymbol{J}^C & \boldsymbol{O} \\ \boldsymbol{O} & m\boldsymbol{I} \end{pmatrix}$$
- the constant 4×4 point transformation matrix from system C with the origin at the centroid C into the system H of the body proposed by *Hartenberg - Denavit:*
  $$\boldsymbol{T}^{HC} = \begin{pmatrix} 1 & \underline{O}^T \\ \boldsymbol{r}_{HC}^H & \boldsymbol{D}_0^{HC} \end{pmatrix} \text{ with the position vector } \boldsymbol{r}_{HC}^H = HC \text{ in system H}$$
  and the 3×3 vector transformation matrix $\boldsymbol{D}_0^{HC}$ from the turned system C to H.
- the variable *Hartenberg - Denavit* 4×4 transformation matrix, or product of them, from system H for the instantaneous position of the body into the system O:
  $$\boldsymbol{T}^{OH} = \begin{pmatrix} 1 & \underline{O}^T \\ \boldsymbol{r}_{OH}^O & \boldsymbol{D}_0^{OH} \end{pmatrix} \text{ with the position vector } \boldsymbol{r}_{OH}^O = OH \text{ in system O}$$
  and the 3×3 vector transformation matrix $\boldsymbol{D}_0^{OH}$ from the turned system H to O.

Sought is now the kinetic wrench of the considered body in system O $\hat{\boldsymbol{F}}^O = \begin{pmatrix} \underline{M} \\ \underline{F} \end{pmatrix}^O$.

8.1. METHOD 1:

The transformation of the kinematic dual vectors to the centroid C of a body, the conversion into a kinetic wrench as a dual vector and its transformation back to the origin O

The product of the given 4×4 matrices yields

$$T^{OC} = T^{OH} T^{HC} = \begin{pmatrix} 1 & \underline{O}^T \\ \underline{r}^O_{OC} & D_0^{OC} \end{pmatrix}. \tag{37}$$

With these results we establish the 6×6 transformation matrices

$$T_{dy}^{OC} = \begin{pmatrix} D_0^{OC} & \tilde{r}^O_{OC} D_0^{OC} \\ O & D_0^{OC} \end{pmatrix} \quad \text{and} \quad T_{ki}^{CO} = \left(T_{dy}^{OC}\right)^T. \tag{38)(39}$$

The kinematic screws are transformed into the system C:

$$\hat{V}^C = \begin{pmatrix} \underline{\omega}^C \\ \underline{v}^C \end{pmatrix} = T_{ki}^{CO} \hat{V}^O \quad \text{and} \quad \hat{A}^C = \begin{pmatrix} \underline{\alpha}^C \\ \underline{a}^C \end{pmatrix} = T_{ki}^{CO} \hat{A}^O. \tag{40)(41}$$

We establish the big ω cross product matrix in system C and obtain as matrix product the ω-dependent conversion matrix of inertia in system C for the velocity dependent part:

$$W^C = \begin{pmatrix} \tilde{\underline{\omega}}^C & O \\ O & \tilde{\underline{\omega}}^C \end{pmatrix} \quad \text{and} \quad M_{\omega V}^C = W^C M_A^C. \tag{42)(43}$$

Now we obtain with eq. (36) the kinetic wrench in the system C:

$$\hat{F}^C = M_A^C \hat{A}^C + M_{\omega V}^C \hat{V}^C \quad \text{and transformed into system O} \quad \hat{F}^O = T_{dy}^{OC} \hat{F}^C \tag{44)(45}$$

8.2. METHOD 2:

Computation of the kinetic wrench in the original coordinate system with transformed conversion matrices of inertia

The formula (35) can also be written with the cross product matrix of the complete dual vector of the velocity screw in the form for the multiplication with a dynamic dual vector:

$$\begin{pmatrix} \underline{M} \\ \underline{F} \end{pmatrix}^C = \begin{pmatrix} J & O \\ O & m \end{pmatrix}^C \begin{pmatrix} \underline{\alpha} \\ \underline{a} \end{pmatrix}^C + \begin{pmatrix} \tilde{\underline{\omega}} & \tilde{\underline{v}} \\ O & \tilde{\underline{\omega}} \end{pmatrix}^C \begin{pmatrix} J & O \\ O & m \end{pmatrix}^C \begin{pmatrix} \underline{\omega} \\ \underline{v} \end{pmatrix}^C. \tag{46}$$

As the expansion delivers $m\tilde{\underline{v}}\underline{v} = \underline{O}$, so $\tilde{\underline{v}}$ can be inserted. But now we transform the whole equation from system C to system O with the complete cross product matrix of the dynamic kind separated by the unit matrix $I_6 = T_{dy}^{CO} T_{dy}^{OC}$:

$$\begin{pmatrix} \underline{M} \\ \underline{F} \end{pmatrix}^O = \underbrace{T_{dy}^{OC} \begin{pmatrix} J & O \\ O & m \end{pmatrix}^C T_{ki}^{CO}}_{M_A^O} \begin{pmatrix} \underline{\alpha} \\ \underline{a} \end{pmatrix}^O + \underbrace{T_{dy}^{OC} \begin{pmatrix} \tilde{\underline{\omega}} & \tilde{\underline{v}} \\ O & \tilde{\underline{\omega}} \end{pmatrix}^C T_{dy}^{CO}}_{\hat{\tilde{V}}_{dy}^O} \underbrace{T_{dy}^{OC} \begin{pmatrix} J & O \\ O & m \end{pmatrix}^C T_{ki}^{CO}}_{M_A^O} \begin{pmatrix} \underline{\omega} \\ \underline{v} \end{pmatrix}^O$$

The 1st conversion matrix of inertia in system O for the acceleration dependent part becomes (47)

$$\boxed{M_A^O = \begin{pmatrix} J^O - m\tilde{r}^O_{OC}\tilde{r}^O_{OC} & m\tilde{r}^O_{OC} \\ -m\tilde{r}^O_{OC} & mI \end{pmatrix}} \qquad J^O = D_0^{OC} J^C \left(D_0^{OC}\right)^T \quad \text{without } \textit{Huygens-Steiner}.$$

With the rule of eq. (25) we obtain the big cross product matrix in system O

$$\boxed{\hat{\tilde{V}}_{dy}^O = \begin{pmatrix} \tilde{\underline{\omega}} & \tilde{\underline{v}} \\ O & \tilde{\underline{\omega}} \end{pmatrix}^O}; \quad \tilde{\underline{v}}^O \equiv \tilde{\underline{v}}_{\underline{O}}^O = O \text{ only if } \vec{v}_C + \vec{r}_{OC} \times \vec{\omega} = \vec{O} \text{ in system O or C.} \tag{48}$$

The 2nd conversion matrix of inertia in system O for the velocity dependent part is

$$\boxed{M^O_{vV} = \hat{V}^O_{dy} M^O_A}. \tag{49}$$

We get the sought kinetic wrench as the sum of the acceleration dependent part and of the velocity dependent part of the kinetic wrench in system O:

$$\boxed{\hat{F}^O = M^O_A \hat{A}^O + M^O_{vV} \hat{V}^O}. \tag{50}$$

Appreciation

I owe many thanks to Prof. Wohlhart and to Prof. Lipkin for very useful discussions and proposals on this theme.

References

[1] Keler, M. L.: Analyse und Synthese der Raumkurbelgetriebe mittels Raumliniengeometrie und dualer Größen. 1959 Forsch.Ing.-Wes.Ausg.B Bd. 25 pp.26-32 and 55-63

[2] Yang, A. T., Freudenstein F.: Application of Dual Number Quaternion Algebra to the Analysis of Spatial Mechanisms. 1964 Journal of Applied Mechanics, vol. 32, Trans. ASME, series E, vol. 86 no.2, pp. 300-308

[3] Dimentberg, F. M.: Metod wintow w prikladnoj mechanike. 1971 Isdatelstwo Mashinostrojenie Moskwa

[4] Duffy J., Habib-Olahi, H. Y.: Eine Bewegungsanalyse räumlicher fünfgliedriger Mechanismen.1973 feinwerktechnik+micronic 77 Heft 5

[5] Rooney, J.: On the three types of complex number and planar transformations. 1978 Environment and Planning B, vol. 5, pp. 89-99

[6] Keler, M., Jucha, J.: Ein Rechenverfahren zur Robotersteuerung bei vorgegebener Greiferbewegung mit sechs Freiheitsgraden. 1985 Forsch.Ing.-Wes.Bd.51, Nr.6, pp. 181-185

[7] Keler, M. L.: Einfluß der Elastizitäten und Gelenkspiele auf die Robotergenauigkeit. 1986, Pergamon Journals, Int.J.Appl.Engng.Ed. vol 2, no. 2, pp. 103-106

[8] Keler, M. L.: Kinematische Analyse räumlicher und ebener Getriebe und Roboter Teil 1, 2, 3, 1989 Werkstattblatt 635-neu, 639-neu, 645-neu, Hoppenstedt-Verlag

[9] Keler, M. L.: The Computation of Elastic Deformations by Dual Vectors. Unpublished papers presented 1994 on the 4th International Workshop on Advances in Robot Kinematics (ARK)July 4-6, Ljubljana - Slovenia and 1995 on the 9th IFToMM Congress in Milano - Italy

[10] Keler, M. L.: Kinematics Using Dual Vectors. 1994 unpublished manuscript of lectures on the University of Florida, Gainesville

[11] Pestel, E.: Technische Mechanik Teil II Kinematik und Kinetik. 1. und 2. Teil 1969 and 1971 Bibliographisches Institut Mannheim

[12] Yang, A. T.: Inertia Force Analysis of Spatial Mechanisms. 1971 Trans. ASME, Journal of Engineering for Industry, Series B, Vol. 93, No. 1, pp. 27-32, Feb.

[13] Wohlhart, K.: Formen des Drallsatzes. 1986 unpublished manuscript Technische Universität Graz

[14] Wohlhart, K.: Dynamik. 1994 unpublished manuscript Technische Universität Graz

[15] Wohlhart, K.: Motor Tensor Calculus in the sense of R. v. Mises 1995 unpublished manuscript Technische Universität Graz

[16] Shoham, M., Brodsky, V.: Analysis of Mechanisms by the Dual Inertia Operator, Computational Kinematics, 1993 ed. by J. Angeles et al. Kluwer Academic Publishers,Netherlands

[17] Lipkin, H.: Extension of Euler's Principal Axes to Articulated Inertia. 1995 Proceedings of the 9th IFToMM Congress vol. 2 pp. 1475 - 1479

DUAL NUMBERS, LIE ALGEBRA AND 6R INVERSE KINEMATICS

KUANGRONG HAO
CERMICS, ENPC
Central 2 La Courtine, 93167 Noisy-Le-Grand, France
Fax : (16)(1) 49 14 35 86. Mail : hao@cermics.enpc.fr

Abstract. The ordinary Euler angles and Briant angles for describing rotations are well known and often used in robotics. However it is sometimes necessary to analyze the product of three general rotations $R_1R_2R_3$. We point out some dual invariants which are independent of the articular parameters and can be calculated, so all the quantities of the form $\{X \mid RY\}$ can be determined for X and Y in the Lie algebra $\mathfrak{D}$. Using these results, the **6R** inverse kinematics problem can be solved.

1. Introduction

In the last years, the inverse kinematics problem with 6 revolute joints gave interesting results. Roth, Rastegar and Scheinman (Roth73) using synthetic arguments showed that in general this problem has at most 32 solutions, their proof was based on arguments from synthetic geometry. Albala and Angeles (Alba79), using indicial notation, presented the first constructive solution for this problem. Later, Duffy and Crane (Duff80) provided a different solution by combining spherical trigonometry and dyalitic elimination to obtain a 32th degree polynomial in the tangent of half-angle of one of the joint variables. Tsai and Morgan (Tsai85) proved that this problem has at most 16 solutions.

In recent years the application of resultant methods (dyalitic elimination) to the inverse kinematic problem of 6 degree-of-freedom robots with 6 revolute joints gave interesting results. In particular for robots of general geometry Lee and Liang (Lee88a) (Lee88b) and Raghavan and Roth (Ragh90) demonstrated that there is an univariate trigonometric polynomial of degree 16 in one of the joint angles, which determines all solutions of the inverse kinematics problem.

Kovács (Kova) found the minimum degree solutions for the inverse kinematics problem by application of the Buchberger algorithm. Combining the geometric criteria with the algebraic procedure, Fanghella (Fang88) and Kecskeméthy and

J. Lenarčič and V. Parenti-Castelli (eds.), Recent Advances in Robot Kinematics, 255-264.
© 1996 *Kluwer Academic Publishers. Printed in the Netherlands.*

Hiller (Hill92) proposed a method suited for detecting and generating closed form solutions of general single-loop spatial mechanisms.

In this work, we present a new approach for three general rotations using dual numbers and Lie algebras. It is different from the classical method based on the matrix computations, the results are developed in free coordinate form, i.e. only the intrinsic properties of mechanism are considered. We can determine some dual invariants which are independent of articular parameters of R_1, R_2 and R_3. Defining an intermediate displacement R, **6R** inverse kinematics problem can be transformed into two **3R** problems. Applying these results to a particular case of **6R** mechanism where there exists at least one $d_i = 0$ (distance between the two adjoin linkages), the dual invariants can be determined only by solving an 8th degree polynomial equation. So all the quantities of the form $\{X \mid RY\}$ can be calculated, therefore, **6R** inverse kinematics problem is resolved and the solutions can be represented in a very compact form. Moreover the solutions are necessary and sufficient.

2. Skew-symmetric fields over the dual number ring Δ

In this section we explain the relationships between the classical screw theory and the algebraic properties of a Lie algebra endowed with additional operations (Chev95). We denote by $\mathcal{E}$ and $\mathbb{E}$ the three-dimensional affine Euclidean space and the associated vector space.

2.1. THE ALGEBRAIC STRUCTURE OF $\mathfrak{D}$

We shall present some useful properties of a skew-symmetric field.

1. **Screws and skew-symmetric vector fields — Vector space structure over real field.**
   A skew-symmetric vector field on $\mathcal{E}$ is a map X from $\mathcal{E}$ to $\mathbb{E}$ such that $X(q) = X(p) + \omega_X \times \overrightarrow{pq}$ for all points p and q, what is equivalent such that for all p and q in $\mathcal{E}$, $\overrightarrow{pq} \cdot X(p) = \overrightarrow{pq} \cdot X(q)$, ω_X is the invariant vector.
   The set of all the skew-symmetric vector fields will be denoted by $\mathfrak{D}$.
   As we see, it is equivalent to define a screw and to define a skew-symmetric field. If o is any fixed origin point the relation $X \to (\omega_X, X(o))$ is a one-one and onto; in fact this relation is an isomorphism between the vector spaces $\mathfrak{D}$ and $\mathbb{E} \times \mathbb{E}$. The vectors ω_X and $X(o)$ are the Plücker vectors.
2. **Lie bracket.** The Lie bracket of two skew-symmetric vector fields X and Y is the skew-symmetric vector field $[X, Y]$ such that

   $$[X,Y](p) = \omega_X \wedge Y(p) - \omega_Y \times X(p), \quad \omega_{[X,Y]} = \omega_X \wedge \omega_Y \quad \text{for all} \quad p \in \mathcal{E}.$$

   We denote also $[X,Y] = \operatorname{ad} X \cdot Y$. Endowed with the Lie bracket, the vector space $\mathfrak{D}$ is a Lie algebra, i.e. $[X, Y]$ is skew-symmetric and the Jacobi identity holds : $[X, [Y, Z]] + [Y, [Z, X]] + [Z, [X, Y]] = 0$.
3. **The Klein form** is the non degenerate symmetric bilinear form defined by :

   $$[X \mid Y] = \omega_X \cdot Y(o) + \omega_Y \cdot X(o) \quad \text{(it does not dependent the point } o).$$

4. **The Killing form and the operator Ω.** For $X \in \mathfrak{D}$, $\Omega X \in \mathfrak{D}$ is defined as the constant vector field such that

$$\Omega X(p) = \omega_X \quad \text{for all} \quad p \in \mathcal{E}.$$

Then Ω is a linear operator in $\mathfrak{D}$ with $\Omega \cdot \Omega = 0$. In the dual number setting, the operator Ω is the multiplication by ϵ. Moreover, the Killing form, a positive degenerate symmetric bilinear form on $\mathfrak{D}$, may be defined from the operator Ω and the Klein form by

$$(X \mid Y) = [X \mid \Omega Y] = [\Omega X \mid Y].$$

5. **The module structure on $\mathfrak{D}$.** Let Δ be the dual number ring, i.e. the "numbers" of the form $\hat{z} = z + \epsilon z^o$ with $z, z^o \in \mathbb{R}$ and $\epsilon^2 = 0$, the real and dual parts of $\hat{z}$ are noted by $Re\hat{z} = z$, $Du\hat{z} = z^o$. Then $\mathfrak{D}$ has a natural module structure over Δ which extends its real vector space structure, the product of $X \in \mathfrak{D}$ by the dual number $\hat{z} \in \Delta$ being defined by :

$$\hat{z}X = zX + z^o\epsilon X = zX + z^o\Omega X.$$

The dual inner product is defined as a dual coefficient combination of the Killing and Klein forms :

$$\{X \mid Y\} = (X \mid Y) + \epsilon[X \mid Y] \qquad \text{for} \qquad X, Y \in \mathfrak{D}.$$

With the Lie bracket we define the dual mixed product by :

$$\{X \mid [Y, Z]\} = \{Y \mid [Z, X]\} = \{Z \mid [X, Y]\}.$$

Using this module structure, it makes sense to speak of Δ -linearity or Δ-bilinearity for properties involving scalars belonging to Δ rather than real scalars. The following important properties, which look like the classical ones for the dot product and cross product of ordinary vectors sum up rather cumbersome calculations on screws in a very compact form and will play a role in the sequel :

– $\{\cdot \mid \cdot\}$ is a symmetric non degenerate Δ-bilinear form on $\mathfrak{D}$. It is completely defined by the relative position between the two elements of $\mathfrak{D}$.

– The dual mixed product is a Δ-trilinear skew-symmetric on $\mathfrak{D}$.

– The formula of double Lie bracket holds :

$$[X, [Y, Z]] = \{X \mid Z\}Y - \{X \mid Y\}Z.$$

The relative position between X and Y can be completely explained by the structure of Lie algebra $\mathfrak{D}$

$$\{X \mid Y\} = \frac{\cos(\theta + \epsilon d)}{\|\omega_X\| \cdot \|\omega_X\|}, \qquad [X, Y] = \frac{\sin(\theta + \epsilon d)}{\|\omega_X\| \cdot \|\omega_X\|} \cdot U. \tag{1}$$

θ and d are respectively the angle and the distance between skew-symmetric fields X and Y, θ is supposed in $[0, \pi]$. U is the screw perpendicular and concurrent to X and Y such that $\|\omega_U\| = 1$.

Proposition 2.1. *(Generalized formula of Olinde-Rodrigues)*
The adjoint operator of $\mathfrak{D}$ represents the screw displacement, X is denoted as its

normed generator (screw-symmetric vector field), θ and d are denoted as rotation and translation :

$C(\theta,d,X) = (\exp(\theta+\epsilon d)X)_* = (\mathbf{1}+\sin\theta ad X + (1-\cos\theta)(ad X)^2)(\mathbf{1}+d ad \Omega X)$

(a) $C(\theta,d,X)$ is a translation, when $\theta = 0$:

$$T = (\exp d\Omega X)_* = \mathbf{1} + dad(\Omega X).$$

(b) $C(\theta,d,X)$ is a rotation, when $d = 0$ (Olinde-Rodrigues Formula (Olind1840)) :

$$R = (\exp\theta X)_* = \mathbf{1} + \sin\theta ad X + (1-\cos\theta)(ad X)^2.$$

(b) $C(\theta,d,X)$ is a displacement with pitch f, when $d = f\theta$:

$H = (\exp(\theta(1+\epsilon f)X)_* = \mathbf{1} + \sin(\theta(1+\epsilon f))ad X + (1-\cos(\theta(1+\epsilon f)))(ad X)^2.$

The property $C(\theta,d,X)\cdot X = X$ is called the invariant property of displacement $C(\theta,d,X)$ for its generator X.

2.2. DEFINITION OF A BASIS OF $\mathfrak{D}$ OVER Δ

If $\{\xi,\eta,\zeta\} \in \mathfrak{D}$ is linearly independent over Δ, then $\{\xi,\eta,\zeta\}$ is a basis of $\mathfrak{D}$ over Δ. The following theorem gives a useful criterion to determine whether set $\{\xi,\eta,\zeta\}$ of $\mathfrak{D}$ is a basis of $\mathfrak{D}$ or not and a formula to calculate the dual coordinates of $x \in \mathfrak{D}$ in basis $\{\xi,\eta,\zeta\}$.

Theorem 2.1. *Let be ξ, η and $\zeta \in \mathfrak{D}$, a necessary and sufficient condition for $\{\ \xi,\ \eta,\ \zeta\ \}$ be a basis of $\mathfrak{D}$ over Δ is that :*

(2) $$(\xi|\,[\eta,\zeta]) \neq 0$$

then $X \in \mathfrak{D}$ can be expressed as a dual linear combination of $\{\xi,\eta,\zeta\}$.

(3) $$X = \hat{x}_1\xi + \hat{x}_2\eta + \hat{x}_3\zeta \qquad \text{for} \quad \hat{x}_i = x_i + \epsilon x_i^o \in \Delta.$$

(4) $$\hat{x}_1 = \frac{\{X|[\eta,\zeta]\}}{\{\xi|[\eta,\zeta]\}} \quad \hat{x}_2 = \frac{\{X|[\zeta,\xi]\}}{\{\xi|[\eta,\zeta]\}} \quad \hat{x}_3 = \frac{\{X|[\xi,\eta]\}}{\{\xi|[\eta,\zeta]\}}$$

From the above theorem, we introduce two corollaries

Corollary 2.1. *Let X, Y and Z be in $\mathfrak{D}$, such that X and Y be linearly independent over Δ, then we can exprime $\{[X,Y]\mid Z\}^2$ by :*

$$\{X\mid X\}\{Y\mid Y\}\{Z\mid Z\} - \{X\mid Y\}^2\{Z\mid Z\} - \{Z\mid X\}^2\{Y\mid Y\}$$
$$-\{Z\mid Y\}^2\{X\mid X\} + 2\{X\mid Y\}\{Y\mid Z\}\{Z\mid X\}$$

3. Presentation of problem 6R

A reference configuration of **6R** mechanism is showed in figure 1. We can describe the mechanism using the Denavit-Hartenberg parameters θ_i, d_i, α_i and a_i for $i = 1,\cdots,6$, where θ_i are articular parameters, the others are constant. T is a given displacement. We assume generators $\xi_i \in \mathfrak{D}$ are given in the reference

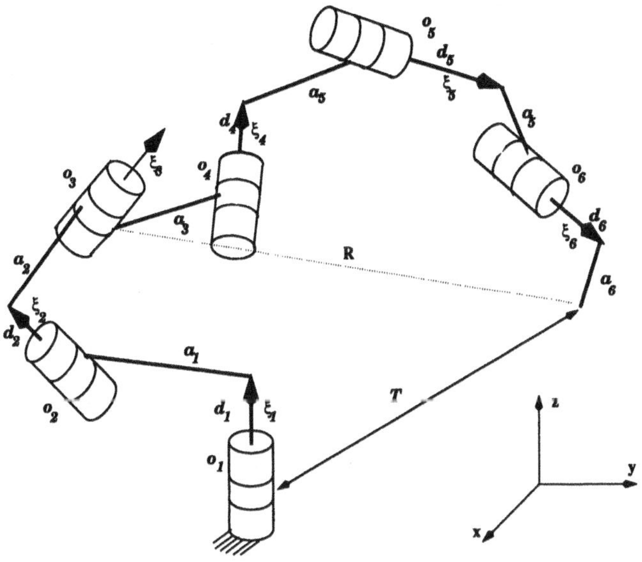

Figure 1. Mechanism 6**R** spatial

configuration. Under assumption $||\omega_\xi|| = 1$, the relationships between Denavit-Hartenberg parameters using dual numbers and Lie algebras properties can be expressed as follows :

$$\{\xi_i \mid \xi_{i+1}\} = \cos\alpha_i - \epsilon a_i \sin\alpha_i$$

the two adjoin generators ξ_i and ξ_{i+1} are supposed linearly independent over Δ.

We can express the closure equation by

$$R_1 R_2 R_3 R_4 R_5 R_6 = T \tag{5}$$

Using proposition 2.1, every rotation $R_i = (\theta_i, d_i, \xi_i)$ around ξ_i is equal to

$$R_i = \mathbf{1} + (\sin\theta_i - d_i\cos\theta_i)\mathrm{ad}\,\xi_i + (1 - \cos\theta_i + d_i\sin\theta_i)(\mathrm{ad}\,\xi_i)^2$$

and its transposition is

$$R_i^\star = \mathbf{1} - (\sin\theta_i - d_i\cos\theta_i)\mathrm{ad}\,\xi_i + (1 - \cos\theta_i + d_i\sin\theta_i)(\mathrm{ad}\,\xi_i)^2$$

Because we suppose there exists at least one $d_i = 0$, with a suitable permutation, we can always assume $d_3 = 0$, then

$$R_3 = \mathbf{1} + \sin\theta_3\mathrm{ad}\,\xi_3 + (1 - \cos\theta_3)(\mathrm{ad}\,\xi_i)^2 \tag{6}$$

It is very difficult to resolve directly equation 5. In fact, we can rewrite it as the following forms

$$T^\star R_1 R_2 R_3 = R \tag{7}$$

$$R_6^\star R_5^\star R_4^\star = R \tag{8}$$

Where R is an unknown displacement, with which we can transform the **6R** problem into a **3R** one, so the objective is to determine intermediate displacement R.

It is obvious that equations 7 and 8 are equivalent to equation 5. In the following section, we will study system $T^{\star}R_1R_2R_3 = R$, some necessary and sufficient conditions will be pointed out, and the displacement R will be determined by the invariants of the form $\{X \mid RY\}$.

4. Determination of invariant $\{X \mid RY\}$

We will prove that product $T^{\star}R_1R_2R_3$ is equivalent to three dual scalar equations.

Proposition 4.1. *The next three conditions are equivalent one another*

$$(a) \; : \left\{ \begin{array}{l} \{\xi_1 \mid R_2\xi_3\} = \{\xi_1 \mid TR\xi_3\} \\ \{\xi_1 \mid TRR_3^{\star}\xi_2\} = \{\xi_1 \mid \xi_2\} \\ \{\xi_2 \mid R_1^{\star}TR\xi_3\} = \{\xi_2 \mid \xi_3\} \end{array} \right.$$

$$(b) \; : R_1R_2R_3 = TR \qquad (c) \; : \left\{ \begin{array}{l} R^{\star}T^{\star}R_1R_2\xi_3 = \xi_3, \\ R^{\star}T^{\star}R_1R_2\xi_2 = R_3^{\star}\xi_2. \end{array} \right.$$

At first, we will determine $\{\xi_1 \mid TR\xi_3\}$ and $\{\xi_1 \mid TR\xi_2\}$, with which we find a dual number set A, all the Δ - bilinear form $\{X \mid RY\}$ is a dual combination of the elements of set A.

Due to proposition 2.1, writing separately the real and dual parts, the first relation of (a) of proposition 4.1 becomes two polynomial equations of second degree for $\tan\dfrac{\theta_2}{2}$.

$$(9) \quad \begin{array}{rl} (Re(2a) + (\xi_1 \mid \xi_3) - r_{13})\tan^2\dfrac{\theta_2}{2} + 2Re(b)\tan\dfrac{\theta_2}{2} + (\xi_1 \mid \xi_3) - r_{13} & = 0 \\ (Du(2a) + [\xi_1 \mid \xi_3] - s_{13} - d_2Re(b))\tan^2\dfrac{\theta_2}{2} & \\ +2\,(Du(b) + d_2Re(a))\tan\dfrac{\theta_2}{2} + [\xi_1 \mid \xi_3] - s_{13} + d_2Re(b) & = 0 \end{array}$$

Where $r_{13} = (\xi_1 \mid TR\xi_3)$ and $s_{13} = [\xi_1 \mid TR\xi_3]$. With

$$(10) \qquad a = \{\xi_1 | [\xi_\phi, [\xi_\phi, \xi_\psi]]\} \quad b = \{\xi_1 | [\xi_\phi, \xi_\psi]\}$$

Because equation 9 contains two polynomials in $\mathbb{R}$ having a common root $\tan\dfrac{\theta_2}{2}$, then their resultant is equal to zero which is a polynomial equation in $\mathbb{R}$ of second degree for r_{13} and s_{13},

$$(11) \qquad A_1r_{13}^2 + A_2s_{13}^2 + A_3r_{13}s_{13} + A_4r_{13} + A_5s_{13} + A_6 = 0$$

On the other hand, equation 9 can be also transformed into the following form :

$$\left\{ \begin{array}{l} Re(b)\sin\theta_2 + Re(a)\cos\theta_2 = r_{13} - (\xi_1 \mid \xi_\psi) - Re(a) \\ (Du(b) + fRe(a))\sin\theta_2 + (fRe(b) - Du(a))\cos\theta_2 = s_{13} - [\xi_1 \mid \xi_\psi] - Du(a) \end{array} \right.$$

we can find the solutions for $\sin\theta_2$ and $\cos\theta_2$ from 2 above equations. Substituting these solutions for $\sin\theta_2$ and $\cos\theta_2$, $\sin^2\theta_2 + \cos^2\theta_2 = 1$ can give a polynomial equation of second degree for r_{13} and d_{13} which is independent of equation 11

$$B_1 r_{13}^2 + B_2 d_{13}^2 + B_3 r_{13} d_{13} + B_4 r_{13} + B_5 d_{13} + B_6 = 0 \tag{12}$$

Eliminating d_{13}, equations 11 and 12 give us

$$C_1 r_{13}^4 + C_2 r_{13}^3 + C_3 r_{13}^2 + +C_4 r_{13} + C_5 = 0 \quad C_i \in \mathbb{R}, \tag{13}$$

so r_{13} can be obtained explicitly as function of C_i.

From the second equation of (a) of proposition 4.1 and equation 6, using the similar method, equation 6 can be transformed into the real and dual parts

(14)

$$\begin{cases} (Re(2a') + r_{12} - (\xi_1 \mid \xi_2)) \tan^2 \frac{\theta_3}{2} - 2Re(b') \tan \frac{\theta_3}{2} + r_{12} - (\xi_1 \mid \xi_2) = 0 \\ (Du(2a') + s_{12} - [\xi_1 \mid \xi_2]) \tan^2 \frac{\theta_3}{2} - 2Du(b') \tan \frac{\theta_3}{2} + s_{12} - [\xi_1 \mid \xi_2] = 0 \end{cases}$$

where $r_{12} + \epsilon s_{12} = \{\xi_1 \mid TR\xi_2\}$, a' and b' are defined by

$$\begin{cases} a' = \{\xi_1 \mid TR[\xi_3, [\xi_3, \xi_2]]\} \\ b' = \{\xi_1 \mid TR[\xi_3, \xi_2]\} \end{cases}$$

which can be further expressed as functions of r_{12} and s_{12} according to corollary 2.1.

Having a common root, the resultant of equation 14 must be equal to zero, this is equivalent to

$$\begin{aligned} & r_{12}^4 + D_1 r_{12}^3 d_{12} + D_2 r_{12}^3 + D_3 r_{12}^2 d_{12}^2 + D_4 r_{12}^2 d_{12} + D_5 r_{12}^2 \\ (15) \quad & +D_6 r_{12} d_{12}^2 + D_7 r_{12} d_{12} + D_8 r_{12} + D_9 d_{12} + D_{10} \quad = \quad 0 \qquad \text{for} \quad D_i \in \mathbb{R}. \end{aligned}$$

Rewriting equation 14 in the form of $\sin\theta_3$ and $\cos\theta_3$, yields

$$\begin{cases} Re(b') \sin\theta_3 + Re(a') \cos\theta_3 = Re(a') + r_{12} - (\xi_1 \mid \xi_2) \\ Du(b') \sin\theta_3 + du(a') \cos\theta_3 = Du(a') + s_{12} - [\xi_1 \mid \xi_2] \end{cases} \tag{16}$$

The condition $\sin^2\theta_3 + \cos^2\theta_3 = 1$ gives the following equation

$$E_1 r_{12}^2 + E_2 d_{12}^2 + E_3 r_{12} d_{12} + E_4 r_{12} + E_5 d_{12} + E_6 = 0 \qquad \text{for} \quad E_i \in \mathbb{R}. \tag{17}$$

Eliminating d_{12}, the resolution of equations 15 and 17 leads to solve an equation of 8th degree for r_{12}

$$F_1 r_{16}^8 + F_2 r_{16}^7 + F_3 r_{16}^6 + F_4 r_{16}^5 + F_5 r_{16}^4 + F_6 r_{16}^3 + F_7 r_{16}^2 + F_8 r_{16}^1 + F_9 = 0. \tag{18}$$

The real coefficients A_i, B_i C_i, D_i, E_i and F_i can be calculated using the symbolic computation software Maple, they are only dependent on the constant parameters. So, r_{12} and s_{13} can be calculated by equations 18 and 17.

Theorem 4.1. *(i) Δ-bilinear form $(X, Y) \to \{X \mid RY\}$, for $X, Y \in \mathfrak{D}$, can be determined with the help of set A of 9 dual numbers* $\{\ \{\xi_1 \mid TR\xi_1\}$, $r_{12} + \epsilon d_{12}$, $\{\xi_1 \mid TR\xi_2\}$, $\{\xi_2 \mid TR\xi_3\}$, $\{\xi_1 \mid TR[\xi_2, \xi_3]\}$, $\{[\xi_1, \xi_2] \mid TR\xi_2\}$, $\{[\xi_1, \xi_2] \mid TR\xi_3\}$, $\{\xi_2 \mid TR[\xi_2, \xi_3]\}$, $\{[\xi_1, \xi_2] \mid TR[\xi_2, \xi_3]\}\ \}$

(ii) The elements of set A are functions of r_{13}, d_{13}, r_{12} and d_{12} which are roots of polynomial equations 13, 12 and 18, 17.

Dual numbers $r_{12}+\epsilon d_{12}$ and $r_{13}+\epsilon d_{13}$ are given above, the other elements of A can be expressed as non-linear functions of these two quantities in accordance with theorem 2.1.

5. Determination of θ_1, θ_2 and θ_3

If we hope to present the solution in a triangular form, i.e. $\theta_2 = f_2(T)$; $\theta_1 = f_1(\theta_2, T)$; $\cdots$, $\theta_6 = f_6(\theta_1,\theta_2,\theta_3,\theta_4,\theta_5,T)$. The vector equation (c) of proposition 4.1 must be used.

Assuming R given by theorem 4.1, then θ_1, θ_2, θ_3 can be calculated by equation (c) of proposition 4.1.

Replacing R_1 with the generalized formula of Olinde-Rodrigues, $R_1R_2\xi_3 = TR\xi_3$ becomes :

$$R_2\xi_3 + \sin(\theta_1+\epsilon d_1)[\xi_1, R_2\xi_3] + (1-\cos(\theta_1+\epsilon d_1))[\xi_1,[\xi_1,R_2\xi_3]] = R\xi_\psi \tag{19}$$

Let us take the killing form with ξ_1, the above equation is equivalent to equation 9, we have directly

$$\tan\frac{\theta_2}{2} = \frac{-Re(b) \pm \sqrt{Re(b)^2 - (Re(a) + (\xi_1 \mid \xi_3) - r_{13})\,((\xi_1 \mid \xi_3) - r_{13})}}{2Re(a) - (\xi_1 \mid \xi_3) + r_{13}} \tag{20}$$

where coefficients a and b are defined by equation 10.

Multiplying equation 19 by $[\xi_1, R_2\xi_3]$, $[\xi_1,[\xi_1,R_2\xi_3]]$, the killing form shows θ_1 is a function of θ_2 :

$$\sin\theta_1 = \frac{(TR\xi_3|[\xi_1,R_2\xi_3])}{1-(\xi_1 \mid TR\xi_3)^2} \qquad \cos\theta_1 = \frac{((TR\xi_3|R_2\xi_3) - (\xi_1 \mid TR\xi_3)^2)}{1-(\xi_1 \mid TR\xi_3)^2} \tag{21}$$

Because $\xi_1, [\xi_1,R_2\xi_3]$, $[\xi_1,[\xi_1,R_2\xi_3]]$ form a basis of $\mathfrak{D}$ over Δ, so conditions 20, 21 are necessary and sufficient for equation 19.

Using equation 7 and the invariant property of R_2 for ξ_2, we have :

$$T^\star R_1\xi_2 = R_3^\star\xi_2$$

Multiplying $[\xi_3,\xi_2]$, $[\xi_3,[\xi_3,\xi_2]]$, the Killing form gives $\sin\theta_3$, $\cos\theta_3$ as follows :

$$\sin\theta_3 = -\frac{(R^\star T^\star R_1\xi_2|[\xi_3,\xi_2])}{\sin^2\alpha_2} \qquad \cos\theta_3 = \frac{((R^\star T^\star R_1\xi_2|\xi_2) - \cos^2\alpha_2)}{\sin^2\alpha_2} \tag{22}$$

Conditions 21, 20 and 22 can be represented as explicit functions of the invariants of theorem 4.1. Because of their complexities, we will introduce the explicit conditions only when the numerical results are wished.

6. Determination of θ_4, θ_5 and θ_6

With the same procedure than the precedent section, we will calculate θ_4, θ_5 and θ_6 under the assumption of θ_1, θ_2 and θ_3 given. The closure equation 5 is equivalent to

$$R_4R_5R_6 = R_3^\star R_2^\star R_1^\star T \tag{23}$$

Multiplying by ξ_6, the above equation becomes

$$R_4R_5\xi_6 = R_3^\star R_2^\star R_1^\star T\xi_6 \tag{24}$$

Multiplying by ξ_4, we can eliminate θ_4, so θ_5 is expressed as a function of θ_1, θ_2 and θ_3

$$\tan\frac{\theta_5}{2} = \frac{-Re(v_1) \pm \sqrt{Re^2(u_1) - Re(2u_1 - w_1)Re(w_1)}}{(2Re(u_1) - Re(w_1))} \tag{25}$$

with :

$$\begin{cases} u_1 = \{\xi_4 \mid [\xi_5, [\xi_5, \xi_6]]\} \\ v_1 = \{\xi_4 \mid [\xi_5, \xi_6]\} \\ w_1 = \{\xi_4 \mid \xi_6 - R_3^\star R_2^\star R_1^\star T\xi_6\} \end{cases} \tag{26}$$

Multiplying equation 19 by $[\xi_4, R_5\xi_6]$ and $[\xi_4, [\xi_4, R_5\xi_6]]$ respectively, θ_4 is a function of θ_1, θ_2, θ_3 and θ_5

$$\begin{cases} \sin\theta_4 &= \dfrac{(R_3^\star R_2^\star R_1^\star T\xi_6 | [\xi_4, R_5\xi_6])}{([\xi_4, R_5\xi_6] \mid [\xi_4, R_5\xi_6])} \\ \cos\theta_4 &= \dfrac{(R_3^\star R_2^\star R_1^\star T\xi_6 | R_5\xi_6) - (\xi_3 \mid R_3^\star R_2^\star R_1^\star T\xi_6)^2}{([\xi_4, R_5\xi_6] \mid [\xi_4, R_5\xi_6])} \end{cases} \tag{27}$$

Equation 23 is equivalent to $R_6^\star T^\star = R_5^\star R_4^\star R_3^\star R_2^\star R_1^\star$, we can express θ_6 as a function of θ_2, θ_3 θ_4 and θ_5

$$\begin{cases} \sin\theta_6 &= \dfrac{(R_5^\star R_4^\star R_3 \star R_2^\star \xi_1 \mid [\xi_6, T^\star\xi_1])}{([\xi_6, T^\star\xi_1] \mid [\xi_6, T^\star\xi_1])} \\ \cos\theta_6 &= -\dfrac{(R_5^\star R_4^\star R_3 \star R_2^\star \xi_1 \mid [\xi_6, [\xi_6, T^\star\xi_1]])}{([\xi_6, T^\star\xi_1] \mid [\xi_6, T^\star\xi_1])} \end{cases} \tag{28}$$

All the solutions 25, 27 and 28 can be represented in explicit function of artcular parameters.

Sections 5 and 6 show that solutions 20, 21 and 22 are necessary and sufficient for equation 7. Conditions 25, 27 and 28 are necessary and sufficient for equation 24, finally all the conditions 20, 21, 22, 25, 27 and 28 are necessary and sufficient for closure equation 5.

7. Conclusion

In this paper, we have studied **6R** inverse kinematics problem through **3R** ones. We have proved the existence of dual invariants. This method can be generalized with resolution of a polynomial equation more than 8th degree when all $d_i \neq 0$.

These results is also convenient to the **4R** closed mechanism, its closure equation is $R_1(\theta_1, \xi_1)R_2(\theta_2, \xi_2)R_3(\theta_3, \xi_3) = R(\theta, \xi)$, if there exist no solutions for the polynomial equations for r_{12} and r_{13}, i.e. R_1, R_2 and R_3 are not independent, we can express θ_1, θ_2 as function of θ_3 or inversely, the mechanism has at least one degree of freedom. Conversely, if there are some solutions for the polynomial equations, i.e. all articular parameter are fixed the mechanism has no degree of freedom.

We can also apply this method to **5R** and **6R** closed mechanisms by defining a suitable intermediate displacement **R**. For example a closed **6R** mechanism of one degree of freedom $R_1(\theta_1, \xi_1)R_2(\theta_2, \xi_2)R_3(\theta_3, \xi_3)R_4(\theta_4, \xi_4)R_5(\theta_5, \xi_5)R_6(\theta_6, \xi_6) = \mathbf{1}$, if θ_3 is considered as the input parameter, we can define $R_1 R_2 R_3 R_4 = R$, $R_6^\star R_5 \star = R$. After had calculated the invariants $\{\xi_1 \mid R\xi_4\}$ and $\{\xi_1 \mid R\xi_2\}$, we can obtained the solution in the form $\theta_2 = f_2(\theta_3)$, $\theta_1 = f_1(\theta_3, \theta_2)$, $\theta_4 = f_4(\theta_3, \theta_2, \theta_1)$, $\theta_5 = f_5(\theta_3, \theta_2, \theta_1)$ and $\theta_6 = f_6(\theta_3, \theta_2, \theta_1)$.

Acknowledgment

The author thanks Professor Dominique P. Chevallier for his advises and interested discussions.

References

H. Albala and J. Angeles. Numerical solution to the input-output displacement equation of the general 7r spatial mechanism. In *Proceedings of the Fifth World Congress on Theory of Machines and Mechanisms*, pages 1008–1011, 1979.

D.P. Chevallier. Coordinate free criteria for testing the linear dependence of the sets of screws. In Sophia Antipolis, editor, *Computational kinematics*. Kluwer Academic Publisher, septembre, 1995.

J. Duffy and C. Crane. A displacement analysis of the general spatial 7r mechanism. *J. of Mech. Mach. Theory*, 15(pp. 153-169), 1980.

P. Fanghella. Kinematics of spatial linkages by group algebra: A structure-based approach. *Mech. Mach. Theory*, 1988.

A. Kecskeméthy and M. Hiller. Automatic closed-form kinematics-solutions for recursive single-loop chains. *ASME - Mechanisms conference Phoenix (Arizona)*, 1992.

P. Kovács. Minimum degree solutions for the inverse kinematics problem by application of the buchberger algorithm. In *Proc. 2nd. Intl. Conf. on Advances in Robot Kinematics*, pages 326–334, Berlin, 1990. Spring.

H. Lee and C. Liang. Displacement analysis of the general spatial 7-link 7r mechanism. *Mech. Mach. Theory*, 23(3, pp.219-226), 1988.

H. Lee and C. Liang. A new vector theory for the analysis of spatial mechanisms. *Mech. Mach. Theory*, 23(3, pp.209-217), 1988.

M. Olinde-Rodrigues. Des lois géométriques qui régissent les déplacements d'un système solide dans l'espace, et de la variation des coordonnées provenant de ces déplacements considérés indépendamment des causes qui peuvent les produire. *Journal de mathématiques pures et appliquées*, 5, 1840.

M. Raghaven and B. Roth. Inverse kinematics of the general 6r manipulators and related linkages. *Mechanism synthesis and analysis : ASME Design Technical Conferences -21st Biennial Mechanisms Conference*, 16-19, pp. 59-65, 1990.

B. Roth, J. Rastegar, and V. Scheinman. On the design of computer controlled manipulators. *On the theory and Practice of Robots and Manipulators*, 1, pp. 93-113, september 1973.

L.W. Tsai and A. Morgan. Solving the kinematics of the most general six and five degree of freedom manipulator by continuation methods. *ASME J. of Mechanisms, Transmissions and Automation in Design*, 107, pp.189-200, 1985.

BACKBONE CURVES TO APPROXIMATE MINIMUM JOINT TORQUE CONFIGURATIONS OF PLANAR MULTIPLE-LINK MANIPULATORS IN PRESENCE OF OBSTACLES

J. LENARČIČ
J. Stefan Institute, University of Ljubljana, SLOVENIA

Abstract. An approach is described whereby, for a given position of the end effector, the target backbone curve for a planar multiple-link manipulator is obtained when a polygonal obstacle is placed in the workspace. The curve approximates an optimum static configuration of the manipulator relative to the minimum joint torques caused by an arbitrary force applied to the end-effector. The proposed approach is numerically effective and can be implemented in real-time procedures.

1. Introduction

Hyper-redundant multiple-link manipulators are more versatile since they are capable to produce an infinity of possible motions that correspond to the prescribed task and simultaneously optimise a given set of performance criteria. Designs of such mechanisms were reported by Ma *et al.* (1992), Chirikjian and Burdick (1993) and many others. However, a practical implementation of multiple-link robots is still far away. In a first place, this is because of enormous difficulties in their design and real-time control associated with an irresolvable numerical complexity if standard algorithms are used. The most elementary inverse kinematics problem can include several thousands of arithmetical operations when the well known pseudoinverse approach is applied. Nevertheless, multiple-link manipulators have a prosperous future in applications where specific working conditions will be provided. For instance, in slow motion tasks in space where the weight of the mechanism, as well as its dynamics and flexibility, can be neglected.

In the area of kinematic analysis and optimisation of hyper-redundant robots the majority of findings in the recent years are due to Chirikjian and Burdick (1990, 1991a, 1991b). They introduced the so-called backbone curve that represents the spine of the robot mechanism. Based on its introduction they minimised the complexity of the numerical treatment, as well as the dependency on the number of degrees of freedom.

J. Lenarčič and V. Parenti-Castelli (eds.), Recent Advances in Robot Kinematics, 265-274.
© 1996 *Kluwer Academic Publishers. Printed in the Netherlands.*

The backbone curve can be used to specify the shape of continuos morphology manipulators. For discrete segmented manipulators, the process of fitting is used to determine the displacements of the actuators to resolve redundancy. An effective modal resolution scheme was found in Burdick and Chirikjian (1994). They proposed a method for finding backbone curve shapes that satisfy a given set of task constraints. In Chirikjian and Burdick (1992) a method is developed to obtain optimum configurations of a hyper-redundant manipulator. The complexity of the problem was significantly reduced by using physically meaningful backbone curve parametrisations associated to a selected optimality criterion. Based on the calculus of variations, their method leads to a differential equation whose solutions are the desired optimum configurations. Yet, relatively complex and time-consuming procedures must be applied that may not be suited for real-time control.

As was shown in Lenarčič (1993), the optimum configurations of planar multiple-link manipulators achieve typical and visually recognisable forms with respect to different optimisation criteria, such as manipulability, kinematic index, or compliance. In Lenarčič (1994), this idea was applied to find the optimum configuration of a planar multiple-link manipulator with respect to a linear criterion of joint torques associated with the gravity of the load and the gravity of the links. In Lenarčič (1995), the results were extended to a more general case where the end points of the configuration are positioned arbitrarily in the workspace. In the present work, we study the situation where an obstacle is placed in the workspace so that it crosses the ideal backbone curve of the optimum configuration related to the minimum joint torques. Therefore, this configuration is not realisable and the objective is to find a curve that avoids the obstacle while it minimises the joint torques.

2. Problem statement, assumptions and practical significance

The horizon of the present work is the control scheme introduced by Burdick and Chirikjian where, in a first stage, we synthesise an optimum continuos backbone curve for a multiple-link manipulator and, in a second stage, we fit this curve with the discrete segmented manipulator. We consider here only the first stage where the problem consists of finding the theoretical optimum for the backbone curve that corresponds to a given end-effector position, avoids the obstacle inserted in the workspace, and minimises the joint torques. Since it is significant that the problem is solved in real-time, standard optimisation procedures cannot be used. The optimisation criterion is defined as the sum of absolute joint torques. In principle, other criteria, could be used. However, as it was demonstrated in Lenarčič (1994, 1995), the utilisation of the mentioned criterion leads to solutions where the optimum configuration can effectively be approximated by simple combinations of linear sections. The included mathematics and algorithms relate to planar multiple-link manipulators although they could be elaborated for spatial mechanisms, too.

The proposed approach is valid for static cases where the manipulator holds or pushes an object in a given position of the end effector. Statics in robotics is presently being considered of minor importance. However, in applications, such as holding and pushing, statics prevails over dynamics. In particular in multiple-link manipulators, where the body of the mechanism is preferably designed of extremely light material with tendon drives, the dynamical effects and even the weight of the links with respect to the load could be neglected so that an optimum control of static external forces would be of acute importance.

Suppose that the *n-R* manipulator moves in the plane $(\vec{x},\vec{y})$. Theoretically, for the purpose of establishing the target backbone curve, the links are infinitesimally small and n is infinite. Therefore, with no additional lost in generality, we can assume identical links and unlimited joint angles. We can also neglect eventual collisions between the links. Let a force F be applied to the end effector. It is assumed that F is vertical (this is not a restriction since the manipulator is symmetric about the first joint axis). When the masses of the links are neglected, the objective is to minimise the criterion of absolute joint torques τ_i

$$f=\frac{1}{n}\sum_{i=1}^{n}|\tau_i|=\frac{F}{n}\sum_{i=1}^{n}|x_n-x_i|=\frac{Fd}{D}\sum_{i=1}^{n}|x_n-x_i| \tag{1}$$

subject to a given end effector position $x_n=x$ and $y_n=y$. Here, d is the length of the links, x_i is the horizontal position of the *i-th* link, and D is the total length of the manipulator. It is assumed that the mechanism is enough long so that it always solves the given task - achieves the desired position, avoids the obstacle and minimises the joint torques. According to Lenarčič (1994), the configuration correspondent to the above optimisation problem is formed of two straight sections as shown in Fig. 1. The first is perpendicular to the force axis and originates in the first joint, while the second coincides with the force axis. The problem posed in the present work is to study how this optimum configuration deforms in the presence of obstacles inside the workspace.

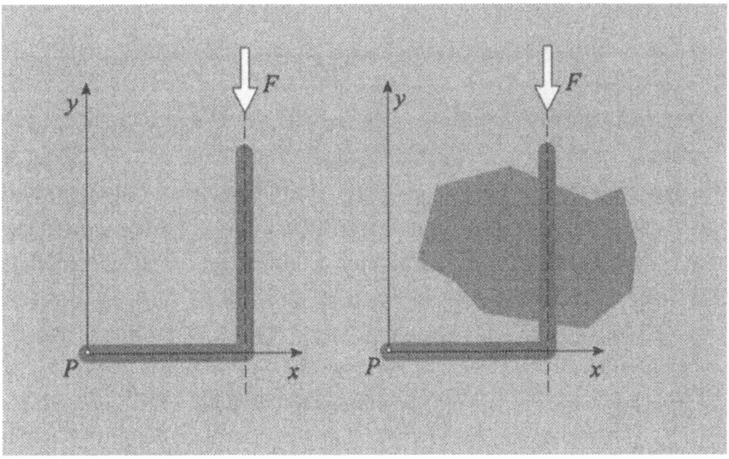

Fig. 1: An obstacle introduced in the workspace of a multiple link manipulator

3. Background

Suppose that a given configuration of the mechanism is composed of straight sections as presented in Fig. 2. Note that a straight section is defined as the sub-chain of the mechanism that possesses equal inclination and all joints are positioned on one side of the force axis. The cost function (1) is additively separable where each f_k relates to one of the straight sections. Let the number of joints in section k be denoted by n_k, the length by D_k, and the cosine of the inclination angle by c_k. If we introduce $n_k = D_k / d$, $n = D / d$, $n \to \infty$, $d \to 0$ and normalise $D = 1$, $F = 1$ we will get

$$f_k = D_k(x_n - x_k + \tfrac{1}{2} D_k c_k) . \quad (2)$$

The aim is then to synthesise the optimum combination of K sections that minimises the cost function f expressed as the sum of f_k. Hence, the complexity of the optimisation problem depends on the number of sections K rather than on the number of degrees of freedom n.

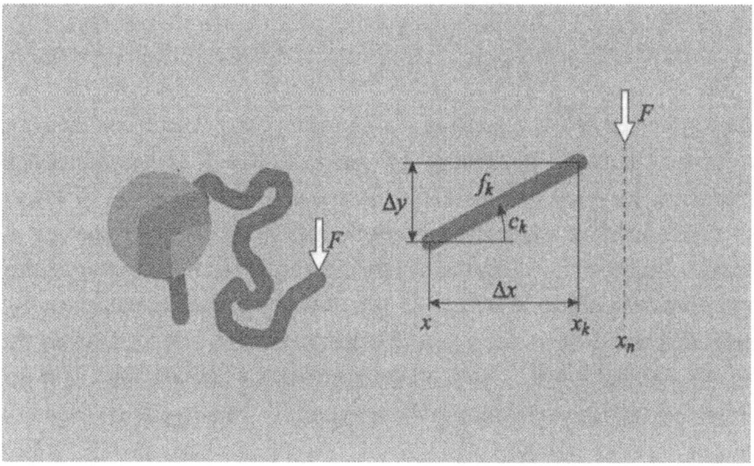

Fig. 2: Manipulator composed of straight sections

According to Lenarčič (1995), the optimum backbone curve that connects a point P and a vertical line L is a horizontal straight line. This can be demonstrated by dividing the mechanism into a series of small straight sections. The cost function of a slant section is always more costly than the cost function of the horizontal section with the same Δx. Consequently, the solution is a series of connected small horizontal sections that form a horizontal line between P and L with the cost function. On the contrary, the optimum configuration between P and a horizontal line L is more complex. We showed in Lenarčič (1995) that this optimum configuration is an near-exponential function that is placed between the vertical line connecting P and L (denominated by (a) in Fig. 3) and the combination of the horizontal line between P and the force axis plus the vertical line along the force axis (denominated by (b) in Fig. 3). The most interesting and profitable property of the optimum solution is seen in Fig. 4

where the relative difference of the cost functions of the configurations (a) and (b) with respect to the cost function of the optimum solution is plotted

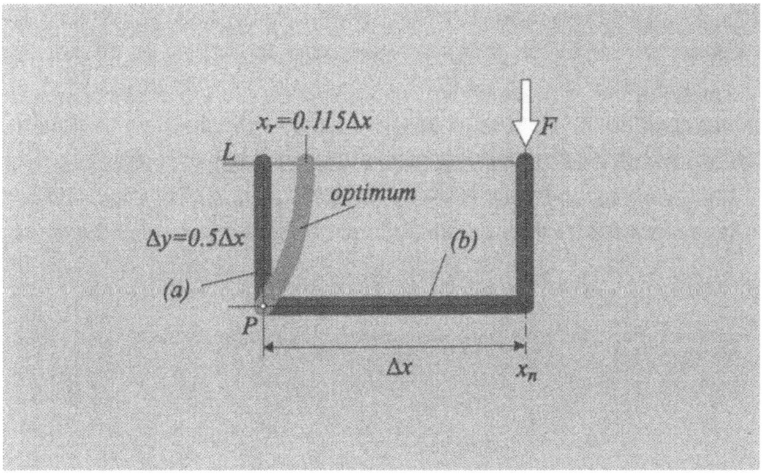

Fig. 3: Optimum configuration when $\Delta y / \Delta x = 0.5$

For the combination (a), this difference is smaller if the ratio $\Delta y / \Delta x < 0.5$, while for the combination (b) it becomes smaller when $\Delta y / \Delta x > 0.5$. It is important to observe that if we take the best of the two, the difference with the optimum configuration will always be smaller than 5%. In practice, it means that we may utilise one of these extremely simple solutions, either (a) or (b), instead of the considerably more complex and time-consuming optimum solution.

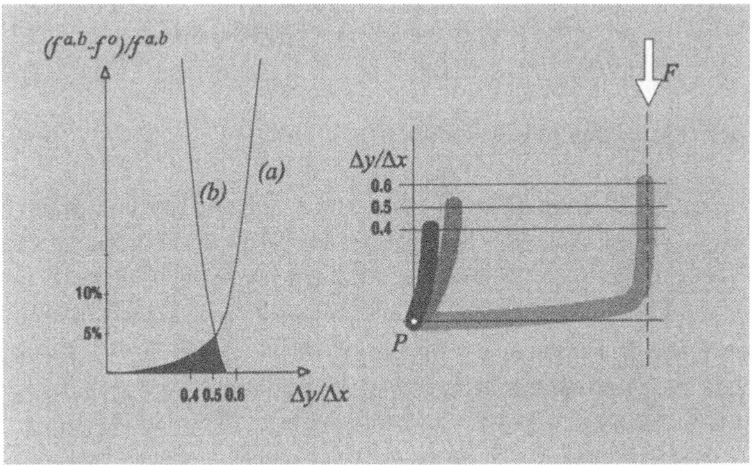

Fig. 4: Dependency of the optimum configuration on $\Delta y / \Delta x$

In this relation and having in mind that the obstacle placed in the workspace possesses a polygonal shape, only few basic structures of straight sections must be analysed. On the left-hand side in Fig. 5 , for instance, it is easy to verify that $f = 2f^h + f^v$ is a convex quadratic function of x that (depending on Δy) achieves its infimum either at $x = x_p$ or at $x = x_n$. Similarly, it is possible to show that in the external triangle on the right-hand side of (above) $f^v + f^h$ is always more costly than f^s without regard to the values of x and Δy. On the other hand, in the internal triangle presented below, $f^v + f^h$ is less expensive than f^s but it is guaranteed only when $x = x_n$. Following these principles we can develop more complex compositions of sections that approximate the optimum configuration with an error that never exceeds 5%.

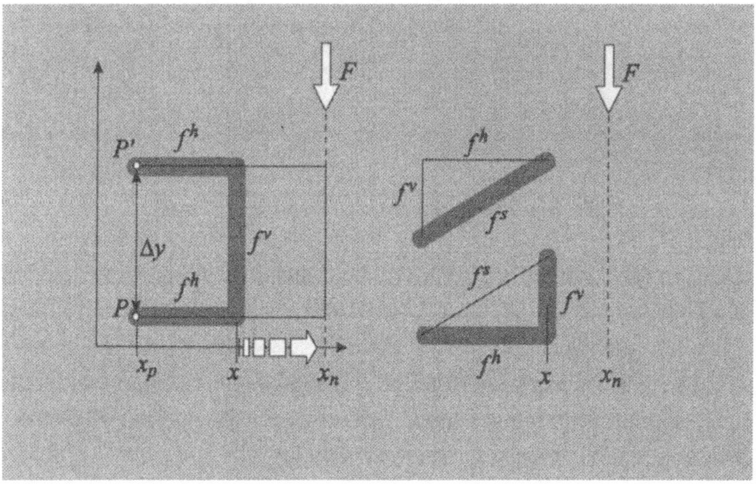

Fig. 5: Simple structures of straight sections

4. Algorithm of obstacle avoidance

Consider now the situation where an object in the form of a convex polygon is inserted in the workspace of a multiple-link manipulator. The critical points or edges of the convex polygon that must be analysed are related to its minimum and maximum x represented by the points $\boldsymbol{w}$ and $\boldsymbol{W}$ and to its minimum and maximum y represented by the points $\boldsymbol{h}$ and $\boldsymbol{H}$. For the sake of simplicity, we presuppose that the obstacle and the mechanism are placed entirely in the area $y > 0$.

When the obstacle is embraced from the right, as shown in Fig. 6, we found that there are three postulates which assign the approximated optimum backbone curves of the manipulator.

Postulate A1:
Sectors of the optimum configuration, where $x > x_n$, that connect the force axis with ***h*** (if $x(\mathbf{h}) > x_n$) and ***H*** (if $x(\mathbf{H}) > x_n$) are horizontal straight sections. They intersect the force axis in ***h'*** and ***H'***.

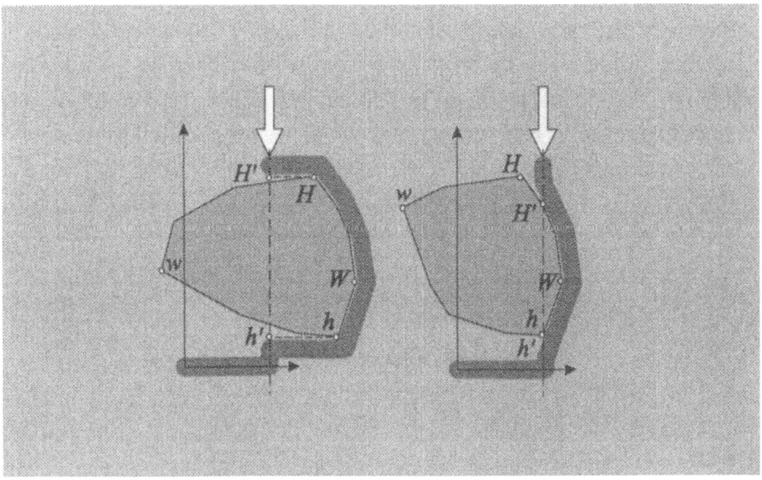

Fig. 6: Obstacle approached from left

Postulate A2:
Sectors of the optimum configuration, where $x > x_n$, between ***h*** (if $x(\mathbf{h}) > x_n$) or ***h'*** (if $x(\mathbf{h}) \leq x_n$) and ***H*** (if $x(\mathbf{H}) > x_n$) or ***H'*** (if $x(\mathbf{H}) \leq x_n$) are straight sections that coincide with the edges of the obstacle.

Postulate A3:
Sectors of the optimum configuration, where $0 \leq x \leq x_n$, between the origin of the manipulator and ***h'***, as well as between ***H'*** and the end of the manipulator, coincide with the ideal optimum.

When the obstacle is embraced from the left, as shown in Fig. 7, we found the following four postulates which assign the approximated optimum backbone curves of the manipulator.

Postulate B1:
Sectors of the optimum configuration, where $x < 0$, that connect $x = 0$ with ***h*** (if $x(\mathbf{h}) < 0$) and **H** (if $x(\mathbf{H}) < 0$) are horizontal straight sections. They intersect $x = 0$ in ***h''*** and ***H''***, respectively. Sectors of the optimum configuration, where $0 \leq x \leq x_n$, that connect the force axis with ***H*** (if $0 \leq x(\mathbf{H}) \leq x_n$) or ***H''*** (if $x(\mathbf{H}) < 0$) is a horizontal straight section. It intersects $x = x_n$ in ***H'***.

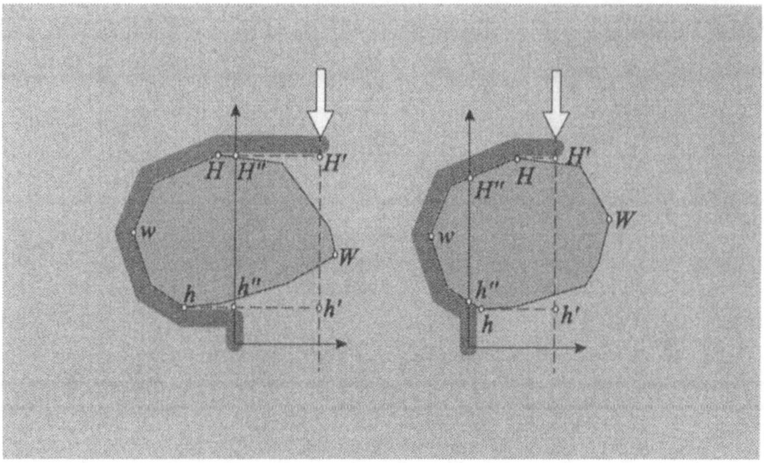

Fig. 7: Obstacle approached from right

Postulate B2:
Sectors of the optimum configuration, where $x < 0$, between $\boldsymbol{h}$ (if $x(\boldsymbol{h}) < 0$) or $\boldsymbol{h''}$ (if $x(\boldsymbol{h}) \geq 0$) and $\boldsymbol{H}$ (if $x(\boldsymbol{H}) < 0$) or $\boldsymbol{H''}$ (if $x(\boldsymbol{H}) \geq 0$) are straight sections that coincide with the edges of the obstacle. Sectors of the optimum configuration, where $0 \leq x \leq x_n$, that connect $\boldsymbol{w}$ (if $x(\boldsymbol{w}) \geq 0$) or $\boldsymbol{H''}$ (if $x(\boldsymbol{w}) < 0$) with $\boldsymbol{H}$ (if $x(\boldsymbol{H}) \leq x_n$) or $\boldsymbol{H'}$ (if $x(\boldsymbol{H}) > x_n$) are straight sections that coincide with the edges of the obstacle.

Postulate B3:
Sectors of the optimum configuration, where $0 \leq x \leq x_n$, that connect the origin of the manipulator with $\boldsymbol{w}$ (if $x(\boldsymbol{w}) \geq 0$) or $\boldsymbol{h''}$ (if $x(\boldsymbol{w}) < 0$) are straight sections composed of combinations reported in Fig. 5.

Postulate B4:
Sectors of the optimum configuration, where $0 \leq x \leq x_n$, that connect $\boldsymbol{H'}$ with the top of the manipulator coincide with the ideal optimum configuration.

To illustrate the above postulates, situations with the obstacle in four different positions are presented in Fig. 8. The numerical complexity of the utilised procedure should be studied for each specific case. However, it is possible to anticipate the number of arithmetical operations for an average case. Let *N* denote the number of edges of the obstacle. When the obstacle is approached from the right, the cost function of a straight section must be calculated approximately *N/2* times (to embrace one half of the obstacle). When the obstacle is approached from the left, the cost function must be calculated about *N/2* times (to embrace the other half of the obstacle), *N/4* times (in relation to the horizontal sections associated with postulate *B3*), and *N/4* times (in relation to vertical sections associated with postulate *B3*). Since a maximum reasonable number of edges N shouldn't be greater than few tens (usually much less), the proposed algorithm is extremely numerically efficient, especially if compared with standard optimisation procedures.

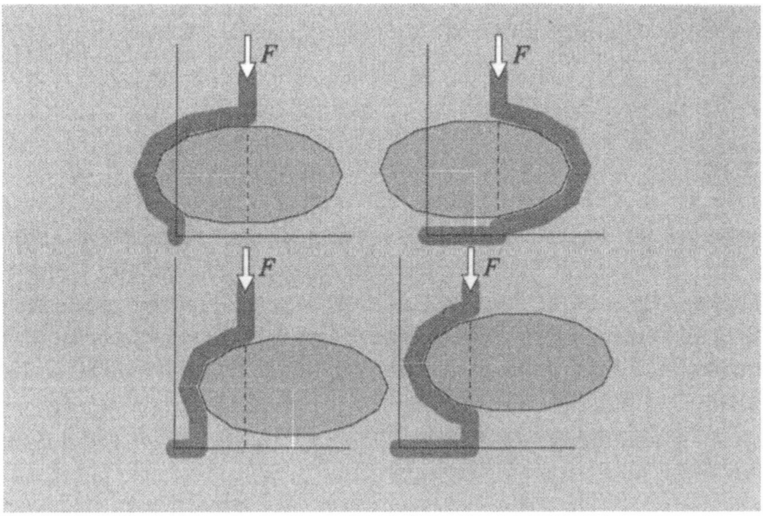

Fig. 8: Target configurations related to an obstacle in different poses

5. Conclusions

We investigated the problem to calculate the optimum backbone curve representing the target configuration of a multiple-link hyper-redundant n-R manipulator when an obstacle of a polygonal form is inserted in the workspace. An approach is described whereby, for a given position of the end effector, the target backbone curve is generated as an approximation the optimum static configuration of the manipulator relative to the minimum joint torques caused by an arbitrary external force. The main idea of the proposed approach is to study the geometrical properties of the optimum configuration when it is divided into a series of straight sections. We present seven basic postulates that approximate the optimum configuration and serve to develop a computer-aided numerical algorithm.

The most apparent advantage of the proposed approach is its extreme simplicity and computational efficiency. The approach is numerically stable, straight forward (non-iterative), and uses simple algebraic relations. However, it can only be applied to obstacles presented as convex polygons with a reasonable number of edges, since its numerical complexity doesn't depend on the number of degrees of freedom but linearly increases with the number of polygon's edges. Even when this number is more than few hundreds, the approximated optimum configuration can be found in real-time on a standard personal computer.

References

Ma, S., Yoshinada, H., Hirose, S. (1992). CT ARM-I: coupled tendon-driven manipulator model I - Design and basic experiments, in *Proc. IEEE Int. Conf. Robotics & Automat.*, Nice, France.

Chirikjian, G.S., Burdick, J.W. (1993). Design, implementation, and experiments with a thirty degree-of-freedom hyper-redundant robot, in *Proc. IEEE Int. Conf. Robotics & Automat.*, Atlanta, GA.

Chirikjian, G.S., Burdick, J.W. (1990). An obstacle avoidance algorithm for hyper-redundant manipulators, in *Proc. IEEE Int. Conf. Robotics & Automat.*, Cincinnati, OH.

Chirikjian, G.S., Burdick, J.W. (1991a). Kinematics of hyper-redundant manipulators, *Advances in Robot Kinematics*, (Lenarčič, J., Stifter, S., Eds.), Springer-Verlag, Vienna.

Chirikjian, G.S., Burdick, J.W. (1991b). Kinematics of hyper-redundant robot locomotion with applications to grasping, in *Proc. IEEE Int. Conf. Robotics & Automat.*, Sacramento, CA.

Chirikjian, G.S., Burdick, J.W. (1994). A Modal approach to hyper-redundant manipulator kinematics, *IEEE Trans. Robotics Automat.*, Vol. 10, No. 3, June 1994, pp. 343-354.

Chirikjian, G.S., Burdick, J.W. (1992). On the determination of kinematically optimal hyper-redundant manipulator configurations, in *Proc. IEEE Int. Conf. Robotics & Automat.*, Nice, France.

Lenarčič, J. (1993). Optimum configurations of planar n-R hyper-redundant mechanism, *Proc. Int. Conf. Advanced Robotics*, Tokyo, Japan, October.

Lenarčič, J. (1994). Minimum joint torque configurations of planar multiple-link manipulator, *Advances in Robot Kinematics and Computational Geometry*, (Lenarčič, J., Ravani, B., Eds.), Kluwer Academic Publishers, Dordrecht, pp. 281-288.

Lenarčič, J. (1995). On minimum joint torque configurations of multiple-link manipulator, *Computational Kinematics*, (Merlet, J.P., Ravani, B., Eds.) Kluwer Academic Publishers, Dordrecht, pp. 133-142.

6. Analysis and simulation

SIMULATING THE MOTION OF A NONHOLONOMIC ROBOT AND ITS TRAILER

A.L. RANKIN, C.D. CRANE, J. DUFFY
Center for Intelligent Machines & Robotics
300 MEB, University of Florida
Gainesville, Florida, U.S.A. 32611

Abstract. This paper describes an algorithm for simulating the motion of a nonholonomic robot vehicle and its trailer. The algorithm assumes that both the vehicle and its trailer move along circular arcs when the incremental time step used to update position and heading is small. The purpose of the simulation is to accurately represent the motion of a Navigation Test Vehicle (NTV) so that the ability of a vehicle's control algorithm to track an explicit path can be evaluated. An auto-tuning algorithm is currently being developed in simulation to autonomously tune the control parameters for a nonholonomic robot vehicle, and this will require an accurate representation of the kinematics of the NTV. An accurate representation of the trailer motion is necessary to develop an autonomous backing-up control algorithm and path tracking control algorithms where the control point is on the trailer. As with the NTV, the positioning subroutine of the simulation program provides the control algorithm with vehicle and trailer pose data at 10 Hz. Locally, the positioning subroutine updates this data at 0.01 second intervals. The simulation currently runs on a Silicon Graphics IRIS Crimson/VGX workstation.

1. Introduction

Computer simulations are recognized and used widely as valuable design and testing tools. The benefits of a robot simulation are numerous. They can be used to develop control algorithms for a robot and verify the correct operation of the robot. The use of a simulation for development and initial testing enables the actual robot to be available for previously debugged testing. Initial simulation testing can be accomplished in an air-conditioned environment and expensive or shared navigation hardware need not be required until preliminary simulation testing is completed. Further, it is easier to detect motion problems in a control algorithm by watching the simulated motion of the robot rather than analyzing a set of printed data.

The simulation discussed in this paper is used to develop and test control algorithms that improve the tracking of paths by a car-like vehicle (which contains an Ackerman steering linkage) and a two-wheeled trailer. An auto-tuning program is currently being developed to autonomously tune the control parameters for a Navigation Test Vehicle (NTV). An accurate

J. Lenarčič and V. Parenti-Castelli (eds.), Recent Advances in Robot Kinematics, 277-286.
© 1996 *Kluwer Academic Publishers. Printed in the Netherlands.*

representation of the trailer's motion is necessary to develop an autonomous backing-up control algorithm and path tracking control algorithms where the control point is the center point of the axle on the trailer.

The NTV is a Kawasaki Mule all-terrain vehicle converted into a robot by placing actuators on the four functions of (1) steering, (2) throttle, (3) brakes, and (4) transmission. The four actuators are placed under the control of an on-board VME computer, operating under the VxWorks operating system (Rankin et al., 1994).

The autonomous navigation subtasks for the NTV are path planning, vehicle positioning, path following, and obstacle avoidance. Path planning is accomplished off-line by an A* search algorithm which outputs a set of via-points that define a collision-free piecewise-linear path (Rankin and Crane, 1994). Vehicle positioning is accomplished by the integration of an inertial navigation system (INS) and a global positioning system (GPS) with a kalman filter algorithm (Crane et al., 1995). Path following is accomplished using PID control to steer the vehicle directly towards a subgoal calculated in real time at a specific distance ahead of the vehicle and on the path (Armstrong et al., 1994). Obstacle avoidance is accomplished using ultrasonic sensors and a reflexive avoidance algorithm (Armstrong and Crane, 1994).

2. Simulation overview

The following list describes the three phases involved in simulating the execution of a specific path by an animated robot vehicle and its trailer.

1) *Plan a safe and achievable path.*
2) *Initialize the current vehicle and trailer pose.*
3) *Draw the vehicle, trailer, planned path, actual path, and the environment until the vehicle is within some tolerance of the goal position and heading.*
    a) *Determine the control parameters from a look-up table.*
    b) *Calculate a command steering angle and vehicle velocity.*
    c) *Determine the new vehicle and trailer pose.*
    d) *Draw the environment.*
    e) *Draw the path to be executed.*
    f) *Draw the vehicle and its trailer at their new pose.*
    g) *Draw a path indicating the actual path of the vehicle or trailer.*

After planning a safe and achievable path and initializing the current vehicle and trailer pose, the steps under phase three are repeated at 10 Hz until the vehicle arrives at the goal position. If the method of steering control involves gains that change as a function of the current vehicle velocity, the gains can be calculated by interpolating the data in a predefined look-up table.

A command steering angle and vehicle velocity are then calculated using controlling equations specific to the method of vehicle control. A subroutine is then called that updates the vehicle and trailer pose, given that the vehicle was directed to use a command steering angle and command vehicle velocity, and assuming 0.1 seconds has expired since the last update. The environment, planned path, actual path, vehicle, and trailer are all drawn in their proper prospective in three dimensions.

It should be noted that while the steps under phase three may not in reality run at 10 Hz, the subroutine that updates the current pose of the vehicle and trailer are told that 0.1 seconds

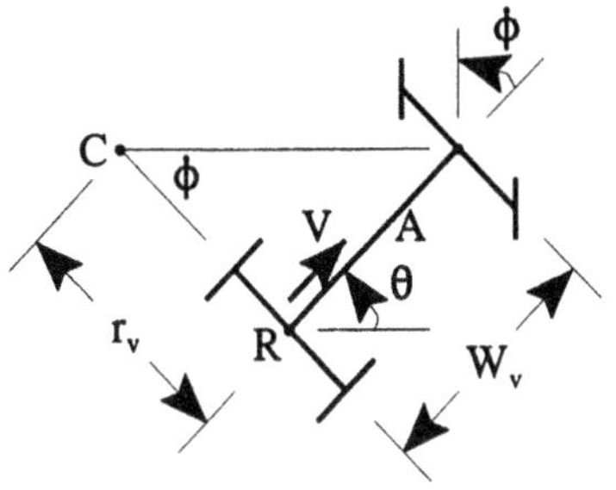

Figure 1. A car-like vehicle with velocity v and steering ange ϕ

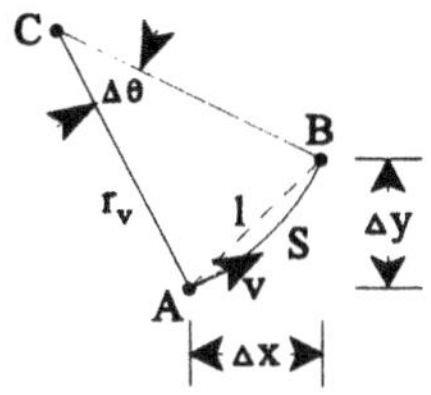

Figure 2. Vehicle moves from point A to point B along circular arc S during Δt

have passed since the previous position update. Thus, position updates "symbolically" occur at 10 Hz in the simulation. The focus of this paper is the methodology used to determine the current vehicle and trailer pose at this rate.

3. Kinematic constraints

Consider the car-like vehicle shown in Figure 1, where the controlled point is chosen to be R, the center of the rear axle of the vehicle. The pose of the vehicle is specified by the three parameters (x,y,θ), where (x,y) are the coordinates of point R and θ is the orientation of the vehicle with respect to the x-axis of a global coordinate system. Hence, the configuration space of the car-like vehicle is three dimensional. Assuming no slip, the instantaneous velocity of the control point R is along the main axis A of the vehicle. The space of achievable velocities at any vehicle configuration is thus two dimensional. The coordinate components of the instantaneous velocity v are

$$\frac{dx}{dt} = v\cos\theta, \qquad \frac{dy}{dt} = v\sin\theta \tag{1}$$

The equations in (1) are combined to yield equation (2), a nonintegratible equation of motion referred to as a *nonholonomic equality constraint.*

$$-\sin\theta\, dx + \cos\theta\, dy = 0 \tag{2}$$

Nonholonomic constraints are generally caused by one or more rolling contacts between rigid bodies, where the relative velocity between the contact points is zero. Car-like robot vehicles are thus typical nonholonomic mechanical systems. The only other kinematic constraint on the motion of a car-like vehicle is a lower bound on the steering angle, ϕ, as

$$|\phi| \le |\phi_{max}| < \frac{\pi}{2} \tag{3}$$

expressed in equation (3). This equation can be rewritten as an inequality (called a *nonholonomic inequality constraint*) involving the differential motions dx, dy, and dθ (Latombe, 1990).

Laumond (1986) has proved that in spite of the kinematic constraints, a car-like robot is fully controllable. In an environment which does not contain obstacles, a car-like robot can achieve any position in the *xy* plane with any orientation.

4. Assumptions

Due to the nonholonomic equality constraint, the space of differential motions (dx, dy, dθ) of the robot at any pose (x,y,θ) is a two dimensional space. The instantaneous motion of the point R is determined by two parameters: the linear velocity of point R, v, and the vehicle's current steering angle, ϕ. Here, it is assumed that the feedback steering angle (obtained from a steering column encoder) can be approximated in simulation by modeling the vehicle as a tricycle. Although the steering angle constantly changes during vehicle execution of a curvilinear path, it is sufficient to assume the vehicle and its trailer travels on circular arcs when the update time, Δt, is small. During this short time interval, the steering angle is considered a constant. In determining the steering angle to use during the next time interval, it is assumed that the steering angular velocity is constant when the current steering angle is not equal to the command steering angle.

It is assumed that no wheel slippage occurs and that perfect velocity tracking occurs, i.e., once the command vehicle velocity is obtained, it is maintained. During each time interval, Δt, the velocity v is considered a constant. In determining the vehicle velocity to use during the next time interval, it is assumed that the acceleration/deceleration of the vehicle is constant when the current vehicle velocity is not equal to the command vehicle velocity.

5. Positioning algorithm

During autonomous navigation, the NTV does not instantaneously achieve a command steering angle or a command vehicle velocity. There is a latency time which is proportional to the difference between the commanded and current control parameter. A commanded steering angle and vehicle velocity may not be achieved before the control algorithm specifies new command values. A simulation of the motion of a car-like vehicle needs to consider these issues. Equation (4) expresses the assumption that the steering wheel angular velocity $\dot{\phi}$ is a constant k_s, when the current steering angle does not equal the command steering angle.

$$\dot{\phi} = \begin{cases} k_s \; for \; (\phi_{cur} \neq \phi_{cmd}) \\ 0 \; for \; (\phi_{cur} = \phi_{cmd}) \end{cases} \tag{4}$$

Equation (5) expresses the assumption that the vehicle acceleration/deceleration is a constant, k_v, when the current vehicle velocity does not equal the command vehicle velocity.

$$\dot{v} = \begin{cases} k_v \; for \; (v_{cur} \neq v_{cmd}) \\ 0 \; for \; (v_{cur} = v_{cmd}) \end{cases} \tag{5}$$

The algorithm for updating the vehicle and the trailer is given below.

```
update_pose(φ_cmd, φ_cur, v_cmd, v_cur)
{
    Δt = 0.1/N
    for I=1 to N
    {
        if (φ_cur < φ_cmd)
            φ_cur = min(φ_cmd, φ_cur + k_s Δt)
        else if (φ_cur > φ_cmd)
            φ_cur = max(φ_cmd, φ_cur - k_s Δt)
        else
            φ_cur = φ_cmd
        if (v_cur < v_cmd)
            v_cur = min(v_cmd, v_cur + k_v Δt)
        else if (v_cur > v_cmd)
            v_cur = max(v_cmd, v_cur - k_v Δt)
        else
            v_cur = v_cmd
        get_new_pose(φ_cur, v_cur)
    }
}
```

The subroutine *update_pose* is called 10 times per second. During each call to this subroutine, the position and heading of the vehicle and trailer is locally updated N=10 times. The update time is Δt=0.01 seconds, i.e., the pose data is locally updated at 100 Hz. The next two sections of this paper will present the equations used in the *get_new_pose* subroutine to update the position and heading of the simulated vehicle and its trailer.

6. Equations for vehicle pose

When the steering angle is nonzero, the vehicle moves around a circular arc centered at C and having a radius of r_v, as shown in Figure 1. The turning radius r_v of the vehicle is related to the steering angle ϕ by the equation

$$r_v = \left| \frac{W_v}{\tan\phi} \right| , \tag{6}$$

where W_v is the *wheelbase*, the distance between the vehicle's front and rear axles. Over the update time step, point A moves to point B along arc S at a constant velocity v, as shown in Figure 2. The change in heading, $\Delta\theta$, can be expressed as a function of the arc length S,

$$\Delta\theta = \frac{S}{r_v} , \tag{7}$$

and the arc length S can be expressed as a function of the time interval and instantaneous velocity v.

$$S = v\Delta t \tag{8}$$

Substituting equations (6) and (8) into equation (7) yields the following expression for $\Delta\theta$ in terms of known values.

$$\Delta\theta = \frac{v\Delta t \tan\phi}{W_v} \tag{9}$$

The length of the line segment, l, can be shown to be

$$l = 2r_v \sin(\frac{\Delta\theta}{2}) \tag{10}$$

When the steering angle is nonzero, the differential changes (Δx,Δy) in the vehicle's position (x,y) are

$$\Delta x = l\cos[\theta + sign(\phi)\frac{\Delta\theta}{2}], \qquad \Delta y = l\sin[\theta + sign(\phi)\frac{\Delta\theta}{2}] \tag{11}$$

When the steering angle is zero, $r_v = \infty$, $\Delta\theta = 0$, and the differential changes (Δx,Δy) in the vehicle's position (x,y) are

$$\Delta x = v\Delta t\cos\theta, \qquad \Delta y = v\Delta t\sin\theta \tag{12}$$

The current pose of the vehicle is updated locally inside the positioning subroutine at 100 Hz using the equations

$$\begin{aligned} \theta_{v,new} &= \theta_{v,old} + \Delta\theta \\ x_{v,new} &= x_{v,old} + \Delta x \\ y_{v,new} &= y_{v,old} + \Delta y \end{aligned} \tag{13}$$

7. Equations for trailer pose

If the nonholonomic vehicle tows a trailer, then a hitch point and some position on the trailer must be updated every time the vehicle position (x,y) is updated. Figures 3 illustrates the four

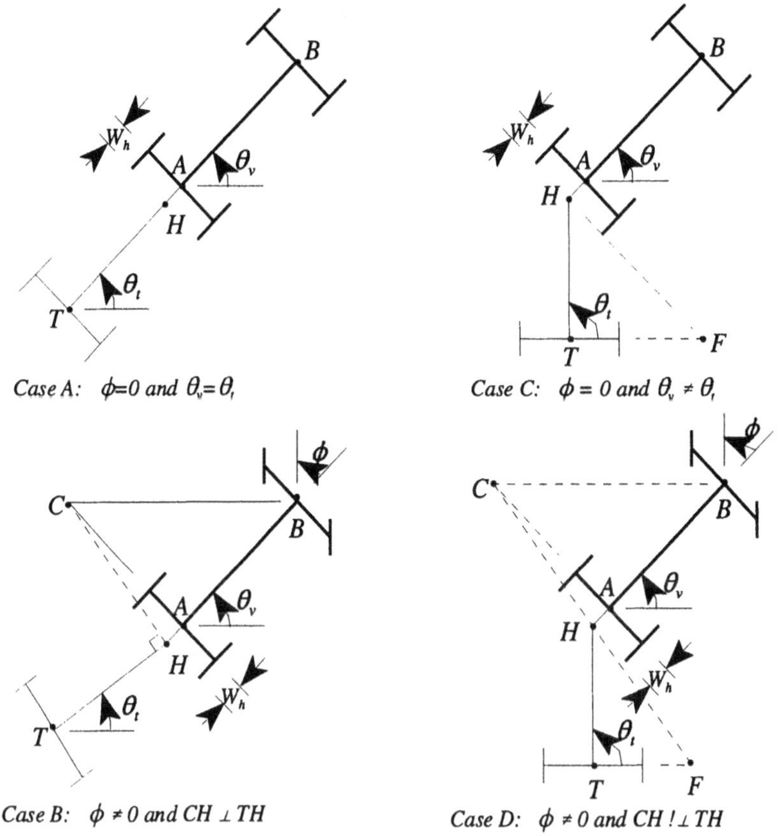

Figure 3. Four general vehicle/trailer configurations

general cases used to determine the motion of the trailer. For all of the cases, the updated hitch point can be calculated given the updated vehicle pose.

$$
\begin{aligned}
x_{h,new} &= x_{v,new} - W_h \cos(\theta_{v,new}) \\
y_{h,new} &= y_{v,new} - W_h \sin(\theta_{v,new})
\end{aligned}
\tag{14}
$$

Case A is the simplest case. Here, the vehicle is traveling straight (ϕ=0) and the trailer heading coincides with the vehicle heading. In case B, the hitch point *H* moves along a

$$
\theta_t = \theta_v - sign(\phi)\tan^{-1}\left[\frac{W_h \tan\phi}{W_v}\right] \tag{15}
$$

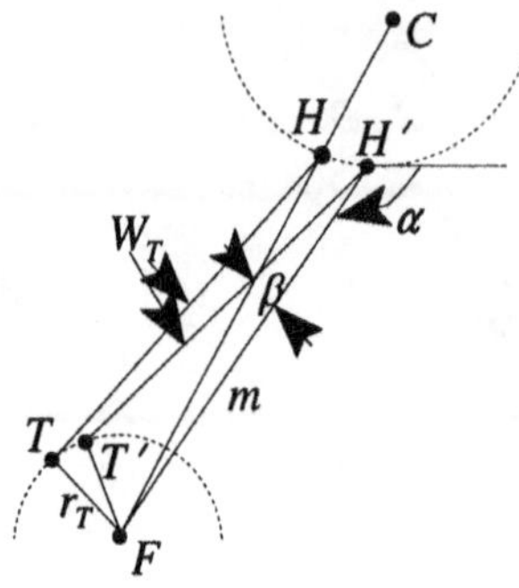

Figure 4. Hitch point H moves to H′ and trailer point T moves to T′

circular arc centered at *C* and having a radius of *CH*. Here, CH is perpendicular to the main axis of the trailer (*TH*) and the instantaneous velocity of point *T* is in the direction of the main axis of the trailer. *CH* is perpendicular to *TH* when θ_t is equal to the expression in equation (15). For case A and B, the new trailer pose is calculated by the equations in (16).

$$\begin{aligned}\theta_{t,new} &= \theta_{t,old}\\ x_{t,new} &= x_{h,new} - W_t\cos(\theta_{t,new})\\ y_{t,new} &= y_{h,new} - W_t\sin(\theta_{t,new})\end{aligned} \tag{16}$$

In case C, the steering angle is zero, but the trailer heading does not coincide with the vehicle heading. Case D is the most general case when the vehicle is executing a turn. Here, the steering angle is nonzero and the line segment *CH* is not perpendicular to *TH*. It is assumed that in both case C and D, the trailer point *T* moves about the circular arc centered at *F* and having radius *TF*. In case C, the point *F* is found using the equations in (17).

$$\begin{aligned}k &= sign(\mathbf{HB} \times \mathbf{HT})\\ x_f &= -\frac{kW_t\sin(\theta_{v,old})}{\sin(\theta_{t,old} - \theta_{v,old})} + x_{h,old}\\ y_f &= \frac{kW_t\cos(\theta_{v,old})}{\sin(\theta_{t,old} - \theta_{v,old})} + y_{h,old}\end{aligned} \tag{17}$$

For case D, point *C* is given by the equations in (18),

$$x_c = x_{v,old} - sign(\phi) r_v \sin(\theta_{v,old}), \qquad y_c = y_{v,old} + sign(\phi) r_v \cos(\theta_{v,old}) \tag{18}$$

and point F can be found by determining the intersection of the line through CH and the line through the axle of the trailer.

As shown in Figure4, the hitch point H moves to the point H′ and the trailer point T moves to the point T′. Once point F is found for case C and D, the algorithm for finding the new trailer position, T′, and its heading, is then as follows:

$$\begin{aligned}
m &= \sqrt{(x_{h,new} - x_f)^2 + (y_{h,new} - y_f)^2} \\
r_t &= \sqrt{(x_{t,old} - x_f)^2 + (y_{t,old} - y_f)^2} \\
\beta &= \cos^{-1}\left[\frac{W_t^2 + m^2 - r_t^2}{2 W_t m}\right] \\
\alpha &= atan2(y_f - y_{h,new}, x_f - x_{h,new}) \\
k &= sign(\boldsymbol{HF} \times \boldsymbol{HT}) \\
x_{t,new} &= x_{h,new} + W_t \cos(\alpha + k\beta) \\
y_{t,new} &= y_{h,new} + W_t \sin(\alpha + k\beta) \\
\theta_{t,new} &= atan2(y_{h,new} - y_{t,new}, x_{h,new} - x_{t_{new}})
\end{aligned} \tag{19}$$

8. Results

The simulation makes the assumption that the vehicle and its trailer travels on circular arcs when the update time is small. Here, the update time is 0.01 seconds, which appears to be adequate for simulating vehicle motion at slow speeds (less than 10 mph). Figure 5 shows the execution of a path used to survey a field for buried munitions (Armstrong et al., 1994).

The simulation is currently being used to develop an auto-tuning algorithm to autonomously tune the control parameters for a nonholonomic robot vehicle. Also, an algorithm is being developed in simulation to control the vehicle/trailer system while path following in reverse. A further area of study is minimizing the lateral error between the path and the trailer control point T during forward motion.

It should be noted that the accuracy with which the current algorithm estimates the vehicle and trailer pose degrades as the current vehicle speed increases. A future improvement, albeit at the cost of program speed, would be to fix the arc length S as a constant, rather that the update time Δt.

References

Rankin, A.L., Crane, C.D., Armstrong, D.G. (1994) Navigation of an autonomous robot vehicle, *ASCE Conference on Robotics for Challenging Environments*, Albuquerque, pp. 44-51.

Rankin A.L., Crane, C.D. (1994) Off-line path planning using an A* search algorithm, *Proceedings of the Florida Conference on Recent Advances in Robotics*, Gainesville, Florida, pp. 218-225.

Crane, C.D., Rankin, A.L., Armstrong, D.G., Wit, J.S., Novick, D.K. (1995) An evaluation of INS and GPS for autonomous navigation, *Proceedings of the 2nd IFAC Conference on Intelligent Autonomous Vehicles*, Espoo, Finland, pp. 205-213.

Armstrong, D.G., Crane, C.D., Rankin, A.L., Nease, A., Brown, E. (1994) Autonomous location of hazardous waste and unexploded ordnance, *Fifth International Symposium on Robotics and Manufacturing*, Maui, Hawaii, pp. 263-268.

Armstrong, D.G., Crane, C.D. (1994) Sonar based obstacle avoidance for outdoor autonomous navigation, *Proceedings of the Florida Conference on Recent Advances in Robotics, Gainesville, Florida*, pp. 137-140.

Latombe, J. (1990) Robot motion planning, Kluwer Academic Publishers, Boston..

Laumond, J. (1986) Feasible trajectories for mobile robots with kinematic and environment constraints, *Proceedings of the International Conference on Intelligent Systems*, Amsterdam, pp. 346-354.

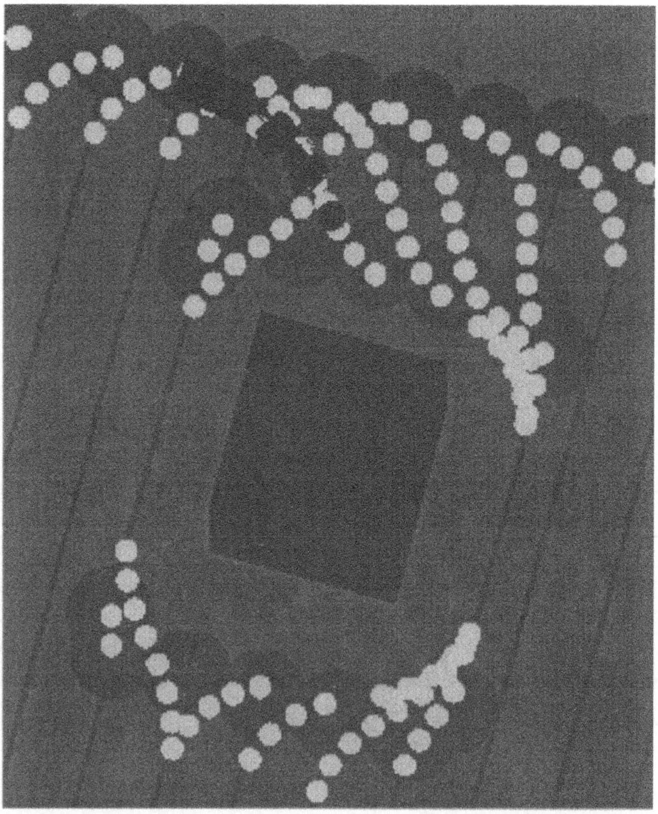

Figure 5. Simulating the survey of a field containing buried munitions

SCREWBALL: A USER INTERFACE FOR SPECIFYING SCREWS

I.A. PARKIN
Basser Department of Computer Science
University of Sydney, N.S.W. 2006, Australia

Abstract. *ScrewBall* is an extension of the *ArcBall* implementation of a notional thumb-ball, visible on the screen and activated by desk-top mouse movements, which provides an acceptable user interface for specifying *rotational* displacements of computer-graphical objects. The *ScrewBall* extension permits specification of *general* displacements — *i.e.* of mixed translations and rotations — by allowing the user to specify a *finite displacement screw* or, equivalently, the vector part of a *unit biquaternion.* The paper describes the mathematical basis of *ScrewBall,* its relationship with *ArcBall,* and the computer-graphical context in which it would be applied.

1. Introduction

In many computer-graphical situations there is a need for the user to specify rotations about axes whose directions are themselves to be specified generally in 3-space. The requirement to specify both axis and rotation in a manner which is reversible and supports intuition is not – given the non-commutability of rotational displacements – easily met. An attractive method, with the merit that it makes use of the ubiquitous *mouse* device, is that proposed under the name *ArcBall* by Shoemake (Shoemake, 1992). Supported by *quaternion* properties in its conception, this method permits the user to compose full-angle rotations of some drawn artefact by the reversible addition of *half-angle* arcs executed on the surface of a *(hemi-)sphere,* which is the *ball* of the method's name.

Beyond this *rotational* requirement, there is often a need to *translate* a drawn artefact in a freely-specified spatial direction; and, not infrequently, to provide a general displacement whose effect is a mixture of rotation and translation executed upon some chosen *screw axis* (Bottema *et al.*, 1990).

J. Lenarčič and V. Parenti-Castelli (eds.), Recent Advances in Robot Kinematics, 287-296.
© 1996 *Kluwer Academic Publishers. Printed in the Netherlands.*

It is desirable, therefore, to extend the *ArcBall* mechanism to encompass translation – but to do so in a manner sympathetic with its established underpinnings. It appears appropriate to direct extension toward user specification of a *finite displacement screw*, or equivalently, the vector part of a *unit biquaternion*, for this embodies the extension to translation of the quaternion on which the properties of *ArcBall* are based.

Such finite displacement screws have attractive properties of linear combination (Parkin, 1992; Huang *et al.*, 1994; Hunt *et al.*, 1995; Parkin, 1996) and may be specified by a user not only for immediate use in displacing an artefact, but also for defining sets of constraints on its future displacement (Parkin, 1995). In more general contexts, a *screw* – being simply a directed *line* with an associated value of *pitch* – finds use in a multiplicity of computer-based analyses in statics and dynamics.

Under the name *ScrewBall*, this paper describes a feasible screw-based extension of the *ArcBall* method. Its mathematical basis, including the relationship between finite displacement screws and unit biquaternions, is first outlined. The computer-based context of *ScrewBall*, and its algorithmic relationship with *ArcBall*, are then described, together with illustration of computer-graphical situations in which it would be applied.

2. Definition of a Screw

We write the general *screw* as a 3-vector of dual numbers, $\hat{\mathbf{S}} \equiv \mathbf{S} + \varepsilon\,\mathbf{S_p}$, in which the real 3-vectors $\mathbf{S}$ and $\mathbf{S_p}$ are respectively the *direction* and *moment* parts of the screw, and ε is the *quasi-scalar* with the property $\varepsilon^2 = 0$. The *magnitude*, *pitch* and *origin radius* of the screw are

$$|\mathbf{S}|\ , \quad p = \frac{\mathbf{S} \bullet \mathbf{S_p}}{\mathbf{S}^2}\ , \quad \mathbf{R} = \frac{\mathbf{S} \times \mathbf{S_p}}{\mathbf{S}^2}\ ,$$

and, in terms of these, the moment part may be written $\mathbf{S_p} = p\,\mathbf{S} + \mathbf{V} \times \mathbf{S}$ for $\mathbf{V}$ being any point on the line of the screw. Alternatively, the screw may be written $\hat{\mathbf{S}} = |\mathbf{S}|\,(1 + \varepsilon p)\,\hat{\mathbf{s}}$, where $\hat{\mathbf{s}}$ represents that line as a specialised screw of unit magnitude and zero pitch. Whenever we refer to the *normalised* instance of a screw $\hat{\mathbf{S}}$, we shall mean that unit line, written with the corresponding lower case letter $\hat{\mathbf{s}}$; and we shall write $\mathbf{s}$ for its unit direction 3-vector. For any two screws $\hat{\mathbf{S}}_1 \equiv \mathbf{S}_1 + \varepsilon\,\mathbf{S}_{\mathbf{p}_1}$ and $\hat{\mathbf{S}}_2 \equiv \mathbf{S}_2 + \varepsilon\,\mathbf{S}_{\mathbf{p}_2}$, the scalar product $\hat{\mathbf{S}}_1 \bullet \hat{\mathbf{S}}_2$ and cross product $\hat{\mathbf{S}}_1 \times \hat{\mathbf{S}}_2$ are defined over their dual elements in the same manner as they are defined over the real elements of real 3-vectors. The scalar product yields a dual number

$$\hat{\mathbf{S}}_1 \bullet \hat{\mathbf{S}}_2 = \mathbf{S}_1 \bullet \mathbf{S}_2 + \varepsilon\,\hat{\mathbf{S}}_1 \otimes \hat{\mathbf{S}}_2 \quad \textit{where} \quad \hat{\mathbf{S}}_1 \otimes \hat{\mathbf{S}}_2 \equiv \mathbf{S}_1 \bullet \mathbf{S}_{\mathbf{p}_2} + \mathbf{S}_{\mathbf{p}_1} \bullet \mathbf{S}_2\ ,$$

in which the binary operation $\hat{\mathbf{S}}_1 \otimes \hat{\mathbf{S}}_2$ measures the *mutual moment* of the screws. In the situation $\mathbf{S}_1 \bullet \mathbf{S}_2 = 0$ where two screws are mutually perpendicular, their mutual moment vanishes if, and only if, the screws intersect. Thus the condition $\hat{\mathbf{S}}_1 \bullet \hat{\mathbf{S}}_2 = 0$, in which the screws are *orthogonal*, implies that each intersects the other at right angles. We note without proof that the cross product screw $\hat{\mathbf{S}}_1 \times \hat{\mathbf{S}}_2$ is sited in the *common perpendicular line* of $\hat{\mathbf{S}}_1$ and $\hat{\mathbf{S}}_2$. If, in particular, $\hat{\mathbf{s}}_1$ and $\hat{\mathbf{s}}_2$ are orthogonal lines, then $\hat{\mathbf{s}}_1 \times \hat{\mathbf{s}}_2$ is also a line and is orthogonal to both $\hat{\mathbf{s}}_1$ and $\hat{\mathbf{s}}_2$.

3. Reference Frames

We represent the typical reference frame by mutually orthogonal unit lines $\hat{\mathbf{x}}_i$, $\hat{\mathbf{y}}_i$, and $\hat{\mathbf{z}}_i$ for which

$$\left.\begin{array}{l} \hat{\mathbf{x}}_i^2 = \hat{\mathbf{y}}_i^2 = \hat{\mathbf{z}}_i^2 = 1 \,, \quad \hat{\mathbf{x}}_i \bullet \hat{\mathbf{y}}_i = \hat{\mathbf{y}}_i \bullet \hat{\mathbf{z}}_i = \hat{\mathbf{z}}_i \bullet \hat{\mathbf{x}}_i = 0 \,, \\ \hat{\mathbf{x}}_i \times \hat{\mathbf{y}}_i = \hat{\mathbf{z}}_i \,, \quad \hat{\mathbf{y}}_i \times \hat{\mathbf{z}}_i = \hat{\mathbf{x}}_i \,, \quad \hat{\mathbf{z}}_i \times \hat{\mathbf{x}}_i = \hat{\mathbf{y}}_i \,. \end{array}\right\} \tag{1}$$

The point of mutual intersection of these axial lines is, of course, the origin of this i-frame. Whenever – in some common reference system – we know the coordinates $\hat{\mathbf{S}} = \mathbf{S} + \varepsilon\, \mathbf{S_p}$ of some general screw and of the i-frame axes $\hat{\mathbf{x}}_i$, $\hat{\mathbf{y}}_i$, and $\hat{\mathbf{z}}_i$, the i-frame coordinates of that screw are given by

$$\hat{\mathbf{S}}_i \equiv \mathbf{S}_i + \varepsilon\, \mathbf{S}_{\mathbf{p}_i} = \begin{bmatrix} \hat{\mathbf{x}}_i^T \\ \hat{\mathbf{y}}_i^T \\ \hat{\mathbf{z}}_i^T \end{bmatrix} \hat{\mathbf{S}} = \begin{bmatrix} \hat{\mathbf{x}}_i \bullet \hat{\mathbf{S}} \\ \hat{\mathbf{y}}_i \bullet \hat{\mathbf{S}} \\ \hat{\mathbf{z}}_i \bullet \hat{\mathbf{S}} \end{bmatrix} = \begin{bmatrix} \mathbf{x}_i \bullet \mathbf{S} \\ \mathbf{y}_i \bullet \mathbf{S} \\ \mathbf{z}_i \bullet \mathbf{S} \end{bmatrix} + \varepsilon \begin{bmatrix} \hat{\mathbf{x}}_i \otimes \hat{\mathbf{S}} \\ \hat{\mathbf{y}}_i \otimes \hat{\mathbf{S}} \\ \hat{\mathbf{z}}_i \otimes \hat{\mathbf{S}} \end{bmatrix} \tag{2}$$

in which the first-written matrix is shown by Equations 1 to be a 3×3 dual orthogonal matrix.

4. Screw Specification of a Finite Displacement

An arbitrary finite displacement of a rigid body can itself be represented by a suitably specified screw. When a rigid body suffers such a displacement there is a directed line $\hat{\mathbf{s}}$ — the *screw axis* — such that the displacement could have been achieved by means of translation through some distance 2σ in the direction of $\hat{\mathbf{s}}$ and by rotation through some angle 2θ measured in a right-handed sense about $\hat{\mathbf{s}}$. The line $\hat{\mathbf{s}}$ is unique except when $\theta = 0$ (a pure translation) in which case *any* line parallel with the translation will serve. We follow Hunt (Hunt, 1987) in defining the *cardinal motion* for the given displacement to be that in which the rotation angle satisfies $-\pi < 2\theta \leq \pi$.

Adopting the standard interpretation of a *dual angle*, thus

$$\hat{\theta} \equiv \theta + \varepsilon\,\sigma \,, \quad \sin\hat{\theta} \equiv \sin\theta + \varepsilon\,\sigma\cos\theta \,, \quad \cos\hat{\theta} \equiv \cos\theta - \varepsilon\,\sigma\sin\theta$$

we use the elements of the cardinal motion, namely the axis $\hat{\mathbf{s}}$, the *half-translation* σ, and the *half-rotation* θ, $-\pi/2 < \theta \leq \pi/2$, to define the (directed) *finite displacement screw* which represents the displacement, viz.

$$\hat{\mathbf{S}} \equiv \sin\hat{\theta}\,\hat{\mathbf{s}} = \sin\theta\,(1 + \varepsilon P_S)\,\hat{\mathbf{s}}\ , \quad P_S \equiv \frac{\sigma}{\tan\theta}\ , \tag{3}$$

of amplitude $\sin\theta$ and pitch P_S whose line is $\hat{\mathbf{s}}$.

We may show that, in displacing 3-space through the dual angle $2\hat{\theta}$, such a finite displacement screw $\hat{\mathbf{S}}$ carries any screw from an initial location $\hat{\mathbf{G}}_i$ into a final location

$$\hat{\mathbf{G}}_f = \hat{\mathbf{G}}_i + 2\hat{\mathbf{S}} \times (\cos\hat{\theta}\,\hat{\mathbf{G}}_i + \hat{\mathbf{S}} \times \hat{\mathbf{G}}_i)\ . \tag{4}$$

The *unit biquaternion* (Bottema *et al.*, 1990) for the same displacement is expressed in terms of the screw $\hat{\mathbf{S}}$ by

$$\hat{\mathbf{q}} = \mathbf{q} + \varepsilon\,\mathbf{q}_0 = \begin{bmatrix} \cos\hat{\theta} \\ \hat{\mathbf{S}} \end{bmatrix} = \begin{bmatrix} \cos\hat{\theta} \\ \sin\hat{\theta}\,\hat{\mathbf{s}} \end{bmatrix}\ , \tag{5}$$

which distinquishes the *cosine component* $\cos\hat{\theta}$ and the *screw component* $\hat{\mathbf{S}}$ of the biquaternion. A biquaternion in which the real part of the cosine component, *viz.* $\cos\theta$, is non-negative – so that $-\pi/2 < \theta \leq \pi/2$ – describes the cardinal motion for the displacement. We observe that the real-part *quaternion* $\mathbf{q}$, referred to by Shoemake (Shoemake, 1992), describes only rotational aspects of the displacement.

5. ScrewBall and Drawing Windows

Consider that – drawn to occupy the screen of a computer terminal – there is a main rectangular window, of primary interest for the user, in which the objects of a kinematic World are depicted as they are seen from a particular viewpoint. Superimposed upon this World window – and providing a tool for manipulation of World objects and the viewpoint itself – there is a secondary *ScrewBall* window.

As two-dimensional coordinates measured in the plane of the screen we shall refer to U (horizontal, to the right) and V (vertically upward). The orthogonal axes on which these are measured meet in an origin at the centre of the screen. When, later, we wish to consider manipulations within the depicted World, we shall consider those axes to be unit orthogonal lines $\hat{\mathbf{u}}_V$ and $\hat{\mathbf{v}}_V$ (as defined in Section 3) whose coordinates – with those of their cross product $\hat{\mathbf{n}}_V = \hat{\mathbf{u}}_V \times \hat{\mathbf{v}}_V$ – are specified in terms of the World reference frame. By this means, the *viewpoint-object* composed of

- the triad of lines $\hat{\mathbf{u}}_V$, $\hat{\mathbf{v}}_V$, and $\hat{\mathbf{n}}_V$ which constitutes the *view-frame*,
- the *view-plane*, being the $\hat{\mathbf{u}}_V\hat{\mathbf{v}}_V$-plane of the screen, and
- the *eye*, which lies on the positive $\hat{\mathbf{n}}_V$-axis at distance e from the screen,

is effectively located, and can be relocated, within the three-dimensional space of the World.

6. The Mouse and its Protocols

Consider that a hand-manipulated *mouse* device is available whose changes of position – in the plane of the table at which the user sits – are the source of two-dimensional data for the computer; and that the mouse carries a number of *buttons* whose depressions provide signals for special control purposes. We assume that movement information from the mouse causes a *cursor* to track across the computer screen so that, apart from the button function, we may ignorc thc mousc and refer only to the coordinates $(U,\ V)$ of the cursor.

It is convenient to describe the composite interaction known as a *mouse-drag* which takes place repeatedly between the computer and a user who, by use of the mouse, is manipulating drawn objects on the screen. The interaction is initiated at *mouse-down* when a mouse-button is depressed and *initial* coordinates $(U_i,\ V_i)$ are passed to the computer. Subsequently – *while the mouse-button remains depressed* – *current* coordinates $(U_c,\ V_c)$ of the most recent cursor position are passed until, at *mouse-up* when the button is released, *final* coordinates $(U_f,\ V_f)$ for the interaction are recorded.

The essence of the interaction is the change which is created between the *initial* and *final* states of the drawing on the screen and of the computer-internal data values which support it. If, as is usual, the images of drawn artefacts are continuously updated throughout the mouse-drag to reflect the most recent positional change of the cursor, the screen is always it its potentially *final* state. It is then appropriate to ignore the distinction between *current* and *final* coordinates.

A *mouse-down* may occur in any window, the interpretation applied to the cursor coordinates being determined by the current *mode of operation* within contexts associated with that window. For example, in Section 9 it is required that the user should have nominated a particular point $\mathbf{Q}$ in the World. The user might achieve this by choosing a *point-selection* mode (by menu) and by then performing a *mouse-down* with the cursor placed at the visible image of some point object within the World window.

7. The ArcBall Paradigm

The *ArcBall* paradigm is provided by a *thumb-ball*, that is, by a spherical ball which can be rotated – by thumb or finger – through an arbitrary angle about any chosen axis through its centre. In place of such a physical ball, the user of *ArcBall imagines* the presence of a three-dimensional sphere. No more of this is shown than the equatorial circle in which the sphere intersects the plane of the screen. When a mouse-drag takes place within this *ArcBall* circle, the position of a *notional* cursor is continuously evaluated as though it were on the surface of the nearer hemisphere at a point (U, V, N) whose coordinates U and V are the same as those of the *actual* cursor in the plane of the screen. This notional cursor on the sphere imitates the thumb on the thumb-ball so that, in the course of the mouse-drag, the imaginary *ArcBall* sphere rotates through an angle θ about an axis of direction $\mathbf{s}$ through its centre.

Details of the interpretation are straightforward. For completeness, consider that the *ArcBall* circle is drawn inside a square *ArcBall* window whose image may be moved or re-sized upon the screen to suit the user's convenience. Measured in the same $\hat{\mathbf{u}}_V$- and $\hat{\mathbf{v}}_V$-units as the cursor coordinates, let the *ArcBall* circle be of radius R_{ball} with centre at (U_{ball}, V_{ball}). The *ArcBall* process is initiated when a mouse-down occurs within the *ArcBall* window. At that instant, and throughout the ensuing mouse-drag, we continuously derive from the cursor coordinates U and V the new coordinates

$$u = U - U_{ball} , \quad v = V - V_{ball} , \tag{6}$$

which, independent of the screen placement of the *ArcBall* circle, measure the coordinates of the cursor relative to its centre. If we find that $u^2+v^2 < R_{ball}^2$, we know that the actual cursor is inside the *ArcBall* circle so we proceed to evaluate the unit vector

$$\mathbf{p} = (l, m, n) , \quad l = u/\ R_{ball} , \quad m = v/\ R_{ball} , \quad n = \sqrt{1-l^2-m^2} , \tag{7}$$

which, relative to the *ArcBall* centre, locates the notional cursor point on the surface of (a unit-radius version of) the *ArcBall* sphere. If, instead, we find that the actual cursor lies on or outside the *ArcBall* circle, we interpret the notional cursor as lying *on* the equatorial circle (in the screen surface) in the same direction from the *ArcBall* centre: we evaluate the appropriate unit vector as

$$\mathbf{p} = (l, m, 0) , \quad l = u/\ u^2+v^2 , \quad m = v/\ u^2+v^2 , \quad l^2+m^2 = 1 . \tag{8}$$

If we denote the unit vectors so derived at *mouse-down*, and in the course of the *mouse-drag*, by

$$\mathbf{p}_i = (l_i, m_i, n_i) \quad and \quad \mathbf{p}_f = (l_f, m_f, n_f) , \tag{9}$$

respectively, the outputs of the *ArcBall* process consist of a unit direction vector $\mathbf{s}$ together with the choice of a particular angle θ of rotation about that direction which are provided by

$$\sin\theta\,\mathbf{s} = \mathbf{p}_i \times \mathbf{p}_f = (l_i, m_i, n_i) \times (l_f, m_f, n_f) , \quad |\mathbf{s}| = 1 . \tag{10}$$

Definition of the the three independent coordinates implicit in this quantity, together with a rotational redundancy of definition about the chosen axis $\mathbf{s}$, exhaust the four degrees of freedom available in a mouse-drag conducted with a 2-dimensional mouse.

8. The ScrewBall Extension

To define the dual analogue quantity $\sin\hat{\theta}\,\hat{\mathbf{s}}$ of Equation 10 – involving both translation and rotation – requires a composite mouse-drag which is punctuated by some mode-altering control event from keyboard or mouse-button. We refer to this composite process as *ScrewBall* and, for consistency, refer to certain artefacts inherited from *ArcBall* as the *ScrewBall* circle, the *ScrewBall* centre, and so on.

Within the *ScrewBall* process, the *ArcBall* process is first used to establish an axis direction $\mathbf{s}$. When, during the *ArcBall* mouse-drag, a specific control event occurs – depression of the keyboard 'Z' key is used at present – the *ScrewBall* extension process is entered. The angular parameter θ obtained from the *ArcBall* process is *discarded* but the 3-vector $\mathbf{s}$ is adopted as the direction of a $\mathbf{z}$-axis at the *ScrewBall* centre which we represent, relative to the $\hat{\mathbf{u}}_V$-, $\hat{\mathbf{v}}_V$-, and $\hat{\mathbf{n}}_V$-directions, by

$$\mathbf{z}= \mathbf{s}= (l_S,\ m_S,\ n_S)\ ,\quad |\mathbf{z}|= l_S^2+m_S^2+n_S^2= 1\ . \tag{11}$$

In place of the *ScrewBall* sphere, the user is now invited to imagine – as reference surface – an infinitely long *cylinder* whose axis is that $\mathbf{z}$-axis and whose surface, of radius R_{ball} , is tangent to the *ScrewBall* sphere in a certain circle of contact centred at the *ScrewBall* centre. To establish coordinates for that cylinder, consider – with the *ScrewBall* centre as origin – that a unit $\mathbf{y}$-axis is sited on the particular diameter of that circle of contact which lies in the plane of the screen so that, with the corresponding unit $\mathbf{x}$-axis directed toward the eye-side of the screen, we have

$$\mathbf{y}= \frac{\mathbf{z}\times(0,\ 0,\ 1)}{\sqrt{1-n_S^2}}= \frac{(m_S,\ -l_S,\ 0)}{\sqrt{1-n_S^2}}\ ,\quad \mathbf{x}= \mathbf{y}\times\mathbf{z}= \frac{(-l_S\,n_S,\ -m_S\,n_S,\ 1-n_S^2)}{\sqrt{1-n_S^2}} \tag{12}$$

Any point $(x,\ y,\ z)$ has a screen projection whose coordinates measured from the *ScrewBall* centre are

$$\begin{bmatrix} u \\ v \end{bmatrix} = \frac{-1}{\sqrt{1-n_S^2}}\left\{x\,n_S\begin{bmatrix} l_S \\ m_S \end{bmatrix} - y\begin{bmatrix} m_S \\ -l_S \end{bmatrix}\right\} + z\begin{bmatrix} l_S \\ m_S \end{bmatrix}\ , \tag{13}$$

so that, on inverting this relationship, we have

$$y= \frac{u\,m_S-v\,l_S}{\sqrt{1-n_S^2}}\quad and\quad -x\,n_S+z\sqrt{1-n_S^2}= \frac{u\,l_S+v\,m_S}{\sqrt{1-n_S^2}}\ . \tag{14}$$

Thus, given coordinates $(u,\ v)$ of the actual cursor – measured from the *ScrewBall* centre as in Equation 6 – we may find the location of a notional cursor (which, when possible, has the same screen projection) at a point $(x,\ y,\ z)$ satisfying $x^2+y^2= R_{ball}^2$ on the surface of the cylinder. Our algorithm is:

- from the first of Equations 14 determine a coordinate y' which, from its definition, measures the signed distance of the actual cursor from the projected image of the cylinder axis;
- if $y'^2 < R_{ball}^2$, the actual cursor lies within the projected horizons of the cylindrical surface, so

$$y= y'\ , \quad x= \sqrt{R_{ball}^2-y'^2}\ , \tag{15}$$

  the positive x value being appropriate to the half-cylinder which is nearer to the eye; if, instead, the actual cursor lies on or outside the projected horizons, interpret the notional cursor as lying *on* the horizon nearer to it, with coordinates

$$y= sign(y')\, R_{ball}\ , \quad x= 0\ ; \tag{16}$$

- finally, by use of the second of Equations 14, determine z .

With the identifications

$$\cos\theta= x/\ R_{ball}\ , \quad \sin\theta= y/\ R_{ball}\ , \quad \sigma= z\ , \quad \mathbf{s}= (l_S,\ m_S,\ n_S)\ , \quad |\mathbf{s}|= 1\ , \tag{17}$$

the *ScrewBall* process provides a unit direction vector $\mathbf{s}$ together with choices of a particular distance of translation σ , $-\infty < \sigma < \infty$, and a particular angle of rotation θ , $-\pi/2 < \theta \leq \pi/2$, as measured along and about that direction. It is to be noted that the user specifies both parameters σ and θ in the course of a single, uninterupted, mouse-drag.

9. Displacing An Object

Like *ArcBall*, the *ScrewBall* process does not provide a spatial location for the axis which it defines. The user must do this in a separate step by nominating, within the kinematic World of interest, some point **Q** through which the axis is to pass. For simplicity we shall assume that this is done before the *ScrewBall* process is used. If, as foreshadowed in Section 5, the directions $\mathbf{u}_V$, $\mathbf{v}_V$, and $\mathbf{n}_V$ of the view-frame axes are known in the coordinates of that World, the vector

$$\mathbf{s}_W= l_S\,\mathbf{u}_V+m_S\,\mathbf{v}_V+n_S\,\mathbf{n}_V\ , \quad |\mathbf{s}_W|= 1\ , \tag{18}$$

is the representation in World coordinates of the *ScrewBall*-defined direction vector $\mathbf{s}$ of Equations 17. The unit *line* of this direction through the point $\mathbf{Q}$, and the *finite displacement screw* constructed on it (as at Equation 3) from the values in Equations 17, then have, in World coordinates, the respective forms

$$\hat{\mathbf{s}}_Q = \mathbf{s}_W + \varepsilon\, \mathbf{Q} \times \mathbf{s}_W \quad \textit{and} \quad \hat{\mathbf{S}}_Q = (\sin\theta + \varepsilon\, \sigma \cos\theta)\, \hat{\mathbf{s}}_W \,. \tag{19}$$

Concatenation of the quantity $\cos\theta - \varepsilon\, \sigma \sin\theta$ provides the *unit biquaternion* of Equation 5.

As required, the screw $\hat{\mathbf{S}}_Q$ may be drawn as an artefact in its own right or, through use of Equation 4, may be employed to displace any other nominated screw $\hat{\mathbf{G}}_i$ to a new location $\hat{\mathbf{G}}_f$. When the latter is done, the screw $\hat{\mathbf{S}}_Q$ need not, of course, be made known to the user, displacement of the nominated screw $\hat{\mathbf{G}}_i$ being made to appear the normal outcome of nominating a point $\mathbf{Q}$ and operating the *ScrewBall* process. However, it is likely to be a significant aid to the user's comprehension of the observed displacement if its screw axis $\hat{\mathbf{s}}_Q$, passing through $\mathbf{Q}$, is displayed while the *ScrewBall* mouse-drag is in progress.

When many lines like $\hat{\mathbf{G}}_i$ are to be simultaneously displaced by the screw $\hat{\mathbf{S}}_Q$, efficiency is gained by reducing Equation 4 to the linear operations of a matrix transformation. To achieve this we first identify *any* unit line $\hat{\mathbf{x}}_i$ which is orthogonal to $\hat{\mathbf{S}}_Q$ and then determine, by use of Equation 4, the unit line $\hat{\mathbf{x}}_f$ into which $\hat{\mathbf{x}}_i$ is displaced by the screw $\hat{\mathbf{S}}_Q$. We evaluate the two unit orthogonal companion lines $\hat{\mathbf{y}}_i = \hat{\mathbf{s}}_Q \times \hat{\mathbf{x}}_i$ and $\hat{\mathbf{y}}_f = \hat{\mathbf{s}}_Q \times \hat{\mathbf{x}}_f$. We can then argue that the triad of unit orthogonal lines $\hat{\mathbf{x}}_i$, $\hat{\mathbf{y}}_i$, and $\hat{\mathbf{s}}_Q$ – to be regarded as a reference frame – is displaced by the screw $\hat{\mathbf{S}}_Q$ into the triad of lines $\hat{\mathbf{x}}_f$, $\hat{\mathbf{y}}_f$, and $\hat{\mathbf{s}}_Q$. It follows, for any other line that is displaced by $\hat{\mathbf{S}}_Q$ from $\hat{\mathbf{G}}_i$ to $\hat{\mathbf{G}}_f$, that the coordinates of $\hat{\mathbf{G}}_i$ in the frame of $\hat{\mathbf{x}}_i$, $\hat{\mathbf{y}}_i$, and $\hat{\mathbf{s}}_Q$ must be identical with those of $\hat{\mathbf{G}}_f$ in the frame of $\hat{\mathbf{x}}_f$, $\hat{\mathbf{y}}_f$, and $\hat{\mathbf{s}}_Q$. That is,

$$\hat{\mathbf{G}}_f = \begin{bmatrix} \hat{\mathbf{x}}_f & \hat{\mathbf{y}}_f & \hat{\mathbf{s}}_Q \end{bmatrix} \begin{bmatrix} \hat{\mathbf{x}}_i^T \\ \hat{\mathbf{y}}_i^T \\ \hat{\mathbf{s}}_Q^T \end{bmatrix} \hat{\mathbf{G}}_i \,. \tag{20}$$

We therefore proceed to evaluate this product matrix from the constructed frame-axial lines, and to use it for the transformation of each general line $\hat{\mathbf{G}}_i$ into its displaced location $\hat{\mathbf{G}}_f$.

10. Displacing the ViewPoint

In modifying the viewpoint from which the World is seen, the user's perception is of displacing some drawn object – and with it, *every* object – to a location in which it can be better viewed. While it is desirable that the *ScrewBall* process should be used to specify the screw $\hat{\mathbf{S}}_Q$ which *appears* to support that perception, there are compelling reasons why the objects should not, themselves, be displaced within the World. These concern both the computational effort and the loss of positional accuracy entailed in (repeatedly) recalculating the World locations of all objects.

It is more efficient and quite as effective to apply the negative of that screw, *viz.* $-\hat{\mathbf{S}}_Q$, to displacement of just the *viewpoint-object* which consists, as described in Section 5, of the triad of axes $\hat{\mathbf{u}}_V$, $\hat{\mathbf{v}}_V$, and $\hat{\mathbf{n}}_V$. When – using algorithms beyond the scope of this paper – the (unaltered) objects of the World are then re-drawn in their new relationships with those (relocated) axes, the effect of the desired change of viewpoint will be achieved.

11. Conclusion

An extension of the *ArcBall* method, under the name *ScrewBall*, has been proposed for the specification of screws at the computer terminal. *ScrewBall* has, in fact, been implemented as a software process and its application, together with associated screw-based evaluation methods like certain of those outlined here, are presently under investigation and refinement.

References

O. Bottema and B. Roth. *Theoretical kinematics.* North-Holland Publishing Company, Amsterdam (1979). Reprinted Dover, New York (1990).

C. Huang and B. Roth. Analytic expressions for the finite screw systems. *Mechanism and Machine Theory* **29**, pp. 207–222 (1994).

K.H. Hunt. Manipulating a body through a finite displacement. *Proc. 7th World Congress on the Theory of Machines and Mechanisms, Sevilla, Spain*, pp. 187–91 (1987).

K.H. Hunt and I.A. Parkin. Finite displacements of points, planes and lines *via* screw theory. *Mechanism and Machine Theory* **30**, pp. 177–192 (1995).

I.A. Parkin. A third conformation with the screw systems: finite twist displacements of a directed line and point. *Mechanism and Machine Theory* **27**, pp. 177–188 (1992).

I.A. Parkin. The application of finite displacement screws to drawing constraints. *Computational Kinematics '95* (eds. J.-P. Merlet and B. Ravani), Kluwer Academic Publishers, pp. 63-72 (1995).

I.A. Parkin. The screws for finite displacement of a rigid body expressed in terms of its symmetry screws. (to appear)

K. Shoemake. ARCBALL: a user interface for specifying three-dimensional orientation using a mouse. *Proceedings of Graphical Interface '92*, pp. 151-156 (1992).

SIMULATION OF FLEXIBLE MANIPULATORS BASED ON THEIR NEWTON-EULER MODEL

F. BOYER
Ecole des Mines de Nantes,
4 rue Alfred Kastler,
44070 Nantes, France.

P. COIFFET
Laboratoire de Robotique de Paris,
10-12 avenue de l'Europe,
78140 Vélizy, France.

Abstract. This paper presents the first part of a research work devoted to simulation, identification and control of flexible robots based on their Newton-Euler model. First, we recall the generalization of the Newton Euler equations for an open chain of flexible links. These equations constitute an original dynamic model because they represent the dynamic behavior of the chain with a mixed formalism since it is both Lagrangian and Eulerian. This formalism gives the linear intrinsic dynamic model of links. Second, we exhibit the constrained nature of these equations. We use a numerical integration method for a constrained system and an updating procedure well suited to our particular formalism. Finally, simulation results on a two links spatial kinematic are reported.

1. Introduction

The symbolic minimal form of flexible manipulators equations are very complex. The main approaches used today to get them are based upon the symbolic calculation of their Lagrange's equations (Cetinkunt and Book 1987, Siciliano and Book, 1988, Lin and Yuan, 1988). A few authors, such as Kim and Haug (1988) or Wehage and Haug (1988), choose to investigate the problem with its constrained form. It consists of breaking off the joints and writting the balance equations of each free isolated body of the chain. Then, constraint equations on velocities are added to assemble the chain topology. With this approach a new difficulty arises. In fact, if we write the balance equations in a Lagrangian way we are led to introduce angular parameters such as Euler's angles which hide the linear intrinsic dynamics of free bodies and lead to a non intuitive formulation of the chain dynamics. For instance, the generalized contraint forces associated with these coordinates are some torques about fictitious axis of intermediate frames, absolutely unconnected with the intrinsic geometry of the links. On the other hand the use of an Eulerian description puts unknown velocities in place of the unknown positional parameters, and thus avoids the use of absolute angular parameters. This formalism is usually applied to the description of motions of rigid manipulators (Khalil and Kleinfinger 1988, Renaud, 1988). In this case it gives the

J. Lenarčič and V. Parenti-Castelli (eds.), Recent Advances in Robot Kinematics, 297-306.
© 1996 *Kluwer Academic Publishers. Printed in the Netherlands.*

compact intuitive equations of Newton-Euler which are namely used for solving the problem of the non linear control (Luh, Walker and Paul 1980). In fact, as far as flexible manipulators are concerned, it is still possible to apply this formalism to describe the rigid non linear component of the motion of each isolated body of the open chain. Nevertheless the elastic motions require the use of the Lagrangian formalism since they require the definition of a reference configuration to measure the current deformation. With such a mixed kinematic description, we obtain a dynamic model which can be considered as a generalization of the Newton-Euler model to the case of flexible manipulators (Boyer and Coiffet, 1996).

2. Kinematic model of a free elastic body

The kinematic model of a flexible link is descibed as the superimposition of two screws:

$$\underline{\mathbf{D}}(M) = \underline{\mathbf{D}}_{\mathbf{r}}(M) + \underline{\mathbf{D}}_{\mathbf{e}}(M) = \begin{Bmatrix} \underline{\mathbf{V}}_{\mathbf{r}} + \underline{\boldsymbol{\omega}}_{\mathbf{r}} \times \underline{\mathbf{r}}(M) \\ \underline{\boldsymbol{\omega}}_{\mathbf{r}} \end{Bmatrix} + \begin{Bmatrix} \underline{\mathbf{V}}_{\mathbf{e}}(M) \\ \underline{\boldsymbol{\omega}}_{\mathbf{e}}(M) \end{Bmatrix} \tag{1}$$

where " $\underline{\ }$ " underlines a vector or a matrix of vectors of the usual space. The first component of (1), $\underline{\mathbf{D}}_{\mathbf{r}}(M)$, is the rigid screw of the velocity field in M. Its resultant $\underline{\boldsymbol{\omega}}_{\mathbf{r}}$ is the rigid Galilean angular velocity of a reference frame driven by the rigid component of the global motion. Its momentum is the rigid Galilean linear velocity. The vector $\underline{\mathbf{r}}(M)$ is the position vector of M in the reference frame of origin O. We have thus defined the Eulerian rigid kinematical variables $\underline{\boldsymbol{\omega}}_{\mathbf{r}}$ and $\underline{\mathbf{V}}_{\mathbf{r}}$ which describe the rigid part of the motion in an Eulerian way. The second component of (1), $\underline{\mathbf{D}}_{\mathbf{e}}(M)$, is the elastic contribution to the motion. Its resultant is the elastic angular velocity $\underline{\boldsymbol{\omega}}_{\mathbf{e}}(M)$ of M, refering to the rigid motion, and its momentum $\underline{\mathbf{V}}_{\mathbf{e}}(M)$ is the linear elastic velocity refering to the same motion. We apply a truncated Rayleigh-Ritz decomposition of the displacement field. This decomposition defines a set of N generalized coordinates q_{ek} such that we can rewrite the elastic screw field:

$$\underline{\mathbf{D}}_{\mathbf{e}}(M) = \sum_{k=1}^{N} \begin{Bmatrix} \underline{\boldsymbol{\Phi}}_{\mathbf{dk}}(M) \\ \underline{\boldsymbol{\Phi}}_{\mathbf{rk}}(M) \end{Bmatrix} \cdot \dot{q}_{ek}$$

Where $\underline{\boldsymbol{\Phi}}_{\mathbf{dk}}(M)$ is the kth shape function of the displacement field and $\underline{\boldsymbol{\Phi}}_{\mathbf{rk}}$ is its rotational counterpart. Thus, our set of kinematical variables is a mixed set of Eulerian and Lagrangian variables: $(\underline{\boldsymbol{\omega}}_{\mathbf{r}}, \underline{\mathbf{V}}_{\mathbf{r}}, q_{e1}, q_{e2} \dots q_{eN})$. In the following no assumption is done on the Raileigh-Ritz functions.

3. Geometric model of an open flexible chain

We consider an open chain of n flexible links $S_1, S_2 \dots S_n$. The chain kinematics is defined by n ponctual joints. Each joint introduces at its geometric center A_j, one degree of freedom which is either linear or angular and parametrized using the joint

Lagrangian variable q_{rj}. In the following we will need to know the geometric model of the chain. We use as many authors do (Chedmail, 1990, Lin and Yuan, 1988..), the terminology of the 4×4 matrices. In fact, any particle of the chain can be located by a string of products of the elementary matrices ${}^{j}\mathbf{T}_{j+1}$ which parametrizes the situation of the reference frame R_{j+1} with respect to that of the previous one R_j. To calculate these matrices (considering that no assumption has been done on the Rayleigh-Ritz functions) we use two sets of frames (Boyer, 1994). The first one is the set of the reference frames already introduced to describe the kinematics of the free elastic bodies. The second set collects frames such that their origins are located at the points A_j's and their "z basis vector" $\underline{\mathbf{z}}_j$'s are along the joint axis. They are thus called joint frames and denoted by Q_j and N_j. The Q_j frame moves with the link S_j, N_j with S_{j+1}. The matrices ${}^{j}\mathbf{T}_{j+1}$ are both dependent on the elastic deformations of S_j and S_{j+1} and of the joint parameter q_{rj} only.

4. Newton Euler model of a flexible manipulator

In (Boyer and Coiffet, 1996) we first applied the D'Alembert's principle to each isolated body of the chain using the mixed formalism. We then wrote the constraint equations on velocities with the same formalism and obtained the following formulation of the dynamics of an open chain of flexible links.

Dynamic model of free bodies: $\forall j = 1,\dots n$

$$\begin{pmatrix} \underline{\underline{\mathbf{M}}}_j & \underline{\underline{\hat{\mathbf{MS}}}}_j & \underline{\mathbf{MS}}_{dej} \\ \underline{\underline{\hat{\mathbf{MS}}}}_j^T & \underline{\underline{\mathbf{I}}}_j & \underline{\mathbf{MS}}_{rej} \\ \underline{\mathbf{MS}}_{dej}^T & \underline{\mathbf{MS}}_{rej}^T & \mathbf{m}_{eej} \end{pmatrix} \cdot \begin{Bmatrix} \underline{\boldsymbol{\gamma}}_{linr,j} \\ \underline{\boldsymbol{\gamma}}_{angr,j} \\ \ddot{\mathbf{q}}_{ej} \end{Bmatrix} + \begin{Bmatrix} \underline{\mathbf{F}}_{inj} + \underline{\mathbf{F}}_{gj} \\ \underline{\mathbf{C}}_{inj} + \underline{\mathbf{C}}_{gj} \\ \mathbf{Q}_{inj} + \mathbf{Q}_{gj} + \mathbf{k}_{eej} \cdot \mathbf{q}_{ej} \end{Bmatrix} = \begin{Bmatrix} \Delta \underline{\mathbf{F}}_j \\ \Delta \underline{\mathbf{C}}_j \\ \Delta \mathbf{Q}_j \end{Bmatrix} \tag{2}$$

Kinematic model of joints: $\forall j = 1,\dots n-1$

$$\underline{\mathbf{D}}_{rj+1} = \underline{\mathbf{D}}_{rj} + \underline{\mathbf{D}}_{lj+1} + \sum_{k=1}^{N_j} \begin{Bmatrix} \underline{\boldsymbol{\Phi}}_{dj,k}(A_j) \\ \underline{\boldsymbol{\Phi}}_{rj,k}(A_j) \end{Bmatrix} \cdot \dot{q}_{ej,k} - \sum_{k=1}^{N_{j+1}} \begin{Bmatrix} \underline{\boldsymbol{\Phi}}_{dj+1,k}(A_{j+1}) \\ \underline{\boldsymbol{\Phi}}_{rj+1,k}(A_{j+1}) \end{Bmatrix} \cdot \dot{q}_{ej+1,k} \cdots$$

$$\cdots + \begin{Bmatrix} \underline{\boldsymbol{\omega}}_{rj} \times \underline{\mathbf{r}}_j(A_{j+1}) - \underline{\boldsymbol{\omega}}_{rj+1} \times \underline{\mathbf{r}}_{j+1}(A_{j+1}) \\ \underline{\mathbf{0}} \end{Bmatrix} \tag{3}$$

This model is a generalization of the well known Newton Euler's model for an open chain of flexible bodies, where we have introduced the following notations in (2) (the detailed formula of these matrices are given in (Boyer and Coiffet, 1996)):

The underscript " $\underline{\underline{\ }}$ " denotes second order tensors and $\underline{\underline{\hat{\mathbf{V}}}}$ denotes the skew symetric tensor associated with the vector $\underline{\mathbf{V}}$. We denote by $\underline{\boldsymbol{\gamma}}_{linr,j}$ and $\underline{\boldsymbol{\gamma}}_{angr,j}$ the linear and angular rigid accelerations. The Rayleigh - Ritz displacement, velocity and acceleration matrices are respectively denoted by $\mathbf{q}_{ej}$, $\dot{\mathbf{q}}_{ej}$ and $\ddot{\mathbf{q}}_{ej}$. The diagonal matrices of inertia and stiffnesses are denoted by $\mathbf{m}_{eej}$ and $\mathbf{k}_{eej}$. The mass, inertia, and unbalance tensors are respectively denoted by: $\underline{\underline{\mathbf{M}}}_j$, $\underline{\underline{\mathbf{I}}}_j$ and $\underline{\underline{\hat{\mathbf{MS}}}}_j$. The row matrices of inertia linear and angular coupling vectors are respectively denoted: $\underline{\mathbf{MS}}_{dej}$,$\underline{\mathbf{MS}}_{rej}$. Their dual column matrices are denoted by ${}^{T}\underline{\mathbf{MS}}_{dej}$ and ${}^{T}\underline{\mathbf{MS}}_{rej}$. The coriofugal forces and torques on the link are respectively denoted by: $\underline{\mathbf{F}}_{inj}$ and $\underline{\mathbf{C}}_{inj}$ and their elastic generalized counterparts $\mathbf{Q}_{inj}$. The gravity forces and torques are denoted by $\underline{\mathbf{F}}_{gj}$ and $\underline{\mathbf{C}}_{gj}$ and their elastic counterparts $\mathbf{Q}_{gj}$, The balance of inter-bodies forces and torques are denoted by $\Delta\underline{\mathbf{F}}_j$ and $\Delta\underline{\mathbf{C}}_j$ and their generalized elastic counterparts $\Delta\mathbf{Q}_j$. Let us also note that the matricial product of (2) handles products of tensors, vectors, and scalars. We also introduced the screws $\underline{\mathbf{D}}_{rj} = \underline{\mathbf{D}}_{rj}(O_j)$, where $\underline{\mathbf{D}}_{rj}(M)$ is the rigid screw field of the link S_j and O_j is its reference point. The screw $\underline{\mathbf{D}}_{lj}$ denotes the ponctual joint screw defined by:

$$\underline{\mathbf{D}}_{lj} = \begin{Bmatrix} \sigma_j . \underline{\mathbf{z}}_j \\ \overline{\sigma}_j . \underline{\mathbf{z}}_j \end{Bmatrix} . \dot{q}_{rj} \, ,$$

where σ_j is the Boolean variable which defines the type of the jth joint. It is equal to 1 if the joint is prismatic, 0 if it is rotational. $\overline{\sigma}_j$ is equal to $1-\sigma_j$.

We are now going to recall the equations of a constrained system using a Lagrangian formalism. Next we will put our model (2-3) in such a canonical form. From this last formulation we will deduce a method of simulation.

5. Analysis of constrained systems in Lagrangian formalism

For a polyarticulated mechanism with fixed obstacles, it is possible, after breaking off its joints and describing its configuration by a set of Lagrangian coordinates $\mathbf{q}$, to exhibit the dynamics of the system under the following constrained form:

$$\mathbf{M}(\mathbf{q}).\ddot{\mathbf{q}} + \mathbf{l}(\mathbf{q},\dot{\mathbf{q}}) = \mathbf{Q}(\mathbf{q}) + \mathbf{H}^{T}(\mathbf{q}).\boldsymbol{\lambda} \tag{4}$$

$$\mathbf{H}(\mathbf{q}).\dot{\mathbf{q}} = 0 \tag{5}$$

where $\mathbf{M}(\mathbf{q})$ is the generalized inertia matrix of the free bodies and $\mathbf{l}(\mathbf{q},\dot{\mathbf{q}})$ groups all the non linear forces acting on each body and induced by their motion. $\mathbf{Q}(\mathbf{q})$ contains all generalized control forces which drive the motions of the mechanism and $\boldsymbol{\lambda}$ is the vector of Lagrange multipliers. Depending on the current configuration, the constraint matrix $\mathbf{H}(\mathbf{q})$ exhibits the constraints induced by the joints. Tacking into consideration a simulation goal, we can find from this canonical form a more suitable equational presentation, by assembling the algebro-differential system (4-5). This is practically achieved by first differentiating the constraint equations (5) with respect to the time:

$$\frac{d}{dt}(\mathbf{H}(\mathbf{q}).\dot{\mathbf{q}}) = \frac{\partial \mathbf{H}}{\partial \mathbf{q}}(\mathbf{q}).\dot{\mathbf{q}}\dot{\mathbf{q}} + \mathbf{H}(\mathbf{q}).\ddot{\mathbf{q}} = \mathbf{0} \tag{6}$$

And by inserting (4) into (6), we can isolate the Lagrange multipliers that we insert into (4) we thus obtain the assembled dynamical equations:

$$\ddot{\mathbf{q}} = -\mathbf{M}^{-1}\left((\mathbf{Q}+\mathbf{I}) + \mathbf{H}^{T}.\left((\mathbf{H}\mathbf{M}^{-1}\mathbf{H}^{T})^{-1}\left(\frac{\partial \mathbf{H}}{\partial \mathbf{q}}\dot{\mathbf{q}}\dot{\mathbf{q}} - \mathbf{H}.\mathbf{M}^{-1}(\mathbf{Q}+\mathbf{I})\right)\right)\right) \tag{7}$$

We have now at our disposal a system directly usable for an explicit numerical integration. In order to apply these results to our specific problem, we are first going to achieve the canonical constrained formulation of (2-3).

6. Constraint equations of an open chain in a mixed formalism

We have to rewrite the constraint equations on velocities (3) under the similar form to (5). For that we first project the recurence on velocities as follows: The kinematic Eulerian variables are projected onto the reference basis, and the constraint equations are then projected onto the joint basis since the constraint forces and torques are naturally known in these basis. This gives six scalar equations per joint. In fact, each joint j introduces only five scalar constraint equations since the sixth introduces the joint speed $\dot{q}_{rj}$ which does not belong to our set of kinematical variables. Thus, we have to eliminate from the projection of (5) one scalar equation per joint. That finally leads us to the folowing system which descibes the constraint equations of the manipulator:

$$\mathbf{H}.\begin{Bmatrix} \mathbf{V_r} \\ \boldsymbol{\omega}_\mathbf{r} \\ \dot{\mathbf{q}}_\mathbf{e} \end{Bmatrix} = \begin{pmatrix} \mathbf{H}_{\mathbf{V_r V_r}} & \mathbf{H}_{\mathbf{V_r}\boldsymbol{\omega}_\mathbf{r}} & \mathbf{H}_{\mathbf{V_r e}} \\ \mathbf{H}_{\boldsymbol{\omega}_\mathbf{r}\mathbf{V_r}} & \mathbf{H}_{\boldsymbol{\omega}_\mathbf{r}\boldsymbol{\omega}_\mathbf{r}} & \mathbf{H}_{\boldsymbol{\omega}_\mathbf{r}\mathbf{e}} \end{pmatrix}.\begin{Bmatrix} \mathbf{V_r} \\ \boldsymbol{\omega}_\mathbf{r} \\ \dot{\mathbf{q}}_\mathbf{e} \end{Bmatrix} = \begin{Bmatrix} \mathbf{LV_{lin}} \\ \mathbf{LV_{ang}} \end{Bmatrix} \tag{8}$$

Where the block matrices are detailed in (Boyer, 1994) and we concatenated the kinematical variables into the following columns:

$$\mathbf{V_r} = col_{j=1,..n}({}^{R_j}\mathbf{V_{rj}}) \ , \ \boldsymbol{\omega}_\mathbf{r} = col_{j=1..n}({}^{R_j}\boldsymbol{\omega}_{\mathbf{rj}}) \ , \ \dot{\mathbf{q}}_\mathbf{e} = col_{j=0..n}(\dot{\mathbf{q}}_{\mathbf{ej}})$$

7. Newton-Euler model of a flexible manipulator in its constrained form

We can now look for the dynamic equations of the chain using a constrained form analogous to (4-5). For that, we re-concatenate the $n+1$ balance equations of (2-3) in order to gather the linear, angular, and elastic balance terms.

Dynamic equilibria of links:

$$\begin{pmatrix} \mathbf{M} & \mathbf{MS} & \mathbf{MS_{de}} \\ \mathbf{MS^T} & \mathbf{J} & \mathbf{MS_{re}} \\ \mathbf{MS_{de}^T} & \mathbf{MS_{re}^T} & \mathbf{M_{ee}} \end{pmatrix} \begin{Bmatrix} \gamma_{\mathbf{linr}} \\ \gamma_{\mathbf{angr}} \\ \ddot{\mathbf{q}}_{\mathbf{e}} \end{Bmatrix} + \begin{Bmatrix} \mathbf{F_{in}} + \mathbf{F_g} \\ \mathbf{C_{in}} + \mathbf{C_g} \\ \mathbf{Q_{in}} + \mathbf{k_{ee}} \cdot \mathbf{q_e} + \mathbf{Q_g} \end{Bmatrix} = \begin{Bmatrix} \Delta\mathbf{F_c} \\ \Delta\mathbf{C_c} \\ \Delta\mathbf{Q_c} \end{Bmatrix} + \begin{pmatrix} \mathbf{H^T_{V_r V_r}} & \mathbf{H^T_{\omega_r V_r}} \\ \mathbf{H^T_{V_r \omega_r}} & \mathbf{H^T_{\omega_r \omega_r}} \\ \mathbf{H^T_{V_r e}} & \mathbf{H^T_{\omega_r e}} \end{pmatrix} \begin{Bmatrix} \mathbf{F_l} \\ \mathbf{C_l} \\ \mathbf{Q_l} \end{Bmatrix} \tag{9}$$

Constraint equations of joints:

$$\begin{pmatrix} \mathbf{H_{V_r V_r}} & \mathbf{H_{V_r \omega_r}} & \mathbf{H_{V_r e}} \\ \mathbf{H_{\omega_r V_r}} & \mathbf{H_{\omega_r \omega_r}} & \mathbf{H_{\omega_r e}} \end{pmatrix} \cdot \begin{Bmatrix} \mathbf{V_r} \\ \boldsymbol{\omega}_\mathbf{r} \\ \dot{\mathbf{q}}_\mathbf{e} \end{Bmatrix} = \begin{Bmatrix} 0 \\ 0 \end{Bmatrix} \tag{10}$$

Here, the intrinsic quantities of (2) are replaced by the matrices of their components after application of suited projections such that: all the tensors and vectors of inertia parameters as the galilean velocities and accelerations, the inertia forces and torques, the gravity forces and torques are written relative to the set of reference frames. The constraint forces and torques, as well as the row vectors of $\mathbf{H}$ are given in the joint frames. Moreover we have separated the constraint forces and torques ($\Delta\mathbf{F_l}$, $\Delta\mathbf{C_l}$ and $\Delta\mathbf{Q_l}$) from the control ones $\Delta\mathbf{F_c}$, $\Delta\mathbf{C_c}$ and $\Delta\mathbf{Q_c}$. The only difficulty to find (9) is to prove that the factor matrix of the constraint forces and torques is the transposition of the constraint matrix. In fact, such a proof is a validation of the previous developments since this property is a theoretical consequence of the non-working nature of constraint wrenchs.

8. Differentiation of constraint equations

In order to assemble the equilibriums in an explicit form easy to simulate, we are led to differentiate with respect to time the constraint equations. This is achieved using the Galilean differentiation of the intrinsic expressions of constraint equations (Boyer, 1994). After projection, we finally obtain the differentiation of the constraint equations:

$$\begin{pmatrix} \mathbf{H_{V_r V_r}} & \mathbf{H_{V_r \omega_r}} & \mathbf{H_{V_r e}} \\ \mathbf{H_{\omega_r V_r}} & \mathbf{H_{\omega_r \omega_r}} & \mathbf{H_{\omega_r e}} \end{pmatrix} \cdot \begin{Bmatrix} \gamma_{\mathbf{linr}} \\ \gamma_{\mathbf{angr}} \\ \ddot{\mathbf{q}}_\mathbf{e} \end{Bmatrix} + \begin{Bmatrix} \boldsymbol{\Psi}_{\mathbf{lin}} \\ \boldsymbol{\Psi}_{\mathbf{ang}} \end{Bmatrix} = \begin{Bmatrix} 0 \\ 0 \end{Bmatrix} \tag{11}$$

Here we have introduced the matrices of components:

$$\gamma_{\mathbf{linr}} = col_{j=1,..n}({}^{\mathbf{R_j}}\mathbf{V_{rj}}) \quad , \gamma_{\mathbf{angr}} = col_{j=1..n}({}^{\mathbf{R_j}}\boldsymbol{\omega}_{\mathbf{rj}}) \ , \ \ddot{\mathbf{q}}_\mathbf{e} = col_{j=0..n}(\dot{\mathbf{q}}_{\mathbf{ej}})$$
$$\boldsymbol{\Psi}_{\mathbf{ang}} = col_{j=1,..n}({}^{\mathbf{N_j}}\boldsymbol{\Psi}_{\mathbf{ang,j}}) \quad , \quad \boldsymbol{\Psi}_{\mathbf{lin}} = col_{j=1..n}({}^{\mathbf{N_j}}\boldsymbol{\Psi}_{\mathbf{lin,j}})$$

Where the ${}^{\mathbf{N_j}}\boldsymbol{\Psi}_{\mathbf{ang,j}}$'s and the ${}^{\mathbf{N_j}}\boldsymbol{\Psi}_{\mathbf{lin,j}}$'s are matrices of nonlinear functions of the velocities. In comparison with the Lagrangian case, the first contribution of (11) is homologous to "$\mathbf{H}(\mathbf{q}).\ddot{\mathbf{q}}$". The second contribution of (11) is homologous to "$\frac{\partial \mathbf{H}}{\partial \mathbf{q}}(\mathbf{q}).\dot{\mathbf{q}}\dot{\mathbf{q}}$".

9. Explicit dynamic system for simulation

Assembling of the constrained model under its mixed formalism becomes obvious. For that, we proceed in the same way as in the Lagrangian case. We finally obtain the explicit differential system:

$$\begin{Bmatrix} \dot{\mathbf{V}}_r \\ \dot{\omega}_r \\ \ddot{\mathbf{q}}_e \end{Bmatrix} = \begin{Bmatrix} \mathbf{V}_r \times \omega_r \\ 0 \\ 0 \end{Bmatrix} + \mathbf{M}_a^{-1} \left(\left(\begin{Bmatrix} \Delta\mathbf{F}_c - \mathbf{F}_{in} - \mathbf{F}_g \\ \Delta\mathbf{C}_c - \mathbf{C}_{in} - \mathbf{C}_g \\ \Delta\mathbf{Q}_c - \mathbf{Q}_{in} - \mathbf{Q}_g - \mathbf{k}_{ee}.\mathbf{q}_e \end{Bmatrix} \right) \right) ..$$

$$..-\mathbf{H}^T.\left((\mathbf{H}.\mathbf{M}_a^{-1}.\mathbf{H}^T)^{-1} \left(\mathbf{H}.\mathbf{M}_a^{-1}. \begin{Bmatrix} \Delta\mathbf{F}_c - \mathbf{F}_{in} - \mathbf{F}_g \\ \Delta\mathbf{C}_c - \mathbf{C}_{in} - \mathbf{C}_g \\ \Delta\mathbf{Q}_c - \mathbf{Q}_{in} - \mathbf{Q}_g - \mathbf{k}_{ee}.\mathbf{q}_e \end{Bmatrix} + \begin{Bmatrix} \Psi_{lin} \\ \Psi_{ang} \end{Bmatrix} \right) \right) \quad (12)$$

Where we have introduced the following columns:

$$\mathbf{V}_r \times \omega_r = col_{j=1,..n}\left({}^{R_j}\mathbf{V}_{rj} \times {}^{R_j}\omega_{rj} \right), \; \dot{\mathbf{V}}_r = col_{j=1,..n}({}^{R_j}\dot{\mathbf{V}}_{rj}) \; , \; \dot{\omega}_r = col_{j=1,..n}({}^{R_j}\dot{\omega}_{rj}) \; ,$$

and $\mathbf{M}_a$ is the mass matrix of the augmented system constituted of the free bodies. The simulation is based on the numerical integration of (12). In order to achieve the simulation process we have to update the geometric configuration of the chain.

10. Updating current geometry and kinematics of the chain

Numerical integration of (12) gives at each step of time the current values of the Eulerian and Lagrangian velocities. To calculate **H** at each time we have to update the geometric model. The updating process that we here propose is based on the following remark: if the rigid kinematical description is Eulerian, nothing prevents us from using the Lagrangian parameters, i.e. the $q_{ek,j}(t)$'s and the $q_{rj}(t)$'s to update the geometrical configuration of the chain. The elastic coordinates are simply obtained by twice integrating the elastic accelerations. Concerning joint variables, we recall that we have previously deleted from the projection of (3), one equation per joint, which was not a constraint equation. In fact these n equations relate joint speeds and Eulerian velocities as follows:

$$\dot{q}_{rj+1} = \sigma_{j+1}.\mathbf{z}_{j+1}^T.\mathbf{LV}_{ang,j+1} + \bar{\sigma}_{j+1}.\mathbf{z}_{j+1}^T.\mathbf{LV}_{lin,j+1}$$

Then we integrate these equalities and find the positional joint parameters which are used with the $q_{ek,j}$ to compute the geometric model. In order to take into account the joint constraints at the level of the velocities we compute once more (but not by integration) the Eulerian velocities from the knowledge of the Lagrangian velocities and using the recurence (3). In such a way we impose to the system a verification of the joint constraints at every level of differentiation: On the accelerations by the way of the Lagrange multipliers, on the velocities by the recurence (3), and on the positions using the geometric model parametrized by the minimal set of Lagrangian coordinates.

11. Simulation results

We here repot a steady-state test initiated by imposing the rigid velocities. Such a test allow us to see the effects of the inertia forces caused by the motions. The considered manipulatoir is a two links spatial kinematic. The links are one meter cylindrical beams with ten centimeters of diameter. They are made of an usual steel. The elastic fields are decomposed on the basis of the embedded-free modes of each link considered as an Euler - Bernouilli beam. Each flexural and twisting field is approximated with one mode. Some dissipative forces are added as modal dampings on each link. The considered kinematic (Fig. 1) has two rotational joints. The first joint axis is parallel to the geometric axis of the first link. The second joint axis is perpendicular to the geometric axis of the second link. The coordinates $q_{x1}, q_{y1}, q_{y2}, q_{z2}$ are the first flexural modal coordinates in directions parallel to the axis noted with their indexes. The coordinate τ_1 is the torsional modal coordinate of the first link.

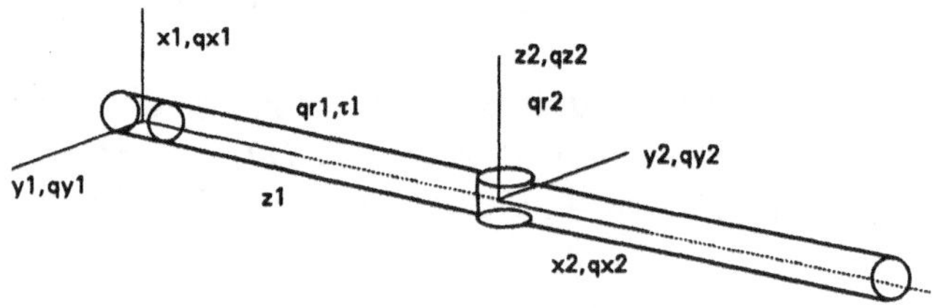

Figure 1. Spatial two links kinematic

STEADY STATE TEST

We initiate the test with the rigid motion.

$$(q_{r1}, q_{r2}) = (0,0) \quad , \quad (\dot{q}_{r1}, \dot{q}_{r2}) = (1\text{rd/s}, 1\text{rd/s})$$

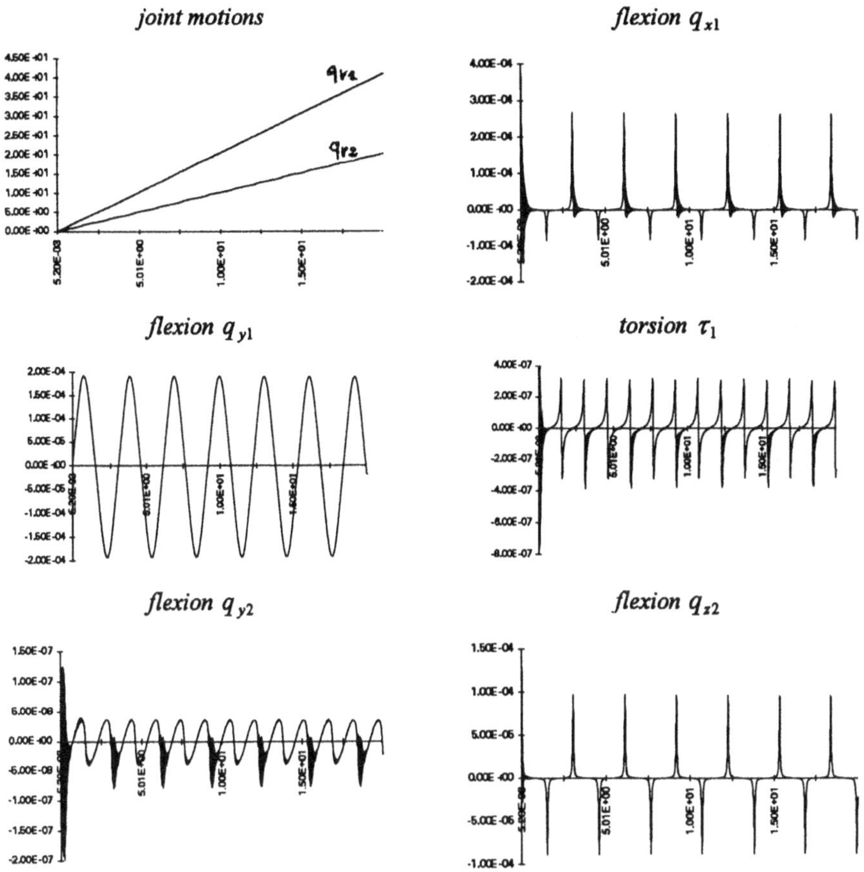

Figure 2. Steady state test of a spatial two links kinematic

COMMENTS OF FIGURE 2

Each of the two links describes a circular uniform motion except that the first one accelerates and slows down around an average fixed velocity. By comparing the two joint responses we note that the velocity of the first link is maximum when the second one passes through a multiple of π - i.e. - when the inertia momentum seen from the first joint is minimum. This well verifies the kinematic momentum conservation. The interpretation of elastic evolutions is more difficult. The flexion q_{y1} is of pure time rigid scale and is driven by the inertia unbalance that causes the motion of the second link. With respect to q_{x1} and q_{z2} we can only say that when the kinematic passes through the configurations such as $q_{r2} = k.\pi$, some inertia forces occur which alternatively push and pull the links and immediately vanish. In some cases, the forces are strong and fast enough to excite the elastic structural modes. This explains the periodic appearance of some fast oscillations in comparison with the slow rigid motion. These oscillations are transmitted to the whole structure by the inertial couplings. We can also remark that if the flexural excitations are symmetric, the torsional one which acts on τ_1 is anti-symmetric. This can be easily understood by considering the symmetries of this kinematic.

12. Conclusions

To our knowledge this method is a new issue to the simulation problem of flexible poly-articulated structures. It permits us to simulate realistic kinematics with just by hand pre-calculations and with a simple programming work. On the other hand, it is computationally time consuming. For the time being, this simulation has been programmed without trying to numerically eliminate the redundant kinematical unknown. Such an improvement could be achieved in a second try. It could be based on the results obtained by authors who consider the simulation problem in its constrained form, from Lagrangian models.

References

Cetinkunt, S., Book, J.W. (1987) Symbolic modelling of flexible manipulators, *IEEE international conference on Robotics Automation*, pp. 2074-2080.

Siciliano, B., Book, W.J. (1988) A singular perturbation approach to control of light weight flexible manipulators, *International Journal of Robotics Research*, Vol.7, No.4.

Lin, L.Y., Yuan, K. (1988) A Lagrange Euler assumed modes approach to modeling flexible robotic manipulators, *Journal of the chinese institute of engineers*, Vol. 11, No. 4, pp. 335-347.

Kim, S.S., Haug, E.J. (1988) A recursive formulation for flexible multibody dynamics, Part 1: Open-loop-systems, *Journal of Computer methods in applied mechanics and engineering*, Vol. 71, No. 3, pp. 293-314.

Wehage, R.A., Haug, E.J. (1982) Generalized coordinate partitionning for dimension reduction in analysis of constrained dynamic systems, *Journal of mechanical design*, Vol. 104, January.

Khalil, W., Kleinfinger, J.F. (1988) Minimum operations and minimum parameters of the dynamic models of tree structure robots, *IEEE Transaction on Robotics and Automation*.

Renaud, M. (1988) Calcul quasi minimal du modèle dynamique inverse d'un robot manipulateur, Technique de la robotique, tome 1, Architectures et Commandes ed. Hermes, Paris.

Luh, J.Y.S., Walker, M.W., Paul, R. (1980) On-line computational scheme for mechanical manipulators, ASME Transaction, *Journal of Dynamic Systems, Measurements and control*, Vol. 102, No 2, pp.69-76.

Boyer, F., Coiffet P. (1996) A generalization of the Newton-Euler models on open chains of flexible links, *Journal of Robotic Systems*, Vol. 13, No. 1, pp. 11-24.

Chedmail, P. (1990) Synthèse de robots et de sites robotisés - Modelisation de robots souples, *Thèse de Doctorat d'état, Université de Nantes, E.NS.M.* Mai.

Boyer, F. (1994), Contribution à la modélisation et à la commande des robots flexibles, *Thèse de Doctorat de robotique de l'Université Pierre et Marie Curie.*

Book, W.J. (1984) Recursive Lagrangian Dynamics of flexible manipulators Arms, *International Journal of Robotics research*, Vol. 3, No 3, pp.87-101.

MODELING OF A BRACED ROBOT WITH FOUR-BAR MECHANISM

J. ZUPANČIČ AND A. KRALJ
Faculty of Electrical Engineering
University of Ljubljana
Tržaška 25, Ljubljana, Slovenia

Abstract.

Kinematic analysis of a braced robot is described. The strategy of bracing is intended for absolute robot end-point pose accuracy and repetability improvement. The principle of mechanical lever is explained with the theoretical point-to-line contact-type of robot bracing. The kinematics of braced robot with true surface-to-surface contact-type are derived: 1) with the local surface geometry of contacting bodies, 2) by an original method of constructing an equivalent four-bar mechanism. Pose uncertainty due to inaccuracy and repetability of free and braced robot is analyzed. An estimation function of accuracy improvement by bracing together with simulation results are presented. Advantages and disadvantages of the proposed method of equivalent four-bar mechanism with regard to the uncertainty reduction are discussed.

1. Introduction

A human operator, when performing precise manipulation tasks, often finds supports for his forearm, wrist or elbow like in many instances watch-makers are practicing. The same simple idea is transferred into robotics. Bracing strategy for robotic manipulators was first presented by Book, Le and Sangveraphunsiri (1984). Asada and West (1984) made a kinematic analysis of the robot with a "jig-hand" attached inside the robot workspace. Significant increase of braced robot stiffness is reported in (West and Asada, 1985).

In some special case, it is possible to increase the robot absolute accuracy and repetability by bracing (Zupančič and Kralj, 1991; Zupančič, 1992). The method is based on the lever effect, which is possible only when rotational component of the lever movement is ensured. This is achieved by pure rolling or rolling/sliding

J. Lenarčič and V. Parenti-Castelli (eds.), Recent Advances in Robot Kinematics, 307-316.
© 1996 *Kluwer Academic Publishers. Printed in the Netherlands.*

combination. Sliding component of motion is ineligible due to nonlinear effects of friction, mechanical wear and problems with contact-point definition.

Three significant points on braced robot segment are observed: the joint with the lower-part of the robot, the contact-point, the joint with the upper-part of the robot. Pose uncertainty of the last joint of the lower-part is transferred and reduced through the contact-point to the first joint of upper-part if two conditions are satisfied: the contact kinematics is included into the braced robot kinematics, the uncertainty of the contact-point is better than the uncertainty of the last joint in lower-part.

Contact kinematics were studied for example by Cai and Roth (1987) and Montana (1988). The movement of two bodies in contact was described by the movement of the contact-point across both surfaces. The method is applicable in cases where the position sensors for observing the contact-point motion on the surfaces are available. The modification concerning the calculation of contact-point for sensorless operation of braced robot is suggested in this paper. Proposed modified method utilizes the local surface geometry at the contact-point on both surfaces in contact.

Because of high computational demands in the minimization procedure for the contact-point calculation an additional simplified method of an equivalent four-bar mechanism is proposed. It is restricted to the rolling contact of braced segment in a limited range of motion. A detailed construction at the contact-point is described. Advantages and disadvantages are discussed.

2. Methods

2.1. A BACKGROUND

Given an object in $\mathbf{R}^3$, we represent the configuration of a rigid body by attaching a Cartesian coordinate frame **O** to some point on the rigid body. Position of the body is specified by the origin of Cartesian frame represented by the vector **p**. We describe the orientation of the body by giving the relative orientation between a coordinate frame attached to the body and a fixed coordinate frame **Ref**. (See Figure 1.)

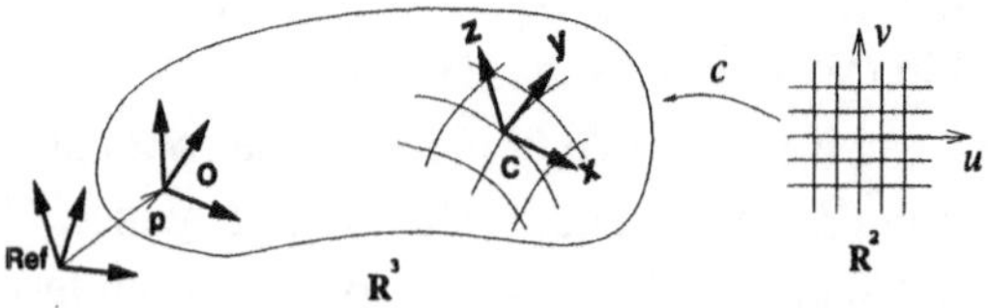

Figure 1. Description of the rigid body with the pose and the shape. [x,y,z] state for normalized Gauss frame at the surface point *c*.

The shape of the same body is described by its surface. We describe the surface of the object using a local coordinate chart, $c : U \subset \mathbf{R}^2 \rightarrow \mathbf{R}^3$. The map c takes a point $(u, v) \in \mathbf{R}^2$ to a point $x \in \mathbf{R}^3$ on the surface of the object, written in the object frame reference, O. Thus, locally, we can describe a point on the surface by specifying the corresponding (u, v). (Murray *et al.*, 1994)

At every nonsingular point of a regular surface normalized Gauss frame can be defined

$$[x\,, y\,, z] = [\frac{c_u}{\| c_u \|}, \frac{c_v}{\| c_v \|}, n] \tag{1}$$

where $c_u = \frac{\partial c}{\partial u}$, $c_v = \frac{\partial c}{\partial v}$ define tangent plane, and $n = (c_u \times c_v)/\| c_u \times c_v \|$ is a unit normal at the surface point.

Consider two objects with surfaces S_o and S_f. Let p_o and p_f be the positions of the point of contact relative to body-fixed frames O and F. Surfaces are described with charts c_o and c_f. Define local surface parameters with $\alpha_o = c_o^{-1}(p_o) = (u_o, v_o)$ and $\alpha_f = c_f^{-1}(p_f) = (u_f, v_f)$. The fifth parameter of 5dof higher contact pair between the bodies is the angle ψ defining the relative rotation around the contact normal between the bodies. We call $\eta = (\alpha_o, \alpha_f, \psi)$ the contact coordinates for S_o and S_f. The scope of higher-pair kinematics is the relative movement of the objects while maintaining in contact. The scope of contact kinematics is the motion of the points of contact across the surfaces of the objects in response to a relative motion of the objects.

2.2. KINEMATICS OF BRACED ROBOT

The kinematics of braced robot can be derived with or without contact kinematics. The idea of using braces for the absolute robot accuracy improvement (Zupančič and Kralj, 1991) requires contact kinematics. First, the lever principle suggested in (Zupančič, 1992) is briefly described. A modified method using local geometry of contacting bodies is explained in the next paragraph.

2.2.1. *Theoretical point-to-line contact-type of robot bracing*

Consider a robot with "line segments" as shown in Figure 2. A cone with the configuration ${}^0\mathbf{S}$ defined in the world coordinates represents the support. Point-to-line type of contact is established at the contact-point c. An arbitrary fixed Cartesian frame is attached at the contact-point, which is described by matrix ${}^S\mathbf{C}$. Because of the surface singularity, a normalized Gauss frame can not be defined at the contact-point. We now define the transformation of joint k, depicted as **Up**, with respect to the frame ${}^S\mathbf{C}$ at the contact-point, instead of defining the same joint transformation with respect to the previous joint *k-1* (like in DH notation, for example).

In our simple case, the frame **Up** is derived by two successive homogeneous transformations. First, the orientation of ${}^S\mathbf{C}$ is changed into new orientation, specified by the direction of the braced line segment. Second, the rotated coordinate system is translated by the value a in the direction of the braced line segment. The

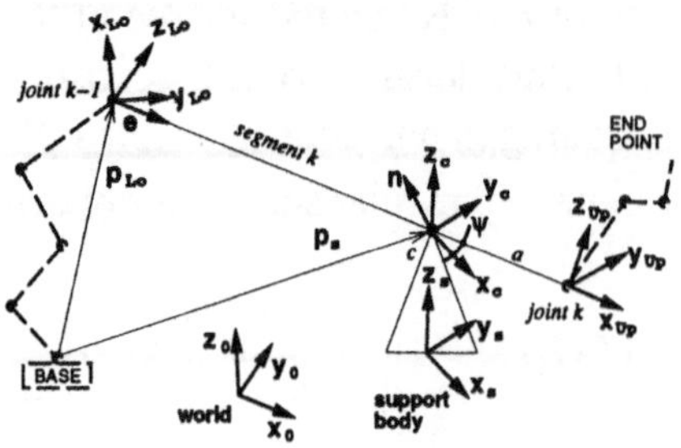

Figure 2. Simplified contact kinematics of bracing the robot with "line segment" against the cone shaped support body. Point-to-line type of contact is modeled.

transformation is described by the matrix product (eq. 2):

$$^{C}\mathbf{Up} \;=\; \mathbf{ROT}(\mathbf{n},\psi)\cdot\mathbf{TRANS}(\mathbf{e},a) \quad , \tag{2}$$

where vectors **e**, **n** and angle ψ define the orientation of braced segment. They are derived by the following equations (eq. 3):

$$\mathbf{e}=\frac{\mathbf{p}_s-\mathbf{p}_{Lo}}{\|\mathbf{p}_s-\mathbf{p}_{Lo}\|} \quad , \quad \psi=\arcsin\left(\frac{\|\mathbf{n}\|}{\|\mathbf{x}_c\|\cdot\|\mathbf{e}\|}\right) \quad , \quad \mathbf{n}=\mathbf{x}_c\times\mathbf{e} \tag{3}$$

The kinematics of braced robot with included (simplified) contact kinematics are represented with the following matrix multiplication (eq. 4):

$$^{0}\mathbf{T}_b \;=\; {}^{0}\mathbf{S}\cdot{}^{S}\mathbf{C}\cdot{}^{C}\mathbf{Up}\cdot{}^{Up}\mathbf{A}_{k+1}\cdots{}^{n-1}\mathbf{A}_n\,, \tag{4}$$

where ${}^{i-1}\mathbf{A}_i$ states for DH matrix of segment i and ${}^{Up}\mathbf{A}_{k+1}$ represents DH matrix of first upper part segment.

2.2.2. *True-type contact-point robot bracing*

Consider a robot with braced segment k as shown in Figure 3. The pose of the support body is specified by ${}^{0}\mathbf{S}=[\mathbf{x}_s,\mathbf{y}_s,\mathbf{z}_s,\mathbf{p}_s]$, where $\mathbf{p}_s$ defines the origin and $\mathbf{x}_s,\mathbf{y}_s,\mathbf{z}_s$ the axes of Cartesian frame with respect to the world frame. The surface of the support body is described by the local chart $s(u_s,v_s)$ in ${}^{0}\mathbf{S}$.

Coordinate system $\mathbf{Lo}=[\mathbf{x}_{Lo},\mathbf{y}_{Lo},\mathbf{z}_{Lo},\mathbf{p}_{Lo}]$ defines the last joint of the lower part of the robot and is derived according to Denavit-Hartenberg (DH) notation. Coordinate system $[\mathbf{x}_{Up},\mathbf{y}_{Up},\mathbf{z}_{Up}]$ is placed at the first joint of the upper part.

The surface of the braced segment is described two fold. First, by the local chart $g_{Lo}(u_{Lo},v_{Lo})$ defined in **Lo** and second, by the local chart $g_{Up}(u_{Up},v_{Up})$ defined in **Up**. The segment and the support are assumed to be convex bodies. The contact-point is calculated with the minimization of distance between the surfaces, while braced segment is approaching the support body. In the minimization procedure the position of the coordinate system **Lo** is constant. The orientation

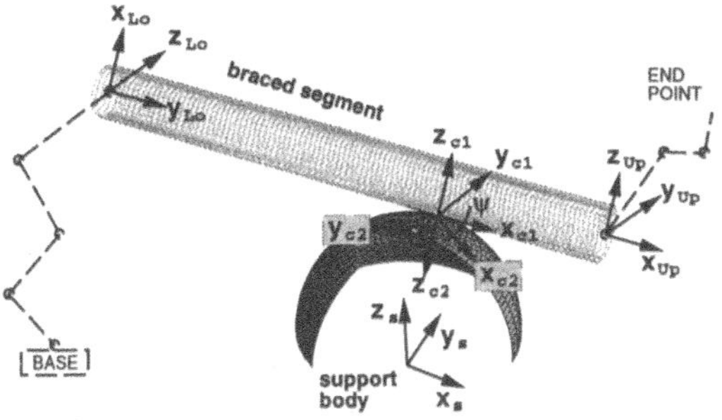

Figure 3. Definition of coordinate systems for braced robot with true-type contact.

of **Lo** is changing according to the joint *k-1* constraints. The contact-point at the support body is parameterized by (u_{sC}, v_{sC}). The same point is parameterized on the surface of braced segment with local charts (u_{LoC}, v_{LoC}) and (u_{UpC}, v_{UpC}).

At the contact-point c normalized Gauss frames are defined for both contacting surfaces (see Figure 3): $[\mathbf{x}_{c1}, \mathbf{y}_{c1}, \mathbf{z}_{c1}]$ states for the support body contact-point Gauss coordinate frame. It is defined by matrix $^{S}\mathbf{C}_1$ (eq. 5):

$$^{S}\mathbf{C}_1 = \begin{bmatrix} \mathbf{x} & \mathbf{y} & \mathbf{z} & s(u_s, v_s) \\ 0 & 0 & 0 & 1 \end{bmatrix} \quad , \quad \mathbf{x} = \frac{c_u}{\| c_u \|} \quad , \quad \mathbf{y} = \frac{c_v}{\| c_v \|} \quad , \quad \mathbf{z} = \frac{c_u \times c_v}{\| c_u \times c_v \|} \tag{5}$$

where (u_s, v_s) are parameters of contact-point parameterized in local support body surface chart.

In Figure 3, $\mathbf{C}_2 = [\mathbf{x}_{c2}, \mathbf{y}_{c2}, \mathbf{z}_{c2}]$ represents two contact-point Gauss frames $^{Lo}\mathbf{G}_c$, $^{Up}\mathbf{G}_c$ at the robot segment surface. It is derived from $^{S}\mathbf{C}_1$, instead of calculating it from the serial kinematic chain "BASE - Lo - Contact-Point". Frame $^{S}\mathbf{C}_1$ is transformed into $\mathbf{C}_2$ with two successive rotation transformations. First, $^{S}\mathbf{C}_1$ is rotated about $\mathbf{z}_{c1}$ axis by $-\psi$ to reach the proper segment direction defined with $\mathbf{x}_{c2}$ axis. Second, the rotation by π about new axis $\mathbf{x}'_{c1}$ is performed to complete the transformation of $^{S}\mathbf{C}_1$ into $^{C1}\mathbf{C}_2$. The following is the matrix equation performing the transformations (eq. 6)

$$^{C1}\mathbf{C}_2 \;=\; \mathbf{Rot}(z_{c1}, -\psi) \cdot \mathbf{Rot}(x_{c1}, \pi) \tag{6}$$

Frames $^{Lo}\mathbf{G}_c$ and $^{Up}\mathbf{G}_c$ represent the same pose in the world frame. Assuming the segment is rigid body, the following matrix equation is describing the geometric relations between distinctive coordinate systems (eq. 7, see also Figure 4).

$$^{Lo}\mathbf{G}_c = {}^{Lo}\mathbf{Up} \cdot {}^{Up}\mathbf{G}_c \quad , \tag{7}$$

where $^{Lo}\mathbf{Up}$ represents the local description of joint *k* pose with respect to body coordinate systems **Lo**. We found the pose of **Up** described in the normalized

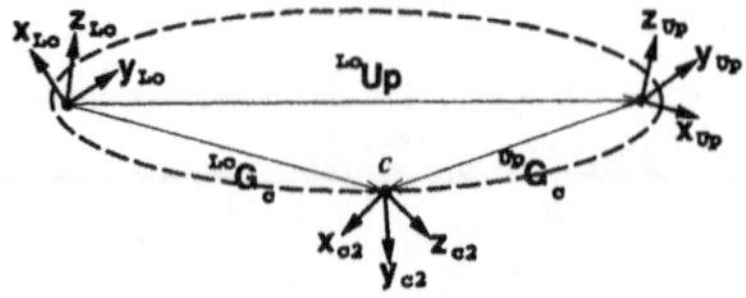

Figure 4. Geometric relations between significant coordinate systems attached onto the braced robot segment.

Gauss frame of the contact-point by inverting the transformation ${}^{\mathrm{Up}}\mathbf{G}_c$ (eq. 8):

$$ {}^{\mathrm{Up}}\mathbf{G}_c^{-1} = {}^{\mathrm{C2}}\mathbf{Up} = {}^{\mathrm{Lo}}\mathbf{G}_c^{-1} \cdot {}^{\mathrm{Lo}}\mathbf{Up} \tag{8} $$

Finally, we define the nominal braced robot end-point pose ${}^{0}\mathbf{T}_{bn}$ by the matrix multiplication (eq. 9):

$$ \begin{aligned} {}^{0}\mathbf{T}_{bn} &= {}^{0}\mathbf{S} \cdot {}^{\mathrm{S}}\mathbf{C}_1 \cdot {}^{\mathrm{C1}}\mathbf{C}_2 \cdot {}^{\mathrm{C2}}\mathbf{Up} \cdot {}^{\mathrm{Up}}\mathbf{A}_{k+1} \cdots {}^{\mathrm{N\text{-}1}}\mathbf{A}_N \\ &= {}^{0}\mathbf{S} \cdot {}^{\mathrm{S}}\mathbf{C}_1 \cdot \mathbf{Rot}(z_{c1}, -\psi) \cdot \mathbf{Rot}(x_{c1}, \pi) \cdot {}^{\mathrm{Lo}}\mathbf{G}_c^{-1} \cdot {}^{\mathrm{Lo}}\mathbf{Up} \cdot {}^{\mathrm{Up}}\mathbf{A}_{k+1} \cdots {}^{\mathrm{N\text{-}1}}\mathbf{A}_N \\ {}^{0}\mathbf{T}_{bn} &= f({}^{0}\mathbf{S}, \eta, {}^{\mathrm{Lo}}\mathbf{Up}, upperDH) \end{aligned} \tag{9} $$

${}^{0}\mathbf{S}$ and ${}^{\mathrm{Lo}}\mathbf{Up}$ are constant matrices, ${}^{\mathrm{S}}\mathbf{C}_1$ and ${}^{\mathrm{Lo}}\mathbf{G}_c^{-1}$ are matrices dependent on the contact coordinates η which are required by the proposed method. Contact parameters u_s, v_s, u_{Lo}, v_{Lo} can be measured or calculated. First option requires special sensor capabilities. In the second case, high computational demands are required for multidimensional minimization procedure. Two restrictions for the contact motion are introduced by the proposed method: rolling motion only, limited range of contact-point motion. An original method for such a motion is described in the following section.

2.3. METHOD OF VIRTUAL FOUR-BAR MECHANISM CONSTRUCTION

Consider the convex robot segment rolling across the convex support. When the joint **Lo** is moved for *d***Lo** in an appropriate direction, the contact-point c is changed into $c + dc$, the joint **Up** is moved for *d***Up**. The instantaneous movement dc is always performed in a tangent plane which is defined by x, y axes of normalized Gauss frame at the contact-point c. The direction and the value of displacement *d***Up** is dependent on the lever crank ratio and the local surface geometry parameters at the contact-point.

The aim of braced segment's kinematics is to define the relation $d\mathbf{Lo}/d\mathbf{Up}$. By introducing an equivalent mechanism for rolling, the contact-point must be defined only at the initial stage. Then, the braced segment together with joints **Lo** and **Up** is modeled with the motion of virtual 4-bar coupler segment. The model is valid for some particular interval of contact-point motion. The lenght of this interval depends on the temporary curvatures of contacting surfaces and geometric parameters of the lever. For simple rolling, only one 4-bar is defined. For more complex rolling trajectories, more 4-bars are constructed across the contact-point trajectory.

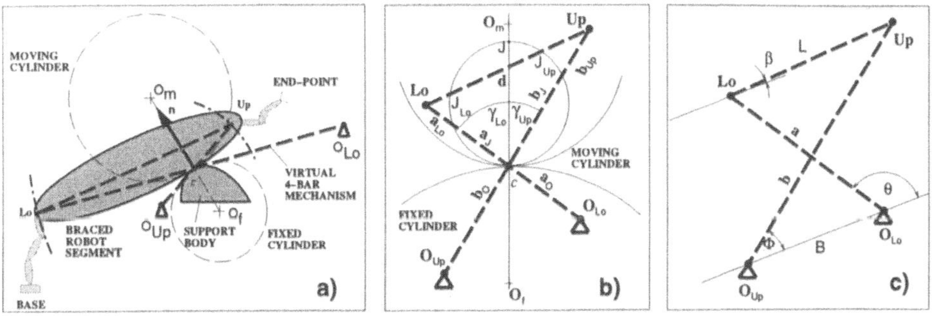

Figure 5. a) Equivalent virtual 4-bar mechanism used for modeling the rolling contact pair; b) Construction prameters of 4-bar mechanism; c) Parameters of 4-bar used in Freundenstein equation

Construction of equivalent 4-bar mechanism is performed in the plane of instantaneous rolling which is defined by the direction of rolling, $\frac{du}{dv}$, and the common normal **n** at the contact-point. Braced robot segment represents the coupler of the virtual 4-bar. Joint **Lo** is equivalent to the joint between the coupler and input segment which is connected to the virtual base in joint O_{Lo}. Robot joint **Up** is virtually connected to the output segment of four-bar mechanism. The virtual four-bar output segment is connected with the joint O_{Up} to the virtual base. (See figure 5a.)

The centers of curvature of the polodes of the 4-bar coincide with the centers of the fixed and moving cylinders which represent the curvatures of the surfaces in contact. (Figure 5b), (Hall, 1986).

Required radii of curvature at each contacting surfaces (R_{seg}, R_{sup}) are calculated at the beginning of the valid 4-bar interval according to the Euler form: $1/R = \cos^2\alpha/R_1 + \sin^2\alpha/R_2$, where α represents the angle between the rolling direction and the x axes of the particular Gauss frame, R_1, and R_2 state for the principal radii of curvature for each surface.

Parameters used in 4-bar construction are defined as follows. The length d represents inflection circle diameter. The distance from c to the intersection of input segment with inflection circle, J_{Lo}, is described with a_J, the part of input segment from c to **Lo** is depicted with a_{Lo} and the part from c to the virtual base $\mathbf{O}_{Lo}$ is represented by a_O. The same definitions for b_J, b_{Up} and b_O are used for the output segment, as shown in Figure 5b. The angles between the contact normal **n** and input and output segments are defined with γ_{Lo} and γ_{Up}. We now write Euler-Savary equations and rearrange them to get the partitions of input/output cranks lengths.

$$a_O = \frac{a_{Lo}^2}{a_{Lo} - d\cos\gamma_{Lo}} - a_{Lo} \quad , \quad b_O = \frac{b_{Up}^2}{b_{Up} - d\cos\gamma_{Up}} - b_{Up} \tag{10}$$

Inflection circle diameter d is derived with the formula $1/d = 1/R_{seg} - 1/R_{sup}$, assuming R_{seg}, R_{sup} represent curvatures in the rolling direction.

When all required 4-bar parameters are known, we apply Freundenstein's equation (eq. 11)(Hall, 1986) into numerical solving of the virtual 4-bar kinematics. Constants $\mathcal{R}_1$, $\mathcal{R}_2$, $\mathcal{R}_3$ define the 4-bar construction.

$$\mathcal{R}_1 \cos\Theta - \mathcal{R}_2 \cos\phi + \mathcal{R}_3 = \cos(\Theta - \phi) \tag{11}$$

$$\mathcal{R}_1 = B/b \qquad \mathcal{R}_2 = B/a \qquad \mathcal{R}_3 = \frac{(B^2 + a^2 - L^2 + b^2)}{2ba} \tag{12}$$

Constants $\mathcal{R}_1$, $\mathcal{R}_2$, $\mathcal{R}_3$ define the 4-bar mechanism. Moving of the joint **Lo** corresponds to the changing of the input angle Θ which provokes the change of the output angle ϕ. (See Figure 5c.) The change of the joint **Up** is as follows: $d\mathbf{Up} = b \cdot d\phi$.

3. Accuracy analysis of robot bracing

Given a robot and a support body in a Cartesian space, we represent three significant configurations (Figure 7a). A nominal configuration of a free robot – $\mathbf{T}_n$, an actual configuration of a free robot – $\mathbf{T}_a$, and an actual configuration of a braced robot – $\mathbf{T}_{ba}$. A nominal configuration of a support body is depicted by $\mathbf{S}_n$. An actual configuration of a support body is represented with $\mathbf{S}_a$. The difference between them is $\Delta\mathbf{S}$. Nominal configurations of a free and a braced robot are the same ($\mathbf{T}_n = \mathbf{T}_{bn}$). A reduction of $\Delta\mathbf{T}_a$ into $\Delta\mathbf{T}_{ba}$ is assumed as the accuracy improvement due to bracing.

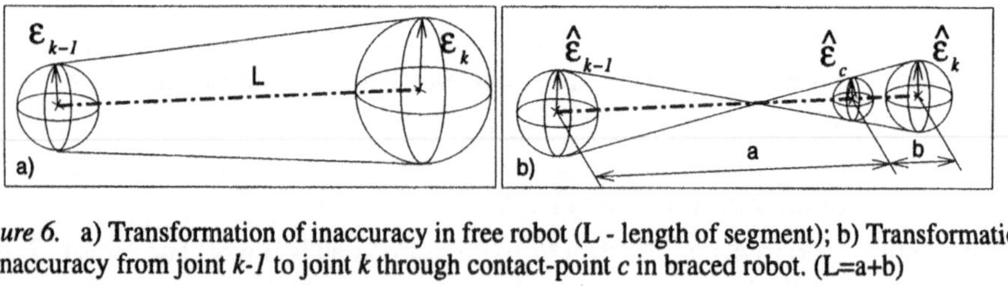

Figure 6. a) Transformation of inaccuracy in free robot (L - length of segment); b) Transformation of inaccuracy from joint *k-1* to joint *k* through contact-point *c* in braced robot. (L=a+b)

3.1. FREE ROBOT

Consider a free robot with n serial links, as shown in Figure 7a. The segment k is connected to the surrounding segments with joints *k-1* and k. Let the joint *k-1* absolute inaccuracy be represented by the sphere with the radius ϵ_{k-1}, as shown

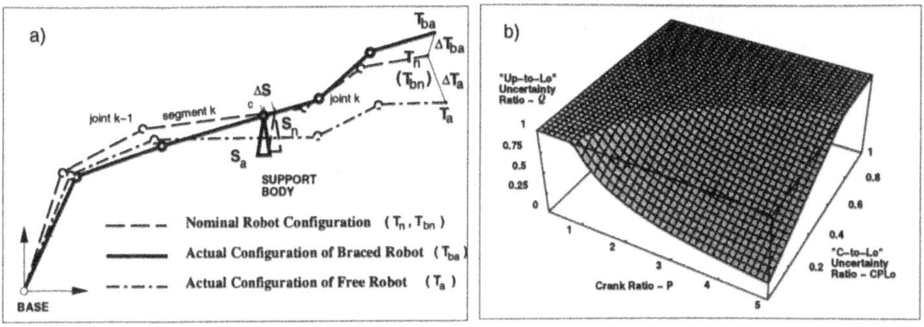

Figure 7. a) The principle of the accuracy improvement by bracing - three significant robot configurations. b) Estimate of accuracy improvement by bracing.

in Figure 6a. The inaccuracy of the joint k is represented with the radius of sphere ϵ_k. In case of free serial kinematic chain, the inequality $\epsilon_k > \epsilon_{k-1}$ is always true because of successive pose error accumulation.

3.2. BRACED ROBOT

Consider a robot bracing with the segment k against the support body at the contact-point c. The inaccuracy sphere of the joint *k-1* has the radius $\widehat{\epsilon}_{k-1}$. The inaccuracy sphere of the contact-point has the radius $\widehat{\epsilon}_c$. Inaccuracy of the joint k is represented by the sphere with the radius $\widehat{\epsilon}_k$.

To estimate an inaccuracy reduction ratio, $\mathcal{Q}$, for the method based on the absolute contact-point definition, we first define the ratio between the lower and the upper lever crank, $P = a/b$, (see Figure 6b). The ratio between the contact-point inaccuracy and the lower-part robot inaccuracy is defined: $CPLo = \| \widehat{\epsilon}_c \| / \| \widehat{\epsilon}_{k-1} \|$. The inaccuracy reduction ratio $\mathcal{Q}$ is estimated with the following function (eq. 13):

$$\mathcal{Q} = \frac{\| \widehat{\epsilon}_k \|}{\| \widehat{\epsilon}_{k-1} \|} = CPLo + \frac{CPLo}{P} + \frac{1}{P} \tag{13}$$

The robot accuracy is improved only when $\mathcal{Q} \leq 1$. In Figure 7b, simulation of the accuracy improvement estimation for the braced robot with bracing parameters: P=[0.2,5]; CPLo=[0.01,1] is shown. On x axis there is the lever ratio,P, on y axis we find the contact/lower-part accuracy ratio, $CPLo$. It is clear, that the accuracy ratio $CPLo$ must be always less then 1. It means that the contact-point must be defined more accurate as the last joint of the lower part of the robot.

4. Discussion

The problem of increasing the absolute robot accuracy was solved by bracing. Accuracy analysis of braced robot shows that absolute robot accuracy is improved by bracing only when absolute contact-point pose is known. An original method for modeling of braced robot kinematics without requirement for the surface position sensor is presented. We found the rolling contact as the most appropriate for the real bracing application. The idea of modeling the rolling contact with virtual 4-bar mechanism is proposed. The method is valid for a limited range of contact-point motion. It is acceptable, because the bracing strategy is meaningful only in fine robot motion.

Compared to the general method, which uses local surface geometry of contacting surfaces, the method of virtual 4-bar simplifies the contact kinematics inside the valid interval. Simplification is due to the modeling of the contact-point trajectory across the contacting surfaces, instead of measuring or calculating it.

In addition, the kinematic model based on the virtual equivalent 4-bar is attractive for optimization of bracing. Contact kinematics of braced robot segment could be presented in the configuration space of 4-bar mechanism. For the specified robot task an optimal pose and shape of the support body should be defined. Optimization procedure regarding relative position of the contact-point with respect to the braced segment's joints and relative curvatures of the contacting surfaces could be easier performed in the configuration space of the virtual 4-bar mechanism.

References

Asada, H. and West, H. (1984) Kinematic analysis and design of tool guide mechanisms for grinding robots, *Proc. ASME Winter Meeting on Computer - integrated manufacturing and robotics*, New Orleans, pp. 1–17

W.J. Book, S. Le, and V. Sangveraphunsiri. (1984) Bracing strategy for robot operation, *Proc. Symposium on the Theory and Practice of Robots and Manipulators*, Udine, Italy, pp. 179–185

Cai, C., Roth, B. (1987) On the spatial motion of a rigid body with point contact, *Proc. 1987 IEEE International Conference on Robotics and Automation*, Raleigh, N.C., U.S.A., pp. 686–695

Hall, A.S. (1986) *Kinematics and Linkage Design*, Waveland Press, Inc., Prospect Heights, IL, U.S.A.

David J. Montana, D.J. (1988) The kinematics of contact and grasp, *The International Journal of Robotics Research* **Vol.7 no.3**, pp. 17–32

Murray, R.M., Li, Z., Sastry, S.S. (1994) *A Mathematical Introduction to Robotic Manipulation*, CRC Press, Inc.

West, H., Asada, H. (1985) Kinematic analysis and mechanical advantage of manipulators constrained by contact with the environment, *Proc. ASME Winter Meeting: Robotics and Manufacturing Automation*, Miami Beach, Florida, **Vol.15**, pp. 175–186

Zupančič, J. (1992) Enhancing robot mechanism performances by using mechanical support – kinematic analysis, *Proc. of the 3rd International Workshop on Advaces in Robot Kinematics*, Ferrara, Italy, pp. 297–303

Zupančič, J., Kralj, A. (1991) Self–suported robot mechanisms, *Proc. 14th IASTED Int. Symp. Manufacturing and Robotics*, Lugano, Switzerland, pp. 118–121

A REMARKABLE CLASS OF OVERCONSTRAINED LINKAGES

O. RÖSCHEL
Institute of Geometry, Technical University Graz
Kopernikusgasse 24, A - 8010 Graz, Austria

Abstract: Given 3 congruently parametrised axial Darboux motions $\zeta_i = \Sigma_i \setminus \Sigma_0$ (i=1,2,3) with pairwise orthogonal and skew axes, which are gained of ζ_0 by rotation of 120, 240 degrees, resp. round a certain axis r_0. Then we are able to show, that each relative motion $\Sigma_i \setminus \Sigma_j$ $(i \neq j;\ i, j = 1,2,3)$ has a two-parametric familiy of (real) points situated on hyperboloids of one sheet $\Phi_{i,j}$, which is moved on spheres centered on an analogue surface $\Phi_{j,i}$. The intersection of two hyperboloids in a system Σ_i splits into a straight line g_i and a cubic circle c_i. It is shown that it is possible to form stiff triangles connecting corresponding points of c_1, c_2, c_3, such that the motions are not disturbed. For the points on the straight lines an analogous result holds: There the stiff triangles degenerate to straight lines, which in Σ_0 determine a regulus on a hyperboloid of rotation at each moment of the motion. Finally as an example two pairs of points on g_i and c_i are linked to get an overconstrained linkage.

1. In the 3-dimensional Euclidean space E_3 we use Cartesian frames $\{O_i;\ x_i, y_i, z_i\}$ $(i = 0,1,2,3)$ to describe points of the given systems Σ_i $(i = 0,1,2,3)$ by their position vectors $\vec{x}_i := (x_i, y_i, z_i)^t$. In Σ_0 we define 3 not intersecting orthogonal axes a_1, a_2, a_3 given by equations

$$a_1 \dots x_0 = A, y_0 = 0, \quad a_2 \dots y_0 = A, z_0 = 0, \quad a_3 \dots x_0 = 0, y_0 = A \qquad (1)$$

with an arbitrary real $A \neq 0$ (see figure 1). Then we may define 3 congruent (even congruently parametrized) axial DARBOUX-motions $\zeta_i = \Sigma_i \setminus \Sigma_0$ (i=1,2,3) with axes a_1, a_2, a_3. They all shall be parametrized by their angle of rotation $t \in [0, 2\pi]$. Then a parametrisation of these motions shall be given by (see O. Bottema-B. Roth (1979) ,p.321)

J. Lenarčič and V. Parenti-Castelli (eds.), Recent Advances in Robot Kinematics, 317-324.
© 1996 *Kluwer Academic Publishers. Printed in the Netherlands.*

$$\zeta_1 : \vec{x}_0(t, \vec{x}_1) := \begin{pmatrix} A + x_1 \cos t - y_1 \sin t \\ x_1 \sin t + y_1 \cos t \\ z_1 + B \sin t \end{pmatrix}, \qquad \zeta_2 : \vec{x}_0(t, \vec{x}_2) := \begin{pmatrix} z_2 + B \sin t \\ A + x_2 \cos t - y_2 \sin t \\ x_2 \sin t + y_2 \cos t \end{pmatrix},$$

$$\zeta_3 : \vec{x}_0(t, \vec{x}_3) := \begin{pmatrix} x_3 \sin t + y_3 \cos t \\ z_3 + B \sin t \\ A + x_3 \cos t - y_3 \sin t \end{pmatrix} \qquad (t \in [0, 2\pi]) \qquad (2)$$

with a further real constant $B \neq 0$. The common time parameter $t \in [0, 2\pi]$ links the 3 systems $\Sigma_1, \Sigma_2, \Sigma_3$. This linkage represents a generalisation of a partial motion of the motions studied in Röschel (1995). We want to discuss properties of the relative motions and show that the motion is not disturbed, if certain stiff rods (with spherical links) connect certain points of each pair of system. $\Sigma_1 \setminus \Sigma_0$ may be moved into $\Sigma_2 \setminus \Sigma_0$ ($\Sigma_3 \setminus \Sigma_0$) (including its parametrisation) by a rotation of 120° (240°) round the axis $r_0 = [(0,0,0), (1,1,1)]$ (see figure 1 for $t = 0$). Therefore $\Sigma_2 \setminus \Sigma_0$ and $\Sigma_3 \setminus \Sigma_0$ are conjugate motions to $\Sigma_1 \setminus \Sigma_0$ with respect to these fixed rotations.

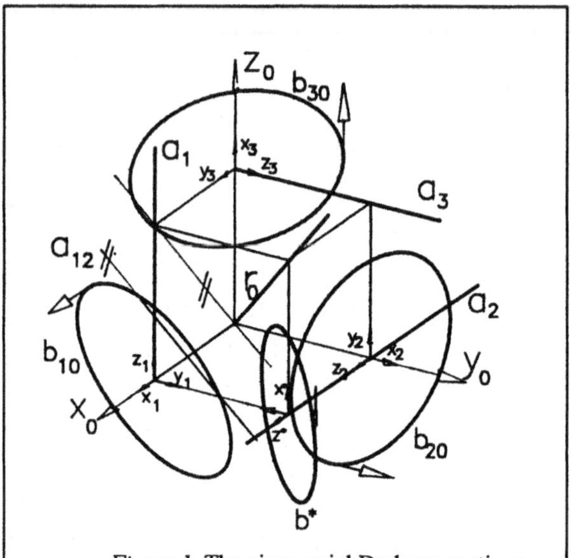

Figure 1. The given axial Darboux motions

2. The relative motions $\Sigma_i \setminus \Sigma_j$ $(i \neq j)$. Our definition of the motions guarantees that for each moment t the positions $\Sigma_j(t)$ in Σ_0 may be gained by reflecting $\Sigma_i(t)$ at a straight line a_{ij} followed by a fixed displacement in Σ_j. Figure 1 shows the situation for i=1, j=2: We have drawn a point path b_{i0} under $\Sigma_i \setminus \Sigma_0$ (i=1,2), which is gained from b_{10} by rotation round r_0. But it may be generated by reflecting the positions of Σ_1 with respect to the straight line a_{12} (results denoted by "*") followed by a fixed screw in Σ_2 (with axis a_2, angle $-\pi/2$ and translation distance $-A$). The situation for $(i, j) \neq (1,2)$ is similar.

Therefore $\Sigma_i \setminus \Sigma_j$ is line-symmetric in the sense of J.Krames with respect to a basic surface Γ_{ji}, which is the path surface of the straight line a_{ij} under the inverse motion Σ_0 / Σ_j. Such motions have been studied by J.Krames (1937b).

Γ_{ji} is a ruled surface of degree 4 (J.Krames (1937b)). In this case in Σ_i there exists a two-parametric family of points with spherical paths under $\Sigma_i \setminus \Sigma_j$. These points in general are situated on a pair of complex conjugate planes (a special singular case of that described by J.Krames (1937b)) and on an orthogonal hyperboloid of one sheet Φ_{ij}, the corresponding centers lie on congruent surfaces, the corresponding hyperboloid will be denoted by Φ_{ji}. We restrict ourselves to the case $i = 1$ - the other cases result from permutations and appropriate changes of some signs. Computation yields

$$\begin{aligned} \Phi_{1,2} \; &\ldots\; 2B\, x_1\, z_1 = A\,(-B^2 + x_1^2 + y_1^2), \\ \Phi_{1,3} \; &\ldots\; 2B\, y_1\, z_1 = A\,(-B^2 + x_1^2 + 2B\,y_1 + y_1^2) \end{aligned} \tag{3}$$

After some algebra the relations $V_{1j}\colon \Phi_{1j} \subset \Sigma_1 \to \Phi_{j1} \subset \Sigma_j \quad (j = 2,3)$ between the points with spherical paths and the centers of their spheres may be written in the following typical cases:

$$V_{1,2}\colon \Phi_{1,2} \to \Phi_{2,1}$$

$$x_2 = \frac{B\,(B^2 - x_1^2 - y_1^2)}{x_1^2 + (y_1 - B)^2},\; y_2 = \frac{-2B^2 x_1}{x_1^2 + (y_1 - B)^2},\; z_2 = \frac{A(x_1 - y_1)}{x_1} \tag{4}$$

and

$$V_{1,3}\colon \Phi_{1,3} \to \Phi_{3,1}$$

$$x_3 = \frac{-2B^2 y_1}{(x_1 + B)^2 + y_1^2},\; y_3 = \frac{B(-B^2 + x_1^2 + y_1^2)}{(x_1 + B)^2 + y_1^2},\; z_3 = \frac{A\,x_1}{y_1} \tag{5}$$

We now show how to find these equations for the example $i = 1,\ j = 2$ (for the other choices of indices it may be done in a similar way):

Two points $\vec{x}_1 = (x_1, y_1, z_1)^t$, $\vec{x}_2 = (x_2, y_2, z_2)^t$ fixed in Σ_1, Σ_2, resp. keep constant distances during the motions ζ_1, ζ_2, iff the difference vector

$$\vec{d}_{1,2}(t, \vec{x}_1, \vec{x}_2) := \vec{x}_0(t, \vec{x}_2) - \vec{x}_0(t, \vec{x}_1) = \begin{pmatrix} z_2 - A \\ A \\ -z_1 \end{pmatrix} + \begin{pmatrix} -x_1 \\ x_2 - y_1 \\ y_2 \end{pmatrix} \cos t + \begin{pmatrix} B + y_1 \\ -x_1 - y_2 \\ x_2 - B \end{pmatrix} \sin t :$$

$$:= \vec{a} + \vec{b}\cos t + \vec{c}\sin t \tag{6}$$

has constant length. This yields

$$0 = \vec{d}_{1,2}(t, \vec{x}_1, \vec{x}_2)\frac{\partial \vec{d}_{1,2}(t, \vec{x}_1, \vec{x}_2)}{\partial t} = \vec{a}\,\vec{c}\cos t - \vec{a}\,\vec{b}\sin t + \vec{b}\,\vec{c}\cos 2t + 0.5\,(\vec{c}^2 - \vec{b}^2)\sin 2t$$

$\forall\, t \in [0, 2\pi]$. This condition is valid, iff

$$0 = \vec{a}\,\vec{b} = A\,x_2 - z_1\,y_2 - x_1\,z_2 + A\,(x_1 - y_1),$$
$$0 = \vec{a}\,\vec{c} = -z_1\,x_2 - A\,y_2 + (B + y_1)\,z_2 + B\,z_1 - A\,(B + x_1 + y_1),$$
$$0 = \vec{b}\,\vec{c} = -x_1\,x_2 + (y_1 - B)\,y_2 - B\,x_1,$$
$$0 = \vec{b}^2 - \vec{c}^2 = 2[(B - y_1)\,x_2 - x_1\,y_2 - B\,(B + y_1)].$$

This is a system of 4 linear equations for x_2, y_2, z_2. It has solutions, if the condition

$$B\,[x_1^2 + y_1^2 + B^2]\,[2\,B\,x_1\,z_1 + A\,(B^2 - x_1^2 - y_1^2)] = 0 \qquad (7)$$

holds. This equation determines the points of Σ_1 which may hold constant distances to certain points of Σ_2. Therefore for $A\,B \neq 0$ these points of Σ_1 belong to a complex conjugate pair of planes and the hyperboloid of one sheet $\Phi_{1,2}$ (3). For the points of $\Phi_{1,2}$ we get the corresponding points in Σ_2 via the map $V_{1,2}$ (4). □

The correspondence $V_{1,2}$ (4) between the x- and y- coordinates (the ground projections $(x_i, y_i, 0), (x_j, y_j, 0)$) *does not depend on the constant* A (therefore for the first two coordinates we may use the results of Röschel (1995)). The equations show, that the ground projection of our relationships **may be gained by an inversion with respect to a circle, followed by a displacement.** The centers of inversion have the coordinates $(0, B, 0)$, $(-B, 0, 0)$ resp. for $V_{1,3}$. The radius of the circles of inversion is $B\sqrt{2}$. We sum up in

Theorem 1: *Given the 3 congruent special axial Darboux-motions* $\Sigma_i \setminus \Sigma_0$ $(i = 1,2,3)$ *in a fixed space* Σ_0 *represented by* (2). *Then each relative motion* $\Sigma_i \setminus \Sigma_j$ *is line-symmetric in the sense of J. Krames and moves a two-parametric family of points of* Σ_i *on spheres centered in* Σ_j. *The corresponding (real) points are (in general) situated on special hyperboloids of one sheet in both systems.*

3. In the case $A = 0$ Röschel (1995) gives a way to generate overconstrained linkages consisting of rods of constant length (with spherical links in each system Σ_i). We now want to extend this procedure to the case $A \neq 0$:

Interesting cases occour, if we start with points P_1 on the intersection of the hyperboloids $\Phi_{1,2}$ and $\Phi_{1,3}$. The curve of intersection is an algebraic curve of order 4 containing the absolute points of the planes z_1 = const. and the point of infinity of the

z_1 – axis. The correspondences $V_{1,i}$ $(i = 2,3)$ map $P_1 \in k_{1,2,3} := \Phi_{1,2} \cap \Phi_{1,3}$ into points $P_i := V_{1,i}(P_1)$. By simple computations it is possible to proof the surprising

Theorem 2: *Given the 3 congruent special axial Darboux-motions $\Sigma_i \setminus \Sigma_0$ in a fixed space Σ_0 represented by (2). Starting with points P_1 on the intersection curve $k_{1,2,3} := \Phi_{1,2} \cap \Phi_{1,3}$ the maps $V_{1,i}$ $(i = 2,3)$ determine corresponding points $P_i := V_{1,i}(P_1)$ $(i=2,3)$. Then for each point $P_1 \in k_{1,2,3} := \Phi_{1,2} \cap \Phi_{1,3}$ the relation $P_i = V_{k,i}(V_{j,k}(V_{i,j}(P_i)))$ is true.*

This means, that for all starting points P_1 on $k_{1,2,3} := \Phi_{1,2} \cap \Phi_{1,3}$ the corresponding points $P_i := V_{1,i}(P_1)$ $(i = 2,3)$ determine a triangle P_1, P_2, P_3 of fixed shape, which does not disturb our special one- parameter motion, if the triangles and the systems $\Sigma_1, \Sigma_2, \Sigma_3$ have spherical links at the vertices P_1, P_2, P_3.

4. Now we are interested in the shape of the **curve of intersection** $k_{1,2,3} := \Phi_{1,2} \cap \Phi_{1,3}$: Suitable geometrical considerations show, that it splits into two parts; a **straight line** g_1 and a **cubic circle** c_1, which may be parametrized by

$$g_1 \;\ldots\ldots\; \left(u,\; B + u,\; \frac{A(u + B)}{B}\right)^t,$$

$$c_1 \;\ldots\ldots\; \left(-\frac{u\,B^2}{u^2 + (u + B)^2},\; -\frac{B^2(u + B)}{u^2 + (u + B)^2},\; \frac{A(u + B)}{B}\right)^t \quad (u \in R) \tag{8}$$

Then the maps $V_{1,2}, V_{1,3}$ transform g_1 and c_1 into image curves with the following parametrisations

$$g_2 := V_{1,2}(g_1) \ldots \left(-\frac{B(u + B)}{u},\; -\frac{B^2}{u},\; -\frac{AB}{u}\right)^t,$$

$$g_3 := V_{1,3}(g_1) \ldots \left(-\frac{B^2}{u + B},\; \frac{Bu}{u + B},\; \frac{Au}{u + B}\right)^t,$$

$$c_2 := V_{1,2}(c_1) \ldots \left(\frac{Bu(u + B)}{(u + B)^2 + B^2},\; \frac{B^2\,u}{(u + B)^2 + B^2},\; -\frac{AB}{u}\right)^t, \tag{9}$$

$$c_3 := V_{1,3}(c_1) \ldots \left(\frac{B^2(u + B)}{u^2 + B^2},\; -\frac{B\,u(u + B)}{u^2 + B^2},\; \frac{Au}{u + B}\right)^t$$

It is remarkable, that the straight line g_1 is mapped into straight lines g_2, g_3, the cubic circle c_1 into cubic circles c_2, c_3. We sum up in

Theorem 3: *Given the 3 congruent special axial Darboux-motions* $\Sigma_i \setminus \Sigma_0$ *in a fixed space* Σ_0 *represented by* (2). *Then the intersection curve* $k_{1,2,3} := \Phi_{1,2} \cap \Phi_{1,3}$ *of the hyperboloids with (real) spherical paths splits into a straight line* g_1 *and a cubic circle* c_1. *Without disturbation of the motions the points* P_1 *on* g_1 *may be linked by sticks with points* $P_i := V_{1,i}(P_1)$ $(i=2,3)$ *on straight lines* g_2, g_3, *the points on* c_1 *with points on cubic circles* c_2, c_3.

Now we start with a point $P_1 \in g_1$: These points are mapped into points $P_2 \in g_2$ and $P_3 \in g_3$, whose paths may be studied in Σ_0. Our difference vectors $\vec{d}_{1,2}(t,u) := \vec{x}_0(t, P_2) - \vec{x}_0(t, P_1)$ (6) and $\vec{d}_{1,3}(t,u) := \vec{x}_0(t, P_3) - \vec{x}_0(t, P_1)$ read

$$\vec{d}_{1,2}(t,u) = \frac{1}{u}\left[\frac{A}{B}\begin{pmatrix} -B(u+B) \\ u\,B \\ -u\,(u+B) \end{pmatrix} - \begin{pmatrix} u^2 \\ (u+B)^2 \\ B^2 \end{pmatrix}\cos t + \begin{pmatrix} u^2 + 2uB \\ B^2 - u^2 \\ -B(2u+B) \end{pmatrix}\sin t\right], \quad (10)$$

$$\vec{d}_{1,3}(t,u) = \frac{1}{u+B}\left[\frac{A}{B}\begin{pmatrix} -B(u+B) \\ u\,B \\ -u\,(u+B) \end{pmatrix} - \begin{pmatrix} u^2 \\ (u+B)^2 \\ B^2 \end{pmatrix}\cos t + \begin{pmatrix} u^2 + 2uB \\ B^2 - u^2 \\ -B(2u+B) \end{pmatrix}\sin t\right].$$

Therefore we have $u\,\vec{d}_{1,2}(t,u) = (u+B)\,\vec{d}_{1,3}(t,u)$ for all $u \in R$. Thus the corresponding points $P_1 \in g_1$, $P_2 := V_{1,2}(P_1)$, $P_3 := V_{1,3}(P_1)$ are collinear during the whole motion! *At each moment* t *in* Σ_0 *the triple of straight lines* g_1, g_2, g_3 *determines a regulus of intersecting straight lines. These straight lines may be materialized as sticks connecting corresponding points on* g_1, g_2, g_3*!* Computation shows that the regulus of sticks for each $t \in [0, 2\pi]$ in Σ_0 is part of a rotational hyperboloid of one sheet with common axis $r_0 = [(0,0,0), (1,1,1)]$, which may degenerate. Of course the shape of the hyperboloids changes with t. In this case we have got a model of a moveable hyperboloid of one sheet, where the generators are formed by sticks linked together by spherical links. This special case is known - a model may be found at institutes of geometry at Vienna and Dresden University of Technology.

5. An example. Now we give an example of a simple overconstrained linkage following the results of chapter 4: Starting with points $u = -0.5B$ and $u = B$ in (8) we gain points

$$\begin{aligned} &P_1 \in g_1 \ldots (-0.5\,B, 0.5\,B, 0.5\,A)^t, \quad R_1 \in g_1 \ldots (B, 2\,B, 2\,A)^t, \\ &Q_1 \in c_1 \ldots (B, -B, 0.5\,A)^t, \qquad S_1 \in c_1 \ldots (-0.2\,B, -0.4\,B, 2\,A)^t \end{aligned} \quad (11)$$

with corresponding points

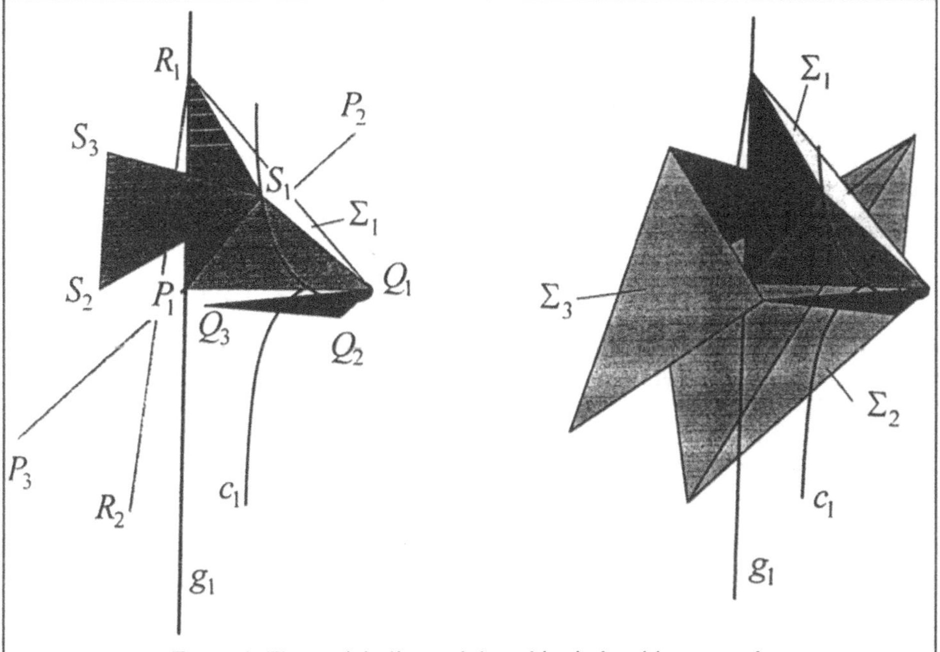

Figure 2: The straight line and the cubic circle with some rods

$$P_2 \in g_2 \;....\; (B, 2\,B, 2\,A)^t, \qquad R_2 \in g_2 \;....\; (-2\,B, -\,B, -\,A)^t, \tag{12}$$
$$Q_2 \in c_2 \;....\; (-\,0.2\,B, -0.4\,B, 2\,A)^t, \quad S_2 \in c_2 \;....\; (0.4\,B, 0.2\,B, -\,A)^t$$

$$P_3 \in g_3 \;....\; (-2\,B, -\,B, -\,A)^t, \quad R_3 \in g_3 \;....\; (-0.5\,B, 0.5\,B, 0.5\,A)^t, \tag{13}$$
$$Q_3 \in c_3 \;....\; (\,0.4\,B, 0.2\,B, -\,A)^t, \quad S_3 \in c_3 \;....\; (B, -\,B, 0.5\,A)^t.$$

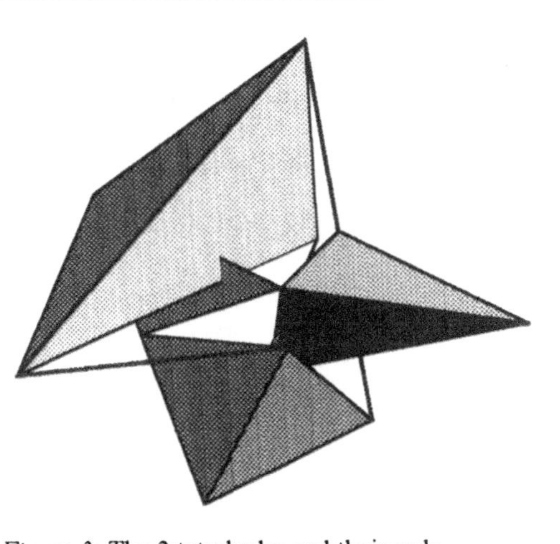

Figure 3: The 3 tetrahedra and their rods

We have got a configuration consisting of 3 tetrahedra, two triangular and two straight line links. The general degree of freedom would be $F = 0$. Adding more sticks according to our rules we would get a linkage with general degree of freedom $F \leq -1$. Figure 2 shows a very simple configuration of this type consisting of the points (11), (12), (13) for $A = B = 100$ and $t = \pi / 6$. As many lines are hidden by the tetrahedra the left side only shows the first tetrahedron, the straight line g_1, the cubic c_1, the triangular and the straight line links. The right side gives the image of the whole configuration. It has to be mentioned that only the edges of the tetrahedra should be materialized - but in order to gain more instructive pictures the author has chosen the materialized version.

Figure 3 shows the situation for the same model for $t = 19\pi/24$. Here the straight line links are situated outside of the 3 tetrahedra. Therefore the two positions only may be gained by a coninous motion, if intersections of some sticks are allowed. But both models built of rods have a one- parametric motion with timeparameter t taken from two different parts of the allowed time - interval.

References:

Bottema, O., Roth, B. (1979) Theoretical kinematics. North-Holland Series. Amsterdam.

Borel, E. (1908) Mémoire sur les déplacements à trajectoires sphériques. Mem. Acad. Sciences 33 (2), 1 - 128.

Bricard, M. (1906) Mémoire sur les déplacements à trajectoires sphériques. Journ. de l' École Polytechnique 2 (11), 1 - 93.

Duporcq, E. (1898) Sur le déplacement le plus général d'une droite dont tous les points décrivent des trajectoires sphériques. Journ. de mathématiques pures et appliquées, **4**, 121 - 136.

Krames, J. (1937a) Zur Bricardschen Bewegung, deren sämtliche Bahnkurven auf Kugeln lie gen (Über symmetrische Schrotungen II). Monatsh. Math. **45**, 407 - 417.

Krames, J. (1937b) Die Borel-Bricard-Bewegung mit punktweise gekoppelten orthogonalen Hyperboloiden (Über symmetrische Schrotungen VI). Monatsh. Math. **146**, 172 - 195.

Röschel, O. (1995) Zwangläufig bewegliche Polyedermodelle I. Math. Pann. **6**/1, 267 - 284.

Röschel, O. (1996) Zwangläufig bewegliche Polyedermodelle II. Studia Sci. Math. Hung. (in print).

Stachel, H. (1991) The HEUREKA-Polyhedron. Proc. Conf. Intiutive Geometry, Szeged Coll. Math. Soc. J. Bolyai, 447 - 459.

Stachel, H. (1992) Zwei bemerkenswerte bewegliche Strukturen. Journal of Geometry **43**, 14-21.

Verheyen, H. F. (1989) The complete set of Jitterbug transformers and the analysis of their motion. Computers Math. Applic. **17**, 203 - 250.

Wohlhart, K. (1993a) Dynamic of the "Turning Tower". Ber. d. IV. Ogolnopolska Konf. Maszyn Wlokienniczych i Dzwigowych, 325 - 332.

Wohlhart, K. (1993b) Heureka Octahedron And Brussels Folding Cube As Special Cases Of The Turning Tower. Syrom 93 II, 303 - 312.

Wohlhart, K. (1995a) Das dreifach plansymmetrische Oktoid und seine Punktbahnen. Math. Pann. 6/2, 243 - 265.

Wohlhart, K. (1995b) New Overconstrained Spheroidal Linkages. Proc. 9th World Congress on the Theory of Machines and Mechanisms, 149 - 154.

HOW TO EXTEND ROBERTS' LAW FOR ECCENTRICALLY DRIVEN, INVERTED SLIDER-CRANKS

E.A.DIJKSMAN (author) & A.T.J.M.SMALS (software)
Faculty of Mechanical Engineering
Eindhoven University of Technology
The Netherlands

Abstract. The extension of Roberts' Law concerns inverted slider-cranks with an eccentricity and a general tracing-point attached to the moving plane of the slider. As both curve-cognates degenerate in this case, the infinite turning-joint has *first* been replaced by a *finite* far-away joint, but such that the coupler-plane also meets *two* accuracy-positions of the erstwhile slider.
One of the curve-cognates may then be adapted or modified until up to *six* accuracy-positions are met. They seem to be sufficient to attain an extraordinary good approximation of the entire branch of the original curve. Besides, even the transmission-angle attains a value about twice as large as the one obtained with the *auxiliary* four-bar representing the *first* approximation.
Cases, where the inverted slider-crank turns into a crank-and-rocker, into a double crank or even into a stretchable four-bar, are shown. The latter corresponds with one of Grashof's border cases for the inverted slider-crank. Roberts' Law then ensures that each inverted slider-crank, having a fully revolving crank, possesses *three* four-bar branch-cognates this way.

1. Introduction

Though Roberts' Law (ref. [1]) describes the general existence of *three* four-bar curve cognates, all able to produce the *same* four-bar coupler curve, its application seems to collapse when applied on a degenerated four-bar such as the inverted slider-crank.
However, a previous paper [4] showed a practical circumvention in the special case where the inverted slider-cranks produced *symmetrical* curves. Then, a very good approximation was obtained based on *stretch-rotation* and *symmetrization*. However, for the more general type of the (eccentric) inverted slider-crank, symmetrization isn't applicable and has to be replaced by another procedure, apparently leading to a similar concurrence between the curves as in the symmetrical case.
In the *eccentric* case, we are going to start from in this paper, also *non*-Grashof types occur. Then, there is no crank making a complete revolution, whereas the curve produced will be singular branched as then only *one* branch occurs as the gradual merging result of two different branches originally produced by Grashof types.
In the *symmetrical* case the replacement four-bar reproduced only *one* branch generated by the then centrically driven inverted slider-crank. Naturally, something alike is to be expected for the eccentric case. Thus, when an eccentrically driven inverted slider-crank is of the Grasfof-type, two branches appear, each possibly replaceable by *one* coupler-branch of a Grashof four-bar. For the *Non*-Grashof type only part of the curve may be reproduced. This part then corresponds with either a forward - or, otherwise, a backward stroke of the "input-rocker" between its extreme positions (Fig. **9**).

J. Lenarčič and V. Parenti-Castelli (eds.), Recent Advances in Robot Kinematics, 325-336.
© 1996 *Kluwer Academic Publishers. Printed in the Netherlands.*

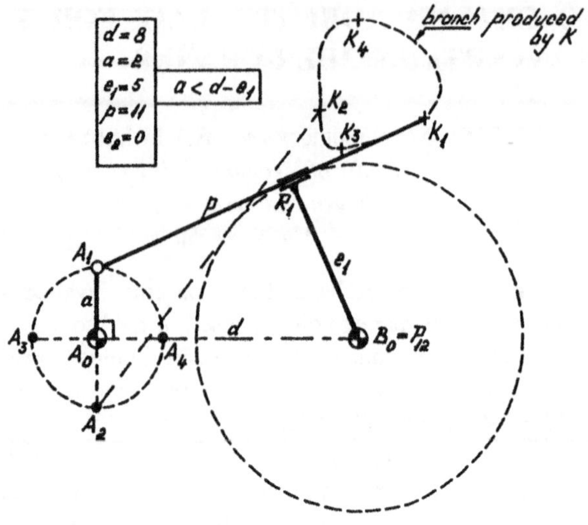

Fig. 1: Initial coupler branch produced by point K attached to the slider of an inverted slider-crank (4 accuracy positions chosen to design a 4-bar branch-cognate with)

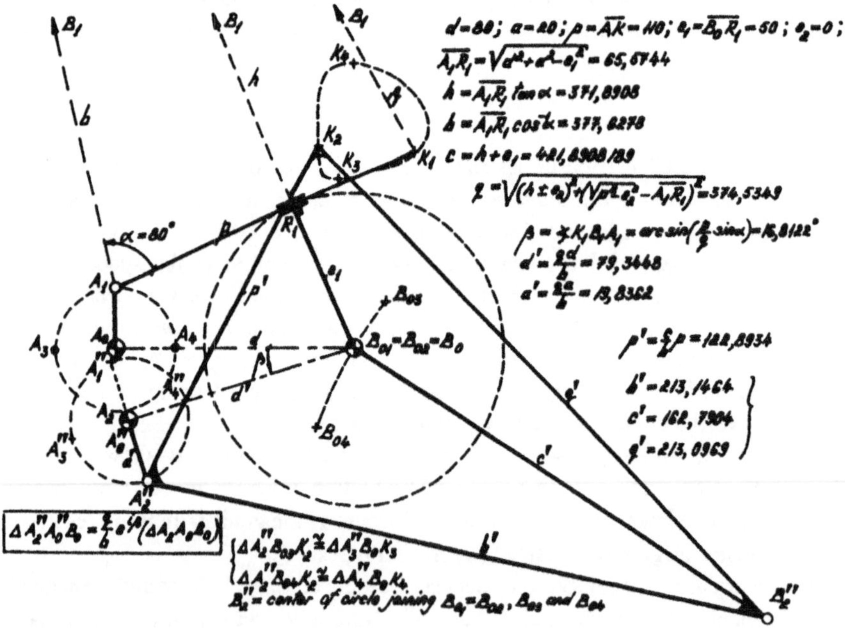

Fig. 2: Design of 4-bar branch cognate (nearly) producing the same coupler branch as the one produced by the eccentric inverted slider-crank (4 position synthesis with position reduction)

2. Investigated Types

It is possible to distinguish between *four* types of eccentric inverted slider cranks. Two of them are of the Grashof type, whereas two others are *Non*-Grashof. For the Grashof-type either $a + e_1 < d$ (Fig. **1**), or $d + e_1 < a$ (Figrs. **3**, **10**).

For the *Non*-Grashof type, the crank-circle about A_0 has to intersect the eccentricity-circle (about B_0) at real points R and $\bar{R}$, leading to the very existence of $\Delta\ A_0RB_0$ meeting the three *Non*-Grashof conditions:

$a < d + e_1$

$d < e_1 + a$

$e_1 < a + d$

Of these, the last one has to be met anyway, otherwise no real mechanism is to be drawn with real tangents to the eccentricity-circle. Thus, as $e_1 < a + d$ represents an almost trivial condition, true also for Grashof linkages, only two cases are left for the *Non*-Grashof type. They are:

case I $\quad d - e_1 < a < d$ (Fig. **9**)
case II $\quad d < a < d + e_1$

Note that for the centric case, for which $e_1 = 0$, the Non-Grashof types disappear. Further, border-cases appear when either $a + e_1 = d$ (Fig. **6**) or $d + e_1 = a$.

Of course, in order to be complete, all occurring cases are of interest when looking for the possible existence of four-bar branch cognates when Grashof linkages are at hand, or "half curve cognates" when *Non*-Grashof inverted slider cranks are presented (Fig. **9**).

3. Design of the 4-bar branch-cognate

The actual design of the four-bar cognate replacing the inverted slider-crank occurs in *two* stages: (Figrs. **2**, **4**, 7, (**9**) and **10**)
In the 1st stage a first approximation is adopted leading to an *auxiliary* four-bar, based on *two* specifically chosen accuracy positions of the coupler-point and on a simultaneous replacement of the slider.
In the 2nd stage, part of Roberts' Configuration is used to find another crank-circle, simultaneously giving the designer the opportunity to improve his first approximation of the branch through two other pairs of specifically chosen accuracy positions.
As a result, only *one* branch of the 4-bar coupler curve approximates the chosen branch, initially produced by the inverted slider-crank. For the other branch one finds another branch-cognate somewhat different from the first.
Accuracy positions, as will be proved, occur in pairs. That is to say, for each accuracy-position another exists leading to the same result. So, basically, the design uses only three accuracy positions, in reality being six.
The first pair is to be allocated for the auxiliary four-bar, whereas the remaining pairs are to be used for the final mechanism found in the 2nd stage.

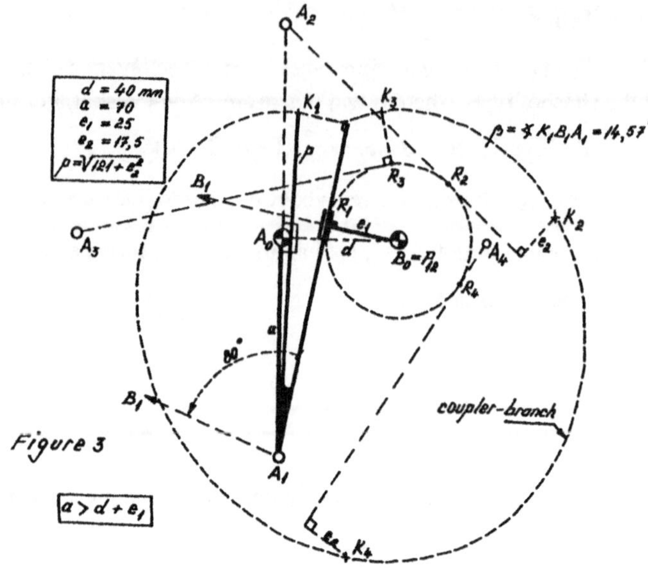

Fig. 3: Original coupler-branch produced by a coupler point K of an inverted slider-crank with eccentricity e_1 (4 accuracy positions are used to obtain the 4-bar branch cognate)

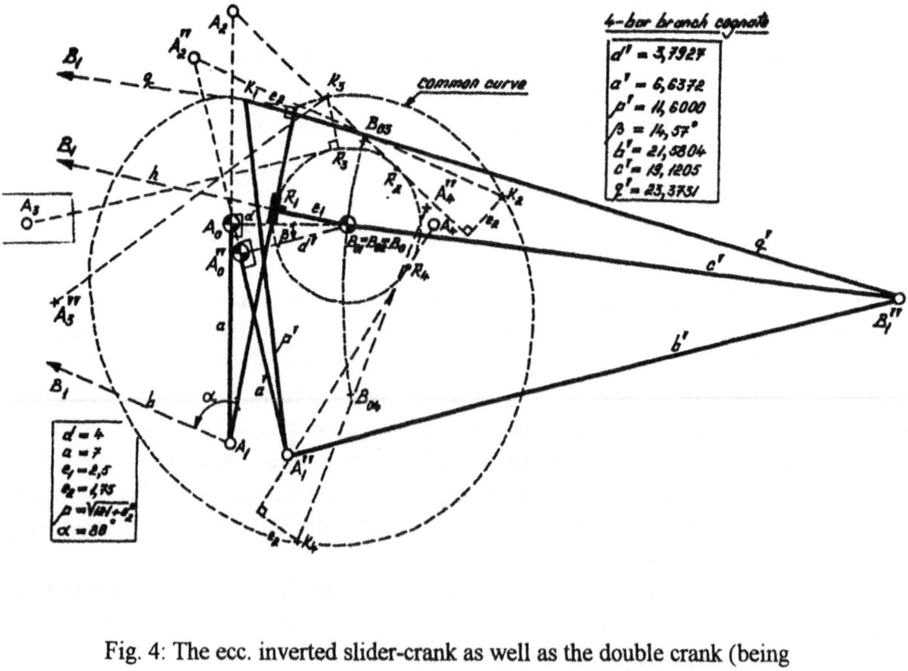

Fig. 4: The ecc. inverted slider-crank as well as the double crank (being the branch cognate) producing the same coupler branch

3.1 THE AUXILIARY FOUR-BAR

In order to simplify matters and to attain a unified method for all cases, the first pair of accuracy-positions are taken to correspond with the perpendicular positions of the input-crank with respect to the frame. (Figrs. **1, 3, 6, 10**). Then, P_{12}, the virtual *rotation center* of the positions, A_1K_1 and A_2K_2, coincides at the center B_0 of the eccentricity-circle. (The mid-normal of B_1"B_2" always joins B_0.)
One further replaces the slider A_1R_1, touching the eccentricity-circle at R_1, by a basically required *large* "circle" with center B_1. Then, the last point, becoming a finite turning-joint, has to comply with the first two accuracy-positions, whence B_1 joins the normal of A_1K_1, simultaneously meeting point $P_{12} = B_0$.

In order to have a large length for $c = \overline{B_1B_0} = h + e_1$ (1)

one chooses the angle $\alpha = \sphericalangle B_1A_1R_1 = 80^0$. (In case $\alpha = 90^0$, one reattains the inverted slider-crank.)
The dimensions of the auxiliary mechanism (A_0-$A_1K_1B_1$-B_0) having the same crank, the same coupler-point, and also the same center - B_0 - as the inverted slider-crank we started from, is then to be established from the initial dimensions: $\overline{A_0B_0} = d$; $\overline{A_1A_0} = a$;

$\overline{B_0R_1} = e_1$; $\overline{A_1K_1} = p$ and e_2, the latter being the shortest distance from K_1 to A_1R_1.

We so find that:

$$\overline{A_1R_1} = \sqrt{d^2 + a^2 - e_1^2} \tag{2}$$

$$\overline{B_1R_1} = h = \overline{A_1R_1} \tan \alpha \tag{3}$$

$$\overline{A_1B_1} = b = \overline{A_1R_1} \cos^{-1} \alpha \tag{4}$$

$$\overline{B_1K_1} = q = \sqrt{(h \pm e_2)^2 + (\sqrt{p^2 - e_2^2} - \overline{A_1R_1})^2} \tag{5}$$

$$\sphericalangle K_1B_1A_1 = \beta = \arcsin[\frac{p}{q} \sin(\alpha - \arcsin\frac{e_2}{p})] \tag{6}$$

Clearly, the design of the auxiliary mechanism is the same for all Grashof-types having a fully revolving crank. Thus,

for $a \le d - e_1$ with $d > e_1$ (Figrs. **1, 6**)
for $a \ge d + e_1$ with $d > e_1$ (Fig. **3**)
and for $a \ge d + e_1$ with $d < e_1$ (Fig. **10**)

Of course, as $\alpha < 90^0$, the curve produced by the auxiliary mechanism will deviate a bit from the initial curve as produced by the eccentric inverted slider-crank. However, the deviation is to be corrected by application of the 2 nd stage leading to the final replacement mechanism.

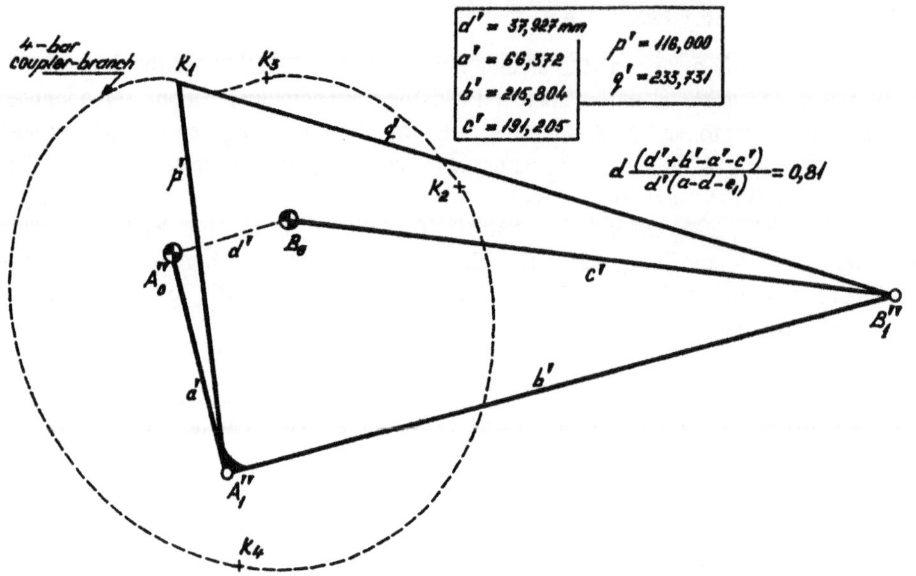

Fig. 5: The double crank as a four-bar branch cognate of an eccentrically driven inverted slider-crank (the 2 mechanisms produce about the same branch with nearly the same input velocity)

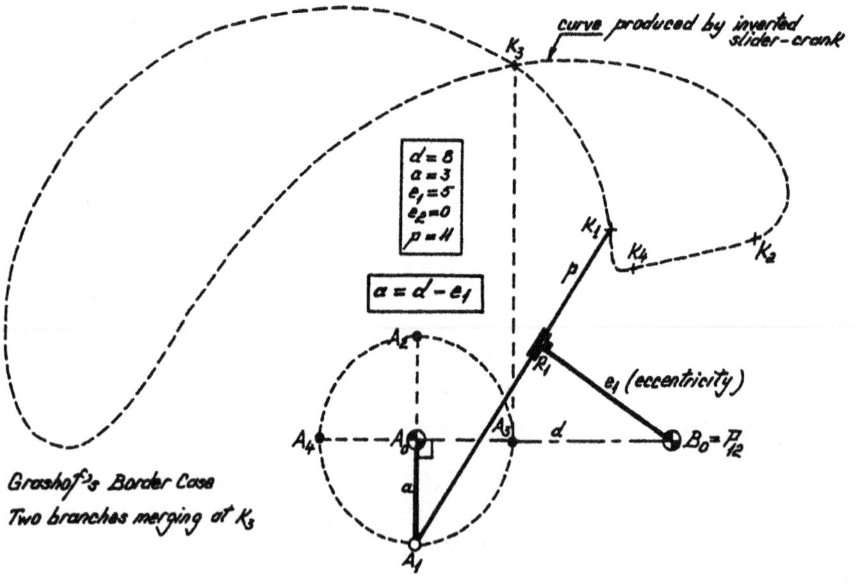

Fig. 6: Eccentrically driven, inverted slider crank (4 accuracy positions to be used for the design of a 4-bar branch cognate)

3.2 FINAL FOUR-BAR BRANCH COGNATE

Naturally, further enlargement of the angle α leads to lower values for the transmission angle $\mu_1 = \sphericalangle B_0B_1A_1$. (For instance, if $\alpha = 90^0$, then $\mu_1 = 0^0$)
As this is undesired with regard to force-transmission, other ways have to be found. Whence, we abide by $\alpha = 80^0$, but look at Roberts' Law and ditto Configuration instead. Though Roberts' Law may then be applied caused by the now finite location of point B_1, the transformation does not change the curve itself.
However, if we consider a particular curve-cognate of Roberts', namely the one having a crank rotating with the same speed as the angular velocity of the initial crank, it becomes feasible to change *three* of its cognate-dimensions, in order to meet *three* different accuracy positions. To carry this out in detail, the following procedure is adopted:

1. First, change the auxiliary four-bar into its Roberts'curve-cognate having a crank $A_0''A''$ rotating with the same speed as the initial crank A_0A,
2. Secondly, replace turning-joint $(B_1'')_{cognate}$ or $(B_2'')_{cognate}$ until B'' meets the three (or six) accuracy-positions being chosen.

In more detail, this would lead to the *stretch-rotation* of $\Delta\, A_1A_0B_0$ about B_0 with the complex multiplication-factor $\frac{q}{b}e^{i\beta}$, yielding

$$\Delta\, A_1''A_0''B_0 = \frac{q}{b}\, e^{i\beta}(\Delta\, A_1A_0B_0)\ . \tag{7}$$

For the 2nd accuracy-position, a similar equation holds:

$$\Delta\, A_2''A_0''B_0 = \frac{q}{b}e^{i\beta}(\Delta\, A_2A_0B_0) \tag{8}$$

Hence, we so obtain the simple formulas:

$$\overline{A_0''B_0} = d' = \frac{q}{b}d \tag{9}$$

$$\overline{A_1''A_0''} = a' = \frac{q}{b}a \tag{10}$$

Further, $\sphericalangle\, B_0A_0''A_1'' = 90^0 = -\sphericalangle\, B_0A_0''A_2''$ (11)

Whereas, similarly according to Roberts' Configuration, also

$$\overline{A_2''K_2} = \overline{A_1''K_1} = p' = \frac{c}{b}p \tag{12}$$

Naturally, the perpendicular bisectors of $\overline{A_1''A_2''}$ and of $\overline{K_1K_2}$ intersect at B_0. Thus,

$$\Delta\, A_2''B_0K_2 \cong \Delta\, A_1''B_{02}K_1 \tag{13}$$

yielding $B_{01} = B_0 = B_{02}$. (14)

Then, with two other chosen accuracy positions from the initially given mechanism, say A_3K_3 and A_4K_4, it becomes quite possible to determine accurate corresponding positions,

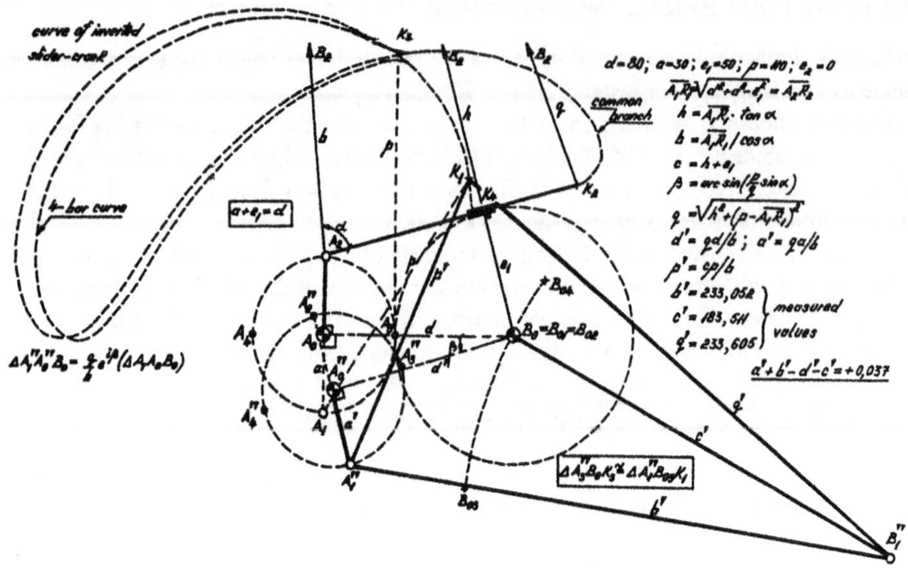

Fig. 7: Design of four-bar Branch-Cognate of inverted slider-crank
(Grashof's Border case $a + e_1 = d$ leading to $a' + b' > d' + c'$)

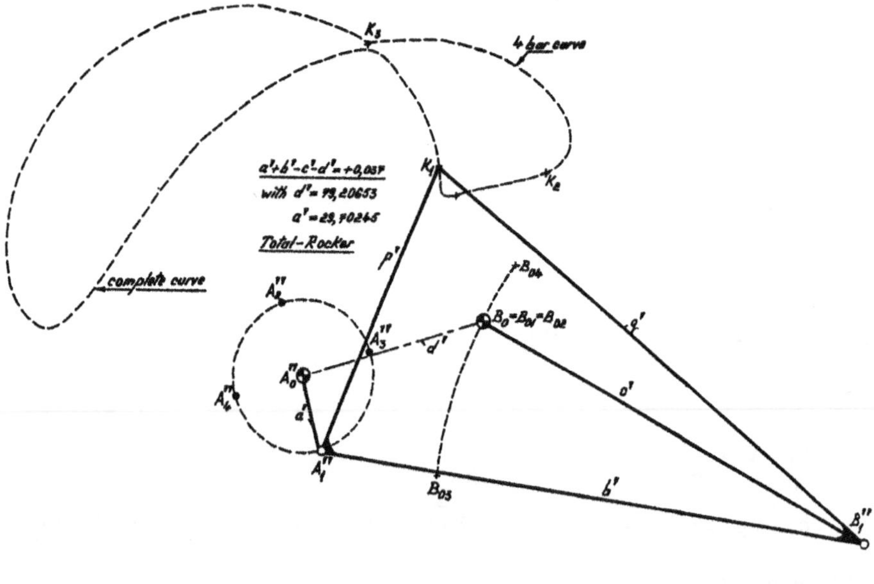

Fig. 8: Four-bar Branch Cognate of inverted slider-crank

A_3'' and A_4'' , at the cognated crank-circle about A_0'' . This may be realized by intersection of circles having radius p' respectively about K_3 and K_4, with the cognated crank-circle about A_0''

When, for instance, A_3 as well as A_4 join the initial fixed link A_0B_0 , we find that both, A_3'' as well as A_4'' , slightly deviates from the line containing the cognated fixed link, that is to say from $A_0''B_0$. Anyway, having determined their exact locations at the cognated crank-circle, we may establish the locations of the points, B_{03} and B_{04} . The way to find these points, will be based on the so-called *attachment-method*, hence, on the equations:

$$\Delta\ A_3''B_0K_3 \equiv \Delta\ A_1''B_{03}K_1 \tag{15}$$

$$\Delta\ A_4''B_0K_4 \equiv \Delta\ A_1''B_{04}K_1 \tag{16}$$

The circle joining the *three* points $B_{01} = B_{02}$, B_{03} and B_{04} then has the required turning-joint B_1'' for its center.

As a result, one obtains the new dimensions:

$$\overline{A_1''B_1''} = b' \quad , \quad \overline{B_0B_1''} = c' \quad \text{and} \quad \overline{K_1B_1''} = q' \tag{17}$$

In the particular case for which the initial branch is *symmetrical*, which occurs when the two eccentricities are zero, B_1'' coincides at the center of a circle joining K_1 , B_0 and A_1'' . In that case $b' = c' = q'$, meaning that the four-bar is of the λ-type (ref.[4]).

The accuracy-positions chosen, corresponded with the initial crank-positions dividing 360^0 into equal parts of 90^0.

A *measure* for the accuracy of the cognate curve may be found by comparing the *Grashof-distances* of the two mechanisms. Then, the best approximation will be the one having its *Grashof-distance-ratio* much nearer the value 1 than with other approximations. For the mechanisms of Fig. **2**, for instance, the auxiliary four-bar yields the Grashof-distance-ratio $\dfrac{a+c-d-b}{a-d+e_1} = 1.57$, whereas the final branch-cognate results into the ratio $\dfrac{d(a'+b'-c'-d')}{d'(a-d+e_1)} = 0.924$. Thus, the latter represents the better mechanism. (Note that the Grashof-distance-ratio equals 1 when two Roberts' 4-bar curve cognates are compared.)

For the auxiliary mechanism of Fig. **4**, we obtain the Grashof-distance-ratio

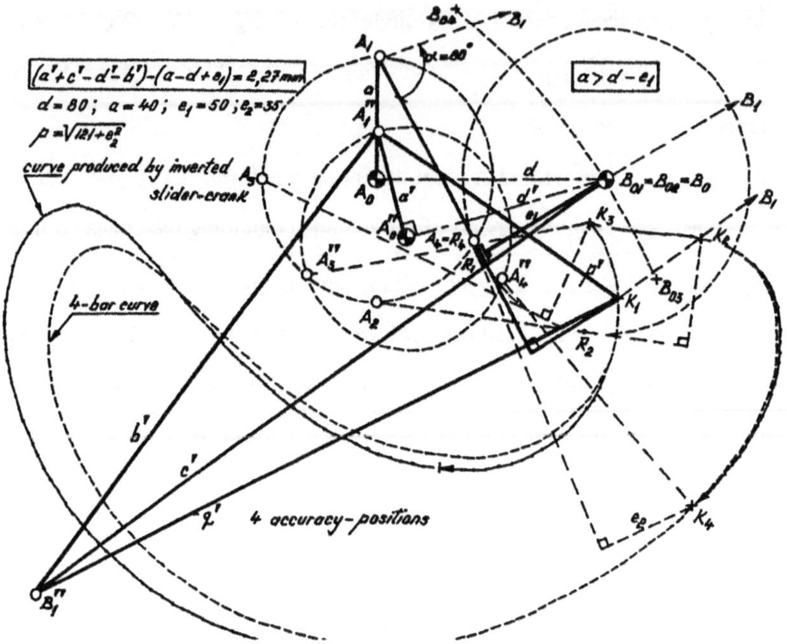

Fig. 9: Non-Grashof Curve-Cognate

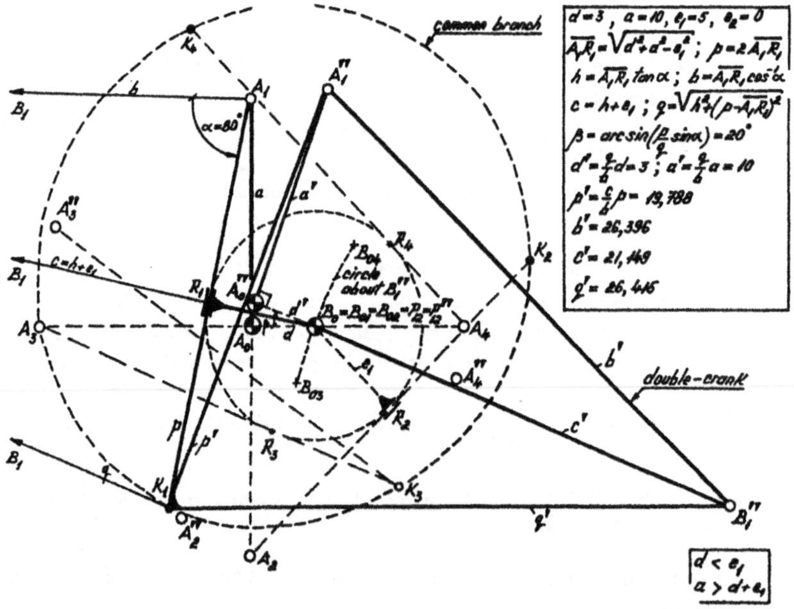

Fig. 10: Iverted slider crank and its 4-bar branch-cognate producing the same branch (4 position-synthesis using a point position reduction method: $B_{01} = B_{02}$)

$\frac{d+c-a-b}{a-d-e_1} = 2.34$, whereas the branch-cognate gives rise to the ratio

$\frac{d(d'+b'-a'-c')}{d'(a-d-e_1)} = 0.81.$

Even for Non-Grashof linkages, namely for those without revolving bars, the method remains applicable. Fig. 9, for instance, yields for the auxiliary mechanism the Grashof-distance-ratio $\frac{a+b-c-d}{a-d+e_1} = 1.65$, whereas the final curve-cognate gives the ratio

$\frac{d(a'+c'-d'-b')}{d'(a-d+e_1)} = 1.35.$

It is quite possible that an other distribution of our accuracy-positions, gives a better Grashof-ratio. Remarkable better results though, are not to be expected. For instance, if we choose point A_4 of Fig. 9 at the other intersection of the crank - and the eccentricity-circle, we obtain the ratio $\frac{d(a'+c'-d'-b')}{d'(a-d+e_1)} = 0.73.$ Now, its distance to 1 is only slightly less than with the 1.35-value belonging to the mechanism demonstrated in Fig. 9. In Grashof's border case, such as the particular one demonstrated in the Figrs. 6 and 7, we rather observe the *difference* between these Grashof-distances.
Then, for the auxiliary mechanism we precisely observe the value

$$(a+c-b-d)-(a+e_1-d) = \sqrt{d^2+a^2-e_1^2}\,.\,(\sin\alpha-1)\cos^{-1}\alpha =$$

$$= -40\sqrt{3}\,\tan\tfrac{1}{2}[(\pi/2)-\alpha] = -40\sqrt{3}\,\tan 5^0 = -6.0614\ .$$

As in this case each *Grashof-distance* should tend to zero, a considerable improvement is obtained with the final branch-cognate as demonstrated in Fig. 8. Then, namely

$$\frac{d}{d'}\,.\,(a'+b'-c'-d')-(a+e_1-d) = +0.037\ .$$

A random *circle* about B_0 intersecting the cognated crank-circle at possible accuracy-positions A''_m and A''_n , yields a virtual rotation-centre $P_{mn} = B_0$, giving $B_{0m} = B_{0n}$. Indeed, accuracy-positions only occur in pairs. Thus, each accuracy-position has to be counted twice, unless of course the cognated crank-positions already are at a circle about B_0, as will be the case for the accuracy-positions, A''_1 and A''_2 .
Generally, the transmission angle $\mu_1 = \sphericalangle A''_1B''_1B_0$ appears to be *about twice* as large as the one obtained with the auxiliary four-bar or with its Roberts' curve-cognate. Complete application of Roberts' Law *would* have lead to a point $(B''_1)_{cognate}$ as the fourth vertex of a linkage parallelogram $B_0B_1K_1(B''_1)_{cognate}$. However, such a point $(B''_1)_{cognate}$ would have been much farther away, giving about half the transmission-angle. All examples demonstrated show this same phenomenon. We conclude, that the 2 nd stage leads to a better approximation of the generated branch, simultaneously giving about twice the transmission angle in comparision to the one obtained in the 1 st stage.

Theoretically, the *circular* curve described by point B_0 of the Stephenson-2 six-bar (A_0 - A - B^* - B_0 - A_0'' - A_1'' - K_1) with respect to the link $A_1''K_1$ will be a Stephenson-2 six-bar curve of order 16, (ref.[5]). Clearly, this curve approximates a circle very neatly, although in reality not more than 16 intersections exist. (We count 2 times 16, minus the (8 times 2) intersections of the circle at the two circular or isotropic points of the curve.)

4. Conclusions

Branches of curves produced by inverted slider-cranks having an eccentricity and/or an eccentrically located coupler point, are to be reproduced by even *three* different four-bar branch-cognates. In the two cases the curves are singular branched, the three four-bar curve-cognates reproduce only part of the curve, namely the part corresponding to either a forward - or, otherwise a backward stroke of the input-rocker. Better transmission angles may be attained at the cost of the accuracy of the reproduction. In most cases though a high accuracy is obtained with acceptable transmission angles. (The accuracy of the replacement-method has been measured with a newly introduced *Grashof-distance-ratio.*)

References

1. Roberts, S. (1875) Three-bar motion in plane space, *Proc.London Math.Soc.* 7, pp.14-23
2. Stoimenov, M., Panteli ć, T. (1979) Possibility of the development of a curved translation and for a member of an extended kinematic chain of a curved sliding mechanism, *Proc. of the 5th World Congress on the Theory of Machines and Mechanisms,* pp 1436-1439, published by the ASME.
3. Dijksman, E.A. (1976) Motion Geometry of Mechanisms, (Chapter 5: Cognate Linkages), *Cambridge University Press.*
   (1981) Cinemática De Mecanismos. (Cap.5: Mecanismos Cognados) *Editorial LIMUSA, S.A.*, Mexico
4. Dijksman, E.A. (1996) How to exchange centric inverted slider cranks with λ-formed four-bar linkages, *Mechanism and Machine Theory* (in press)
5. Primrose, E.J.F., Freudenstein, F., Roth, B. (1967) Six-Bar Motion II The Stephenson-1 and Stephenson-2 Mechanisms, *Archive for Rational Mechanics and Analysis,* Volume **24**,Nr.1, pp.55-72.

7. Performance of parallel mechanisms

ON SELF-MOTIONS OF A CLASS OF PARALLEL MANIPULATORS

A. KARGER
Charles University Prague, Mathematical Institute,
Sokolovská 83 18600 Praha 8, Czech Republic,
email: karger@karlin.mff.cuni.cz

AND

M. HUSTY
Montan University Leoben, Inst. of Applied Geometry,
Fr. Josefstr.18, A-8700 Leoben, Austria,
email: husty@unileoben.ac.at

Abstract: In this paper we shall introduce an algorithm to determine self-motions of a special class of platform mechanisms. This class of mechanisms has a geometry that generalizes the geometry of existing flight simulator platforms. The algorithm uses the Study representation of the Lie group of space congruences and the quadratic constaint equation which decribes the condition that a point of the moving system is bound to move on a sphere.

1. Introduction

Within the last years a lot of research has been done on platform mechanisms. They have been introduced by Gough and were later on used by Stewart (1965) as early models of flight simulators. Now this type of kinematic structure seems to be widely accepted for the design of commercially used flight simulators. The geometry of this device can be described as follows: a base-platform is connected to a mobile (upper) platform with six legs. The six legs are driven by linear actuators and connected to the platform by spherical joints (Fig.1). Most of the research on this platform mechanisms was devoted to the kinematic and dynamic analysis (direct analysis, workspace) as well as the development of control strategies. Special types of platform mechanisms have been investigated, and their direct analysis was presented in many papers (for an overview see Merlet (1990)). The direct analysis of the most general type of platform has been solved in

J. Lenarčič and V. Parenti-Castelli (eds.), Recent Advances in Robot Kinematics, 339-348.
© 1996 *Kluwer Academic Publishers. Printed in the Netherlands.*

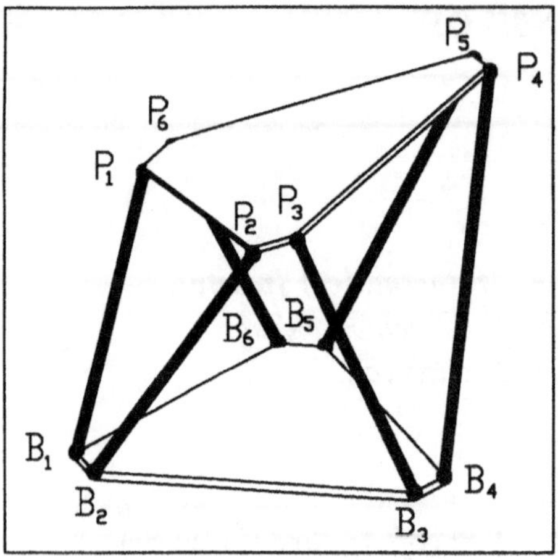

Figure 1. Platform Mechanism

Husty (1996), where a special algorithm has been developed, which also will be used in this paper, that deals with the singularity analysis of a special class of such structures. Architecture dependent singularities have been presented in Ma and Angeles (1992), where it was shown that special architectures lead to rank deficient Jacobians within regions of the workspace or even within the whole workspace. This means that the platform gains at each point of its workspace an infinitesimal degree of freedom. Merlet (1992) has pointed out that singularity analysis of platform mechanisms is closely related to linear dependencies of lines (legs) and special line structures like complices, congruences, ruled surfaces which are well known from line geometry (Grassmann geometry in Merlet's paper). Using old results from Borel and Bricard it was shown in Husty and Zsombor-Murray (1994) that it is possible to construct platform mechanisms which have continuous singularities and therefore move even when all actuators are locked. The class of manipulators investigated in this paper derives its geometry from existing flight simulators. As shown in Fig. 1 platform and base are planar and the joints on both of them are on hexagons which are obtained from equilateral triangles that are cut with circles centered at the centroid of each triangle (Figs. 2 and 3). Both

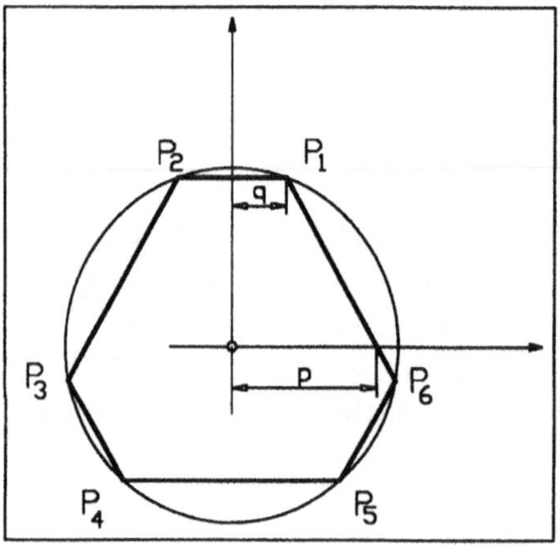

Figure 2. Platform Coordinate System

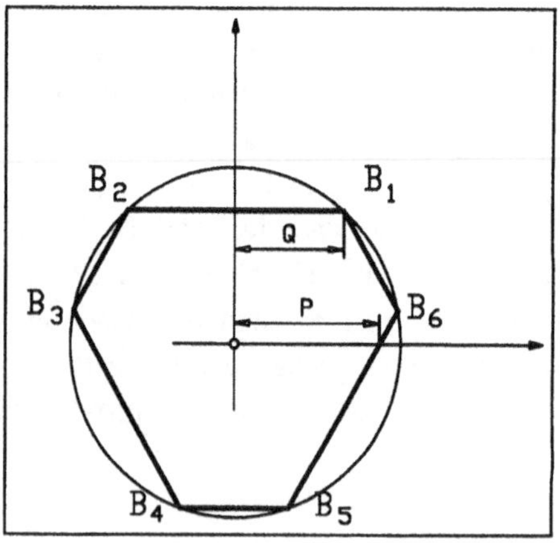

Figure 3. Base Coordinate System

platform and base are completely described by the two homogenous design parameters p, q and P, Q. Using the design parameters the coordinates of the joint centers of platform and base are found as

$$(1)\quad \begin{aligned} P_1 &= [q, \sqrt{3}(4p/3 - q)], & P_2 &= [-q, \sqrt{3}(4p/3 - q)], \\ P_3 &= [-2p + q, \sqrt{3}(q - 2p/3)], & P_4 &= [2(q - p), -2p\sqrt{3}/3], \\ P_5 &= [2(p - q), -2p\sqrt{3}/3], & P_6 &= [2p - q, \sqrt{3}(q - 2p/3)], \end{aligned}$$

$$(2)\quad \begin{aligned} B_1 &= [2(P - Q), 2P\sqrt{3}/3], & B_2 &= [2(Q - P), 2P\sqrt{3}/3], \\ B_3 &= [-2P + Q, \sqrt{3}(2P/3 - Q)], & B_4 &= [-Q, \sqrt{3}(Q - 4P/3)], \\ B_5 &= [Q, \sqrt{3}(Q - 4P/3)], & B_6 &= [2P - Q, \sqrt{3}(2P/3 - Q)]. \end{aligned}$$

It should be noted that by special choice of the design parameters the platform or the base or both of them can be equilateral triangles.
In this paper part of the classification of all self-motions of the above described class of platform manipulators will be shown. This classification is based on Study representation of the euclidean congruences and on the equation which represents the constraint that one point of the moving system (platform) is bound to move on a sphere. At first a basic set of six nonlinear equations will be derived. Then we will show how to find from this set of equations special architectures and configurations of platform mechanisms which allow self-motions.

2. Basic Equations

Let $\mathbf{A}$ be the matrix representation of a space congruence

$$(3)\qquad \mathbf{A} = \begin{pmatrix} 1 & 0 & 0 & 0 \\ a_1 & a_{11} & a_{12} & a_{13} \\ a_2 & a_{21} & a_{22} & a_{23} \\ a_3 & a_{31} & a_{32} & a_{33} \end{pmatrix}$$

where $\mathbf{a} = (a_1, a_2, a_3)^T$ represents the translation vector and a_{ij} are the components of the rotation matrix. We shall understand the matrix $\mathbf{A}$ as the matrix of the correspondence between the coordinates in the moving platform and the fixed base. According to this correspondence a point $\mathbf{p} = (1, a, b, c)^T$ in the platform system will have coordinates $\mathbf{P} = (1, X, Y, Z)^T$,

$$(4)\qquad \mathbf{P} = \mathbf{A}.\mathbf{p}$$

Let $\mathbf{W} = (1, A, B, C)^T$ be a general point in the base (fixed system). The condition that the point $\mathbf{p}$ is on the sphere with center at $\mathbf{W}$ and radius r is expressed by the equation

$$(5)\qquad (X - A)^2 + (Y - B)^2 + (Z - C)^2 - r^2 = 0.$$

Expanding eq.(5) yields

$$X^2+Y^2+Z^2-2(XA+YB+ZC)+R_0=0 \tag{6}$$

where $R_0 = A^2 + B^2 + C^2 - r^2$. Furtheron the transformation matrix **A** will be parametrized by Study parameters in the following form

(7)

$$\mathbf{A} = \begin{pmatrix} t_0 & 0 & 0 & 0 \\ t_1 & x_0^2+x_1^2-x_2^2-x_3^2 & 2(x_1x_2-x_0x_3) & 2(x_1x_3+x_0x_2) \\ t_2 & 2(x_1x_2+x_0x_3) & x_0^2-x_1^2+x_2^2-x_3^2 & 2(x_2x_3-x_0x_1) \\ t_3 & 2(x_1x_3-x_0x_2) & 2(x_2x_3+x_0x_1) & x_0^2-x_1^2-x_2^2+x_3^2 \end{pmatrix}$$

where

$$t_0 = x_0^2+x_1^2+x_2^2+x_3^2, \qquad t_1 = 2y_0x1 - 2y_1x_0 - 2y_3x_2 + 2y_2x_3$$
$$t_2 = 2y_0x2 - 2y_2x_0 - 2y_1x_3 + 2y_3x_1, \; t_3 = 2y_0x3 - 2y_3x_0 - 2y_2x_1 + 2y_1x_2.$$

Substitution of eqn.(7) into eq.(6) yields after some manipulations and rearrangements (see Husty (1996)) the required constraint equation. This equation will not be displayed here because we will take into account that in the discussed cases platform and base are planar, which simplifies the equation considerably. Let $X = [A, B, 0]$ be a general point connected with the base, $x = [a, b, 0]$ be a general point connected with the upper platform and their distance be r. Then we obtain the following constraint equation:

$$\begin{aligned} h = {} & R_0(x_3^2+x_0^2+x_1^2+x_2^2) + 4Ay_1x_0 + 4Ay_3x_2 + 4By_2x_0 - 4Ay_2x_3 \\ & - 4By_0x_2 - 4Ay_0x_1 + 4By_1x_3 - 4y_3bx_1 - 4y_2x_3a + 4y_0x_1a + 4y_1x_3b \\ & - 4y_2x_0b - 4y_1x_0a + 4x_2y_0b + 4x_2y_3a - 4Abx_1x_2 + 4Abx_0x_3 \\ & - 4Bax_1x_2 - 4Bax_0x_3 - 4By_3x_1 + 2Aa(-x_1^2+x_2^2+x_3^2-x_0^2) \\ & + 2Bb(-x_0^2+x_1^2-x_2^2+x_3^2) + 4(y_0^2+y_1^2+y_2^2+y_3^2) = 0. \end{aligned} \tag{8}$$

In eq.(8) $x_0, x_1, x_2, x_3, y_0, y_1, y_2, y_3$ are the Study parameters. They have to satisfy the fundamental relation

$$U_7 = x_0y_0 + x_1y_1 + x_2y_2 + x_3y_3 = 0, \tag{9}$$

the equation of the *Study quadric*. It should be noted that $x_0, x_1, x_2, x_3, y_0, y_1, y_2, y_3$ are homogeneous coordinates of points in a seven-dimensional projective space in which we represent the space congruences on the Study quadric (kinematic mapping, see Ravani and Roth (1984)). As a point in a projective space can be represented by any nonzero multiple of its coordinates, we leave the possible normalizing condition $x_0^2+x_1^2+x_2^2+x_3^2=1$

out, because we will use more suitable normalizing conditions for representing the motion and hence normalize the point conveniently during the discussion of different cases. We substitute the points A_i, B_i with R_i into eq.(8) and we obtain six equations $h_1, ..., h_6$, which are not displayed here, because they are relatively long. But we show some of their linear combinations, which are much simpler and which give more information. The following denotations are introduced:

$$R_1 - R_2 + R_3 - R_4 + R_5 - R_6 = 2S_1, \quad R_3 - R_4 = S_3, \quad R_5 - R_6 = S_4,$$
$$R_1 + R_2 + R_3 + R_4 + R_5 + R_6 = 2S_2, \quad R_4 - R_6 = S_5,$$
$$P - Q = W, \quad p - q = w, \quad qQ - 4Ww = X, \quad QqWw \neq 0$$
$$U_1 = h_1 - h_2 + h_3 - h_4 + h_5 - h_6, \quad U_2 = h_1 + h_2 + h_3 + h_4 + h_5 + h_6,$$
$$U_3 = h_1 - h_5, \quad U_4 = h_1 - h_2, \quad U_5 = h_3 - h_4, \; U_6 = h_1 - h_3.$$

Then we have the equations:

$$U_1 = -8\sqrt{3}x_0x_3X + S_1(x_0^2 + x_1^2 + x_2^2 + x_3^2) = 0,$$
$$(10) \qquad U_2 = 12Q(y_0^2 + y_1^2 + y_2^2 + y_3^2) + QS_2(x_0^2 + x_1^2 + x_2^2 + x_3^2) +$$
$$4(XQ + 8QWw + 4Q^2w + 4WX + 16W^2w)(x_3^2 - x_0^2).$$

The remaining equations $U_3, ..., U_7$ are linear in y_0, y_1, y_2, y_3 and can be used to compute these unknowns. Let us denote a_{ij} the coefficient in U_i at y_j,where $i = 3, ..., 7$, $j = 0, ..., 3$, and $\mathbf{B}$ be the matrix of a_{ij}, the coefficients of linear unknowns in $U_3, ..., U_7$. Now we compute the determinants of 4×4 - submatrices of $\mathbf{B}$ which will be denoted by:

$$M_3 = \det(U_4, U_5, U_6, U_7), \; M_4 = \det(U_3, U_5, U_6, U_7),$$
$$M_5 = \det(U_3, U_4, U_6, U_7), \; M_6 = \det(U_3, U_4, U_5, U_7),$$

Furtheron we set: $H_1 = M_3 + M_6$, $H_2 = M_3 - M_6$. Now the determinants are computed and we get:

$$(11) \quad H_1 = XQ^2[(x_3x_2 - x_0x_1)(X + 4Ww) + 2WQ(x_3x_2 + x_0x_1),$$
$$H_2 = XQ^2\sqrt{3}[2WQ(x_1x_3 - x_0x_2) + (x_0x_2 + x_1x_3)(X + 4Ww)],$$
$$(12) \qquad M_3 - M_5 = XQ^3\sqrt{3}[x_1x_3(2w + Q) + x_0x_2(2w - Q)]$$
$$M_4 = \sqrt{3}Q(2w - Q)x_0x_1 + (2X - Q^2 + 2wQ - 4QW + 8Ww)x_0x_2,$$
$$+(2X + Q^2 + 2wQ + 4QW + 8Ww)x_1x_3 - \sqrt{3}Q(Q + 2w)x_2x_3.$$

Only if the rank of $\mathbf{B}$ is four we can compute the linear unknowns from $U_3, ..., U_7$, if it is less (which means that there is a dependency among them) we need additional equations to solve for them. We show at first the

Lemma: The rank of **B** is less then 4 iff $X = 0$ or $x_0 = x_3 = 0$ or $x_1 = x_2 = 0$.

Proof. Let the rank of **B** be less then 4. This means that $M_3 = M_4 = M_5 = M_6 = 0$. Let $X \neq 0$. We see from eqns.(11) that $H_1 = 0, H_2 = 0$ is a system of two linear equations with a non-trivial solution. This means that the determinant of this system must be equal zero and we get $x_0 x_3 (x_1^2 + x_2^2) = 0$. Let $x_0 = 0, x_3 \neq 0, x_1^2 + x_2^2 \neq 0$. Then eqns.(11) yield $X = -2WQ - 4Ww$, and eqns.(12) yield $Q = -2w$ and $X = 0$, which is a contradiction. The case $x_3 = 0$ is similar.

3. Discussion of Some Cases

Due to the limitations of space we cannot present the full classification of all self-motions of the class of manipulators in discussion. We can only give an idea of the procedure and some examples. The main idea of the following is to search for design conditions or position conditions which cause one (or even more) of the equations $U_1, ...U_7$ to vanish. Then the remaining equations describe a curve (or a surface, threedimensional manifold) in the seven-dimensional image space, which corresponds to a one (or higher) degree of freedom self-motion of the platform. Examining U_1 in eq. (10) yields for example: U_1 will vanish if the two conditions $X = 0$ and $S_1 = 0$ hold simultaneously. $X = 0$ is a condition on the design of the platform mechanism and it means that the upper platform is similar to the base. $S_1 = 0$ is a (rather weak) positioning condition and means that one of the leg lengths cannot be chosen arbitrarily. Hence if a platform fullfills the two conditions above it will go into a self-motion. This result extends the results obtained in Ma and Angeles (1992) where it was shown that the Jacobian of the manipulator vanishes in a significant subregion of the workspace if platform and base are similar. Now we go on more systematically. In the following we will consider only the case $S_1 = 0$, that means we have a configuration in which $R_1 - R_2 + R_3 - R_4 + R_5 - R_6 = 0$.

3.1. THE CASE $X = 0$ (AND $S_1 = 0$!).

As discussed above for all platforms within this case we have $U_1 = 0$ and self-motions in all positions for which $S_1 = 0$. But from the Lemma we know that the rank of **B** is at most 3. So what still remains is to discuss several subcases according to the rank of **B**.

3.1.1. *Rank of* **B** *is equal to one*

A discussion of the subdeterminants shows that there are only two possibilities:

1. $x_1 = x_2 = x_3 = y_0 = S_3 = S_4 = S_5 = 0,\ x_0 = 1,\ S_2 = 3R_6,\ Q = 2w.$

We get one more design condition ($Q = 2w$), that is both platform and base are congruent, and conditions on the legs: they have to be of equal length. The only remaining equation is

$$(13) \qquad U_2 = 12(y_0^2 + y_1^2 + y_2^2) + 3R_6 + 32(wW - w^2 - W^2) = 0.$$

2. $x_0 = x_1 = x_2 = y_3 = S_3 = S_4 = S_5 = 0,\ x_3 = 1,\ S_2 = 3R_6,\ Q = -2w$. From this set of conditions we get the same design and configuration conditions as in the case above. The remaining equation is

$$(14) \qquad U_2 = 12(y_1^2 + y_2^2 + y_3^2) + 3R_6 - 32(wW + w^2 + W^2) = 0.$$

In both equations (13 and 14) we set $y_1 = t$ and $y_2 = s$ and we can compute $y_0(t, s)$ or $y_3(t, s)$. This means we have obtained two-parametric translatory motions in both cases. The difference between the two cases is that the first contains the identity of the transformation group and the second does not contain the identity.

3.1.2. *Rank of* **B** *is equal to two*

Subdeterminants of order 3 for equations $U_4 = U_5 = U_7 = 0$ yield $x_1 = x_2 = 0$, subdeterminants of order 3 for equations $U_3 = U_6 = U_7 = 0$ yield $x_0 = Q + 2w = 0$, or $x_3 = Q - 2w = 0$. We obtain the previous case, for which the rank of **B** is equal to one.

3.1.3. *Rank of* **B** *is equal to three*

The rank deficiency of **B** by one means that we can express three of the linear unknowns from $U_3, ..., U_7$. For instance we express y_0, y_1, y_2 from $U_4 = U_5 = U_7 = 0$ and y_3 from U_2 (note that y_3 is not linear in U_2). Two quadratical equations for the unknowns x_0, x_1, x_2, x_3 remain. Those equations Q_1 and Q_2 are obtained in their simpliest form by adding and subtacting U_3 and U_6: $Q_1 = U_3 - U_6,\ \ Q_2 = U_3 + U_6$.

$$(15) \quad Q_1 = -64\sqrt{3}x_1x_2Ww(Q + 2W) + (x_0^2 + x_1^2 + x_2^2 + x_3^2)(2S_5W$$
$$-QS_4 - QS_3 - 2WS_4) = 0, Q_2 = (QS_4 - QS_3 - 12R_6W - 2WS_4 + 4S_2W$$
$$-4WS_3 - 6S_5W)(x_0^2 + x_1^2 + x_2^2 + x_3^2) + 96Ww(Q + 2W)(x_1^2 - x_2^2) = 0$$

In this case we set $x_0^2 + x_1^2 + x_2^2 + x_3^2 = 1$ and it is easy to verify that Q_1 and Q_2 are two hyperbolas (each of them can split into two lines in special cases) in the x_1x_2-parameter-plane which always have two real and two complex intersections. We set $x_3 = t$, compute the intersections and now the self-motion is determined. This rather general case is illustrated with an example.

Example 1: The platform is given by the design constants $p = 5, q = 4$ and the base is given by $P = 2, Q = 1$. It is easy to verify that the design condition $X = 0$ is fullfilled, which is significant for the case. Not to get too complicated formulas we set $R_1 = R_2 = 2, R_3 = R_4 = R_5 = R_6 = 0$. Substituting the design parameters into eq.(8) we obtain h_i from which U_i are computed. Then we verify that all M_i vanish, and compute from U_4, U_5 and U_7 the Study parameters y_0, y_1, y_2. Q_1 and Q_2 have already a very simple form:

$$(16) \qquad Q_1: \ -192x_2\sqrt{3}x_1 = 0, \quad Q_2: \ 36(x_1^2 - x_2^2) + 1 = 0.$$

Computing the intersection we get two real solutions $l_1 : x_1 = 0, x_2 = \frac{1}{6}$ and $l_2 : x_1 = 0, x_2 = \frac{-1}{6}$. We take the first solution and set (taking into account $x_0^2 + x_1^2 + x_2^2 + x_3^2 = 1$):

$$x_0 = \tfrac{1}{6}\sqrt{35}\cos(t), \quad x_3 = \tfrac{1}{6}\sqrt{35}\sin(t).$$

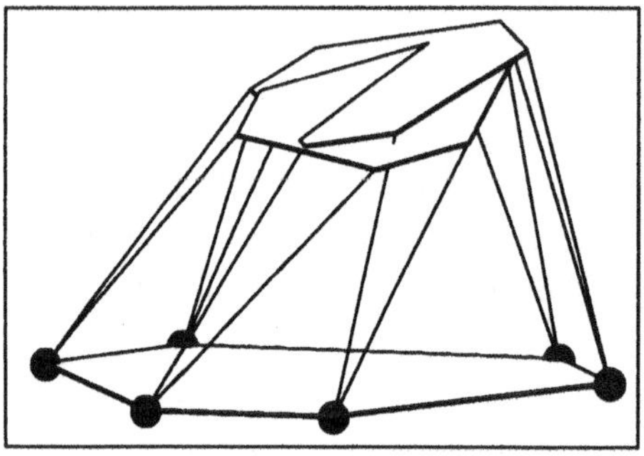

Figure 4. Two Positions of Example 1

Now the motion is determined and back substitution into eqn.(7) yields the transformation matrix **A**. Fig.4 shows the platform mechanism at positions $t = 0$ and $t = \frac{\pi}{6}$. It should be mentioned that there exists also a degenerate case, which is obtained from eqns.(15) by imposing the design constraints $Q = -2W$. Then both equations (15) vanish and we have obtained a three-parametric self-motion. But a short inspection of the geometrical meaning of the design constraint shows that we have no proper platform mechanism any more: the base is degenerated into an equilateral triangle and so is the upper platform. Moreover we have only three legs, because each two of them coincide.

3.2. THE CASE $X \neq 0$ ($S_1 = 0$!).

In this case the classification becomes already rather complicated, so we only discuss one case.

3.2.1. *Rank of* **B** *is equal to four*

As $R_1 - R_2 + R_3 - R_4 + R_5 - R_6 = S_1 = 0$, we must have $x_0x_3 = 0$. This yields two possible cases. We discuss one of them.
Let $X \neq 0, x_0 = 0, x_3 = 1$. This case is much more complicated then previous ones. To avoid very large formulas, which we get if we continue in the same way as above, we shall use different denotations. Therefore

let us return to original equations $h_1 = 0, ..., h_6 = 0$, which were obtained from eq.(8). Into these equations we shall substitute as follows: $x_0 = 0, x_3 = 1, y_3 = -x_1y_1 - x_2y_2, R_6 = R_1 + R_3 + R_5 - R_2 - R_4, P = Q + W, p = (2QW + 8W^2 - QY - 4YW - X)/(4W), q = -2W + Y, R_4 = R_1 + S_1, R_3 = R_2 + S_2, R_2 = R_1 + S_4, S_2 = S_1 + T, R_5 = R_1 + U$. Now we continue similarly as before. For the determinants M_3, M_4, M_5, M_6 we obtain

$$(17)\ M_3 = Y(x_1 + \sqrt{3}x_2), M_6 = Y(\sqrt{3}x_2 - x_1), M_4 = \sqrt{3}(QY + X)x_2 \\ -(QY + 4YW + X)x_1, M_5 = (YW + QY + X)x_1 + \sqrt{3}YWx_2.$$

Let $Y \neq 0$ and $x_1 + \sqrt{3}x_2 \neq 0$. As $M_3 \neq 0$, we express y_0, y_1, y_2 from equations $U_4 = U_5 = U_6 = 0$ and substitute into equations $U_2 = 0$ and $U_3 = 0$. We obtain two equations, from both of them the factor $1 + x_1^2 + x_2^2$ splits and as the result we obtain an equation of degree 3 of length 51 ($U_3 = 0$) and one equation of degree 8 of length 3089 ($U_6 = 0$) in variables x_1, x_2. U_3 is an expression of the following form (for convenience we shall substitute $x_1 = k, x_2 = m$ in the rest of the paper),

$$(18)\qquad U3 = (p_1k + p_2m\sqrt{3})(1 + k^2 + m^2) + 8kG(1 + 4m^2)$$
$$\text{where}\quad p_1 = p_2 - 8G - 2p_3,\quad p_3 = 2YW(S_1 + T) + (X + QY)T,$$
$$p_2 = 2YW(T + S_1) - (X + QY)S_4,$$
$$G = -4X(Y - W)(Q + W) + 6QYW(Y - Q + 2W) + 3Y^2Q^2 + X^2.$$

We have to find out under which conditions plane curves (in the plane x_1, x_2) determined by equations $U_3 = 0, U_6 = 0$ have a common part, which means that equations $U_3 = 0$ and $U_6 = 0$ must have a common factor. This is possible only if U_3 is contained in U_2 or the cubic curve determined by $U_3 = 0$ splits into factors. Let us consider the second possibility. A rather complicated discussion yields the following example: If a platform mechanism fullfills additionally to the characteristic conditions for this case the design condition $Q = -2W$, which gives a triangular shape of the base and the two positioning constraints

$$R_1 = -\frac{1}{12}(128W^2Y^4 + 8Y^2X^2 - 64WY^3X - 72W^3YT + 448W^4Y^2 \\ +128Y^2W^2X - 12WXYT - 512W^3Y^3 + 36W^2Y^2T + 3W^2T^2)/(W^2Y^2)$$
$$S_4 = -(24W^2Y + 2YX - 8WY^2 - WT)/W,$$

then U_3 and U_6 have the common factor $k - m\sqrt{3} = 0$ and the platform allows a self-motion. The computation of the motion shows that it is a pure rotation. Fig. 5 shows base and platform in a top view and Fig. 6 shows two positions of the platform in an axonometric view. The axis of rotation

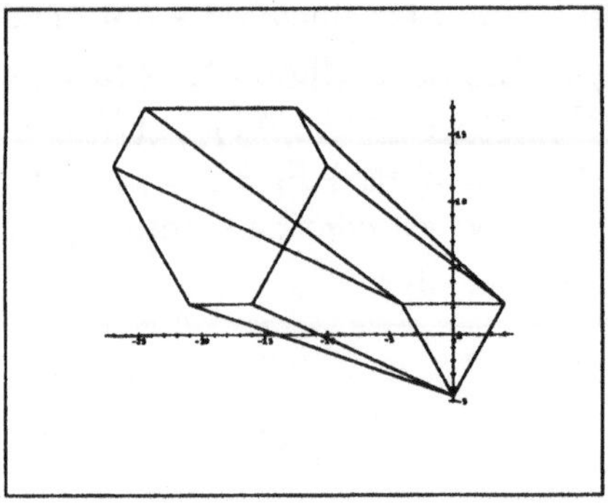

Figure 5. Top View

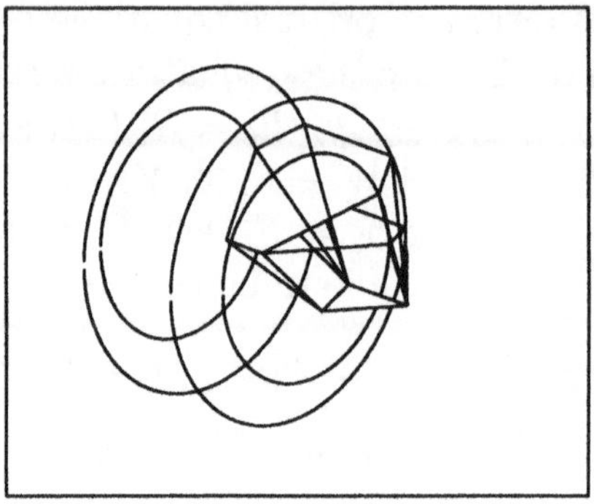

Figure 6. Axonometric View

is in the plane of the base and contains one side of the base and one side of the upper platform.

4. Conclusions

In this paper we have discussed examples of self-motions of a class of platform mechanisms. The geometry of this class was derived from the geometry of existing flight simulator platforms. We have discussed design parameters and positioning parameters which cause self-motions of such platforms. It is believed that informations on the possibility of an unwanted behaviour of these mechanism are of great importance for the designer to avoid critical design.

5. Acknowledgements

The research work reported here was made possible by Grant No. GACR 201/93/0496 of Czech Grant Agency.

References

Ma, O., and Angeles, J., (1992), Architecture Singularities of Parallel Manipulators, Int. Journal of Rob. and Automation,**vol. 7**, no. 1, pp. 23–29.

Husty, M.,(1996) An Algorithm for Solving the Direct Kinematics of General Stewart-Gough Platforms, Mech. Mach.Theory, (in print).

Merlet, J.-P., (1990), *Les robots parallèles*, Hermes, Paris.

Merlet, J.-P., (1992), Singular Configurations of Parallel Manipulators and Grassmann Geometry, The Int. Journal of Robotics Research, **vol. 8**, No. 5, pp. 45–56.

Husty, M. and Zsombor-Murray, P., (1994), A Special Type of Singular Stewart-Gough Platform, in *Advances in Robot Kinematics and Computational Geometry*, eds. Lenarcic, J. and Ravani, B., Kluwer academic publishers, Dordrecht, pp. 449–458.

Ravani, B. and Roth, B.,(1984), Mappings of Spatial Kinematics, J. of Mech. Trans. and Auto. in Design, **vol. 106**, pp. 341–347.

Stewart, D., (1965), A Platform With Six Degrees of Freedom, Proc. of the Inst. of Mech. Eng., **180 (1)**, pp. 371–386.

ON THE DESIGN OF PARALLEL MANIPULATORS FOR A PRESCRIBED WORKSPACE: A PLANAR QUATERNION APPROACH

A. P. MURRAY*, F. PIERROT**, P. DAUCHEZ** and J. M. McCARTHY*
* *Department of Mechanical and Aerospace Engineering*
University of California, Irvine
Irvine, California 92717-3975, USA
** *LIRMM*
UMR 9928 CNRS / Université Montpellier II
161 rue Ada, 34392 Montpellier Cedex 5, France

Abstract. In this paper we present a technique for designing parallel planar manipulators capable of reaching any set of points in a prescribed workspace. The manipulator consists of a platform connected to ground by *RPR* chains. The set of positions and orientations available to the end-effector of a single *RPR* chain is mapped into the space of planar quaternions to obtain a quadratic manifold. The coefficients of this *constraint manifold* are functions of the locations of the base and platform *R* joints and the distance between them. The structure of these equations is used to define a solution technique that yields the desired chains. Parallel manipulators that can reach the prescribed workspace are assembled from these chains. An example shows the determination of three *RPR* chains that construct a manipulator that reaches the prescribed workspace.

1. Introduction

The fundamentals of parallel robots are presented in [9]. The modeling and control of such machines have been widely studied and experimental results have been obtained [1][6][13]. Even though parallel robots are already in use, the design of these systems presents many challenges and stays a focus of research. This design process includes considerations of symmetry of the manipulator, isotropy of the Jacobian, velocity and force capabilities, singularity avoidance, redundancy, and size of the workspace to name some examples [4][5][7][10][11]. Designing a parallel manipulator capable of reaching a prescribed workspace is another challenge. When it is desired that the workspace definition include *both* positions *and* orientations of the end-effector, the design problem becomes more complex.

As a first step towards solving this problem, we consider the design of a planar parallel manipulator, constructed of *RPR* chains (*R* and *P* denoting revolute and prismatic joints, respectively). The design problem consists of finding such chains that reach the desired workspace. The configuration space of the moving platform is mapped to a subset of planar quaternion space, and the desired workspace is then defined by a generalized cube in this space. The goal is to determine the dimensions of the manipulator such that the set of end-effector positions and orientations encloses the desired workspace.

The set of end-effector configurations of this manipulator can be viewed as the

J. Lenarčič and V. Parenti-Castelli (eds.), Recent Advances in Robot Kinematics, 349-358.
© 1996 *Kluwer Academic Publishers. Printed in the Netherlands.*

intersection of the positions *and* orientations reachable by each of the RPR chains that connect it to ground. The configuration space for an individual RPR chain is a quadratic algebraic manifold in planar quaternion space, called its *constraint manifold*. The quadratic form of this manifold contains the coordinates of the base and platform R joints and the displacement of the P joint as free parameters.
To represent the workspace, we identify points along its boundaries as well as inside the generalized cube. The requirement that each of these points lies on the constraint manifold of an RPR chain defines a set of constraint equations for the design parameters. It is possible to exploit the structure of these equations and use its coefficients directly. The result is called the *coefficient representation* of constraint manifold, and the design problem reduces to searching intersections of annular regions in two different planes. Any set of N points within these intersections defines an N-chain parallel manipulator that solves the problem.
This represents a preliminary result in the direction of kinematic design of spatial parallel manipulators for a specific workspace. The methodology is generalized by using dual quaternions and the constraint manifold of an SPS chain (S denoting a spherical joint). At its most basic form, this can be viewed as an extension of Burmester theory [12].

2. Dual and planar quaternions

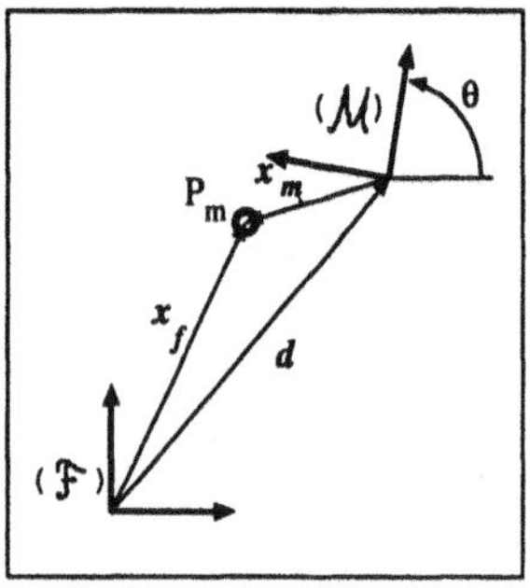

Figure 1: A planar displacement

Clifford [3] introduced a product operation for vectors in a metric product space that defines a geometric algebra, called its Clifford algebra. Complex numbers, quaternions, planar quaternions and dual quaternions are obtained using Clifford's construction [8]. The Clifford algebra associated with spatial displacements encodes the rotation and translation of the displacement in an eight dimensional vector. The components of this vector are defined by the Plücker coordinates of the screw axis and the rotation and translation around and along this axis. Dual quaternions that represent planar displacements have only four non-zero components and form a planar quaternions subalgebra. Their components are defined by the rotation and translation of the planar displacement.
The advantage of using planar quaternions in the formulation of our problem is that we construct a set of simple *quadratic* equations.

The planar displacement shown in Figure 1 is given by:

$$\boldsymbol{x}_f = \boldsymbol{R}\boldsymbol{x}_m + \boldsymbol{d} \tag{1}$$

where:

- $\boldsymbol{x}_f$ is the coordinate vector of a point in a fixed frame $(\mathcal{F})$
- $\boldsymbol{x}_m$ is the coordinate vector of the same point in a moving frame $(\mathcal{M})$
- $\boldsymbol{R}$ is a 2×2 rotation matrix
- $\boldsymbol{d}$ is a 2×1 displacement vector

Classical representations use three parameters (namely x, y, and θ) to specify the location of $(\mathcal{M})$ in $(\mathcal{F})$ by defining $\boldsymbol{R}$ and $\boldsymbol{d}$ as follows:

$$\boldsymbol{R}(\theta) = \begin{bmatrix} \cos(\theta) & -\sin(\theta) \\ \sin(\theta) & \cos(\theta) \end{bmatrix}$$

$$\boldsymbol{d}(x, y) = \begin{bmatrix} x \\ y \end{bmatrix}$$

An alternate set of parameters for expressing a planar displacement is the *planar quaternion* [8]:

$$Q = \begin{pmatrix} q_1 \\ q_2 \\ q_3 \\ q_4 \end{pmatrix} = \begin{pmatrix} \frac{1}{2}\left(x \cos\frac{\theta}{2} + y \sin\frac{\theta}{2}\right) \\ \frac{1}{2}\left(y \cos\frac{\theta}{2} - x \sin\frac{\theta}{2}\right) \\ \sin\frac{\theta}{2} \\ \cos\frac{\theta}{2} \end{pmatrix}$$

The planar displacement in equation (1) expressed in terms of Q is:

$$\boldsymbol{R}(Q) = \begin{bmatrix} q_4^2 - q_3^2 & -2q_3q_4 \\ 2q_3q_4 & q_4^2 - q_3^2 \end{bmatrix} \tag{2}$$

and

$$\boldsymbol{d}(Q) = \begin{bmatrix} 2(q_1q_4 - q_2q_3) \\ 2(q_1q_3 + q_2q_4) \end{bmatrix} \tag{3}$$

Note the property of planar quaternions that:

$$q_3^2 + q_4^2 = 1 \tag{4}$$

3. Constraint manifold of an RPR chain

The set of positions available to a moving body constrained by joints, either revolute or prismatic, is an algebraic submanifold in the space of planar quaternions called a *constraint manifold* [2]. The constraint manifold of the end-effector of a parallel manipulator is the intersection of the constraint manifolds of the individual chains connecting the end-effector to the base. To perform the kinematic design of a parallel manipulator, each individual chain is designed to encompass the desired workspace. The constraint manifold of a single, general *RPR* chain is now determined to accomplish this task.

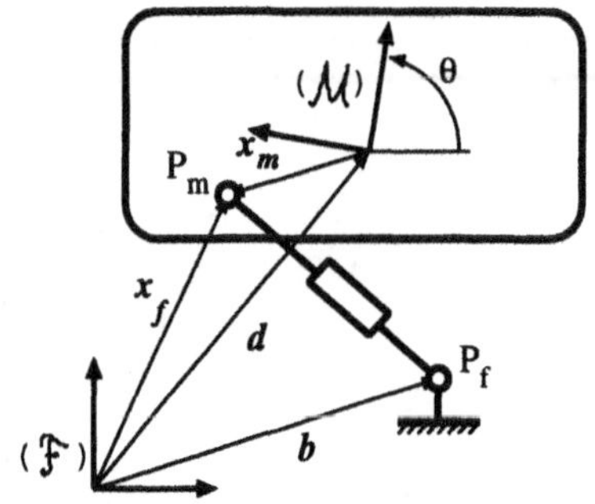

Figure 2: A planar *RPR* chain

A planar *RPR* chain is shown in Figure 2. The point P_f is the location of the fixed pivot, and P_m is the location of the moving pivot. The location of the fixed pivot in the *fixed* frame, $\mathcal{F}$, is specified by the vector $\boldsymbol{b} = (u, v)^T$. The location of the moving pivot in the *moving* frame, $\mathcal{M}$, is specified by the vector $\boldsymbol{x_m} = (\lambda, \mu)^T$. The location of the moving pivot in the *fixed* frame is $\boldsymbol{x_f}$. The distance between the two pivots is l and can be expressed as:

$$||P_m P_f|| = l$$

then:

$$(\boldsymbol{b} - \boldsymbol{x_f}).(\boldsymbol{b} - \boldsymbol{x_f}) = l^2$$

Substituting equation (1) into this leads to:

$$(\boldsymbol{b} - \boldsymbol{R}\boldsymbol{x_m} - \boldsymbol{d}).(\boldsymbol{b} - \boldsymbol{R}\boldsymbol{x_m} - \boldsymbol{d}) = l^2 \tag{5}$$

Substituting equations (2) and (3) into equation (5), and using identity (4) determines the constraint manifold:

$$\begin{aligned} 4(A_0 q_1^2 + A_1 q_2^2 + A_2 q_3^2 + A_3 q_4^2 + A_4 q_1 q_2 + A_5 q_1 q_3 \\ + A_6 q_1 q_4 + A_7 q_2 q_3 + A_8 q_2 q_4 + A_9 q_3 q_4) = l^2 \end{aligned} \tag{6}$$

where

$$\left|\begin{array}{c} A_0 \\ A_1 \\ A_2 \\ A_3 \\ A_4 \\ A_5 \\ A_6 \\ A_7 \\ A_8 \\ A_9 \end{array}\right| = \left|\begin{array}{c} 1 \\ 1 \\ \frac{1}{4}((\lambda+u)^2 + (\mu+v)^2) \\ \frac{1}{4}((\lambda-u)^2 + (\mu-v)^2) \\ 0 \\ -(\mu+v) \\ \lambda - u \\ \lambda + u \\ \mu - v \\ \mu u - \lambda v \end{array}\right| \tag{7}$$

A relationship between the design parameters (λ, μ, u and v) and four of the ten coefficients of the quadratic equation (6) is derived from equation (7):

$$\left|\begin{array}{c} \lambda \\ \mu \\ u \\ v \end{array}\right| = \frac{1}{2}\begin{bmatrix} 0 & 1 & 1 & 0 \\ -1 & 0 & 0 & 1 \\ 0 & -1 & 1 & 0 \\ -1 & 0 & 0 & -1 \end{bmatrix}\left|\begin{array}{c} A_5 \\ A_6 \\ A_7 \\ A_8 \end{array}\right| \tag{8}$$

Notice that the quadric corresponding to an *RPR* chain is distinguished from the general quadric by the coefficients A_i satisfying the following:

$$A_0 = 1 \tag{9}$$
$$A_1 = 1 \tag{10}$$
$$A_4 = 0 \tag{11}$$
$$2A_9 - A_5A_6 - A_7A_8 = 0 \tag{12}$$
$$4A_2 - A_7^2 - A_5^2 = 0 \tag{13}$$
$$4A_3 - A_6^2 - A_8^2 = 0 \tag{14}$$

Equations (9)-(14) are called the *coefficient representation* of an *RPR* chain. The relationships in the coefficient representation allow the constraint manifold of equation (6) to be written as:

$$(2q_1 + A_5q_3 + A_6q_4)^2 + (2q_2 + A_7q_3 + A_8q_4)^2 = l^2 \tag{15}$$

Equation (15) is the desired form for the constraint manifold.

4. Using the constraint manifold for design

To design *RPR* chains capable of reaching the desired workspace, we must solve for the constant parameters A_5, A_6, A_7, A_8 and varying parameter l such that equation (15) is true for all Q in the workspace. Notice that if the distance, l, between the revolute joints is not bounded, any workspace is achievable. Since these are not realistic solutions to the design problem, we eliminate them from consideration by placing bounds on the extension of the prismatic joint: $l_{min} \leq l \leq l_{max}$.

Also notice that, without loss of generality, the workspace is assumed to contain points where the orientation $\theta = 0$. In this case:

$$\text{if } \theta = 0 \Longrightarrow \left|\begin{array}{l} q_3 = 0 \\ q_4 = 1 \end{array}\right. \Longrightarrow$$
$$(2q_1 + A_6)^2 + (2q_2 + A_8)^2 = l^2 \tag{16}$$

The design process proceeds as follows. The prescribed workspace is discretized into p points. Let n be the number of points in the $\theta = 0$ subset of the workspace.

The planar quaternion Q that corresponds to each point in this subset is calculated. For each value of Q, equation (16) is written for $l = l_{min}$ and $l = l_{max}$:

$$(2q_1 + A_6)^2 + (2q_2 + A_8)^2 = l_{min}^2 \quad (17)$$

$$(2q_1 + A_6)^2 + (2q_2 + A_8)^2 = l_{max}^2 \quad (18)$$

Equations (17) and (18) define an *annulus* in the (A_6, A_8) plane. n values of Q produce n annuli. The intersection of these annuli is available as possible values for A_6 and A_8. Note that this intersection defines a necessary but not sufficient condition on the values of A_6 and A_8. k possible $(A_{6_i}, A_{8_i}), (i = 1 \ldots k)$, points are now chosen.

For every choice (A_{6_i}, A_{8_i}) and for each of the remaining $p - n$ workspace points, equation (15) is written for $l = l_{min}$ and $l = l_{max}$:

$$(2q_1 + A_5q_3 + A_{6_i}q_4)^2 + (2q_2 + A_7q_3 + A_{8_i}q_4)^2 = l_{min}^2 \quad (19)$$

$$(2q_1 + A_5q_3 + A_{6_i}q_4)^2 + (2q_2 + A_7q_3 + A_{8_i}q_4)^2 = l_{max}^2 \quad (20)$$

Equations (19) and (20) define an *annulus* in the (A_5, A_7) plane: the $p - n$ workspace points produce $p - n$ annuli. The intersection of these annuli is available as possible values of A_5 and A_7 to be used with A_{6_i} and A_{8_i}; note that no intersection may exist. The procedure allows us to determine r_i possible choices for A_5 and A_7.

Selecting r_i points for each (A_{6_i}, A_{8_i}) finally produces $\sum_{i=1}^{k}(r_i)$ acceptable choices for (A_5, A_6, A_7, A_8); each corresponding to an *RPR* chain whose pivots locations are determined using equation (8). Assembling N such chains produces an N-chain planar manipulator capable of reaching the n prescribed points of the workspace within the desired limits on l.

We implement the following procedure as a convenient step for automatic selection of values (A_5, A_6, A_7, A_8). Any viable choice for the pair (A_6, A_8) must lie within *every* circle specified by equation (18) for the $\theta = 0$ subset of the workspace. Define:

$$\begin{aligned} A_6^+ &= -2\min(q_1)l_{max} \\ A_6^- &= -2\max(q_1)l_{max} \\ A_8^+ &= -2\min(q_2)l_{max} \\ A_8^- &= -2\max(q_2)l_{max} \end{aligned}$$

The region in the plane where acceptable values lie are: $A_6^- < A_6 < A_6^+$ and $A_8^- < A_8 < A_8^+$. Note that if either $A_6^+ < A_6^-$ or $A_8^+ < A_8^-$, there exist no choices for (A_6, A_8) that satisfy the design specification. A similar procedure using equation (20) determines limits on A_5 and A_7 for a given choice of (A_6, A_8).

5. Case study

As an application of our approach we design a 3-chain fully parallel non-redundant manipulator. Its workspace is defined as: $0 < x < 0.5$, $0 < y < 0.25$, $-30 < \theta < 30$ (deg) and is discretized as shown in Figure 3 with 18 points distributed throughout the worskspace.

	1	2	3	4	5	6	7	8	9
x	0	0	.25	.25	.5	.5	0	0	.25
y	0	.25	0	.25	0	.25	0	.25	0
θ	0	0	0	0	0	0	30	30	30
	10	11	12	13	14	15	16	17	18
x	.25	.5	.5	0	0	.25	.25	.5	.5
y	.25	0	.25	0	.25	0	.25	0	.25
θ	00	00	00	-80	-00	-00	-80	-00	-80

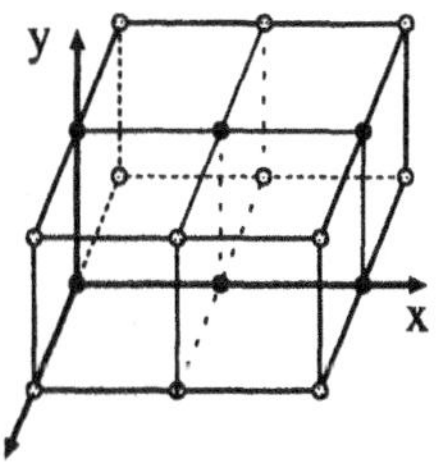

Figure 3: Discretized workspace

Following the procedure described in Section 4, points 1 through 6 determine the 6 annuli in the (A_6, A_8) plane shown in Figure 4:

- the large solid circles are determined by equation (18),
- the dashed circles are determined by equation (17),
- the square is the boundary inside which acceptable values for (A_6, A_8) must lie,
- the crosses, belonging to the intersection of the n annuli, represent possible values for (A_6, A_8) that could produce satisfactory designs. Note that a 10×10 grid is used here to define candidate values of (A_6, A_8) within the boundary.
- the small circles are points not belonging to the intersection. These choices are rejected for values of A_6 and A_8.

For every acceptable point in the (A_6, A_8) plane (represented by a cross in figure 4), we look for suitable (A_5, A_7) points. Again, a boundary region is determined automatically, and the region is searched for candidate values on a (4×4) grid. This generates 246 possible chains, after only few seconds of computation on an *Silicon Graphics Indigo* station running *Matlab*. We selected as examples the 6 chains shown in Figure 4.

Among the 246 possible chains, the user chooses 3 in order to build the manipulator. Any set of 3 chains guarantees that every point chosen to represent the desired workspace is actually reachable within the desired limits on the prismatic joint. There is *no* guarantee that other points in the workspace are reachable within the desired limits or that the mechanism is free of singular positions within the workspace.

As a first example, chains 1, 2 and 4 are assembled as a manipulator. This mechanism is shown in Figure 5 in position $(x = 0, y = 0, \theta = 0)$. The *Matlab* function allows the animation of the manipulator between the desired workspace points. A cursory visual inspection for singular positions can be performed. Figure 5 shows the mechanism moving through a singular position at the workspace point ($x = 0.5, y = 0.25, \theta = 30$).

Trial-and-error is used to determine an acceptable solution. As a second example chains 1, 2 and 3 are now combined to form the mechanism in Figure 6, shown in the initial position. A second position, ($x = 0, y = 0, \theta = 30$), of this manipulator is shown in Figure 6.

This pair of examples displays the ability to rapidly construct a large number of manipulators capable of reaching the discretized workspace. They also indicate the need for the ability to reject poor manipulator design due to singularities, collisions, lack of mechanical advantages, etc.

6. Conclusion

In this paper we formulated an initial step in the design of parallel manipulators for a prescribed workspace. Setting the problem in the context of planar quaternions produces a set of quadratic equations. The consideration of these quadratic equations for the portion of the workspace parallel to the fixed frame produces annular regions. Discretizing the workspace produces a finite number of annular regions. Candidate values are chosen from the intersection of these annular regions and a second set of annuli is generated. The result is an automated procedure for generating the sets of *RPR* chains that combine to form a parallel manipulator encompassing the desired workspace. This procedure only guarantees results at the points of the discretized workspace. Implementation of ideas to eliminate mechanisms with singular configurations or to guarantee continuity of the workspace is a direction for future work, as is the extension of this approach to the design of more general parallel robots.

References

[1] P. Bégon, F. Pierrot, and P. Dauchez. High-precision, high-speed insertion with a 6-dof parallel robot. In *24th International Symposium on Industrial Robots*, pages 145–152, November 1993.

[2] O. Bottema and B. Roth. *Theoretical Kinematics.* Dover Publications, Inc., New York, 1979.

[3] W. K. Clifford. *Mathematical Papers.* R. Tucker, London, 1882.

[4] C. Gosselin and J. Angeles. The optimum kinematic design of a planar three-degree-of-freedom parallel manipulator. *ASME Journal of Mechanisms, Transmission, and Automation in Design*, 110:35–41, March 1988.

[5] C. Gosselin and J. Angeles. A global performance index for the kinematic optimisation of robotic manipulators. *ASME Journal of Mechanical Design*, 113(3):220–226, 1991.

[6] C. Gosselin and J.F. Hamel. The agile eye: a high-performance three-degree-of-freedom camera-orienting device. In *IEEE International Conference on Robotics and Automation*, pages 781–786, May 1994.

[7] R.B. Hertz and P.C. Hughes. *Forward kinematics of a 3-dof variable-geometry-truss manipulator*, pages 241–250. Kluwer Academic Publishers, 1993.

[8] J. M. McCarthy. *Introduction to Theoretical Kinematics.* MIT Press, Cambridge, Massachusetts, 1986.

[9] J.-P. Merlet. *Les Robots Parallèles.* Hermès Ed., Paris, 1990.

[10] F. Pierrot, M. Uchiyama, and K. Unno. HEXA: a fully-parallel 6-dof Japanese-French robot. In *1st Japanese-French congress of Mechatronics*, pages 1–8, October 1992.

[11] C. Reboulet, C. Lambert, and N. Nombrail. A parallel redundant manipulator Speed-R-Man and its control. In *4th International Symposium on Robotics and Manufacturing*, pages 285–291, November 1992.

[12] B. Roth. On screw axes and other special lines associated with spatial displacements of rigid body. *ASME Journal of Engineering for Industry*, pages 102–110, February 1967.

[13] M. Uchiyama and al. Experiment on dynamic control of a hexa-type parallel robot. In *First World Automation Congress, WAC '94*, pages 281–286, August 1994.

Acknowledgements

Pierre Dauchez and François Pierrot are grateful to Mike McCarthy for welcoming them at the University of California, Irvine, during their sabbatical leaves. This visit was possible thanks to financial support provided by the *North Atlantic Treaty Organization*, the *National Science Foundation* and the *Centre National de la Recherche Scientifique.*

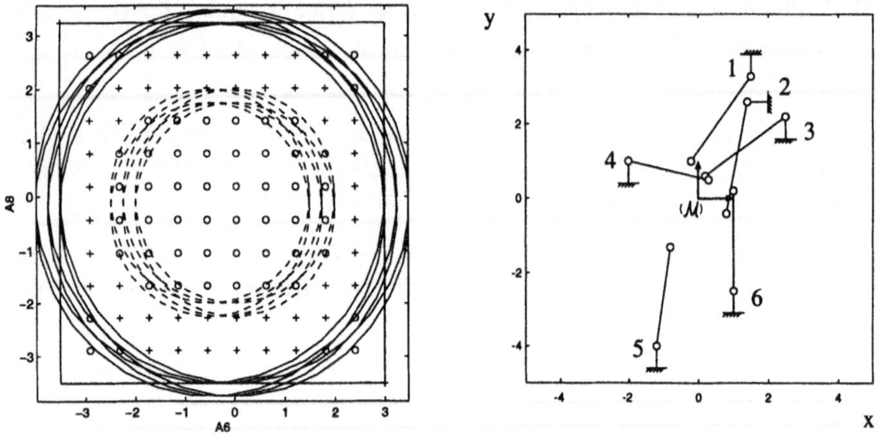

Figure 4: Annuli in (A_6, A_8) space, and 6 possible chains

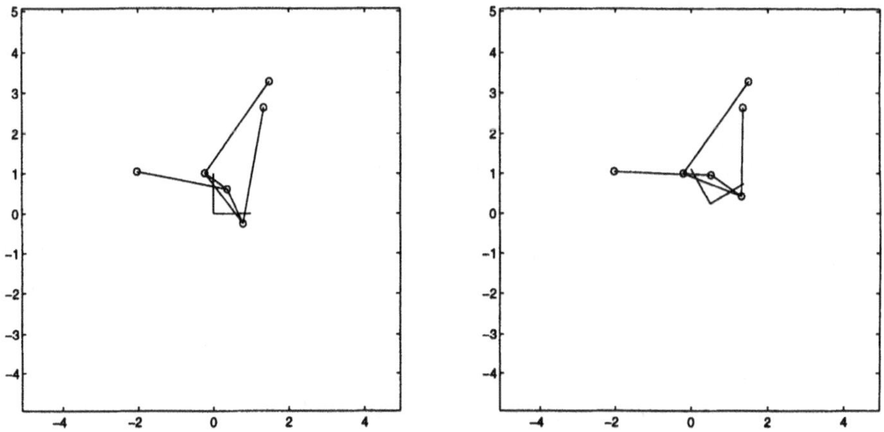

Figure 5: Chains 1, 2 and 4 assembled: two positions (one is singular)

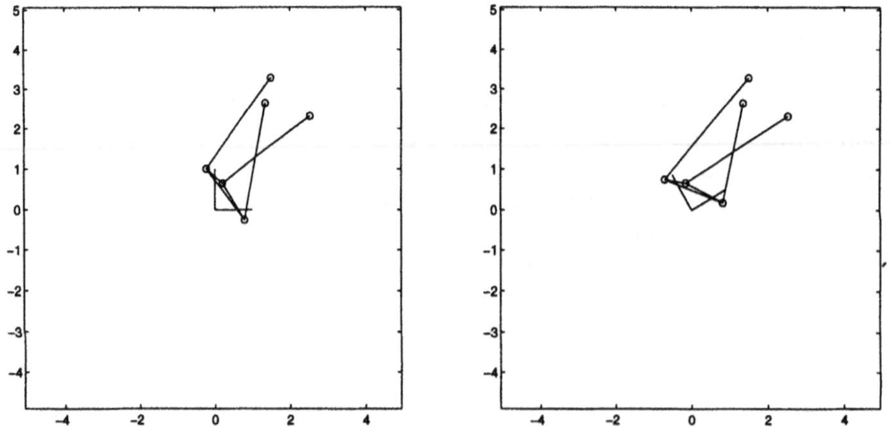

Figure 6: Chains 1, 2 and 3 assembled: two positions

KINEMATOTROPIC LINKAGES

K. WOHLHART
Institut für Mechanik, Technische Universität Graz
Kopernikusgasse 24, A–8010 Graz, Austria
e-mail: wohlhart@mech.tu–graz.ac.at

Abstract. In the main part of this paper there is described a new spatial mechanism which exhibits the very peculiar property that a variation of the position parameter(s) can change its mobility. Although many linkages of this type might exist, only two are to be found in the relevant literature. Linkages of this type will be labelled here as *kinematotropic linkages.* In passing a singularity position (in which a certain transitory infinitesimal mobility is attained) these mechanisms permanently change their global mobilities.

1. Introduction

Mobility of a mechanism does not depend on its topology alone. The Grübler–Kutzbach formula [1, 2] only allows one to determine the minimum degree of freedom of a given mechanism. A mechanism with L fundamental loops and joints which admit relative mobility f to the two rigid bodies they connect, possesses a global minimum degree of freedom $F_{\min}$ given by:

$$F_{\min} = \Sigma f - 6L.$$

By changing the system parameters while retaining the topological structure of the mechanism, higher mobility might be attained. In this case one speaks of paradoxical linkages, although it can only be a minor surprise that dimensions and relative orientations of the joint–axes affect the mobility of the linkage. More surprising seems to be the fact that a simple variation of the position parameter can lead to a permanent change of the mobility of the mechanism. Mechanisms which show this "kinematotropy" are the subject of the present paper and we shall call them kinematotropic linkages. To characterize their peculiar feature the following remarks might be expedient. It is common knowledge, that the degrees of freedom of a mechanism in its various closure–modes differ from one another. The model for the cyclohexane molecule is a example for this: it is rigid in its wan configuration and flexible in the boat form [3]. It is furthermore well known that the infinitesimal mobility of a mechanism in a singularity position differs from the permanent mobility in an adjacent full region of the position parameters, an issue comprehensively discussed in [4]. A planar four–bar linkage, forming a parallelogram, can be brought into a completely flattened position, out of which it can be "opened" in two different ways so that it either again forms a parallelogram or, alternatively an antiparallelogram. In both cases the degree of freedom is $F=1$, i.e., the passage through the flattened position does not change the mobility of the linkage. The flattened position is a singularity position with

J. Lenarčič and V. Parenti-Castelli (eds.), Recent Advances in Robot Kinematics, 359-368.
© 1996 *Kluwer Academic Publishers. Printed in the Netherlands.*

transitory mobility $F = 2$, which allows bifurcation into two different configurations with mobility $F = 1$. This special 4R linkage does not show kinematotropy. But the three linkages we are going to describe show kinematotropy.

2. The Wunderlich Mechanism

In 1954 W. Wunderlich published a paper [5] in which he describes a planar twelve–bar mechanism with six parallelogram or antiparallelogram loops which can be arranged in four different closure modes, all of them movable with either one or two degrees of freedom. What makes this mechanism especially remarkable is the fact that by passing a singularity position the mechanism might change its movability. In that case the singularity position is, therefore, not a simple bifurcation position, but a sort of mobility turning position. Fig. 1 shows Wunderlich's mechanism in two different positions. In the position in which it is has two parallelogram and four antiparallelogram loops, the mechanism is movable with mobility 1 because the two position angles ψ and ϕ are related by the equation:

$$(\tan\frac{\psi}{2})^2 k_2[1-k_1(\tan\frac{\phi}{2})^2]+\tan\frac{\psi}{2}\tan\frac{\phi}{2}(1+k_1+k_2+k_1k_2)+k_1(\tan\frac{\phi}{2})^2-1=0,$$

wherein $k_1=(a+b)/(a-b)$, $k_2=(c+b)/(c-b)$ are the system parameters. In the position, however, in which Wunderlich's mechanism has four parallelogram and two antiparallelogram loops, its mobility is 2, as ψ and ϕ can be chosen arbitrarily. The passage from one position to the other goes smoothly through a singularity position ($\phi = 0$). The Wunderlich Mechanism, therefore, belongs to the category of kinematotropic linkages.

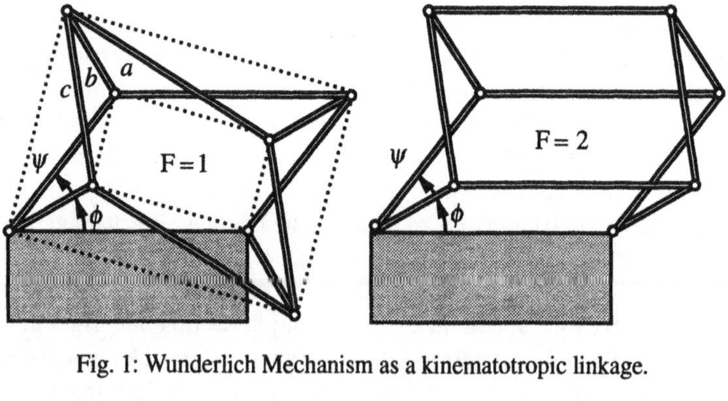

Fig. 1: Wunderlich Mechanism as a kinematotropic linkage.

3. The Wren Platform

In [6] and in [7] a movable Gough–Stewart platform has been described whose elementary mechanism is a model which, according to [6], Christopher Wren in 1645 used to demonstrate that the one–sheeted hyperboloid of revolution contains (two series of) straight lines. In Fig. 2 (a minimum of) five bars of equal length (h) connect via spherical joints two rigid bodies, the base body and the platform. The midpoints of the spherical joints are arranged identically on equal circles (a) in the base body as well as in the platform. Out of the singular position in which the axes of the bars are orthogonally directed to the planes of the two circles, the mechanism can be led either into a region where it has a finite mobility 1 or into a region in which it moves with finite mobility 2. In the first case the two angles ψ and ϕ are related by $2a\sin(\phi/2) = h\sin\psi$, while in the second case the position angles φ and σ are independent. It is easy to prove that in both cases the infinitesimal mobility of Wren's platform is 3, i.e., the local mobility does not change. But, as its finite mobility depends on the position parameters, the platform has to be reckoned among the category of the kinematotropic linkages.

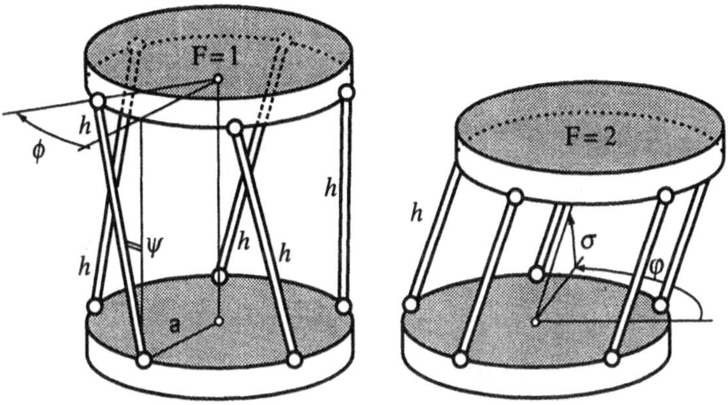

Fig. 2: Wren's Platform Mechanism.

In the following a new spatial mechanism will be described and analyzed which shows kinematotropy, i.e., its degree of freedom depends on the position parameter(s). To bring a kinematotropic mechanism from one region into an other region where it moves with a different global degree of freedom, it has to be led through a singularity position. The infinitesimal transitory movability in this singularity position generally differs from the mobilities in the two adjacent regions. A singularity position as a rule, is a bifurcation position. The term "bifurcation position", should be reserved for a position of a mechanism where only paths with mobility 1 meet together. In the cases of Wunderlich's mechanism and of Wren's platform the singularity positions evidentially are not bifurcation positions.

4. A new kinematotropic linkage (The Queer Square)

The linkage we are going to describe is a collapsible RR(4R)(4R)RR–mechanism with three fundamental loops. The group of joints (4R) represents a parallelogram–joint Π which has been introduced in [8]. Fig. 3 shows this mechanism in three positions. In the following exposition we shall prove that in the first position it is movable with two degree of freedom, in the second, the flattened position, it has a local (infinitesimal) mobility 5, and in the third position the mechanism moves with mobility 1. From the flattened position the mechanism can be unfolded into two positions in which it has different mobilities and belongs, therefore, to the category of kinematotropic linkages.The mechanism has three fundamental loops and its ten plates are held together by twelve rotary joints. From the Grübler–Kutzbach formula we obtain for the minimum of the global degrees of freedom: $F_{\min} = \Sigma f + 6L = 12 - 6.3 = -6$. I.e., the mechanism is highly overconstrained. If it is movable, then it is only because of its special dimensions. Evidently, the bodies of the mechanism are subject to one general constraint: the absolute angular velocity of each body has no component in the direction of $\mathbf{e}_3$. Therefore, the Grübler–Kutzbach formula should be applied in the form $F_{\min} = \Sigma f - 5.L = 12 - 5.3 = -3$. We see, however, that even by taking into account the (visible) general constraint, overconstraint is indicated. Although it is evident that the global degree of freedom in the first position of the mechanism is 2 and in the second 1, we shall make sure of it by determining the dimension of the nullspace of the corresponding Jacobian matrices [9].

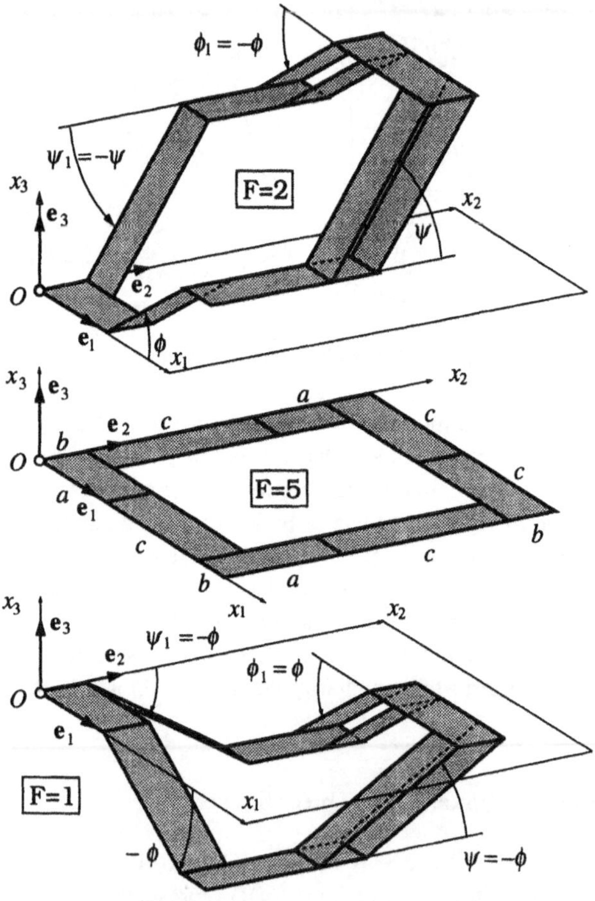

Fig. 3: The new kinematotropic linkage.

The rotary joint axes are given by the unit vectors $\mathbf{n}_\alpha$ and the position vectors $\mathbf{x}_\alpha$ leading from the origin O to a point on each axis (Fig. 4 and 5). With the unit motor $\widehat{\mathbf{n}}_\alpha$ and, the relative angular velocity ω_α of the two bodies the rotary joint connects, the corresponding relative speed motor $\widehat{\mathbf{s}}_\alpha$ is given by [10]:

$$\widehat{\mathbf{s}}_\alpha = \widehat{\mathbf{n}}_\alpha \omega_\alpha = \begin{bmatrix} \mathbf{n}_\alpha \\ \mathbf{x}_\alpha \times \mathbf{n}_\alpha \end{bmatrix} \omega_\alpha . \tag{1}$$

The three infinitesimal closure conditions can then be written as follows:

$$\widehat{\mathbf{n}}_1\omega_1 + \widehat{\mathbf{n}}_2\omega_2 + \widehat{\mathbf{n}}_3\omega_3 + \widehat{\mathbf{n}}_4\omega_4 + \widehat{\mathbf{n}}_7\omega_7 + \widehat{\mathbf{n}}_8\omega_8 + \widehat{\mathbf{n}}_{11}\omega_{11} + \widehat{\mathbf{n}}_{12}\omega_{12} = \widehat{\mathbf{o}}, \tag{2}$$

$$\widehat{\mathbf{n}}_3\omega_3 + \widehat{\mathbf{n}}_4\omega_4 + \widehat{\mathbf{n}}_5\omega_5 + \widehat{\mathbf{n}}_6\omega_6 = \widehat{\mathbf{o}}, \tag{3}$$

$$\widehat{\mathbf{n}}_7\omega_7 + \widehat{\mathbf{n}}_8\omega_8 + \widehat{\mathbf{n}}_9\omega_9 + \widehat{\mathbf{n}}_{10}\omega_{10} = \widehat{\mathbf{o}}. \tag{4}$$

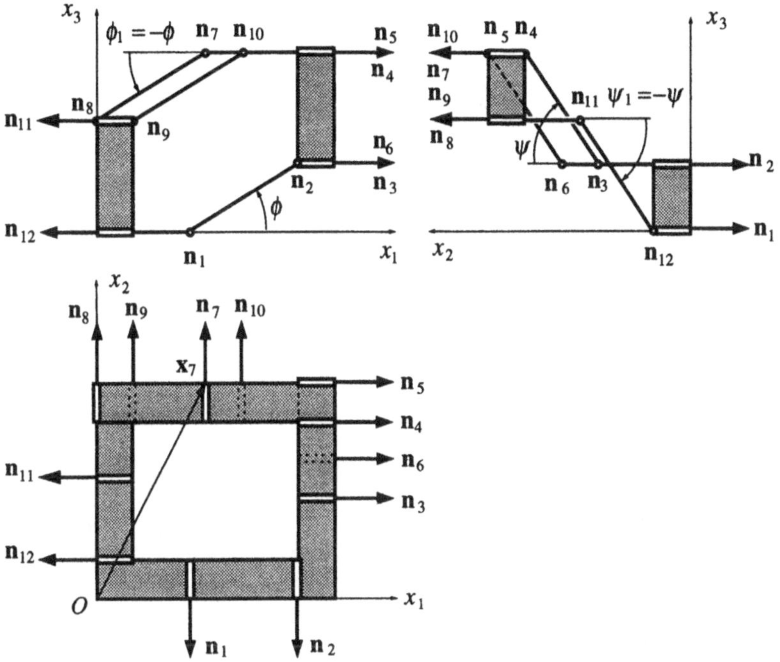

Fig. 4: The new mechanism in the first position.

With the column matrix ω which collects all the relative angular velocities

$$\omega^{\mathrm{T}} = \boxed{\omega_1 \quad \omega_2 \quad \omega_3 \quad \omega_4 \quad \omega_5 \quad \omega_6 \quad \omega_7 \quad \omega_8 \quad \omega_9 \quad \omega_{10} \quad \omega_{11} \quad \omega_{12}} \tag{5}$$

and the Jacobian matrix $\mathbf{J}$ given by

$$\mathbf{J} = \begin{array}{|cccccccccccc|} \widehat{\mathbf{n}}_1 & \widehat{\mathbf{n}}_2 & \widehat{\mathbf{n}}_3 & \widehat{\mathbf{n}}_4 & \widehat{\mathbf{o}} & \widehat{\mathbf{o}} & \widehat{\mathbf{n}}_7 & \widehat{\mathbf{n}}_8 & \widehat{\mathbf{o}} & \widehat{\mathbf{o}} & \widehat{\mathbf{n}}_{11} & \widehat{\mathbf{n}}_{12} \\ \widehat{\mathbf{o}} & \widehat{\mathbf{o}} & \widehat{\mathbf{n}}_3 & \widehat{\mathbf{n}}_4 & \widehat{\mathbf{n}}_5 & \widehat{\mathbf{n}}_6 & \widehat{\mathbf{o}} & \widehat{\mathbf{o}} & \widehat{\mathbf{o}} & \widehat{\mathbf{o}} & \widehat{\mathbf{o}} & \widehat{\mathbf{o}} \\ \widehat{\mathbf{o}} & \widehat{\mathbf{o}} & \widehat{\mathbf{o}} & \widehat{\mathbf{o}} & \widehat{\mathbf{o}} & \widehat{\mathbf{o}} & \widehat{\mathbf{n}}_7 & \widehat{\mathbf{n}}_8 & \widehat{\mathbf{n}}_9 & \widehat{\mathbf{n}}_{10} & \widehat{\mathbf{o}} & \widehat{\mathbf{o}} \end{array} \tag{6}$$

the three conditions (2), (3) and (4) can be written in matrix form:

$$\mathbf{J}.\omega = 0. \tag{7}$$

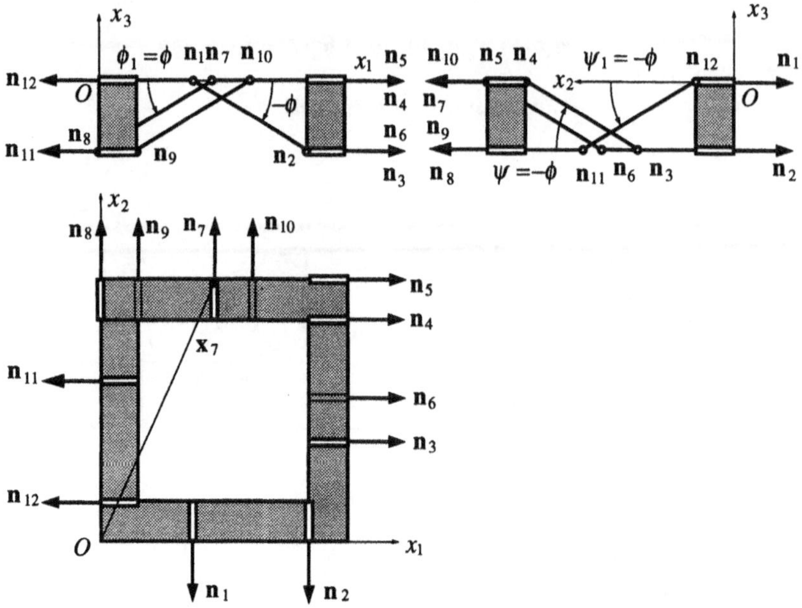

Fig. 5: The new mechanism in the second position.

The Jacobian matrix $\mathbf{J}$ is a function of the position angles ϕ, ψ, ϕ_1 and ψ_1. With the position vectors $\mathbf{x}_\alpha$ and the unit vectors $\mathbf{n}_\alpha$ given in the coordinate system $(O, \mathbf{e}_1\mathbf{e}_2\mathbf{e}_3)$:

$$\mathbf{x}_1 = a\,\mathbf{e}_1$$

$$\mathbf{x}_2 = (a + c\cos\phi)\,\mathbf{e}_1 + c\sin\phi\,\mathbf{e}_3$$

$$\mathbf{x}_3 = c(1+\cos\phi)\,\mathbf{e}_1 + (c-b)\,\mathbf{e}_2 + c\sin\phi\,\mathbf{e}_3$$

$$\mathbf{x}_4 = c(1+\cos\phi)\,\mathbf{e}_1 + (c-b+c\cos\psi)\,\mathbf{e}_2 + c(\sin\phi+\sin\psi)\,\mathbf{e}_3$$

$$\mathbf{x}_5 = c(1+\cos\phi)\,\mathbf{e}_1 + c(1+\cos\psi)\,\mathbf{e}_2 + c(\sin\phi+\sin\psi)\,\mathbf{e}_3$$

$$\mathbf{x}_6 = c(1+\cos\phi)\,\mathbf{e}_1 + c\,\mathbf{e}_2 + c\sin\phi\,\mathbf{e}_3$$

$$\mathbf{x}_7 = c\cos\phi\,\mathbf{e}_1 + c(1+\cos\psi)\,\mathbf{e}_2 + c(\sin\phi+\sin\psi)\,\mathbf{e}_3$$

$$\mathbf{x}_8 = c(\cos\phi - \cos\phi_1)\,\mathbf{e}_1 + c(1+\cos\psi)\,\mathbf{e}_2 + c(\sin\phi+\sin\psi+\sin\phi_1)\,\mathbf{e}_3$$

$$\mathbf{x}_9 = (c(\cos\phi-\cos\phi_1)+b)\,\mathbf{e}_1 + c(1+\cos\psi)\,\mathbf{e}_2 + c(\sin\phi+\sin\psi+\sin\phi_1)\,\mathbf{e}_3$$

$$\mathbf{x}_{10} = (c\cos\phi + b)\,\mathbf{e}_1 + c(1+\cos\psi)\,\mathbf{e}_2 + c(\sin\phi+\sin\psi)\,\mathbf{e}_3$$

$$\mathbf{x}_{11} = c(\cos\phi-\cos\phi_1)\,\mathbf{e}_1 + (b+c\cos\psi)\,\mathbf{e}_2 + c(\sin\phi+\sin\psi+\sin\phi_1)\,\mathbf{e}_3$$

$$\mathbf{x}_{12} = c(\cos\phi-\cos\phi_1)\,\mathbf{e}_1 + (b+c\cos\psi - c\cos\psi_1)\,\mathbf{e}_2 + c(\sin\phi+\sin\psi+\sin\phi_1+\sin\psi_1)\,\mathbf{e}_3$$

$$\mathbf{n}_1 = -\mathbf{e}_2, \quad \mathbf{n}_2 = -\mathbf{e}_2, \quad \mathbf{n}_3 = \mathbf{e}_1, \quad \mathbf{n}_4 = \mathbf{e}_1, \quad \mathbf{n}_5 = \mathbf{e}_1, \quad \mathbf{n}_4 = \mathbf{e}_6,$$
$$\mathbf{n}_7 = \mathbf{e}_2, \quad \mathbf{n}_8 = \mathbf{e}_2, \quad \mathbf{n}_9 = \mathbf{e}_2, \quad \mathbf{n}_{10} = \mathbf{e}_2, \quad \mathbf{n}_{11} = -\mathbf{e}_1, \quad \mathbf{n}_{12} = -\mathbf{e}_1 \tag{8}$$

the closure condition $\mathbf{x}_0 = \mathbf{x}_{12} - b\,\mathbf{e}_2 = 0$ yields for the four angles:

$$\cos\phi - \cos\phi_1 = 0, \quad \cos\psi - \cos\psi_1 = 0, \quad \text{and} \quad \sin\phi + \sin\phi_1 + \sin\psi + \sin\psi_1 = 0, \tag{9}$$

and the non–trivial solutions of these set of equations are:

$$\phi_1 = -\phi, \quad \psi_1 = -\psi \tag{10}$$

and

$$\psi = -\phi, \quad \phi_1 = \phi, \quad \psi_1 = \psi. \tag{11}$$

5. Numerical results

The solutions (10) and (11) correspond to the positions of the mechanism shown at the top and bottom in Fig. 3, respectively. Written in components the Jacobian $\mathbf{J}$, defined by (6), is a 18×12 matrix, and the set of homogeneous equations (7) is solvable only if this matrix has a rank smaller than twelve.

With $\phi_1 = -\phi, \ \psi_1 = -\psi$ one obtains for any numerical set of system parameters a, b, c, ϕ and ψ a Jacobian matrix with the rank $R = 10$. I.e., the dimension of the nullspace of this matrix is $D = 2$. Therefore, the degree of freedom of the mechanism is in fact in this region permanently: $F = 2$. A non–numerical proof should be possible but is not easy to establish.

With $\psi = -\phi, \ \phi_1 = \phi, \ \psi_1 = \psi$, however, the rank of the Jacobian matrix becomes, for any numerical set of the system parameters a, b, c, and ϕ: $R = 11$. I.e., the nullspace of the Jacobian matrix has dimension $D = 1$ and the mobility of the mechanism, therefore, in this realm is $F = 1$. The only (unit)vector of the nullspace is given by

$$\omega_0^{\mathrm{T}} = \frac{1}{\sqrt{12}} \boxed{1 \ \ -1 \ \ -1 \ \ 1 \ \ -1 \ \ 1 \ \ 1 \ \ -1 \ \ 1 \ \ -1 \ \ -1 \ \ 1}$$

The mechanism can be flattened in four configurations; they are characterized by: $\phi = \phi_1 = \psi = \psi_1 = 0$, or $\phi = \phi_1 = 0, \ \psi = \psi_1 = \pi$, or $\phi = \phi_1 = \pi, \ \psi = \psi_1 = 0$, or finally by $\phi = \phi_1 = \pi, \ \psi = \psi_1 = \pi$. In the first flattened configuration the mechanism forms a square with side lengths $2c \ (= a + b + c)$, and it is this position from which the mechanism can be unfolded into either of two realms with different mobilities. The local mobility of the mechanism in the flattened position turns out to be in the first configuration $F = 5$, in the second and the third $F = 2$, and in the fourth configuration $F = 1$.

6. Workspaces of point P

Three plates of the Queer Square mechanism are moving translatorially if the plate which carries the coordinate system $(x_1, \ x_2, \ x_3)$ is held fixed on ground (Fig. 3). To find the regions in which the point P is possibly moving we introduce a second coordinate system $(y_1, \ y_2, \ y_3)$ fixed on one of this three plates (see Fig. 6 and 7). By actuating both angles ϕ and ψ simultaneously we get for the relation between the position vectors x and y:

$$x = a\{0,1,0\} + c\{\cos\phi, \cos\psi, \sin\phi + \sin\psi\} + y. \tag{12}$$

Point P with the given position vector $y = \{y_1, y_2, y_3\}$ in (y_1, y_2, y_3) is then moving with two degrees of freedoms in its workspace, an algebraic surface of order four:

$$x_3(x_2, x_1) = y_3 + \sqrt{c^2 - (x_2 - a - y_2)^2} + \sqrt{c^2 - (x_1 - y_1)^2}. \tag{13}$$

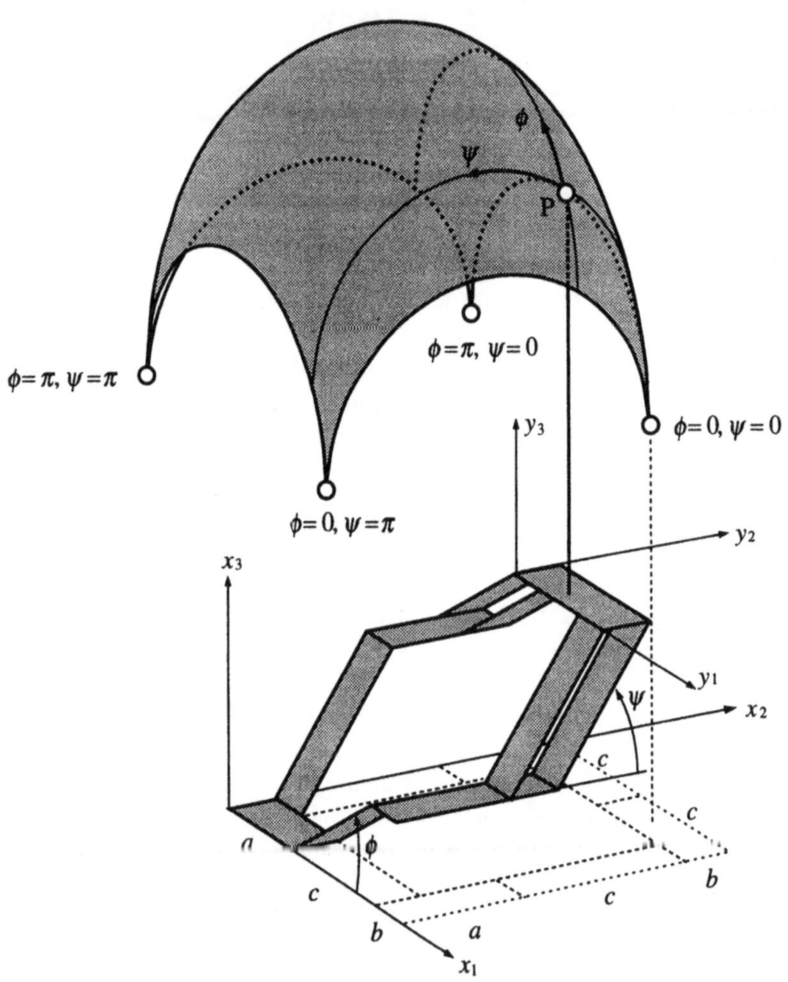

Fig. 6: Workspace of point P ($F = 2$).

But with $\psi \Rightarrow \phi$ and $\phi \Rightarrow -\phi$ (Fig. 7) formula (12) yields:

$$x(\phi) = a\{0,1,0\} + c\{\cos\phi, \cos\phi, 0\} + y \tag{14}$$

and the workspace of P reduces to a straight line in the plane $x_3 = y_3$. The two workspace are in contact at the singularity position $\phi = 0 \wedge \psi = 0$. The four accessible singularity positions of the queer square mechanism are shown in the Fig. 8 and Fig. 9.

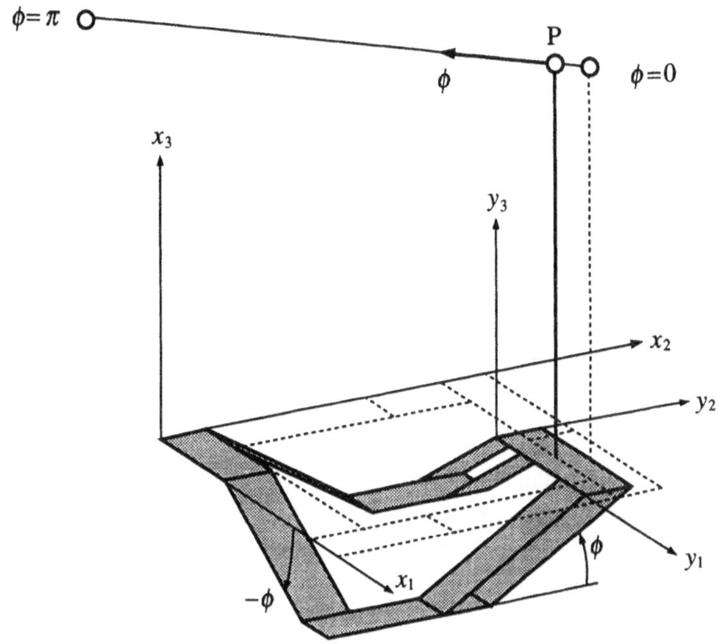

Fig. 7: Workspace of point P ($F = 1$).

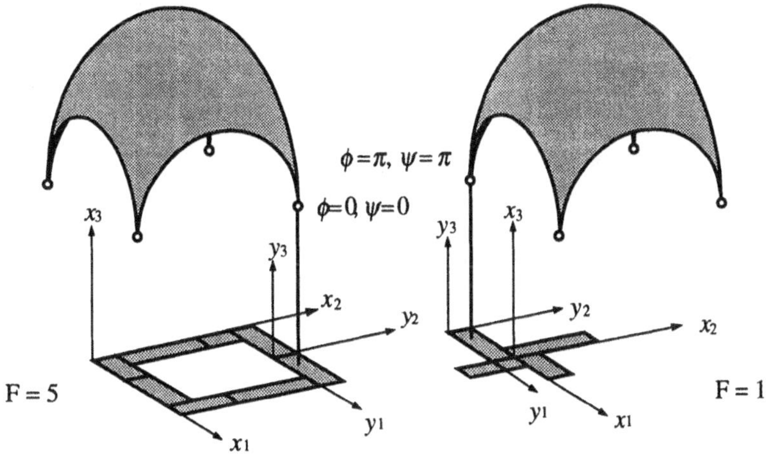

Fig. 8: Flattened positions: $(\phi = 0 \wedge \psi = 0)$, $(\phi = \pi \wedge \psi = \pi)$

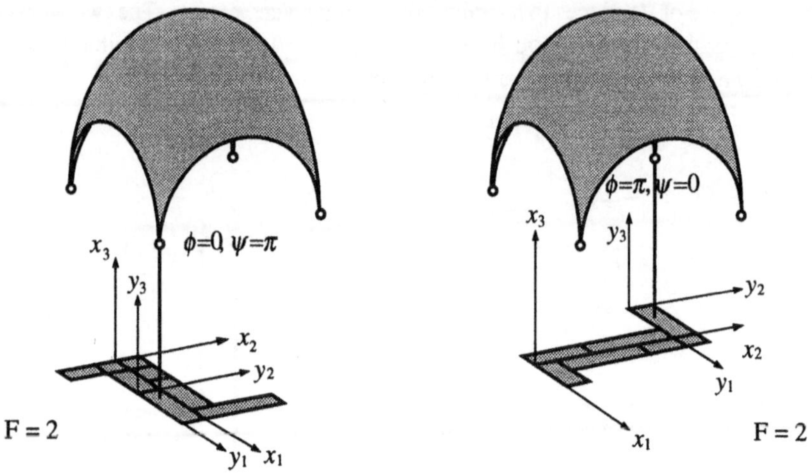

Fig. 9: Flattened positions: ($\phi = 0 \wedge \psi = \pi$), ($\phi = \pi \wedge \psi = 0$)

Summary

Some linkages show “kinematotropy”, i.e., from a singularity position they can open up into regions with different global mobilities. In the paper three examples of such kinematotropic linkages are given, one of them is new. The question whether there exist many or only few kinematotropic linkages, can not be answered.

References

[1] Grübler, M., *Getriebelehre*, Springer, Berlin 1917.

[2] Kutzbach, K. *Mechanische Leistungverzweigung*, Maschinenbau, Der Betrieb, Band 8, S. 710, 1929.

[3] Baker, J. E., *Limiting Positions of the Bricard Linkage and their Possible Relevance to the Cyclohexan Molecule*, Mech.Mach.Theory, Pergamon, Vol.21, No.3, pp.253–260, 1986.

[4] Hunt, K. H., *Kinematic Geometry of Mechanisms*, Clarendon Press, Oxford, 1978.

[5] Wunderlich, W., *Ein merkwürdiges Zwölfstabgetriebe*, Österreichisches Ingenieurarchiv, Band 8, Heft 2/3, S. 224–228, 1954.

[6] Husty, M. L. and Zsombor Murray, P., *A Special Type of Singular Stewart–Gough Platform*, Advances in Robot Kinematics and Computational Geometry, Kluwer, p. 449–548, 1994.

[7] Wohlhart, K., *Das dreifach plansymmetrische Oktoid und seine Punktbahnen*, Mathematica Pannonica, Band 6, Heft 2, S. 243–265, 1995.

[8] Wohlhart, K., *Displacement Analysis of the General Spatial Parallelogram–Manipulator*, 3rd International Workshop on Advances in Robot Kinematics, Ferrara, S. 104–111, 1992.

[9] Angeles, J., *Rational Kinematics*, Section. 5.7, p. 99–102, Springer–Verlag, New–York, Berlin Heidelberg, 1988.

[10] Wohlhart, K., *Motortensorcalculus*, Computational Kinematics, J. P. Merlet and B. Ravani (eds), Kluwer Academic Publishers p. 93–102, 1995.

THE INTEREST OF REDUNDANCY FOR THE DESIGN OF A SPHERICAL PARALLEL MANIPULATOR

CLAUDE REBOULET AND SYLVIE LEGUAY*
*CERT-ONERA - SUP'AERO**
Département d'automatique 2 avenue Edouard Belin
31055 TOULOUSE cedex (FRANCE)
e-mail : reboulet@cert.fr - leguay@cert.fr

Abstract. A new kinematic design of a parallel spherical wrist with actuator redundancy is presented. The particularity of this type of parallel manipulator is that all the actuator axes are collinear, thus permitting unlimited rotation around any axis inside a given workspace. Detailed kinematic analysis has shown that actuator redundancy not only removes singularities but also increases workspace while improving dexterity.

Optimisation of the structure was guided by a global criterion of dexterity. Comparison with other, non-redundant, structures of the same type, carried out with the help of a conditioning measure has shown a significant increase in dexterity in the workspace.

1. INTRODUCTION

A wrist is intended to modify the orientation of an end effector around any arbitrary direction in space. An ideal wrist would have constant performance independent of its configuration and would not penalize a given direction. This property, known as isotropy is, unfortunately, not always guaranteed for many existing mechanisms. Their performance is often mediocre if, in certain configurations, the actuators are obliged to carry out large-scale movements for each change in orientation of the end effector. This isotropic property is however necessary in many applications, as it is generally desirable to keep the same characteristics (acceleration, sensitivity, precision, torque, etc.) for any direction in space and whatever the config-

J. Lenarčič and V. Parenti-Castelli (eds.), Recent Advances in Robot Kinematics, 369-378.
© 1996 *Kluwer Academic Publishers. Printed in the Netherlands.*

uration. There is no reason for the manipulator to have preferred configurations in which it would have better properties.

The solution to this problem depends, firstly, on the choice of an appropriate mechanical concept. Secondly, the geometrical parameters of the structure must be optimised. To do this, it is necessary to define an objective function.

For the mechanical concept, it is possible to choose between serial and parallel chain kinematics. The serial chains, which are characterised by three consecutive rotational joints, generally have a very large usable workspace but very bad isotropy. They have singular points at which the end effector cannot be rotated in certain directions. By adding a fourth joint, this problem of singularities can be avoided. In this case, it is a question of kinematic redundancy.

The other mechanical concept examined in this paper is a parallel structure whose main advantages are its lightness and its rigidity. The use of these chains in a robotic context is relatively recent [14], [9], [20], [15]. Their lightness, due mainly to the possibility of bringing the actuators as close as possible to the base, is interesting in many applications.
Among these applications can be mentioned those requiring a high rate of acceleration [18], [4], [17]; those requiring effort control [20], [15] and, finally, teleoperation applications with a force-reflecting hand controller [19]. In this latter case, the master-slave coupling must be transparent, which calls for the design of a master system with very low inertia, and the most isotropic possible.

The use of these structures as wrists is even more recent [2], [5], [7], [8], [16].
This article presents a new kinematic design of a parallel spherical mechanism with actuator redundancy. Actuator redundancy has been studied in particular by [8] who showed its interest for increasing the dexterity of a parallel mechanism. Actuator redundancy means that the mechanism has more actuators than are strictly necessary to produce a given trajectory without increasing mobility. In this case, actuator rates are uniquely determined by the specified trajectory of the end actuator but actuator forces are undetermined. This type of redundancy is dual to the kinematic redundancy in serial chain manipulators. It is used to increase dexterity.

Many authors have analysed the dexterity of a manipulator and compared the different performance indices [11]. Amongst the principal indices which have been used, [1] uses a quadratic form index to evaluate the instantaneous characteristics of a manipulator robot. It is based on an inertia ellipsoid. [24] uses the square root of JJ^T. Other indices have also been proposed for use with redundant robots [23], [3]; see also [22] who uses a dynamic manipulability ellipsoid. Condition numbers [10] must not be forgotten ; they are certainly the most interesting as there is a direct physical significance as has been shown by [21] and by [12].

The purpose of this article, after a description of this new mechanism and a short presentation of the kinematic equations of the structure, is to define a global dexterity criterion corresponding to a form of dexterity integration over the whole workspace. This criterion, used in conjunction with geometrical constraints concerning the workspace, allow the structure to be optimised. Finally, the results are compared with other, similar, non redundant structures.

1.1. DESCRIPTION OF THE MECHANISM

This is a new type of spherical parallel manipulator with actuator redundancy. It is composed of two pairs of sub arms. The first pair, (1) and (2), is attached to a point $P_{1,2}$ on the moving platform as is shown in Figure 1a. The second pair (3), (4) is attached to $P_{3,4}$. The axes of each articulation (passive and active) are concentric to a point O. The particularity of this structure is that it has four collinear actuators.

This structure, similar to the non redundant mechanism studied by [5], [7], [2], differs in its conception as, in particular, there are only two attachment points on the moving platform. The structure, shown here in its nominal configuration, is symmetrical to a plane. The authors chose an angle of 90° for $P_{1,2}OP_{3,4}$. This choice, in addition to simplifying modelisation, corresponds to the best possible condition number, for obvious reasons which are not presented here. This leads to the basic concept of the mechanism depicted in Figure 1a.

One of the interests of this concept with four collinear motors is that it permits unlimited rotation around any axe inside a given cone. It would be advantageous to have as wide an open cone as possible so a minimal opening for the cone has been introduced as one of the constraints. Another aim, made possible by actuator redundancy, is to eliminate parallel-type redundancies in the workspace, as is shown in [13], and to standardize dexterity in all this space.

2. FORMULATION OF THE KINEMATIC EQUATION

2.1. INVERSE KINEMATIC PROBLEM

The inverse kinematic problem for this manipulator consists in finding the joint variables corresponding to a given orientation of the moving platform. As there are two admissible positions for each sub arm, this gives 16 solutions. Practically, there is only one solution because of the risk of collisions between the sub arms.

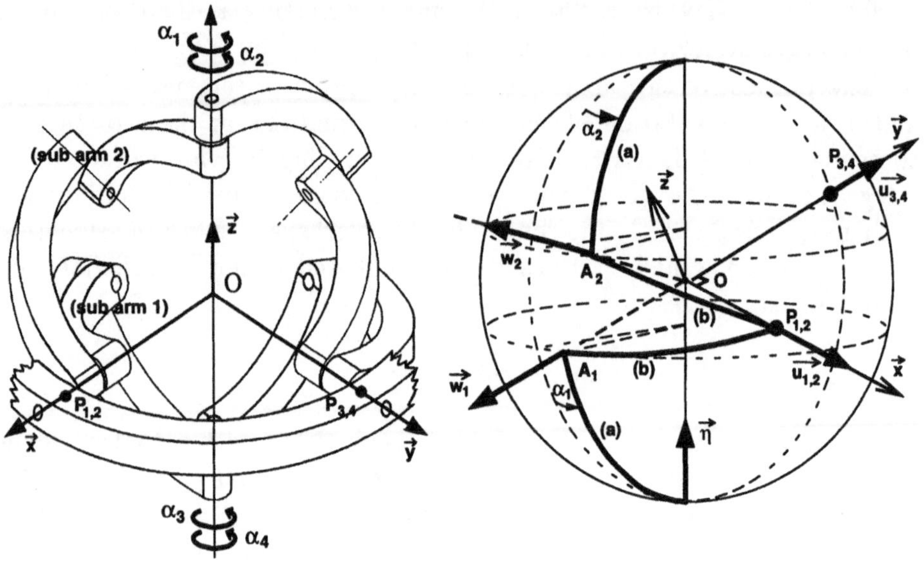

Figure 1. The spherical wrist and its geometrical parameters

In Figure 1b, $\vec{w_i}$ represents the unitary vector of segment $\vec{OA_i}$.
$\vec{u_i}$ represents the unitary vector of segment $\vec{OP_i}$.
In the proposed concept, points P_1 and P_2 are merged as $P_{1,2}$ and points P_3 and P_4 as $P_{3,4}$.
With : $\vec{u_i} = [x_i, y_i, z_i]^T$; x_i, y_i, z_i, components of $\vec{u_i}$ in the fixed base, are functions of three Euler angles ψ, θ, φ defining the orientation of the moving platform.
The components of vectors $\vec{w_i}$ are given by :

$$\vec{w_i} = \begin{bmatrix} \sin a \cos \alpha_i \\ \sin a \sin \alpha_i \\ (-1)^i \cos a \end{bmatrix} \tag{1}$$

The solution to the inverse kinematic problem is obtained by solving the equation :

$$\vec{u_i} \cdot \vec{w_i} = \cos b \tag{2}$$

a (respect. b) represents the angle of the first segment (respect. of the second segment) of each sub arm, equation (2) is represented as :

$$X_i \cos \alpha_i + Y_i \sin \alpha_i = Z_i \tag{3}$$

And :

$$\begin{aligned} X_i &= x_i \sin a \\ Y_i &= y_i = \sin a \\ Z_i &= \cos b - (-1)^i z_i \cos a \end{aligned}$$

Solving this classic equation easily gives solutions for the inverse kinematic problem :

$$\alpha_i = \beta_i \pm \arccos\left(\frac{Z_i}{d_i}\right) \tag{4}$$

with

$$\begin{aligned} d_i &= \sqrt{X_i^2 + Y_i^2} \\ \beta_i &= \arctan 2\left(\frac{Y_i}{X_i}\right) \end{aligned} \tag{5}$$

2.2. JACOBIAN MATRIX

The Jacobian matrix can be found by diffentiation of equation (2) as shown by [7], which gives :

$$\dot{\vec{u_i}} \cdot \vec{w_i} + \vec{u_i} \cdot \dot{\vec{w_i}} = 0 \tag{6}$$

however : $\dot{\vec{u_i}} = \vec{\omega} \wedge \vec{u_i}$, $\vec{\omega}$ being the angular velocity of the moving platform. Equation (6) can be rewritten as

$$b_i \dot{\alpha}_i + (\vec{u_i} \wedge \vec{w_i}) \cdot \vec{\omega} = 0 \tag{7}$$

with : $b_i = \vec{\eta} \cdot (\vec{w_i} \wedge \vec{u_i})$ which gives the 4x3 Jacobian matrix :

$$\dot{\vec{\alpha}} = J.\vec{\omega} \tag{8}$$

where $\frac{-(\vec{u_i} \wedge \vec{w_i})^T}{b_i}$ is the ith line of the Jacobian matrix.
$\dot{\alpha}$ is the 4x1 vector of joint rates.
[13] shows that the 4x3 matrix with lines composed of vectors : $(\vec{u_i} \wedge \vec{w_i})^T$ has no longer any singularity.

It is also interesting to note that, because of the existence of four collinear motors axes , it is better to choose a Euler representation where the first rotation ψ takes place around an axis common to all the motors. This choice permits a kinematic analysis in function of two angles only : θ and φ.

3. KINEMATIC ANALYSIS AND OPTIMAL DESIGN

The notion of optimality is difficult to incorporate in the design process because of the lack of global objective functions. There are no criteria with a sufficiently synthetic power to include simultaneously the principal objective functions for the design of a robot manipulator. It is particularly difficult to combine the geometrical and dynamic aspects.

The primary goal of the design is generally the maximisation of the usable workspace which is frequently a drawback in parallel mechanisms. It is also desirable that inside this workspace, the mechanism has the best possible conditioning. Theses two aims are often contradictory.

In the case of the present mechanism, it is possible to choose parameters a and b, characteristic of the structure, in such a way that it is perfectly isotropic in normal working conditions. Unfortunately, the usable workspace is then very limited. It may, therefore, be interesting to accept isotropy degraded at nominal configuration in order to dispose of a larger workspace. It is evident that isotropy as the only criterion is insufficient for designing a mechanism in most applications.

Let us consider the condition number which measures the ratio between the largest and the smallest singular values :

$$k(J) = \frac{\sigma_{max}}{\sigma_{min}}$$

As this criterion has only a local character, it is incapable of finding an optimal design. It must, therefore, be combined with other objective functions or integrated into an objective function in order to achieve more global dexterity indices.
Amongst the global criteria used, the criterion defined in [6] is particularly interesting.

$$D_g = \frac{\int_w \frac{1}{k(J)} dw}{\int_w dw} \tag{9}$$

This criterion represents integrated dexterity over the whole workspace and standardised by the volume of this workspace. Several other global criteria are shown in [12].

The problem of global criteria is linked to the difficulty of interpreting the work volume. The notion of workspace for a rotation-producing system is not as clear as it is for translations. This is, precisely, the difficulty with using such criteria. Integration in a three Euler angle space would obviously be meaningless as we would not give the same weighting to each element of a given work volume. For this reason, it is necessary to use quadratic invariants, also known as quaternions.
Let : $\begin{array}{l} q_0 = \cos\frac{\alpha}{2} \\ \vec{q} = \sin\frac{\alpha}{2}\vec{u} \end{array}$ with $\vec{u}$, axis of rotation and α the angle of rotation.
In the four dimensional quaternions space, an orientation corresponds to a point of the unit hypersphere. We will, therefore, integrate on the unit

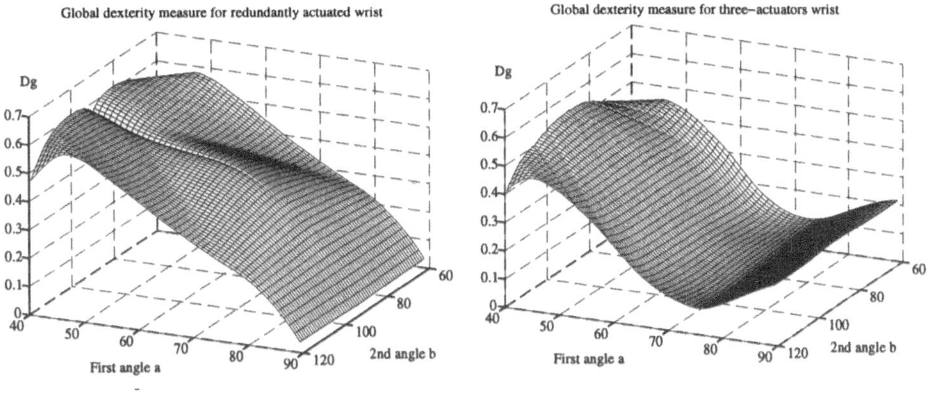

Figure 2. The global dexterity measure

hypersphere in the quaternions space before coming back to Euler angles for which the limits of the workspace are well-known.

In [13], the authors gave a new expression for the previous criterion, which, given the choice of the Euler angles, takes the form :

$$D_g = \frac{1}{B} \iint_{\theta,\varphi} \frac{1}{k(J)} \sin\theta d\theta d\varphi \tag{10}$$

in which B represents : $\int_{\theta,\varphi} \sin\theta d\theta d\varphi$
The interest of this new criterion is the simplicity with which it can be formulated.

Figure 2a shows the value of this criterion as a function of the two characteristic parameters of the structure, a and b. It is interesting to compare the results with the case of a non redundant equivalent spherical wrist like the one examined by [7], [5], [2]. Figure 2b shows the value of this criterion in function of the two characteristic parameters, a and b.

The graphs in Figure 2a for the redundant case and in Figure 2b for the non redundant permit the determination of the couples of optimal parameters in the sense of the criterion mentioned above. From this analysis, notable behavioural differences between the two structures can be seen. However, this single criterion is not sufficient for fixing the definitive choice of coupling for parameters a and b.

Another criterion which seems important for this type of mechanism is its capacity to allow unlimited rotation around any axes situated within a given cone. It is, therefore, interesting to be able to specify a minimal opening for the cone and to compare performance between the different structures under the same conditions.

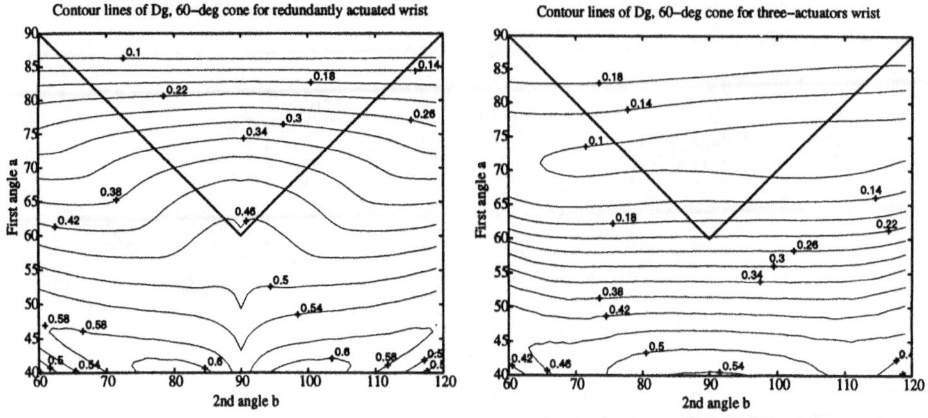

Figure 3. Contour line of global dexterity measure

In fact, the limitations on the workspace come from serial type singularities when two links of a sub-arm are aligned. This additional constraint concerning the minimal opening can be expressed mathematically as :

$$\begin{aligned} a + b &\geq \lambda + \tfrac{\pi}{2} \\ a - b &\geq \lambda - \tfrac{\pi}{2} \end{aligned} \qquad (11)$$

Figures 3a (respect 3b) show the contour lines of the preceding criterion and the geometrical constraints represented by equations (11), where λ was fixed at 60° thus permitting unlimited rotation inside a 60° cone.
It can be seen that the value of the criterion satisfying this constraint is much higher in the case of the redundant robot.

The preceding study permits a choice of values for the parameters a and b. It is thus interesting to visualise the condition number in each case. The rotation matrix of the moving platform is described by three Euler angles (ψ, θ, φ) where ψ is a rotation around axis z, θ is the rotation around the new axis x and φ is a rotation around the new axis z.
The rotation matrix R with respect to the base coordinates is given as a function of the three Euler angles (ψ, θ, φ).

Thanks to this choice, it is possible to represent the condition number as a function of only two Euler angles : θ and φ. As has been shown, this is enough to cover the entire workspace as the surfaces are independent of angle ψ (which has the same axis common to all the motors).

Figures 4a, for the redundant system, and 4b, for the non redundant system show the dexterity measure (the inverse of the condition number) for the two Euler parameters and chosen optimal values of a and b. θ represents the angle of inclination of axis z of the moving platform and φ represents rotation around this axis. A detailed analysis of the contour

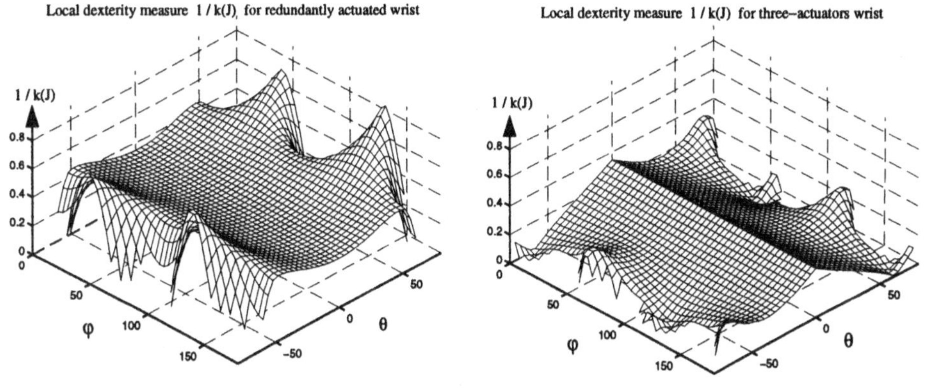

Figure 4. The local dexterity measure 1/k(J)

lines in each case shows a noticeable degradation in performance for non redundant mechanisms.

It can also be seen that for the point corresponding to $\theta = 0$, both types of mechanism accept the same value for the dexterity criterion. When θ is increased, when the moving platform axis is inclined, performance improves for the redundant mechanism whereas it tends to worsen for non redundant mechanism.

4. CONCLUSION

A new kinematic design of a parallel spherical wrist with actuator redundancy has been presented. Kinematic analysis has shown that actuator redundancy not only removes singularities but also improves dexterity in an enlarged workspace. The method of optimisation adopted is based on a global criterion which covers both the notion of isotropy and the workspace volume. Supplementary constraints concerning the workspace volume have been introduced, mainly the possibility of unlimited rotation around any axis situated within a 60° cone (a property which is not common in most known parallel mechanism). The structure which has been presented guarantees dexterity superior to 1/3 over the whole workspace. Comparisons with other equivalent non redundant structures were carried out with the aid of condition number surfaces thus allowing a visualisation of dexterity. They show a notable improvement in performance in terms of uniformity of dexterity for the redundant structure.

Acknowledgements
The authors wish to thank Roger Damien and Roger Isabelle for their help with the construction of the first prototype.

References

1. H. Asada. A geometrical representation of manipulator dynamics and its applications to arm design. *ASME Journal of Dyna., Syst., Meas. and Control*, 105:131–135, 1983.
2. H. Asada and J.A. Cro Granito. Kinematic and static characterisation of wrist joints and their optimal design. In *the IEEE Int Conf on Robotics and Automation, Saint Louis (USA)*, pages 244–250, 1985.
3. P.H. Chang. Development of a dexterity measure for kinematically redundant manipulators. In *Proc. American Control Conference*, pages 496–506, 1989.
4. R. Clavel. Delta a fast robot with parallel geometry. In *the International Symposium in Industrial Robots*, pages 91–100, 1988.
5. D.J. Cox and D. Tesar. The dynamic model of a three of freedom parallel robotic shoulder module. In *4th Int Conf on Advanced Robotics, Colombus (USA)*, jun 1989.
6. C. Gosselin and J. Angeles. A new performance index for the kinematic optimization of robotic manipulators. In *the 20 th ASME Mechanism Conf*, pages 441–447, 1988.
7. C. Gosselin and J. Angeles. The optimal kinematic design of a spherical three degree of freedom parallel manipulator. *J. of Mech., Transm. and Auto. in Design (ASME)*, 8:202–207, 1989.
8. V. Hayward and R. Kurtz. Modeling of a parallel wrist mechanism with actuator redundancy. *Advanced in Robot Kinematics*, 1991.
9. H. Inoue, Y. Tsusaka, and T. Fukuizumi. Parallel manipulator. In *the 3rd International Symposium on Robotics Research, Gouvieux (France)*, pages 321–327, 1986.
10. E. Isaacson and H.B. Keller. *Analysis of numerical methods.* New-York, Wiley, 1966.
11. C.A. Klein and B.E. Blaho. Dexterity measures for the design and control of kinematically redundant manipulators. *Int Journal of Robotics Research*, 6(2):72–83, 1987.
12. R. Kurtz and V. Hayward. Multiple-goal kinematic optimization of a parallel spherical mechanism with actuator redundancy. *IEEE Trans. on Robotics and Automation*, 8(5):644–653, 1992.
13. S. Leguay and C. reboulet. New design of a redundant spherical parallel manipulator. In *ISRAM 96, Montpellier (to appear)*, jun 1996.
14. H. McCallion, G.R. Johnson, and D.T. Pham. A compliant device for inserting a peg in a hole. *The industrial Robot*, pages 81–87, 1979.
15. J.P. Merlet. Parallel manipulator, part i : Theory, design, kinematics, dynamics and control. Technical Report 646, INRIA (France), 1987.
16. J.P. Merlet. Parallel manipulator, part iv : Mode d'assemblage et cinematique directe sous forme polynomiale. Technical Report 1135, INRIA (France), 1989.
17. F. Pierrot, P. Dauchez, and A. Fournier. Hexa a fast six-dof fully parallel robot. In *the 5th International Conference on Advanced Robotics, Pisa (Italy)*, pages 1158–1163.
18. C. Reboulet, C. Lambert, and N. Nombrail. A parallel redundant manipulator : Speed-r-man and its control. In *the 4th ISRAM, Santa-Fe*, pages 285–291, 1992.
19. C. Reboulet, Y. Plihon, and Y. Briere. Interest of the dual hybrid control scheme for teleoperation with time delays. In *ISER 95, Stanford*, pages 311–316, 1995.
20. C. Reboulet and A. Robert. Hybrid control of a manipulator equipped with an active compliant wrist. In *the 3rd Symp on Robotics Research*, pages 237–241, 1985.
21. J.K. Salisbury and J. Craig. Articulated hands ; force control and kinematic issues. *International Journal on Robotics Resarch*, 1(1):4–17, 1982.
22. S. Tadokara, I. Kimura, and T. Takamori. A measure for evaluation of dynamic dexterity based on a stochastic interpretation of manipulator design. In *Proc. IEEE Robotics and Automation*, pages 509–514, 1991.
23. T. Yoshikawa. Analysis and control of robot manipulators with redundancy. In MIT Press, editor, *The 1st International Symposium on Robotics Research, Cambridge (USA)*, pages 735–747, 1984.
24. T. Yoshikawa. Manipulability of robotic mechanis. *Int Journal of Robotics Research*, 4(2):3–9, 1985.

FORWARD KINEMATICS AND MOBILITY CRITERIA OF ONE TYPE OF SYMMETRIC STEWART–GOUGH PLATFORMS

P. DIETMAIER
Institut für Mechanik, Technische Universität Graz
Kopernikusgasse 24, A–8010 Graz, Austria
e-mail: dietmaier@mech.tu–graz.ac.at

Abstract. In this paper a special type of symmetric Stewart–Gough platform is analyzed where the six attachment points in each of the two bodies are arranged symmetrically with respect to a centerpoint. The lengths of the six actuators, which connect the two bodies, are given, and the relative position of the two bodies to each other has to be computed (forward kinematics). In a preliminary step a particular symmetric platform with three pairs of equal actuator lengths is studied. It is shown that such a platform can pose in up to 24 different configurations, and that it can become finitely mobile if the geometric dimensions fulfill certain conditions which are provided in the paper. For the more general type of symmetric platform the forward kinematic analysis reveals again that there might exist up to 24 possible postures. A final numerical example even shows that all 24 postures might be real.

1. Introduction

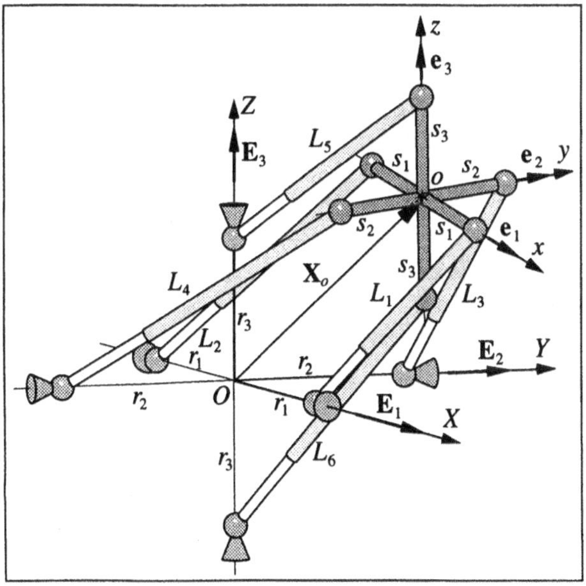

Fig. 1: The symmetric Stewart–Gough platform with attached local coordinate frames.

The object treated in the paper at hand consists of two rigid bodies connected by six actuators with spherical pairs at both ends. Such a kinematic structure is usually referred to as Stewart–Gough platform [2, 10]. When dealing with these platforms one of the kinematic tasks is the so called forward kinematics, where the lengths of the actuators and the positions of the attachment points relative to the rigid bodies are given, and all possible configurations of the structure have to be computed. Some researchers using different techniques have solved this task for the general case in exemplary

J. Lenarčič and V. Parenti-Castelli (eds.), Recent Advances in Robot Kinematics, 379-388.
© 1996 *Kluwer Academic Publishers. Printed in the Netherlands.*

form (e. g. [8]). Furthermore, it has been proved that a general Stewart–Gough platform may have at most 40 configurations [6, 7, 9]. Besides the work on the general case, there have been several publications on closed–form solutions for platforms where some of the attachment points coincide (e. g. [4, 5]). An overview and a classification of this particular kind of platforms can be found in [1]. Although these particular platforms may have 16, 24, 32 or 40 configurations, examples so far have only revealed at most 16 real assembly positions. One of the goals of this paper was to find a Stewart–Gough platform for which it would be easy to find more than 16 real configurations.
In contrast to prior work, this paper provides a solution to the forward kinematic problem of a Stewart–Gough platform where the attachment points are arranged in a symmetrical way and none of them coincide. For a special case of this type of platform the conditions under which this particular platform has an infinite number of configurations, or in other words, becomes mobile, will be investigated.
Fig. 1 shows the symmetric Stewart–Gough platform under investigation. In this drawing the base of the platform is represented by the six cones which are either assumed to be fixed in space or to be rigidly connected to each other — although this is not shown in the drawing. The platform, on the other hand, is represented by three cylinders arranged like a cross and which are again rigidly connected. These two rigid bodies are connected by six actuators or prismatic joints. At one end the actuators are attached to one of the cones and at the other end to one of the endpoints of the rigid cross by means of spherical joints. The attachment points are arranged in pairs symmetrically to the point O in the base, and symmetrically to the point o on the platform. Furthermore, the lines connecting each of the three pairs of attachment points in the base and on the platform are perpendicular to each other, and the distances of attachment points from the centre points (r_i, s_i) as well as the actuator lengths (L_i) are given.
For the purpose of our calculation we attach to the base an orthogonal coordinate frame with the origin O and the unit direction vectors $\mathbf{E}_i$ and a second orthogonal coordinate frame $(o, \mathbf{e}_i)$ to the platform. With the additional use of the vector $\mathbf{X}_o$, which points from the origin O of the base coordinate frame to the origin o of the coordinate frame on the platform, we can express the constraints imposed by the actuators in the following six scalar equations:

$$[\mathbf{X}_o + (s_i\mathbf{e}_i - r_i\mathbf{E}_i)] \circ [\mathbf{X}_o + (s_i\mathbf{e}_i - r_i\mathbf{E}_i)] = L_{2i-1}^2, \qquad i = 1,2,3 \tag{1}$$

and

$$[\mathbf{X}_o - (s_i\mathbf{e}_i - r_i\mathbf{E}_i)] \circ [\mathbf{X}_o - (s_i\mathbf{e}_i - r_i\mathbf{E}_i)] = L_{2i}^2, \qquad i = 1,2,3 \tag{2}$$

with "$\circ$" denoting a vector dot–product. In addition to these conditions it has to be formulated that the unit direction vectors of the platform constitute an orthonormal vector basis. This can be stated by the following three scalar equations and one vector equation:

$$\mathbf{e}_1 \circ \mathbf{e}_1 = 1, \quad \mathbf{e}_2 \circ \mathbf{e}_2 = 1, \quad \mathbf{e}_1 \circ \mathbf{e}_2 = 0, \quad \text{and} \quad \mathbf{e}_3 = \mathbf{e}_1 \times \mathbf{e}_2 \tag{3}$$

with "$\times$" denoting a vector cross–product. These equations constitute the starting point for our calculations.

2. The symmetrically actuated platform

Before analyzing the general symmetric platform we will first examine a particular type of this platform. The additional restrictions which characterize this particular platform are of such kind that the two actuators connected to each pair of symmetrical attachment points must have the same length. However, this means for the platform that it can no longer be positioned arbitrarily in space. For the platform this is a significant drawback, but for the forward kinematics of the general symmetric platform this is one of the possible cases. Furthermore, this approach to the general case will yield a new type of mobile (singular) Stewart–Gough platform.

With the additional relationships between the lengths of the actuators ($L_{2i} = L_{2i-1}$ for $i = 1,2,3$) the six Equs. 1 and 2 can be simplified to:

$$[\mathbf{X}_o \pm \mathbf{v}_i]\circ[\mathbf{X}_o \pm \mathbf{v}_i] = \mathbf{X}_o \circ \mathbf{X}_o \pm 2\mathbf{X}_o \circ \mathbf{v}_i + \mathbf{v}_i \circ \mathbf{v}_i = L_i^2, \quad i = 1,2,3 \tag{4}$$

with $\mathbf{v}_i = s_i \mathbf{e}_i - r_i \mathbf{E}_i$. The task of the forward kinematic now consists in solving these six equations together with the six scalar Equs. 3 for the three coordinates of the vector $\mathbf{X}_o$ and the nine coordinates of the unit vectors $\mathbf{e}_i$ represented in the base coordinate frame. From Equs. 4 it follows right away that the three relations

$$\mathbf{X}_o \circ \mathbf{v}_i = 0, \qquad i = 1,2,3 \tag{5}$$

must hold true. These relations can be satisfied if either $\mathbf{X}_o = \mathbf{0}$ or if the three vectors $\mathbf{v}_i$ become linearly dependent.

For the case $\mathbf{X}_o = \mathbf{0}$ the forward kinematic analysis becomes only a matter of determining the relative orientation of the platform with respect to the base. When $\mathbf{X}_o = \mathbf{0}$ Equs. 4 reduce to the three equations $\mathbf{v}_i \circ \mathbf{v}_i = s_i^2 - 2 s_i r_i (\mathbf{e}_i \circ \mathbf{E}_i) + r_i^2 = L_i^2$, (for $i = 1,2,3$) from which one relative coordinate for each of the unit vectors $\mathbf{e}_i$ can be computed as

$$e_{ii} = \mathbf{e}_i \circ \mathbf{E}_i = \frac{r_i^2 + s_i^2 - L_i^2}{2 s_i r_i}, \qquad i = 1,2,3. \tag{6}$$

The remaining six coordinates of the unit vectors $\mathbf{e}_i$ are then obtained from the constraints of Equs. 3. Therefore, in a first step, we solve the third scalar equation and the vector equation of (3) for the coordinates e_{12}, e_{13}, e_{31}, and e_{32}, which making use of $\mathbf{e}_2 \circ \mathbf{e}_2 = 1$ yields the following results:

$$e_{12} = \frac{e_{11}e_{22} - e_{33}}{e_{21}}, \qquad e_{13} = \frac{e_{22}e_{33} - e_{11}e_{22}^2 - e_{11}e_{21}^2}{e_{21}e_{23}},$$
$$e_{31} = \frac{e_{11}e_{22} - e_{33}e_{22}^2 - e_{33}e_{23}^2}{e_{21}e_{23}}, \quad \text{and} \quad e_{32} = \frac{e_{22}e_{33} - e_{11}}{e_{23}}. \tag{7}$$

Substitution of these four coordinates in $\mathbf{e}_1 \circ \mathbf{e}_1 = 1$ by the respective expressions of (7) and again making use of the relation $\mathbf{e}_2 \circ \mathbf{e}_2 = 1$ results in

$$e_{21}^4 + (e_{11}^2 + e_{22}^2 - e_{33}^2 - 1)e_{21}^2 + (e_{33} - e_{11}e_{22})^2 = 0. \tag{8}$$

From this polynomial equation of degree four in the unknown e_{21} we obtain four solutions for this direction vector coordinate, and from

$$\mathbf{e}_2 \circ \mathbf{e}_2 = 1 \;\Rightarrow\; e_{23} = \pm\sqrt{1 - e_{21}^2 - e_{11}^2} \tag{9}$$

for each of the four solutions of e_{21} two associated solutions for e_{23}. Finally, for each of the eight sets (e_{21}, e_{23}) the last four corresponding coordinates are obtained from Equs. 7 in a unique way.Therefore, the case $\mathbf{X}_o = \mathbf{0}$ yields eight configurations with coinciding origins, but different orientations.

A second set of configurations is obtained from the demands $\mathbf{X}_o \neq \mathbf{0}$ and $\mathbf{X}_o \circ \mathbf{v}_i = 0$ (for

$i = 1,2,3$), which means that the three vectors $\mathbf{v}_i$ have to be linearly dependent. This circumstance can be expressed by

$$(\mathbf{v}_1 \times \mathbf{v}_2) \circ \mathbf{v}_3 = 0 \tag{10}$$

or other similar conditions. Evaluating this equation, and taking into account that the $\mathbf{e}_i$'s and $\mathbf{E}_i$'s are unit vectors of two orthogonal bases we obtain a linear relation between the three coordinates e_{ii} which can be stated as

$$(s_1 s_2 s_3 - r_1 r_2 r_3) + (s_1 r_2 r_3 - r_1 s_2 s_3) e_{11} + (r_1 s_2 r_3 - s_1 r_2 s_3) e_{22} + (r_1 r_2 s_3 - s_1 s_2 r_3) e_{33} = 0. \tag{11}$$

Two additional equations with the three e_{ii} as the sole unknowns are obtained by subtracting one of the Equs. 4 — which for $\mathbf{X}_o \circ \mathbf{v}_i = 0$ are just three different equations instead of 6 — from the other two equations, yielding e. g.

$$L_2^2 - L_1^2 + r_1^2 - r_2^2 + s_1^2 - s_2^2 + 2 r_2 s_2 e_{22} - 2 r_1 s_1 e_{11} = 0,$$

and

$$L_3^2 - L_1^2 + r_1^2 - r_3^2 + s_1^2 - s_3^2 + 2 r_3 s_3 e_{33} - 2 r_1 s_1 e_{11} = 0. \tag{12}$$

In general, the three e_{ii} can now be found by solving the system of linear equations given by Equ. 11 and Equs. 12. The remaining coordinates of the three direction vectors $\mathbf{e}_i$ can then be computed the same way as for the case $\mathbf{X}_o = \mathbf{0}$ described above. This computation results again in eight different orientations of the platform. In order to obtain also the remaining three coordinates of the position vector $\mathbf{X}_o$, we can first express two of the coordinates via $\mathbf{X}_o \circ \mathbf{v}_i = 0$ as homogeneous, linear functions of the third coordinate, and insert the result into one of the Equs. 4. This yields a quadratic equation for one of the coordinates of $\mathbf{X}_o$. So, for each of the eight orientations of the platform there are two associate positions of the centre point, which results in sixteen configurations for the case $\mathbf{X}_o \neq \mathbf{0}$. Together with the eight configurations with coinciding origins the symmetrically actuated platform has a total of 24 configurations. These configurations are arranged symmetrically with respect to the three planes spanned by the coordinate axes of the base. The 24 configurations of this particular type of platform can therefore also be viewed as three groups of eight configurations, with each group representing a particular configuration in all octants of the base coordinate frame. This may justify calling this type of Stewart–Gough platforms symmetrical.

In the process of computing the configurations for the case $\mathbf{X}_o \neq \mathbf{0}$ a system of linear equations (Equ. 11 and Equs. 12) for the three e_{ii} had to be solved, for which these equations were assumed to be linearly independent. We now assume that the system dimensions are such that these three equations become linearly dependent, which means that we can express two of the three e_{ii} by the remaining third one. Consequently, the configurations of this particular platform will then depend on the choice of one of the e_{ii}'s, which in turn means that this platform is finitely mobile although the lengths of the actuators are fixed. The conditions for the finite mobility of a symmetric Stewart–Gough platform are thus obtained by the demand of linear dependency of the respective system of linear equations (Equ. 11 and Equs. 12), which are

$$s_1 s_2 s_3 (r_1^2 r_2^2 + r_1^2 r_3^2 + r_2^2 r_3^2) - r_1 r_2 r_3 (s_1^2 s_2^2 + s_1^2 s_3^2 + s_2^2 s_3^2) = 0 \text{ and}$$

$$\begin{aligned} &[r_2 s_2 (r_1 r_2 s_3 - s_1 s_2 r_3) + r_3 s_3 (r_1 s_2 r_3 - s_1 r_2 s_3)] L_1^2 - \\ &r_3 s_3 (r_1 s_2 r_3 - s_1 r_2 s_3) L_2^2 - r_2 s_2 (r_1 r_2 s_3 - s_1 s_2 r_3) L_3^2 + (r_1 s_2 s_3 - s_1 r_2 r_3)(r_2^2 s_3^2 + s_2^2 r_3^2) - \\ &(r_1^2 + s_1^2)[r_2 s_2 (r_1 r_2 s_3 - s_1 s_2 r_3) + r_3 s_3 (r_1 s_2 r_3 - s_1 r_2 s_3)] = 0. \end{aligned} \tag{13}$$

(Note: These conditions do not restrict the actuator lengths if e.g. $s_i = r_i$, $i = 1,2,3$.)

For a symmetric Stewart–Gough platform which complies with these restrictions and where no two of the r_i's or s_i's are zero it is ensured that this platform is mobile with one degree of freedom.
In this context it is of some interest to find out if it's possible to attach an additional rigid actuator to the mobile platform such that its mobility is not restricted, since there already exist examples of special Stewart–Gough platforms where this is possible (e. g. [3]). For a given point on a symmetric, mobile platform we can show that it moves on a spatial algebraic curve of degree 16. In order to find the attachment points for a rigid actuator we have to find those points on the moving platform which generate spatial curves which lie on the surface of a sphere. For a general symmetric, mobile platform moving in one of the possible (four) modes, it can be shown that the primary six attachment points are the only ones which move on the surface of a sphere.
Finally we want to know what happens to the degree of freedom of a symmetric, mobile Stewart–Gough platform if one of the six rigid actuators is removed. In order to answer this question we choose a numerical example with dimensions according to Tb. 1. Furthermore, we pick a point with coordinates $x = 0.15$, $y = 0.1$, and $z = 0.07$ on the platform. If all six rigid actuators are in use, this particular mobile platform can move in four different modes. For each of these modes the chosen point on the platform traces a closed loop in space. The normal projections onto the $X-Z$–plane of these loops are labelled 1 to 4 and appear as thick lines in Fig. 2.

i	r_i	s_i	L_i
1	1	0.9	1
2	0.85	0.8	0.8
3	0.9	1.026198	0.933955

Tb. 1: Dimensions of a mobile symmetric Stewart–Gough platform.

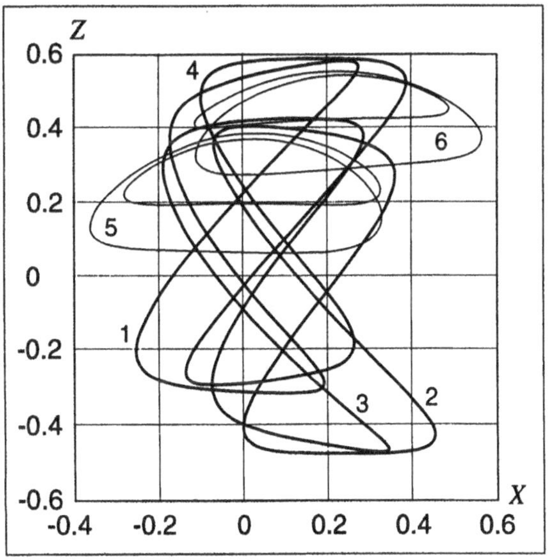

Fig. 2: Paths of the point ($x = 0.15$, $y = 0.1$, $z = 0.07$) with and without the sixth actuator.

If we now remove the sixth actuator (L_6 in Fig. 1) from this particular mobile platform it turns out that the altered platform with five actuators is still mobile with only one degree of freedom, and does not have two degrees of freedom as one could expect. However, the platform with five actuators can now move in two additional modes, labelled 5 and 6 in Fig. 2.
This example demonstrates that the possibility to attach an additional rigid actuator to a mobile platform so that it stays mobile with one degree of freedom not just depends on the geometric dimensions of a five legged platform, but may also depend on the mode the five legged platform is moving in.

3. The general symmetric Stewart–Gough platform

While the solution of the forward kinematics of the symmetrically actuated platform could, in fact, be reduced to the solution of two quadratic equations, this can not be expected for the general case of this type of symmetric Stewart–Gough platform where all the six actuators are of different length. What can be expected, however, is the same number of possible configurations which finally are obtained by solving a corresponding polynomial equation of degree 24.

In order to perform this calculation we start again with Equs. 1 and 2. Our goal is to eliminate, step by step, all but one of the unknown coordinates of the position vector $\mathbf{X}_o$ and the direction vectors $\mathbf{e}_i$. In a preparatory step we subtract Equs. 2 from Equs. 1, which yields three equations with the coordinates of the vector $\mathbf{X}_o$ appearing only in linear form:

$$4\mathbf{X}_o \circ (s_i \mathbf{e}_i - r_i \mathbf{E}_i) = L_{2i-1}^2 - L_{2i}^2, \quad i = 1,2,3. \tag{14}$$

Two additional equations, which do not contain the coordinates of $\mathbf{X}_o$, are obtained by the following linear combinations between Equs. 14, 1 and 2:

$$\begin{aligned}
&[4\mathbf{X}_o \circ (s_1\mathbf{e}_1 - r_1\mathbf{E}_1) - L_1^2 + L_2^2] - [4\mathbf{X}_o \circ (s_2\mathbf{e}_2 - r_2\mathbf{E}_2) - L_3^2 + L_4^2] - \\
&2[\,2\mathbf{X}_o \circ (s_1\mathbf{e}_1 - r_1\mathbf{E}_1) + (s_1\mathbf{e}_1 - r_1\mathbf{E}_1) \circ (s_1\mathbf{e}_1 - r_1\mathbf{E}_1) - \\
&\quad 2\mathbf{X}_o \circ (s_2\mathbf{e}_2 - r_2\mathbf{E}_2) - (s_2\mathbf{e}_2 - r_2\mathbf{E}_2) \circ (s_2\mathbf{e}_2 - r_2\mathbf{E}_2) - L_1^2 + L_3^2] = 0, \\
\text{and} \quad &[4\mathbf{X}_o \circ (s_1\mathbf{e}_1 - r_1\mathbf{E}_1) - L_1^2 + L_2^2] - [4\mathbf{X}_o \circ (s_3\mathbf{e}_3 - r_3\mathbf{E}_3) - L_5^2 + L_6^2] - \\
&2[\,2\mathbf{X}_o \circ (s_1\mathbf{e}_1 - r_1\mathbf{E}_1) + (s_1\mathbf{e}_1 - r_1\mathbf{E}_1) \circ (s_1\mathbf{e}_1 - r_1\mathbf{E}_1) - \\
&\quad 2\mathbf{X}_o \circ (s_3\mathbf{e}_3 - r_3\mathbf{E}_3) - (s_3\mathbf{e}_3 - r_3\mathbf{E}_3) \circ (s_3\mathbf{e}_3 - r_3\mathbf{E}_3) - L_1^2 + L_5^2] = 0.
\end{aligned} \tag{15}$$

Taking into account that the vectors $\mathbf{e}_i$ are unit vectors, Equs. 15 can be simplified and solved for e_{22}, and e_{33} as linear functions of the unknown e_{11}:

$$\begin{aligned}
e_{22} &= \frac{L_1^2 + L_2^2 - L_3^2 - L_4^2 - 2r_1^2 + 2r_2^2 - 2s_1^2 + 2s_2^2 + 4r_1 s_1 e_{11}}{4r_2 s_2}, \\
\text{and} \quad e_{33} &= \frac{L_1^2 + L_2^2 - L_5^2 - L_6^2 - 2r_1^2 + 2r_3^2 - 2s_1^2 + 2s_3^2 + 4r_1 s_1 e_{11}}{4r_3 s_3}.
\end{aligned} \tag{16}$$

Next we insert these two expressions into Equs. 14 and solve these modified three equations for the coordinates of the vector $\mathbf{X}_o$ (a system of linear equations). Using Equs. 3 the result can then be expressed in the following structural form:

$$\begin{aligned}
X_o &= (a_{0x} + a_{1x}e_{11} + a_{2x}e_{12} + a_{3x}e_{13} + a_{4x}e_{21} + a_{5x}e_{31})/(d_0 + d_1 e_{11}), \\
Y_o &= (a_{0y} + a_{1y}e_{11} + a_{2y}e_{12} + a_{3y}e_{21} + a_{4y}e_{23} + a_{5y}e_{32})/(d_0 + d_1 e_{11}), \\
Z_o &= (a_{0z} + a_{1z}e_{11} + a_{2z}e_{13} + a_{3z}e_{23} + a_{4z}e_{31} + a_{5z}e_{32})/(d_0 + d_1 e_{11}).
\end{aligned} \tag{17}$$

In a next step we use the relations $\mathbf{e}_1 \circ \mathbf{e}_2 = 0$, and $\mathbf{e}_3 = \mathbf{e}_1 \times \mathbf{e}_2$ (from Equs. 3) together with Equs. 16 to express also the coordinates e_{21}, e_{23}, e_{31}, and e_{32} as functions of the coordinates of the direction vector $\mathbf{e}_1$. The structure of these relations can be expressed by:

$$
\begin{aligned}
e_{21} &= \frac{a_{0,21} + a_{1,21}e_{11} + a_{2,21}e_{11}^2}{e_{12}}, \\
e_{23} &= -\frac{(a_{0,21} + a_{1,21}e_{11} + a_{2,21}e_{11}^2)\,e_{11} + (a_{0,23} + a_{1,23}e_{11})\,e_{12}^2}{e_{12}e_{13}}, \\
e_{31} &= \frac{a_{0,31} + a_{1,31}e_{11} + a_{2,31}e_{11}^2}{e_{12}}, \\
e_{32} &= -\frac{(a_{0,31} + a_{1,31}e_{11} + a_{2,31}e_{11}^2)\,e_{11} + (a_{0,32} + a_{1,32}e_{11})\,e_{13}^2}{e_{12}e_{13}}.
\end{aligned} \tag{18}
$$

With these relations and together with Equs. 16 and 17 the coordinates of the position vector $\mathbf{X}_o$ and the two direction vectors $\mathbf{e}_2$ and $\mathbf{e}_3$ can be expressed as functions of the three coordinates of the direction vector $\mathbf{e}_1$ only. Therefore it is also possible to express the restrictions $\mathbf{e}_1 \circ \mathbf{e}_1 = 1$, and $\mathbf{e}_2 \circ \mathbf{e}_2 = 1$ as well as one of the Equs. 1 or 2 as algebraic functions of the coordinates of $\mathbf{e}_1$ only. Thus, These three independent algebraic equations can, after some simplification, be written as

$$
\begin{aligned}
& e_{11}^2 + e_{12}^2 + e_{13}^2 = 1, \\
& f_1(1, e_{11}, e_{11}^2, e_{11}^3, e_{11}^4, e_{12}^2, e_{11}e_{12}^2, e_{11}^2e_{12}^2, e_{12}^4) = 0, \\
\text{and}\quad & f_2(1, e_{11}, e_{11}^2, e_{11}^3, e_{11}^4, e_{11}^5, e_{11}^6, e_{12}, e_{11}e_{12}, e_{11}^2e_{12}, e_{11}^3e_{12}, e_{11}^4e_{12}, e_{11}^5e_{12}, e_{12}^2, e_{11}e_{12}^2, e_{11}^2e_{12}^2, \\
& e_{11}^3e_{12}^2, e_{11}^4e_{12}^2, e_{11}^5e_{12}^2, e_{12}^3, e_{11}e_{12}^3, e_{11}^2e_{12}^3, e_{11}^3e_{12}^3, e_{12}^4, e_{11}e_{12}^4, e_{11}^2e_{12}^4, e_{11}^3e_{12}^4, e_{12}^5, e_{11}e_{12}^5, e_{12}^6, \\
& e_{13}, e_{11}e_{13}, e_{11}^2e_{13}, e_{11}^3e_{13}, e_{11}^4e_{13}, e_{11}^5e_{13}, e_{12}e_{13}, e_{11}e_{12}e_{13}, e_{11}^2e_{12}e_{13}, e_{11}^3e_{12}e_{13}, e_{11}^4e_{12}e_{13}, \\
& e_{12}^2e_{13}, e_{11}e_{12}^2e_{13}, e_{11}^2e_{12}^2e_{13}, e_{11}^3e_{12}^2e_{13}, e_{12}^3e_{13}, e_{11}e_{12}^3e_{13}, e_{11}^2e_{12}^3e_{13}, e_{12}^4e_{13}, e_{11}e_{12}^4e_{13}, e_{12}^5e_{13}) = 0
\end{aligned} \tag{19}
$$

with f_i being a linear function of the respective power products. Since the unknown coordinate e_{13} only appears linearly in f_2 it is possible without difficulties to eliminate e_{13} from f_2 and the first condition of Equs. 19, and obtain a second equation which only relates the two unknowns e_{11}, and e_{12}. In this derived relation and with the help of f_1 we can further replace all powers of e_{12} greater than 3 by lower ones, and obtain an algebraic equation

$$
\begin{aligned}
& f_3(1, e_{11}, e_{11}^2, \ldots, e_{11}^{13}, e_{11}^{14}, e_{12}, e_{11}e_{12}, e_{11}^2e_{12}, \ldots, e_{11}^{11}e_{12}, e_{11}^{12}e_{12}, \\
& \qquad e_{12}^2, e_{11}e_{12}^2, e_{11}^2e_{12}^2, \ldots, e_{11}^{11}e_{12}^2, e_{11}^{12}e_{12}^2, e_{12}^3, e_{11}e_{12}^3, e_{11}^3e_{12}^3, \ldots, e_{11}^9e_{12}^3, e_{11}^{10}e_{12}^3) = 0
\end{aligned} \tag{20}
$$

with f_i being again a linear function of its arguments.
In a final step we eliminate the unknown coordinate e_{12} from f_1 of Equs. 19 and Equ. 20 by using the resultant method. This yields a determinant of dimension 7×7 with polynomial elements up to a degree of 12, which, when evaluated, gives the following polynomial equation for the last unknown e_{11}

$$
(a_{0,21} + a_{1,21}e_{11} + a_{2,21}e_{11}^2)^8\,(a_{0,31} + a_{1,31}e_{11} + a_{2,31}e_{11}^2)^8\,p_{24}(e_{11}) = 0. \tag{21}
$$

The first two factors of this equation are the numerators of e_{21} and e_{31} of Equs. 18. These two factors do not contain any valid solutions but were introduced by the elimination process, and could have been partially avoided. The third factor on the other hand, a polynomial of degree 24, contains all the valid solutions for e_{11}. Furthermore, for each of the 24 solutions all the other unknowns are obtained in a unique way by solving linear equations only. The general symmetric Stewart–Gough platform has therefore a maximum of 24 different configurations.

$r_1 = 0.9$	$r_2 = 0.85$	$r_3 = 0.8$
$s_1 = 0.9$	$s_2 = 1.0$	$s_3 = 0.77$
$L_1 = 1.0, L_2 = 1.01$	$L_3 = 1.1, L_4 = 1.09$	$L_5 = 0.9, L_6 = 0.89$

Tb. 2: Dimensions of a general symmetric Stewart–Gough platform.

To illustrate this fact, and to show also that all configurations of such a platform might even be real, we have chosen a particular platform of this type with dimensions according to Tb. 2. The coordinates of the vectors $\mathbf{X}_o$, $\mathbf{e}_1$, $\mathbf{e}_2$ and $\mathbf{e}_3$ as obtained for each of the configurations by the above described procedure, are brought together in Tb. 3. Corresponding to each row of this table Fig. 3 shows a sketch for each of these 24 real configurations with a viewpoint similar to Fig. 1. Furthermore, the sketches are arranged in such a way that similar configurations are side by side. Therefore it is easier to recognize that all 24 configurations are in fact different. The chosen mechanical design shows also that none of the parts interfere in any of the configurations.

4. Conclusions

For a type of symmetric Stewart–Gough platforms, where the attachment points on the platform and in the base are each located symmetrically to a centre point, the forward

i	$X_{o,i}$	$Y_{o,i}$	$Z_{o,i}$	$e_{11,i}$	$e_{12,i}$	$e_{13,i}$	$e_{21,i}$	$e_{22,i}$	$e_{23,i}$	$e_{31,i}$	$e_{32,i}$	$e_{33,i}$
1	0.04864	0.04839	0.04202	0.38051	-0.26261	0.88671	0.90314	0.31172	-0.29524	-0.19887	0.91316	0.35578
2	-0.04818	0.05010	-0.04940	0.38100	-0.90290	-0.19900	0.26302	0.31219	-0.91288	0.88637	0.29547	0.35643
3	-0.05224	-0.06393	0.06186	0.38308	-0.26479	-0.88495	0.90190	0.31417	0.29642	0.19954	-0.91169	0.35917
4	0.06752	0.05249	0.05646	0.38300	0.90195	-0.19952	-0.26471	0.31409	0.91174	0.88501	-0.29638	0.35905
5	-0.05915	0.05409	-0.06373	0.38298	0.26470	0.88502	-0.90195	0.31408	0.29637	-0.19951	-0.91175	0.35904
6	-0.06313	-0.05990	0.04752	0.38258	0.90215	0.19941	-0.26436	0.31370	-0.91198	-0.88529	0.29619	0.35851
7	-0.17080	0.17569	0.18440	0.43456	0.30715	-0.84665	-0.87583	0.36323	-0.31777	0.20993	0.87961	0.42686
8	-0.19261	0.17100	0.16815	0.43492	-0.87565	0.20998	0.30743	0.36357	0.87938	-0.84636	-0.31790	0.42733
9	-0.41431	0.30537	0.41011	0.64385	-0.74361	0.18029	0.45694	0.56267	0.68892	-0.61373	-0.36118	0.70206
10	-0.35191	0.32216	0.45306	0.64373	0.45686	-0.61391	-0.74370	0.56255	-0.36117	0.18035	0.68906	0.70190
11	-0.45631	-0.30171	0.46600	0.69528	0.70389	0.14536	-0.48928	0.61168	-0.62166	-0.52649	0.36110	0.76968
12	-0.36907	-0.32353	0.52086	0.69267	-0.48766	-0.53140	0.70599	0.60920	0.36119	0.14759	-0.62535	0.76626
13	0.39178	-0.30679	0.57544	0.73376	0.51323	0.44519	-0.67135	0.64835	0.35908	-0.10435	-0.56235	0.82029
14	0.48089	-0.26803	0.51942	0.73015	-0.67454	-0.10899	0.51095	0.64491	-0.56835	0.45366	0.35929	0.81554
15	0.45868	0.28407	0.48761	0.70296	0.69762	-0.13845	-0.49404	0.61900	0.61054	0.51163	-0.36078	0.77979
16	0.39063	0.31796	0.53668	0.71091	-0.49897	0.49563	0.69103	0.62657	-0.36040	-0.13072	0.59870	0.79023
17	0.47868	0.23861	-0.56875	0.75277	0.65387	0.07606	-0.52547	0.66646	-0.52888	-0.39651	0.35816	0.84528
18	0.41605	0.26741	-0.60660	0.75464	-0.52671	-0.39127	0.65208	0.66825	0.35810	0.07285	-0.52538	0.84774
19	0.49153	-0.21542	-0.58691	0.76692	-0.63986	0.04917	0.53516	0.67995	0.50128	-0.35417	-0.35813	0.86389
20	0.41194	-0.24987	-0.63266	0.76688	0.53513	-0.35431	-0.63990	0.67991	-0.35813	0.04926	0.50136	0.86383
21	-0.39458	-0.28507	-0.59214	0.73922	-0.51669	0.43195	0.66645	0.65356	-0.35877	-0.09693	0.55308	0.82747
22	-0.47983	-0.26711	-0.54256	0.74438	0.66174	-0.08943	-0.51999	0.65847	0.54408	0.41893	-0.35850	0.83425
23	-0.39354	0.29968	-0.55336	0.71657	0.50248	0.48379	-0.68626	0.63196	0.36009	-0.12480	-0.59003	0.79768
24	-0.46088	0.28620	-0.51464	0.72168	-0.68190	-0.11912	0.50566	0.63684	-0.58202	0.47274	0.35980	0.80440

Tb. 3: The 24 sets of coordinate of a general symmetric Stewart–Gough platform.

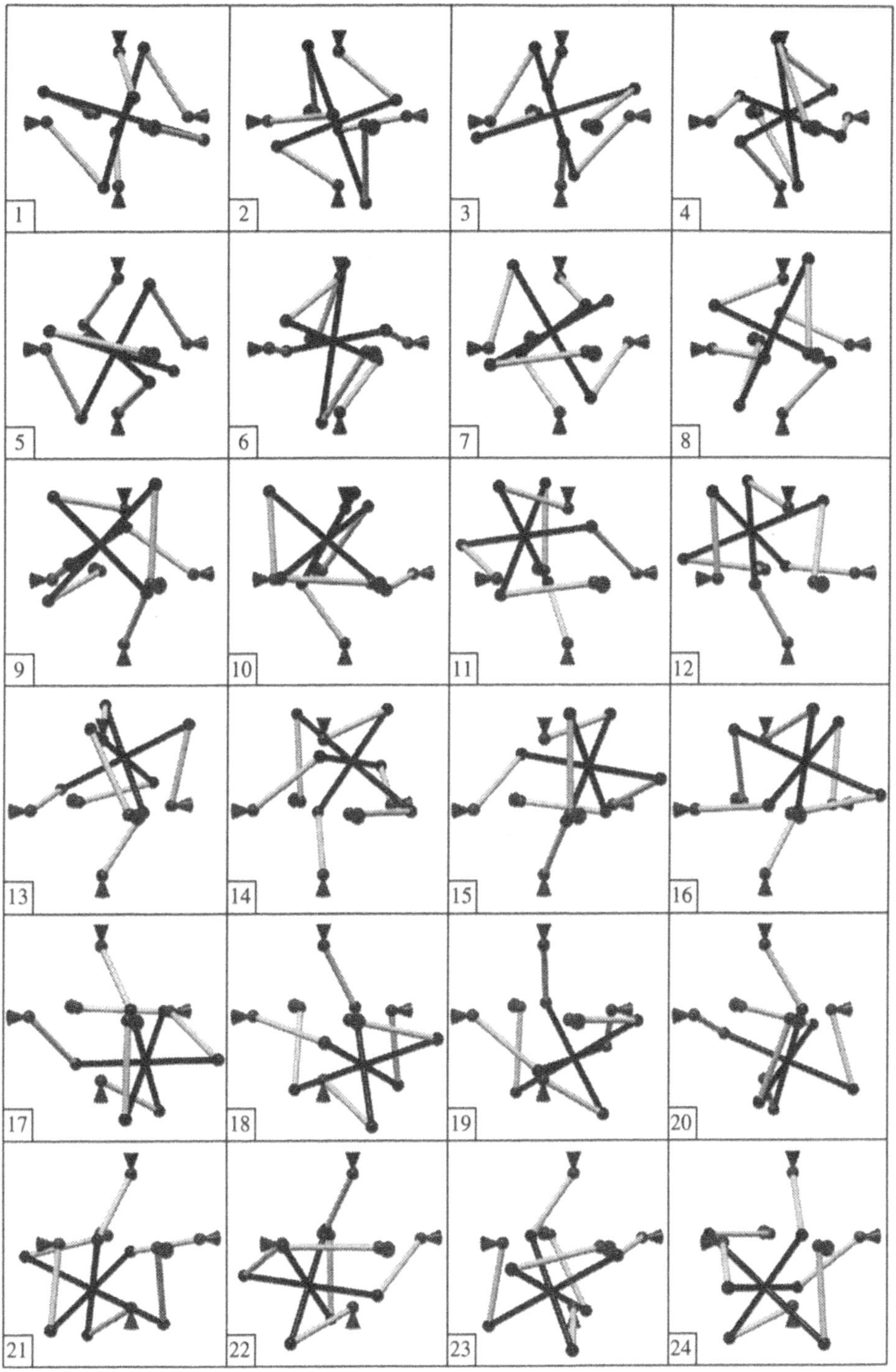

Fig. 3: The 24 postures of a particular general symmetric Stewart–Gough platform.

kinematics was performed. For a special sub–group of these platforms, which are characterized by symmetrically equal actuator lengths, it was shown that the forward kinematics can be solved remarkably easily and that such platforms can have up to 24 configurations. The calculation process for the forward kinematics further revealed, that there might exist symmetric Stewart–Gough platforms with six rigid actuators which in contrast to the general rule might be finitely mobile with one degree of freedom. The conditions for the mobility of such platforms were given in analytical form, and the existence of such platforms was demonstrated by way of a numerical example. Additionally, the matter of how mobility changes if a rigid actuator is added or removed, was also discussed. Finally, an analytical solution for the forward kinematics of the general symmetric Stewart–Gough platform was given, and a numerical example demonstrated that there are not just 24 possible configurations of such a platform but that all of them can be real and that they can be realized in practice.

References

[1] Faugère, J. C. and Lazard, D., 1995: *Combinatorial Classes of Parallel Manipulators*, Mech. Mach. Theory 30, No. 6, pp. 765–776.

[2] Gough, V. E., 1956–1957: *Contribution to discussion to papers on research in automobile stability and control in tyre performance, by Cornell staff*, Proc. AutoDiv. Inst. Mech. Eng. 1956, pp. 392–395.

[3] Husty, M. L. and Zsombor–Murray P., 1994: *A Special Type of Singular Stewart–Gough Platform*, Advances in Robot Kinematics and Computational Geometry, Kluwer Academic Publisher, Dortrecht, pp. 449–458.

[4] Innocenti, C., 1992: *Forward kinematics in analytical form of the topologically–symmetric 4–4 fully–parallel mechanism*, Robotics and Manufacturing, Vol. 4, ASME Press, New York, pp. 411–418.

[5] Innocenti, C. and Parenti–Castelli V., 1993: *Closed–form direct position analysis of a 5–5 parallel mechanism*, ASME J. of Mechanical Design, Vol. 115, No. 3, pp. 515–521.

[6] Lazard, D., 1993: *On the representation of rigid–body motions and its application to generalized platform manipulators*, Computational Kinematics (J. Angeles, P. Kovac and G. Hommel, eds), Kluwer Academic Publisher, Dortrecht, Boston, London, pp. 175–181.

[7] Mourrain, B., 1993: Proc. of ISSSAC'93 (Kiew) (M. Bronstein, ed.), ACM Press, pp. 173–182.

[8] Raghavan, M., 1993: *The Stewart platform of general geometry has 40 configurations*, ASME J. of Mech. Design, Vol. 115, pp. 277–282.

[9] Ronga, F. and Vust, Th., 1992: *Stewart platforms without computer?*, Preprint. Université de Genève.

[10] Stewart, D., 1965: *A platform with 6 degrees of freedom*, Proc. Institution of Mechanical Engineers, London, Vol. 180, Part 1,No. 15, pp. 371–386.

SINGULARITY ANALYSIS AND REPRESENTATION OF SPATIAL SIX-DOF PARALLEL MANIPULATORS

BORIS MAYER ST-ONGE AND CLÉMENT M. GOSSELIN
Département de Génie Mécanique, Université Laval
Québec, Québec, Canada, G1K 7P4

Abstract. In this paper, the singularity loci of general six-degree-of-freedom spatial parallel manipulators are studied and a graphical representation of these loci in the manipulator's workspace is obtained. The algorithm presented is based on analytical expressions of the determinant of the Jacobian matrix. As mentioned elsewhere, two different types of singularities can occur when parallel manipulators are actuated. Both types are considered here and it is shown that one of the two types leads to a trivial description while the second type is more challenging. Moreover, it is shown that, for a given orientation of the platform, the singularity locus in the Cartesian space is represented by a polynomial of degree four. Examples illustrating these results are given. A graphical representation in the Cartesian space is also obtained.

1. Introduction

The drawbacks of conventional serial robot arms have motivated researchers to investigate alternative manipulator geometries. Among other architectures, parallel manipulators have been given considerable attention. Indeed, their closed-loop architecture leads to high load-carrying capacity, high precision and very good stiffness. Therefore, they can be used in many applications where these properties are of primary importance while a limited workspace is acceptable. Manipulators of this type have been designed or studied for several years. The first prototype was designed in 1947 by Gough for testing tires (Gough, 1956–57). Later, Stewart (1965) proposed to use a parallel mechanism as a flight simulator. It is only in the seventies

J. Lenarčič and V. Parenti-Castelli (eds.), Recent Advances in Robot Kinematics, 389-398.
© 1996 *Kluwer Academic Publishers. Printed in the Netherlands.*

though, that Hunt (1978) introduced the concept of parallel manipulator and suggested, in his book, to use this type of mechanism in robotics.

One of the important limitations of parallel mechanisms is that they may lead to singular configurations in which the stiffness of the mechanism is lost. This property has attracted the attention of several researchers and is a crucial issue in a context of analysis and design of parallel manipulators. Gosselin and Angeles (1990) have shown that singularities of closed-loop architecture can be classified in three differents groups. This method is based on the determination of the roots of the determinant of the manipulator's Jacobian matrix. Their classification has been further refined (Zlatanov *et al.*, 1994) where detailed physical interpretations are provided. For the spatial six-degree-of-freedom manipulator, Merlet (1988; 1989) used a method based on Grassmann geometry to characterize the singular configurations. Thus, singularities have been studied for sereval types of six-degree-of-freedom parallel manipulators from a geometric perspective. However, in a context of design, it is also useful to obtain analytical expressions for the singular configurations in order to be able to determine the singularity loci and to superimpose them graphically on the workspace of the manipulator. Analytical expressions for the singularity loci of planar and spherical three-degree-of-freedom parallel manipulators have been obtained (Gosselin and Wang, 1995; Sefrioui and Gosselin, 1993; Sefrioui and Gosselin, 1994; Sefrioui and Gosselin, 1995). However, such expressions are more difficult to obtain for mechanisms with more than three degrees of freedom because of the complexity of the determinant. A custom algorithm is introduced in this paper in order to obtain expressions for the singularity loci of spatial six-degree-of-freedom parallel manipulators which can be used in a practical context of design.

The objective of this work is two-fold: *i*) to obtain analytical expressions of the singularity loci of spatial six-degree-of-freedom parallel manipulators and *ii*) to use these expressions in order to obtain a graphical representation of the singularity loci in the Cartesian space. Such graphical representations are of great interest in the context of design since they allow one to obtain a complete picture of the location of the singular configurations in the workspace. For a given application, it is therefore possible to verify if the singularities will be avoided or not.

2. Kinematic analysis

A spatial six-degree-of-freedom parallel manipulator with prismatic actuators is represented schematically in Fig. 1. The kinematics of this type of manipulator has been studied extensively by sereval authors (Fichter, 1986; Hunt, 1983; Merlet, 1987). It consists of a mobile platform $P_1P_2P_3P_4P_5P_6$

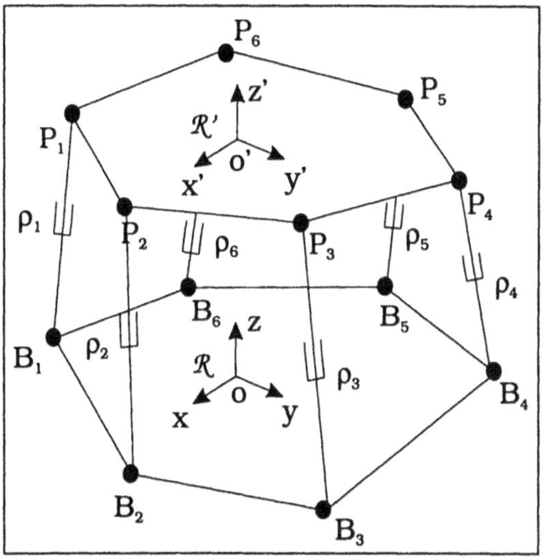

Figure 1. Six-degree-of-freedom parallel manipulator.

and a fixed base $B_1B_2B_3B_4B_5B_6$. The points B_i and P_i, $i = 1,..,6$, are arbitrarily located in 3-dimensional Cartesian space. They are connected via prismatic actuators and the variables representing the distance between B_i and P_i will be noted ρ_i, $i = 1,...,6$. The position and orientation of the platform $P_1P_2P_3P_4P_5P_6$ in space are controlled by adjusting the length of the six legs. The attachments to the platform, at points P_i, $i = 1,...,6$, are universal joints and those at the base, at points B_i, $i = 1,...,6$, are spherical joints. Hence, the mechanism has six degrees of freedom.

For the present analysis, a fixed reference frame $\mathcal{R}$ $(Oxyz)$ is attached to the base of the manipulator, and a moving frame $\mathcal{R}'$ $(O'x'y'z')$ is attached to the platform of the manipulator (Fig. 1). Additionally, the center of the spherical joint connecting the ith leg to the base is noted B_i with coordinates (b_{ix}, b_{iy}, b_{iz}) which are constant when expressed in frame $\mathcal{R}$. Similarly, the center of the joint connecting the ith leg to the platform is noted P_i with coordinates $(p'_{ix}, p'_{iy}, p'_{iz})$ which are constant when expressed in frame $\mathcal{R}'$. The Cartesian coordinates of the platform are given by its position and its orientation. Its position is defined by the position of point O' expressed in frame $\mathcal{R}$ noted (x, y, z), and its orientation by the orientation of frame $\mathcal{R}'$, attached to the platform, with respect to the fixed frame $\mathcal{R}$. Let $\mathbf{Q}$ be the rotation matrix and $\mathbf{r}$ be the translation vector relating frame $\mathcal{R}'$ to the fixed reference $\mathcal{R}$. From the foregoing, the position coordinates of point P_i — the joint connecting the ith leg to the platform — expressed in frame $\mathcal{R}$ and noted (p_{ix}, p_{iy}, p_{iz}), are given by

$$\mathbf{p}_i = \mathbf{r} + \mathbf{Q}\mathbf{p}'_i, \quad i = 1, ..., 6 \tag{1}$$

where $\mathbf{p}_i = [p_{ix}, p_{iy}, p_{iz}]^T$, $\mathbf{r} = [x, y, z]^T$ and $\mathbf{p}'_i = [p'_{ix}, p'_{iy}, p'_{iz}]^T$. By writing this equation in scalar form, one obtains

$$p_{ix} = x + q_{11}p'_{ix} + q_{12}p'_{iy} + q_{13}p'_{iz}, \quad i = 1, ..., 6 \tag{2}$$

$$p_{iy} = y + q_{21}p'_{ix} + q_{22}p'_{iy} + q_{23}p'_{iz}, \quad i = 1, ..., 6 \tag{3}$$

$$p_{iz} = z + q_{31}p'_{ix} + q_{32}p'_{iy} + q_{33}p'_{iz}, \quad i = 1, ..., 6 \tag{4}$$

where q_{ij} is the ijth element (ith row, jth column) of matrix $\mathbf{Q}$. Finally, since the length of the ith segment is the distance between the points P_i and B_i, one can write

$$\rho_i^2 = (p_{ix} - b_{ix})^2 + (p_{iy} - b_{iy})^2 + (p_{iz} - b_{iz})^2, \quad i = 1, ..., 6 \tag{5}$$

Hence, eqs. (5) and (1) represent the explicit solution of the inverse kinematic problem. In order to obtain the Jacobian matrix of this manipulator, the joint velocity vector, noted $\dot{\boldsymbol{\rho}}$, is defined as $\dot{\boldsymbol{\rho}} = [\dot{\rho}_1, \dot{\rho}_2, \dot{\rho}_3, \dot{\rho}_4, \dot{\rho}_5, \dot{\rho}_6]^T$ while the vector of Cartesian velocity is defined as $\mathbf{t} = [x, y, z, \omega_1, \omega_2, \omega_3]^T$, where $\boldsymbol{\omega} = [\omega_1, \omega_2, \omega_3]^T$ is the angular velocity vector of the platform. It is introduced by the relation $\dot{\mathbf{Q}} = \boldsymbol{\Omega}\mathbf{Q}$ where $\boldsymbol{\Omega}$ is the angular velocity tensor ($\boldsymbol{\Omega} = \mathbf{1} \times \boldsymbol{\omega}$). Thus, the Jacobian matrices will be obtained in invariant form. The differentiation of eq. (5) with respect to time then leads to:

$$\mathbf{At} = \mathbf{B}\dot{\boldsymbol{\rho}} \tag{6}$$

where $\mathbf{A}$ and $\mathbf{B}$ are 6 x 6 Jacobian matrices which are written as:

$$\mathbf{A} = \begin{bmatrix} a_{11} & a_{12} & a_{13} & a_{14} & a_{15} & a_{16} \\ \vdots & & & & & \vdots \\ a_{61} & a_{62} & a_{63} & a_{64} & a_{65} & a_{66} \end{bmatrix} \tag{7}$$

and

$$\mathbf{B} = \mathrm{Diag}[\rho_1, ..., \rho_6] \tag{8}$$

with, after simplification

$$a_{i1} = p_{ix} - b_{ix}, \quad i = 1, ..., 6 \tag{9}$$

$$a_{i2} = p_{iy} - b_{iy}, \quad i = 1, ..., 6 \tag{10}$$

$$a_{i3} = p_{iz} - b_{iz}, \quad i = 1, ..., 6 \tag{11}$$

$$a_{i4} = -(p_{iy} - b_{iy})(q_{31}p'_{ix} + q_{32}p'_{iy} + q_{33}p'_{iz}) + (p_{iz} - b_{iz})(q_{21}p'_{ix} + q_{22}p'_{iy} + q_{23}p'_{iz}), \quad i = 1, ..., 6 \tag{12}$$

$$a_{i5} = (p_{ix} - b_{ix})(q_{31}p'_{ix} + q_{32}p'_{iy} + q_{33}p'_{iz}) - (p_{iz} - b_{iz})(q_{11}p'_{ix} + q_{12}p'_{iy} + q_{13}p'_{iz}), \quad i = 1, ..., 6 \tag{13}$$

$$a_{i6} = -(p_{ix} - b_{ix})(q_{21}p'_{ix} + q_{22}p'_{iy} + q_{23}p'_{iz}) + (p_{iy} - b_{iy})(q_{11}p'_{ix} + q_{12}p'_{iy} + q_{13}p'_{iz}), \quad i = 1, ..., 6 \tag{14}$$

3. Singularity loci

Based on the two Jacobian matrices defined in eq. (6), Gosselin and Angeles (1990) have suggested a classification of singularities encountered in closed kinematic chains and three types of singularities having different physical interpretations were found. These singularities occur, respectively, when (*i*) matrix **B** is singular or (*ii*) matrix **A** is singular or (*iii*) the positioning equations degenerate. In this paper, the singularity loci associated with the singularities of type I and II of the spatial six-degree-of-freedom parallel manipulator are studied. The third type of singularities, i.e., architectural singularities, are assumed to be avoided by a proper choice of the kinematic parameters.

Singularities of type I occur when

$$\det(\mathbf{B}) = 0 \tag{15}$$

Mathematically, this condition leads to:

$$\rho_i = 0, \quad i = 1 \ or \ 2 \ or \ ... \ 6 \tag{16}$$

However, since the actuators do not have an infinite range of motion, this type of singularity will in fact occur when one of the actuators reaches one of its limits, i.e., when $\rho_i = \rho_{i,min}$ or $\rho_i = \rho_{i,max}$ for i=1 or 2 ... 6. This is in fact the boundary of the Cartesian workspace.

Singularities of type II occur when the Jacobian matrix **A** is singular, i.e., when

$$\det(\mathbf{A}) = 0 \tag{17}$$

where **A** is given in equation (7). This type of singularities is associated with configurations in which the platform of the manipulator gains extra degree(s) of freedom and is therefore uncontrollable. One of the objectives of the present work is to obtain a graphical representation of the singularity loci in the Cartesian space. Therefore, Jacobian matrix **A** will be written as a function of the Cartesian position coordinates x, y and z. For a given orientation of the platform, the elements of the Jacobian matrix **A** can be written as

$$a_{i1} = G_i + x, \quad i = 1, ..., 6 \tag{18}$$
$$a_{i2} = H_i + y, \quad i = 1, ..., 6 \tag{19}$$
$$a_{i3} = K_i + z, \quad i = 1, ..., 6 \tag{20}$$
$$a_{i4} = L_i y + M_i z + N_i, \quad i = 1, ..., 6 \tag{21}$$
$$a_{i5} = P_i x + Q_i z + R_i, \quad i = 1, ..., 6 \tag{22}$$
$$a_{i6} = S_i x + T_i y + U_i, \quad i = 1, ..., 6 \tag{23}$$

with, after simplification

$$G_i = q_{11}p'_{ix} + q_{12}p'_{iy} + q_{13}p'_{iz} - b_{ix}, \quad i = 1, ..., 6 \tag{24}$$

$$H_i = q_{21}p'_{ix} + q_{22}p'_{iy} + q_{23}p'_{iz} - b_{iy}, \quad i = 1, ..., 6 \tag{25}$$

$$K_i = q_{31}p'_{ix} + q_{32}p'_{iy} + q_{33}p'_{iz} - b_{iz}, \quad i = 1, ..., 6 \tag{26}$$

$$L_i = -(K_i + b_{iz}), \quad M_i = H_i + b_{iy}, \quad i = 1, ..., 6 \tag{27}$$

$$N_i = -b_{iy}L_i - b_{iz}Mi, \quad i = 1, ..., 6 \tag{28}$$

$$P_i = -L_i, \quad Q_i = -(G_i + b_{ix}), \quad i = 1, ..., 6 \tag{29}$$

$$R_i = b_{ix}L_i - b_{iz}Q_i, \quad i = 1, ..., 6 \tag{30}$$

$$S_i = -M_i, \quad T_i = -Q_i, \quad i = 1, ..., 6 \tag{31}$$

$$U_i = b_{ix}M_i + b_{iy}Q_i, \quad i = 1, ..., 6 \tag{32}$$

Using the latter expressions, it is possible to expand eq. (17) to obtain a polynomial expression in x, y and z which represents the singularity locus in the Cartesian space for a given orientation of the platform. This polynomial expression is of degree 4. However, this expression, written in terms of coefficients G_i to U_i is complex and takes about 450 pages. By using a simplified kinematic modeling, i.e., by locating the reference frames at joint connections, it is possible to reduce the length of the equation to 55 pages. Although this represents a significant reduction over the general case, it is still too long and too complex, especially in a context of interactive kinematic analysis, where this procedure must be repeated several times. Thus, a more efficient approach will be introduced in the following section.

4. General algorithm

The objective of this section is to reduce the size — and hence the computation time — of the expression obtained for the determinant. When using a symbolic software tool like *Mathematica* to find this expression, it is not possible to control the algorithm used to calculate the determinant. Hence, a custom algorithm has been developed. The method consists in expanding the determinant by the cofactors of the first row. The advantage of using this method is that the computation of the determinant is divided in several steps, which allows intermediate simplifications. The determinant of matrix $\mathbf{A}$ can be expanded as a combination of the elements of row i and the cofactors of row i, i.e.,

$$\det(\mathbf{A}) = a_{i1}M_{i1} + a_{i2}M_{i2} + ... + a_{in}M_{in} \tag{33}$$

where the *cofactor* M_{ij} is the determinant of matrix $\mathbf{A}_{ij}$ with the proper sign, i.e.,:

$$M_{ij} = (-1)^{i+j}\det(\mathbf{A}_{ij}) \tag{34}$$

and where matrix $\mathbf{A}_{ij}$ is formed by deleting row i and column j of matrix $\mathbf{A}$. These formulas express $\det(\mathbf{A})$ as a combination of determinants of order $(n-1)$. According to the definition, the determinant of a 6×6 Jacobian matrix $\mathbf{A}$ expanded using the first row is then written as:

$$\begin{aligned}\det(\mathbf{A}) &= a_{11}\det(\mathbf{A}_{11}) - a_{12}\det(\mathbf{A}_{12}) + a_{13}\det(\mathbf{A}_{13}) - \\ &\quad a_{14}\det(\mathbf{A}_{14}) + a_{15}\det(\mathbf{A}_{15}) - a_{16}\det(\mathbf{A}_{16}) \end{aligned} \tag{35}$$

where, for instance, matrix $\mathbf{A}_{11}$ is obtained by deleting row one and column one of $\mathbf{A}$. Hence six 5×5 determinants are obtained, which will be expanded by the cofactors of the first row. This method is applied until 3×3 determinants are obtained. At this level, it is possible to write the 3×3 determinants as polynomial equations, as described above with the Jacobian matrix $\mathbf{A}$ as a function of x, y and z. For instance, the 3×3 matrix associated with rows 4, 5 and 6 and columns 3, 4 and 5 can be written as

$$\mathbf{A}_{345} = \begin{bmatrix} a_{43} & a_{44} & a_{45} \\ a_{53} & a_{54} & a_{55} \\ a_{63} & a_{64} & a_{65} \end{bmatrix} = \begin{bmatrix} (K_4 + z) & (L_4 y + M_4 z + N_4) & (P_4 x + Q_4 z + R_4) \\ (K_5 + z) & (L_5 y + M_5 z + N_5) & (P_5 x + Q_5 z + R_5) \\ (K_6 + z) & (L_6 y + M_6 z + N_6) & (P_6 x + Q_6 z + R_6) \end{bmatrix} \tag{36}$$

and its determinant is written as

$$\begin{aligned}\det(\mathbf{A}_{345}) &= F_1 + F_2 x + F_3 y + F_4 xy + F_5 z + F_6 xz + F_7 yz + F_8 xyz + \\ &\quad F_9 z^2 + F_{10} xz^2 + F_{11} yz^2 + F_{12} z^3 \end{aligned} \tag{37}$$

where coefficients F_1 to F_{12} are functions of coefficients K_i to R_i, with $i = 4, 5, 6$, and have between 6 and 18 terms. Following the above procedure, the expression of the determinant will be substantially simplified since each coefficient F_i is multiplied sereval times in the computation of the 6×6 determinant.

Starting with a 6×6 matrix, 20 different 3×3 cofactor matrices can be found, 15 of dimension 4×4 and 6 of dimension 5×5. Thus, the global algorithm can be written as

1. *Determination of the 20* 3×3 *determinants*
   As illustrated above, the 20 3×3 determinants are functions of coefficients G_i to U_i, with $i = 4, 5, 6$. Hence, coefficients F_i can be obtained for each of the 3×3 determinants, which leads to a total of 201 coefficients. Each coefficient has between 6 and 18 terms.
2. *Determination of the 15* 4×4 *determinants*
   The 4×4 determinants are combinations of row 3 and cofactors of row 3 (3×3 determinants). Hence, they are functions of coefficients G_3 to U_3 and F_1 to F_{201}. Coefficients E_i are now introduced in order to simplify the expression of the determinants. 243 such coefficients are defined which have between 4 and 11 terms.

3. *Determination of the* $6\ 5 \times 5$ *determinants*
   Similarly, the 5×5 determinants are combinations of row 2 and cofactors of row 2 (4×4 determinants). Coefficients G_2 to U_2 and E_1 to E_{243} are used and 141 coefficients D_i are introduced. Each coefficient has between 5 and 13 terms.
4. *Determination of the* 6×6 *determinant*
   Finally, the 6×6 determinant is expressed as a function of coefficients G_1 to U_1 and D_1 to D_{141}. A polynomial equation in x, y and z is obtained. Variables x, y and z have a maximum degree of three while the total degree of the terms in the polynomial equation is four. This expression can be written as

$$\begin{aligned}&C_1x^3y + C_2x^3z + C_3x^3 + C_4x^2y^2 + C_5x^2yz + C_6x^2y + C_7x^2z^2 +\\&C_8x^2z + C_9x^2 + C_{10}xy^3 + C_{11}xy^2z + C_{12}xy^2 + C_{13}xyz^2 + C_{14}xyz +\\&C_{15}xy + C_{16}xz^3 + C_{17}xz^2 + C_{18}xz + C_{19}x + C_{20}y^3z + C_{21}y^3 +\\&C_{22}y^2z^2 + C_{23}y^2z + C_{24}y^2 + C_{25}yz^3 + C_{26}yz^2 + C_{27}yz +\\&C_{28}y + C_{29}z^3 + C_{30}z^2 + C_{31}z + C_{32} = 0\end{aligned} \tag{38}$$

   where coefficients $C_i, i = 1, 2, ..., 32$ are functions of coefficients G_1 to U_1 and D_1 to D_{141} and have between 6 and 15 terms.

Hence, the result is a polynomial of degree 4 (eq. 38) which is a function of x, y and z obtained through a cascade of 617 coefficients (C_i, D_i, E_i and F_i). The length of each coefficient is acceptable (maximum 3 lines) while the total length of all coefficients is 13 pages. This substantial reduction in length is due to the fact that, with a custom algorithm, it is possible to simplify the expressions at each of the steps of the expansion of the determinant. The expressions for the intermediate coefficients introduced above are not given here because of space limitation but are available from the authors, upon request. Finally, it is pointed out that, by using a simplified kinematic modeling, i.e., locating the reference frames at some joints, the polynomial equation is exactly the same. In fact, only some coefficients C_i and D_i will be equal to zero. Therefore, the general case is used. The significant reduction of the length of the expression obtained with the custom algorithm leads to an important reduction of the computational time, thereby yielding useful interactive programs.

5. Numerical example

In order to illustrate the application of the algorithm described above, the INRIA prototype, which is a six-degree-of-freedom parallel manipulator, will now be studied (Merlet, 1987). Its geometric properties are summarized in Table 1. The orientation used here is $\phi = -2°$, $\theta = 30°$ and $\psi = -87°$, where ϕ, θ and ψ are three Euler angles (zyz) used to compute matrix $\mathbf{Q}$.

TABLE 1. Geometric properties of the INRIA prototype (all lengths are given in mm)

i	b_{ix}	b_{iy}	b_{iz}	p'_{ix}	p'_{iy}	p'_{iz}
1	92.58	99.64	23.10	30.00	73.00	-37.10
2	132.58	30.36	23.10	78.22	-10.52	-37.10
3	40.00	-130.00	23.10	48.22	-62.48	-37.10
4	-40.00	-130.00	23.10	-48.22	-62.48	-37.10
5	-132.58	30.36	23.10	-78.22	-10.52	-37.10
6	-92.58	99.64	23.10	-30.00	73.00	-37.10

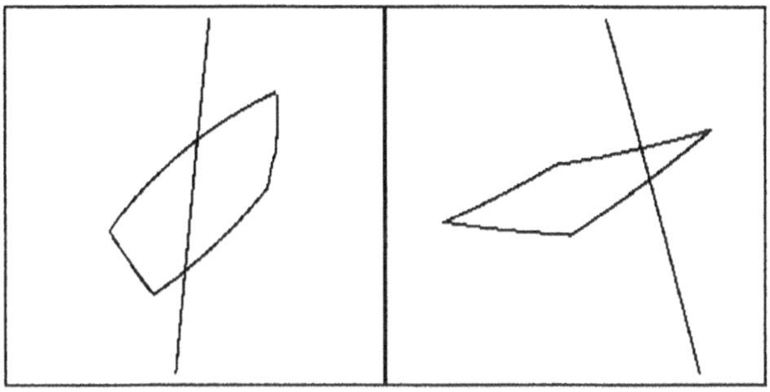

(a) (b)

Figure 2. Section of the workspace obtained (a) in xy plane, with $z = 512$ mm (b) in xz plane, with $y = 65$ mm, showing the singularity locus.

Two differents examples are shown. The first example (Figure 2a), represents a section of the workspace in a plane parallel to the xy plane, when z is equal to 512 mm. It is clear that the singularity locus passes through the workspace. The second example (Figure 2b), represents a section of the workspace in a plane parallel to the xz plane, when y is equal to 65 mm. Again, it is clear that the singularity locus passes through the workspace.

The examples given above illustrate how the expressions for the singularity locus can be used in a context of analysis and design of parallel manipulators. Indeed, an interactive tool based on the above algorithm is of great interest since it is allows the designer to assess the properties of potential kinematic designs.

6. Conclusion

The analytical determination and the graphical representation of the singularity loci of spatial six-degree-of-freedom parallel manipulators has been presented in this paper. The method used here is based on the derivation of an explicit expression for the determinant of the Jacobian matrix of the manipulator. This expression is then used to obtain a graphical representation of the singularity loci. It has been shown that the curve obtained in the three-dimensional Cartesian space, i.e., for a given value of the orientation of the platform, is a polynomial of degree four. An example of application to an architecture of manipulator has been presented. The graphical superposition of the singularity loci and the Cartesian workspace is a very powerful tool in the analysis and design of parallel manipulators.

Acknowledgments - This work was completed under research grants from the Natural Sciences and Engineering Research Council of Canada (NSERC) and le Fonds pour la Formation des Chercheurs et l'aide à la Recherche (FCAR) of Québec. The authors also wish to acknowledge the help of Louis Perreault who wrote and implemented the software allowing the graphical representation of the singularity loci.

References

Fichter, E. F., (1986), A Stewart platform based manipulator: General theory and practical construction, *The International Journal of Robotics Research*, **Vol. 5, No. 3**, pp. 157–181.

Gosselin, C. and Angeles, J., (1990), Singularity analysis of closed-loop kinematic chains, *IEEE Transactions on Robotics and Automation*, **Vol. 6, No. 3**, pp. 281–290.

Gosselin, C. and Wang, J., (1995), Singularity loci of planar parallel manipulators, *Proceedings of the Ninth World Congress on the Theory of Machines and Mechanisms*, Milano, Italy, **Vol. 3**, pp. 1982–1986.

Gough, V. E., (1956–57), Contribution to discussion to papers on research in automobile stability and control and in tyre performance, by Cornell staff, *Proceedings of Auto. Div. Instn mech. Engrs*, 392.

Hunt, K. H., (1978), *Kinematic geometry of mechanisms*, Oxford, Great Britain: Oxford University Press.

Hunt, K. H., (1983), Structural kinematics of in-parallel-actuated robot arms, *ASME Journal of Mechanisms, Transmissions and Automation in Design*, **Vol. 105, No. 4**, pp. 705–712.

Merlet, J.-P., (1987), Parallel manipulator, part 1: Theory, design, kinematic, dynamics and control, INRIA research Report No. 646, France.

Merlet, J.-P., (1988), Parallel manipulator, part 2: Singular configurations and Grassmann geometry, INRIA research Report No. 791, France.

Merlet, J.-P., (1989), Singular Configurations of Parallel Manipulators and Grassmann Geometry, *The International Journal of Robotics Research*, **Vol. 8, No. 5**, pp. 45–56.

Sefrioui, J. and Gosselin, C., (1993), Singularity analysis and representation of planar parallel manipulators, *Journal of Robotics and Autonomous Systems*, **Vol. 10**, pp. 209–224.

Sefrioui, J. and Gosselin, C., (1994), Étude et représentation des lieux de singularité des manipulateurs parallèles sphériques à trois degrés de liberté avec actionneurs prismatiques, *Mechanism and Machine Theory*, **Vol. 29, No. 4**, pp. 559–579.

Sefrioui, J. and Gosselin, C., (1995), On the quadratic nature of the singularity curves of planar three-degree-of-freedom parallel manipulators, *Mechanism and Machine Theory*, **Vol. 30, No. 4**, pp. 533–551.

Stewart, D., (1965), A platform with six degrees of freedom, *Proceedings of the institution of mechanical engineers*, **Vol. 180, No. 5**, pp. 371–378.

Zlatanov, D., Fenton, R.G. and Benhabib, B., (1994), Singularity analysis of mechanisms and robots via a motion-space model of the instantaneous kinematics, *Proceedings of the IEEE International Conference on Robotics and Automation*, **Vol. 2**, pp. 980–991.

8. Analysis of parallel mechanisms

KINEMATICS OF A THREE-DOF PLATFORM WITH THREE EXTENSIBLE LIMBS

LUNG-WEN TSAI
Mechanical Engineering Department
and
Institute for Systems Research
University of Maryland
College Park, MD 20742, U.S.A

Abstract. A three degree-of-freedom manipulator that has a fairly large translational workspace is presented. The mechanism consists of a fixed base, a moving platform, and three extensible limbs. Each limb consists of a prismatic joint and two universal joints connecting the moving platform to the fixed base. Both the direct and inverse kinematics are investigated. The inverse kinematics problem yields two equal and opposite limb lengths for each limb while the direct kinematics problem is reduced to a second-degree polynomial in one unknown. Further, the workspace and singular conditions of the manipulator are discussed.

1. Introduction

A parallel manipulator typically consists of a moving platform that is connected to a fixed base by several limbs. Generally, the number of limbs is equal to the number of degrees of freedom (DOF). Each limb is controlled by one actuator and all limbs are fully in parallel with each other. Parallel mechanisms can be found in many applications such as vehicle and airplane simulators (Stewart, 1965), adjustable articulated trusses (Reinholtz and Gokhale, 1987), mining machines (Arai et al., 1991), pointing devices (Gosselin and Hamel, 1994), fine positioning devices, and off-shore drilling platform. Recently, it has also been developed as a high precision milling machine (Giddings and Lewis, 1995).

The design of parallel manipulators can be dated back to 1947 when Gough and Whitehall (1962) devised a six-linear jack system for use as a universal tire test machine. Stewart (1965) presented his platform ma-

J. Lenarčič and V. Parenti-Castelli (eds.), Recent Advances in Robot Kinematics, 401-410.
© 1996 *Kluwer Academic Publishers. Printed in the Netherlands.*

nipulator for use as an aircraft simulator in 1965. Hunt (1983) made a systematic study of the parallel manipulator structures. Since then, parallel manipulators have been studied extensively by many other researchers (Mohamed and Duffy, 1985; Fichter, 1986; Grffis and Duffy, 1989; Innocenti and Parenti-Castelli, 1990; Nanua et al., 1990). Most of the six-DOF manipulators studied to date consist of six extensible limbs connecting a moving platform to a fixed base by spherical joints. Other variations of the Stewart platform have also been proposed (Kohli et al, 1988; Hudgens and Tesar, 1988; Pierrot et al., 1991; Tsai and Tahmasebi, 1993).

In general, parallel manipulators have the advantages of high stiffness, low inertia, large payload capacity, and simple inverse kinematics. However, they do suffer the following shortcomings:

(a) The direct kinematics is rather difficult and sometimes impossible to solve.
(b) Position and orientation of the moving platform are coupled.
(c) Their workspace is relatively small.
(d) Spherical joints are difficult to manufacture precisely at low cost.

Note that the only six-limbed, six-DOF parallel manipulators for which closed-form direct kinematic solutions have been reported in the literature are special forms of the Stewart platform (Nanua et al., 1990; Grffis and Duffy, 1989; Innocenti and Parenti-Castelli, 1990; Lin et al., 1994; Zhang and Song, 1994). In these special forms, pairs of concentric spherical joints may present design and manufacturing problems. As to the general Stewart platform, researchers have to resort to numerical techniques for the solutions. Innocenti and Parenti-Castelli (1993) developed an exhaustive mono-dimensional search algorithm to find the direct kinematics of the general Stewart platform. Raghavan (1993) showed that the general Stewart platform has 40 direct kinematics solutions by applying the continuation method.

To overcome the above shortcomings, parallel manipulators with fewer than six-DOF ($F < 6$) have been investigated. These $F < 6$ manipulators can maintain all the advantages of a parallel manipulator and, if properly designed, they also have the advantage of having closed-form direct kinematics solutions. For examples, Sternheim (1987) and Clavel (1988) designed a four-DOF high-speed robot called the "Delta" robot. Lee and Shah (1987) analyzed a three-DOF parallel manipulator. Although, these two robots possess closed-form direct kinematics solutions, the Delta robot contains twelve (12) spherical joints while Lee and Shah's manipulator contains three (3) spherical joints. In addition, the position and orientation of Lee and Shah's manipulator are coupled. Recently, Tsai et al. (1996) introduced a novel three-DOF translational platform which is made up of only revolute joints. It possesses pure translational motion and has a closed-form solution for the direct kinematics.

In this paper, a new three-DOF parallel manipulator which is made up of prismatic and universal joints is presented. This manipulator also has a fairly large translational workspace and a closed-form direct kinematics solution. In what follows, we first describe the construction of the mechanism. The sufficient conditions for achieving a pure translational motion of the platform are developed. Then, the inverse and direct kinematics are solved in closed forms. Finally, the workspace and singular conditions of the manipulator are discussed.

2. Mechanism Description

A schematic of the parallel manipulator is shown in Fig. 1, where the fixed base is labeled as link 0 and the moving platform is labeled as link 1. Three identical limbs connect the moving platform to the fixed base by universal joints at points B_i and A_i, $i = 1, 2$, and 3, respectively. Each limb consists of an upper member and a lower member (links 2 and 5 for the first limb, 3 and 6 for the second limb, and 4 and 7 for the third limb) that are connected together by a prismatic joint, P_i. The base connected axes of the universal joints, A_i, $i = 1, 2$, and 3, are co-planar. Similarly, the moving-platform connected axes of the universal joints, B_i, $i = 1, 2$, and 3, are also co-planar. The axes of the two universal joints that are attached to the upper and lower members of each limb are parallel to each other and are perpendicular to the axis of the prismatic joint. The three limbs are preferably, but not necessarily, separated by 120 degrees at the points of connection with the moving and base platforms. Ball screws or hydraulic jacks can be used to vary the lengths of the prismatic joints and, therefore, to control the position of the moving platform.

Let F denote the degrees of freedom of a mechanism, n the number of links, j the number of joints, f_i the degrees of freedom associated with the ith joint, and $\lambda = 6$ the motion parameter. Then, the mobility equation can be written as:

$$F = \lambda(n - j - 1) + \sum_i f_i. \tag{1}$$

For the manipulator shown in Fig. 1, we have $n = 8, j = 9$, and $f_i = 2$ for the universal joints and 1 for the prismatic joints. Applying Eq. (1) to the manipulator, yields

$$F = 6(8 - 9 - 1) + 6 \times 2 + 3 \times 1 = 3. \tag{2}$$

Hence, the manipulator is a three-DOF mechanism. Since the limbs are connected to the moving platform and the fixed base by universal joints,

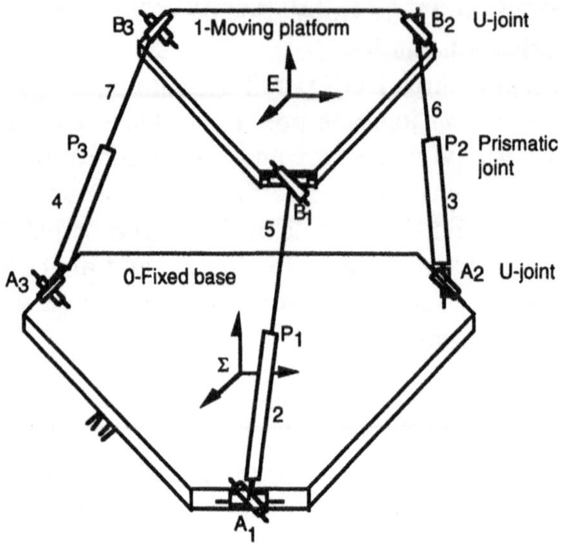

Figure 1. A three-DOF parallel Manipulator

no bending moments will transmit to the limbs. That is the force acting on each limb is directed along the longitudinal axis of the limb and the only moment acting on each limb is a twisting moment. Hence, these limbs can be made of hollow cylindrical rods to produce a light weight, high stiffness, and high speed manipulator.

3. Translational Motion

To facilitate the analysis, two coordinate systems Σ and E are attached to the fixed base and moving platform, respectively, as shown in Fig. 1. Figure 2 shows the geometry of limb i, where w_1 denotes the based attached axis of the lower universal joint, w_5 denotes the moving platform attached axis of the upper universal joint, w_2 and w_4 denote the axes of the universal joints that are respectively attached to lower and upper members of the limb, u_1 is a unit vector pointing from O to A_i, u_2 is a unit vector normal to the plane defined by w_1 and w_2, u_4 is a unit vector pointing from A_i to B_i, u_5 is a unit vector normal to the plane defined by w_4 and w_5, and u_6 a unit vector pointing from Q to B_i. The angles θ_{1i} is measured from u_1 to u_2 about the w_1 axis, θ_{2i} is measured from u_2 to u_4 about the w_2 axis, θ_{4i} is measured from u_4 to u_5 about the w_4 axis, and θ_{5i} is measured from u_5 to u_6 about the w_6 axis.

Let $\mathbf{a}_i = (a_{ix}, a_{iy}, a_{iz})^T$ and $\mathbf{b}_i = (b_{ix}, b_{iy}, b_{iz})^T$ be the position vectors of points A_i and B_i in coordinate systems Σ and E, respectively. Also let

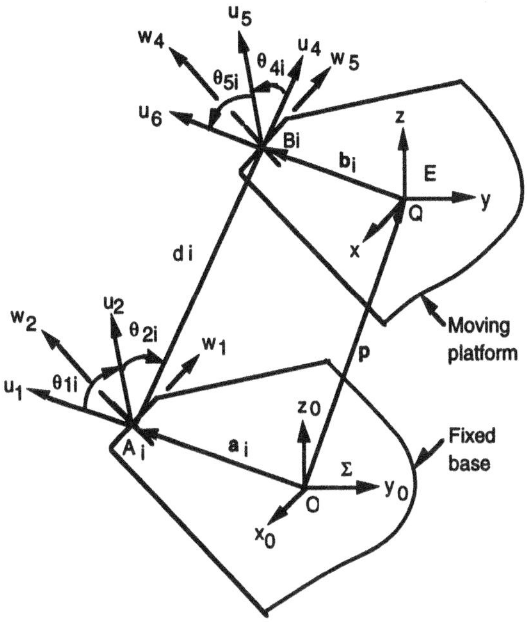

Figure 2. Joint angles and link lengths of limb i

$\mathbf{p} = (p_x, p_y, p_z)^T$ be the position vector of the origin Q of E in Σ, and R be the 3×3 orthogonal matrix describing the orientation of E in Σ. Then, for the ith limb of the manipulator, we may write

$$(\mathbf{p} + R\mathbf{b}_i - \mathbf{a}_i) \cdot (\mathbf{p} + R\mathbf{b}_i - \mathbf{a}_i) = d_i^2, \tag{3}$$

where d_i is the length of the ith limb.

Using the joint angles defined in Fig. 2, it can be shown that the rotation matrix becomes an identity matrix, $R = I$, if the following two conditions are satisfied:

$$\theta_{1i} + \theta_{5i} = 0, \tag{4}$$

and

$$\theta_{2i} + \theta_{4i} = 0. \tag{5}$$

Hence, if the universal joints of the limbs are arranged in such a way that Eqs. (4) and (5) are satisfied, the manipulator will possess a three-DOF translational motion within its workspace. This unique characteristic is very useful in many applications such as a non-conventional $X - Y - Z$ positioning device.

With matrix $R = I$, Eq. (3) reduces to

$$(\mathbf{p} - \mathbf{e}_i) \cdot (\mathbf{p} - \mathbf{e}_i) = d_i^2, \tag{6}$$

where $\mathbf{e}_i = \mathbf{a}_i - \mathbf{b}_i$.

4. Inverse Kinematics

For the inverse kinematics problem, the position vector $\mathbf{p}$ of E with respect to Σ is given and the limb lengths d_i, $i = 1, 2$, and 3, are to be found. Solving Eq. (6) for d_i, yields

$$d_i = \pm\sqrt{(\mathbf{p} - \mathbf{e}_i) \cdot (\mathbf{p} - \mathbf{e}_i)}, \quad \text{for } i = 1, 2, 3. \tag{7}$$

Hence, corresponding to a given position of the moving platform, there are two solutions for each limb length, provided that d_i falls within the range of the corresponding limb length. However, the negative limb length is physically not feasible.

5. Direct Kinematics

For the direct kinematics problem, the limb lengths d_i, $i = 1, 2$ and 3, are given and the position vector of the moving platform, $\mathbf{p}$, is to be found. Expanding Eq. (6), yields

$$\mathbf{p}^2 - 2\mathbf{p} \cdot \mathbf{e}_i + \mathbf{e}_i^2 = d_i^2, \quad \text{for } i = 1, 2, 3. \tag{8}$$

Equation (8) written three times for $i = 1, 2$, and 3, yields three equations in the unknown vector $\mathbf{p}$. Each equation represents a sphere of radius d_i centered at a point defined by the vector $\mathbf{e}_i = \mathbf{a}_i - \mathbf{b}_i$. The intersection of these three spheres yields the solutions to the direct kinematics problem. For the general case, there are two solutions, since the intersection of two of the spheres forms a circle which will generally intersect the third sphere in two locations. Five cases are possible:

(a) Two solutions. The two solutions are realized at the intersection of three spheres.
(b) One double root. Two of the spheres may be tangent to each other. Hence, a double root is possible only if the point of contact lies on the third sphere.
(c) No real solution. The three spheres do not intersect at a common point.
(d) Singular solution. Two of the spheres are concentric. Hence, an infinite number of solutions is possible only if these two spheres coincide, i.e. they are of equal radii.
(e) Singular solution. All three spheres are concentric. Hence, an infinite number of solutions is possible only if the three spheres coincide.

Algebraically, we may eliminate $\mathbf{p}^2$ by subtracting Eq. (8) for $i = 1$ from Eq. (8) for $i = 2$, and $i = 3$, respectively. This results in two linear equations:

$$\mathbf{p} \cdot (\mathbf{e}_2 - \mathbf{e}_1) = k_2, \tag{9}$$

$$\mathbf{p} \cdot (\mathbf{e}_3 - \mathbf{e}_1) = k_3, \tag{10}$$

where
$k_2 = (\mathbf{e}_2^2 - \mathbf{e}_1^2 - d_2^2 + d_1^2)/2$ and
$k_3 = (\mathbf{e}_3^2 - \mathbf{e}_1^2 - d_3^2 + d_1^2)/2$.
We note that Eq. (9) represents a plane containing the circle of intersection between the two spheres defined by limbs 1 and 2. Similarly, Eq. (10) represents another plane containing the circle of intersection between the two spheres defined by limbs 1 and 3. These two planes intersect in a straight line which can be found as follows. Solving Eqs. (9) and (10) for p_x and p_y in terms of p_z, we obtain

$$p_x = \frac{\delta_x}{\delta}, \tag{11}$$

$$p_y = \frac{\delta_y}{\delta}, \tag{12}$$

provided $\delta \neq 0$, where
$\delta_x = (e_{2y} - e_{1y})[p_z(e_{3z} - e_{1z}) - k_3] - (e_{3y} - e_{1y})[p_z(e_{2z} - e_{1z}) - k_2]$,
$\delta_y = (e_{3x} - e_{1x})[p_z(e_{2z} - e_{1z}) - k_2] - (e_{2x} - e_{1x})[p_z(e_{3z} - e_{1z}) - k_3]$, and
$\delta = (e_{2x} - e_{1x})(e_{3y} - e_{1y}) - (e_{3x} - e_{1x})(e_{2y} - e_{1y})$.

Equations (11) and (12) are the parametric representation of a straight line. In general, this line will intersect either one of the spheres defined by Eq. (8) in two points. Substituting Eqs. (11) and (12) into Eq. (8) for $i = 1$, yields

$$c_0 p_z^2 + c_1 p_z + c_2 = 0 \tag{13}$$

where
$c_0 = \delta^2$,
$c_1 = -2\delta^2 e_{1z}$, and
$c_2 = {\delta_x}^2 + {\delta_y}^2 - 2\delta\delta_x e_{1x} - 2\delta\delta_y e_{1y} + \delta^2(e_1^2 - d_1^2)$.
Equation (13) yields:

(a) Two real roots, provided $c_1^2 - 4c_0c_2 > 0$,
(b) One double root, provided $c_1^2 - 4c_0c_2 = 0$, and
(c) No real root, provided $c_1^2 - 4c_0c_2 < 0$.

Corresponding to each p_z, Eqs. (11) and (12) yield a unique solution for p_x and p_y, respectively. We note that the two manipulator postures obtained from Eq. (13) are always mirror image of each other about the plane defined by points A_1, A_2, and A_3.

6. Discussions

From Eq. (7), we note that the workspace is bounded by the maximum extension of the limbs. Let d_{im} be the maximum length of limb i. Then,

$$d_{im} \geq \sqrt{(\mathbf{p}-\mathbf{e}_i)\cdot(\mathbf{p}-\mathbf{e}_i)}, \quad \text{for } i = 1, 2, 3. \tag{14}$$

That is the workspace of the manipulator is bounded by the intersection of the three spheres defined by the maximal extension of the limbs.

As mentioned in the previous section, the direct kinematics solution is determined by the intersection of three spheres. In the general case, Eq. (13) yields two, one, or zero real solution. A singular condition occurs when $\delta = 0$. When $\delta = 0$, the two planes defined by Eqs. (9) and (10) do not intersect. This condition occurs when $\mathbf{e}_i = \mathbf{e}_j = 0$ for any $i \neq j$. Under this condition, the two spheres generated by limbs i and j are concentric and the third sphere intersects them in two parallel planes. In general, two parallel planes do not intersect unless they are linearly dependent, i.e. $d_i = d_j$. Hence, if $\mathbf{e}_i = \mathbf{e}_j = 0$ and $d_i = d_j$ for any $i \neq j$, the three spheres intersect in a circle, resulting in an infinite number of solutions. Furthermore, if $\mathbf{e}_1 = \mathbf{e}_2 = \mathbf{e}_3 = 0$ and $d_1 = d_2 = d_3$, then the three spheres coincide with one another and there are infinite number of solutions lying on the entire surface of the three spheres.

From the above discussions, it becomes clear that in the design of such a manipulator, the condition $\mathbf{e}_i = \mathbf{a}_i - \mathbf{b}_i = 0$ should be avoided. This can be realized by making the size of the moving platform different from that of the fixed base.

7. Summary

A light weight, high stiffness, and high speed parallel manipulator with three translational degrees of freedom is introduced. First, the kinematic structure of the mechanism is described. Second, the sufficient conditions for the moving platform to possess a pure translational motion are developed. Third, the inverse and direct kinematics of the manipulator are investigated. It is shown that the inverse kinematics yields two equal and opposite limb lengths for each limb, but only one is physically possible. The direct kinematics problem has been reduced to a second-degree polynomial in one unknown. Hence, corresponding to each set of limb lengths, there are generally two possible manipulator postures. Finally, the workspace and singular conditions of the manipulator are discussed. It is shown that singular conditions can be avoided by making the size of the moving platform different from that of the fixed base.

8. References

1. Arai, T., Cleary, K., Homma, K., Adachi, H. and Nakamura, T., 1991, "Development of Parallel Link Manipulator for Underground Excavation Task," 1991 Int'l Symposium on Advanced Robot Technology, pp. 541-548.
2. Clavel, R., 1988, "A Fast Robot With Parallel Geometry," 18th Int'l Symposium on Industrial Robots, Sydney, Australia, pp. 91-100.
3. Fichter, E.F., 1986, "A Stewart Platform Based Manipulator: General Theory and Practical Construction," Int'l Journal of Robotics Research, Vol. 5, pp. 157-182.
4. Giddings and Lewis, 1995, Giddings and Lewis Automation Technology, P.O. Box 590, Fond Dulac, WI.
5. Gosselin, C. and Angeles, J., 1989, "The Optimum Kinematic Design of a Spherical Three-Degree-of-Freedom Parallel Manipulator," ASME Journal of Mechanisms, Transmissions, and Automation in Design, Vol. 111, pp. 202-207.
6. Gosselin, C. and Hamel, J., 1994, "The Agile Eye: A High-Performance Three-Degree-of-Freedom Camera-Orienting Device," IEEE Intl. Conference on Robotics and Automation, pp. 781-786.
7. Gough V.E. and Whitehall, S.G.,, 1962, "Universal Tyre Test Machine," Proc. 9th Int'l Tech. Congress, F.I.S.I.T.A., 177 (Institution of Mechanical Engineers).
8. Grffis, M. and Duffy, J., 1989, "Forward Displacement Analysis of a Class of Stewart Platforms," Journal of Robotic Systems, Vol. 6, pp. 703-720.
9. Hudgens, J.C. and Tesar, D., 1988, "A Fully-Parallel Six Degree-of-Freedom Micromanipulator: Kinematic Analysis and Dynamic Model," Trends and Developments in Mechanisms, Machines, and Robotics, Proc. of the 20th ASME Biennial Mechanisms Conf., DE-Vol. 15.3, pp. 29-37.
10. Hunt, K.H., 1983, "Structural Kinematics of In-Parallel-Actuated Robot-Arms," ASME Journal of Mechanisms, Transmissions, and Automation in Design, Vol. 105, pp. 705-712.
11. Innocenti, C. and Parenti-Castelli, V., 1990, "Direct Position Analysis of the Stewart Platform Mechanism," Mechanism and Machine Theory, Vol. 25, pp. 611-612.
12. Innocenti, C. and Parenti-Castelli, V., 1993, "Forward Kinematics of the General 6-6 Fully Parallel Mechanism: An Exhaustive Numerical Approach Via a Mono-Dimensional Search Algorithm," ASME Journal of Mechanical Design, Vol. 115, pp. 932-937.

13. Kohli, D., Lee, S.H., Tsai, K.Y. and Sandor, G.N., 1988, "Manipulator Configurations Based on Rotary-Linear (R-L) Actuators and Their Direct and Inverse Kinematics," ASME Journal of Mechanisms, Transmissions, and Automation in Design, Vol. 110, pp. 397-404.
14. Lee, K. and Shah, D.K., 1987, "Kinematic Analysis of a Three Degrees of Freedom In-Parallel Actuated Manipulator," Proc. IEEE Int'l Conf. on Robotics and Automation, Vol. 1, pp. 345-350.
15. Lin, W., Crane, C.D. and Duffy, J., 1994, "Closed Form Forward Displacement Analysis of the 4-5 In Parallel Platforms," ASME Journal of Mechanical Design, Vol. 116, pp. 47-53.
16. Mohamed, M.G. and Duffy, J., 1985, "A Direct Determination of the Instantaneous Kinematics of Fully Parallel Robotic Manipulators," ASME Journal of Mechanisms, Transmissions, and Automation in Design, Vol. 107, pp.226-229
17. Nanua, P., Waldron, K.J. and Murthy, V., 1990, "Direct Kinematic Solution of a Stewart Platform," IEEE Trans. on Robotics and Automation, Vol. 6, pp. 438-444.
18. Pierrot, F., Fournier, A. and Dauchez, P., 1991, "Toward a Fully Parallel 6 DOF Robot for High-Speed Applications," Proc. of the 1991 IEEE Int'l Conf. on Robotics and Automation, pp. 1288-1293.
19. Raghavan, M., 1993, "The Stewart Platform of General Geometry Has 40 Configurations," ASME Journal of Mechanical Design, Vol. 115, pp. 277-282.
20. Reinholz, C. and Gokhale, D., 1987, "Design and Analysis of Variable Geometry Truss Robot," Proc. 9th Applied Mechanisms Conference, Oklahoma State University.
21. Sternheim, F., 1987, "Computation of the Direct and Inverse Kinematic Model of the Delta 4 Parallel Robot," Robotersysteme 3, pp. 199-203.
22. Stewart, D., 1965, "A Platform with Six Degrees of Freedom," Proc. Institute of Mechanical Engr., London, England, Vol. 180, pp. 371-386.
23. Tsai, L.W., Walsh, G.C. and Stamper, R., 1996, "Kinematics of a Novel Three DOF Translational Platform," accepted for presentation at the IEEE 1996 Int'l Conf. on Robotics and Automation, Minneapolis, MN.
24. Tsai, L.W. and Tahmasebi, F., 1993, "Synthesis and Analysis of a New Class of Six-DOF Parallel Mini-manipulators," Journal of Robotic Systems, Vol. 10, No. 5, pp. 561-580.
25. Zhang, C. and Song, S.M., 1994, "Forward Position Analysis of Nearly General Stewart Platforms," ASME Journal of Mechanical Design, Vol. 116, pp. 54-60.

A CLASS OF FULLY PARALLEL MANIPULATORS WITH CLOSED-FORM FORWARD POSITION KINEMATICS

HERMAN BRUYNINCKX AND JORIS DE SCHUTTER
Katholieke Universiteit Leuven, Dept. Mechanical Engineering
Celestijnenlaan 300B, B-3001 Heverlee, Belgium

Abstract. This paper describes a class of fully parallel manipulators, with planar base and end effector platforms, that have closed-form solutions to their forward position kinematics. "Closed-form" means that only linear, quadratic, and quartic equations have to be solved.

1. Introduction

Parallel manipulators have received growing interest over the last decade, due to some superior characteristics with respect to serial manipulators: rigidity, dynamic characteristics, accuracy, etc., Fichter (1986), Hara and Sugimoto (1989), Hunt (1983), Kerr (1989), Merlet (1989, 1992), Nair and Maddocks (1994), Shi and Fenton (1994). Fully parallel manipulators consist of six serial kinematic chains ("legs") between a fixed "base" and a moving "end effector." Each leg has six joints, but only one is actuated. In most designs, the actuated joint is prismatic (Fig. 1).

The inverse position kinematics for a fully parallel manipulator finds the lengths of the six legs when the end effector position and orientation are given. As is well known, this problem turns out to be simple, and yields unique solutions. The forward position analysis, on the other hand, tries to find the end effector pose when the leg lengths are given. In general, this problem involves highly non-linear equations, and closed-form solutions are only known for a couple of special designs. Moreover, "closed-form" most often means that polynomials of order eight or sixteen have to be solved, Griffis and Duffy (1989), Husain and Waldron (1994), Innocenti and Parenti-Castelli (1989), Nanua *et al.* (1990). "Real" closed-form solutions—i.e., involving only *linear*, *quadratic* and/or *quartic* equations—have been found for very few manipulator designs only. In fact, to the best

J. Lenarčič and V. Parenti-Castelli (eds.), Recent Advances in Robot Kinematics, 411-420.
© 1996 *Kluwer Academic Publishers. Printed in the Netherlands.*

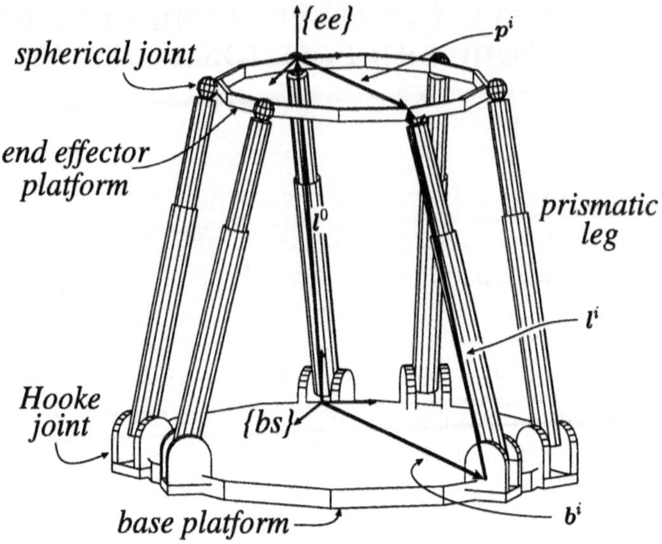

Figure 1. Typical construction of a fully parallel manipulator with planar base and end effector platforms.

of the authors' knowledge, only *three* such designs have been reported in the literature: (i) the five attachment points on either the base or the end effector are collinear, Zhang and Song (1992), (ii) the base and end effector have the same form but a different size, i.e., they are "similar," Lee and Roth (1993), Sreenivasan *et al.* (1994), Wang (1992), or (iii) the legs are arranged in a "321" structure, Nanua and Waldron (1990). All these designs have *planar* base and end effector platforms, i.e., the attachment points lie in the same plane on both the base and the end effector platforms. On the other hand, closed-form solutions are possible if *redundant sensors* are incorporated into the parallel manipulators, Baron and Angeles (1994), Cheok *et al.* (1993), Han *et al.* (1995), Merlet (1993). Examples of redundant sensors are: extra "passive" legs, that are not actuated but whose lengths are measured, or angular encoders on some of the passive revolute joints in one or more legs.

This paper derives the (real) closed-form forward position kinematics for a new, large class of fully parallel manipulators. The members of this class are determined by *four* geometrical parameters: given an (almost) arbitrary design of the base platform, these parameters are used to design the shape of the planar end effector platform. The above-mentioned class of platforms with similar base and end effector, Lee and Roth (1993), Sreenivasan *et al.* (1994), Wang (1992), is a special case of the more general class presented in this paper.

Overview of the Paper. Section 2 describes the mathematical representation of a general manipulator's geometry, and of its loop closure equations. Section 3 then explains which relations between the design parameters of the manipulator give rise to closed-form solutions of the forward kinematics, and how these solutions are constructed. Finally, Section 4 gives some examples.

2. Geometry

Figure 1 shows a general fully parallel manipulator with six legs, and planar base and end effector platforms. The right-handed, orthogonal reference frames $\{bs\}$ and $\{ee\}$ are fixed to the base and to the end effector, respectively. The unit vectors along the axes of these frames are e_x, e_y, e_z, and f_x, f_y, f_z, respectively. Only the prismatic joints along the legs are actuated, and only their joint positions ("leg lengths") are measured. Their attachment points to the base are denoted by the position vectors $b^i, i = 0, \ldots, 5$, and their attachment points to the end effector by the vectors $p^i, i = 0, \ldots, 5$. Only the coordinates of these vectors with respect to their local reference frames $\{bs\}$ and $\{ee\}$ will be needed. These coordinates are denoted as $(b^i_x\ b^i_y\ b^i_z)^T, i = 0, \ldots, 5$, and $(p^i_x\ p^i_y\ p^i_z)^T, i = 0, \ldots, 5$, respectively. The legs are represented by the vectors $l^i, i = 0, \ldots, 5$. The reference frames $\{bs\}$ and $\{ee\}$ are chosen to have their origins at the base and end effector attachment points of the zeroth leg. Hence, $b^0_x = b^0_y = b^0_z = 0$, and $p^0_x = p^0_y = p^0_z = 0$. If one chooses the z axes of $\{bs\}$ and $\{ee\}$ perpendicular to the planes of the platforms, all z coordinates of b^i and p^i are zero.

There are always five kinematic loops through leg zero:

$$l^0 + p^i - b^i = l^i, \quad i = 1, \ldots, 5. \tag{1}$$

The dot product of these closure equations with themselves yields

$$-p^i \cdot l^0 + p^i \cdot b^i + b^i \cdot l^0 = \frac{1}{2}\left((b^i)^2 + (p^i)^2 + (l^0)^2 - (l^i)^2\right) \triangleq L^i, \quad i = 1, \ldots, 5. \tag{2}$$

The five L^i are completely known since they depend on the *constant* design parameters b^i and p^i, and on the *measured* leg lengths l^i. Going to coordinate representation yields

$$b^i_x l^0_x + b^i_y l^0_y - p^i_x l^{0'}_x - p^i_y l^{0'}_y + b^i_x p^i_x r_{xx} + b^i_x p^i_y r_{xy} + b^i_y p^i_x r_{yx} + b^i_y p^i_y r_{yy} = L^i, \tag{3}$$

with

$$l^0_i = e_i \cdot l^0, \quad l^{0'}_i = f_i \cdot l^0, r_{ij} = e_i \cdot f_j. \tag{4}$$

Or, rearranging in matrix form,

$$\boldsymbol{D x} = \begin{pmatrix} \boldsymbol{D_p} & \boldsymbol{D_o} \end{pmatrix} \begin{pmatrix} \boldsymbol{x_p} \\ \boldsymbol{x_o} \end{pmatrix} = \boldsymbol{c}, \tag{5}$$

with

$$D_p = \begin{pmatrix} b_x^1 & b_y^1 & -p_x^1 & -p_y^1 \\ b_x^2 & b_y^2 & -p_x^2 & -p_y^2 \\ b_x^3 & b_y^3 & -p_x^3 & -p_y^3 \\ b_x^4 & b_y^4 & -p_x^4 & -p_y^4 \\ b_x^5 & b_y^5 & -p_x^5 & -p_y^5 \end{pmatrix}, \quad D_o = \begin{pmatrix} b_x^1 p_x^1 & b_x^1 p_y^1 & b_y^1 p_x^1 & b_y^1 p_y^1 \\ b_x^2 p_x^2 & b_x^2 p_y^2 & b_y^2 p_x^2 & b_y^2 p_y^2 \\ b_x^3 p_x^3 & b_x^3 p_y^3 & b_y^3 p_x^3 & b_y^3 p_y^3 \\ b_x^4 p_x^4 & b_x^4 p_y^4 & b_y^4 p_x^4 & b_y^4 p_y^4 \\ b_x^5 p_x^5 & b_x^5 p_y^5 & b_y^5 p_x^5 & b_y^5 p_y^5 \end{pmatrix}, \tag{6}$$

$$x_p = \begin{pmatrix} l_x^0 & l_y^0 & l_x^{0'} & l_y^{0'} \end{pmatrix}^T \quad x_o = \begin{pmatrix} r_{xx} & r_{xy} & r_{yx} & r_{yy} \end{pmatrix}^T, \tag{7}$$

$$c = \begin{pmatrix} L^1 & L^2 & L^3 & L^4 & L^5 \end{pmatrix}^T. \tag{8}$$

The 5×8 matrix D contains only constant *design-dependent* parameters, i.e., functions of the relative positions of the attachment points on both base and end effector. The 8×1 vector x contains unknown, *pose-dependent* parameters. The right-hand side 5×1 vector c contains the known (but pose-dependent) L^i. D and x are split into position-dependent and orientation-dependent subparts, denoted by the subscripts "p" and "o," respectively.

The 5×8 set of equations (5) is linear in the design parameters of the platforms, but nonlinear in their relative pose parameters. This set of equations is to be extended with the three independent and nonlinear orthonormality constraints on the first two columns of the rotation matrix $R = (r_{ij})$:

$$\begin{aligned} 0 &= r_{xx}r_{xy} + r_{yx}r_{yy} + r_{zx}r_{zy} && (9) \\ 1 &= r_{xx}^2 + r_{yx}^2 + r_{zx}^2 && (10) \\ 1 &= r_{xy}^2 + r_{yy}^2 + r_{zy}^2. && (11) \end{aligned}$$

The major contribution of this paper is to show that it is possible to impose two more linear constraints on the design parameters in D_p, in such a way that Eq. (5) can be solved with only linear, quadratic and quartic polynomials. These linear constraints have the following form:

$$P_x = P_{xx}B_x + P_{xy}B_y, \qquad P_y = P_{yx}B_x + P_{yy}B_y. \tag{12}$$

B_x and B_y are the first and second columns in D_p; P_x and P_y are (the negatives of) its third and fourth columns. The constants P_{xx}, P_{xy}, P_{yx} and P_{yy} can be chosen by the designer: once the base platform geometry is chosen (i.e., B_x and B_y), these four parameters completely determine the geometry of the end effector platform. (The designs with similar platforms, Lee and Roth (1993), Sreenivasan *et al.* (1994), Wang (1992), are just special cases of (12), i.e., with $P_{xx} = P_{yy} = k$, and $P_{xy} = P_{yx} = 0$.) The relations (12) imply a third relationship between the design parameters: multiplying each row of the first relation by the corresponding element of P_y, each row of the the second relation by the corresponding element in P_x, and subtracting gives

$$P_{yx}B_x.P_x - P_{xx}B_x.P_y + P_{yy}B_y.P_x - P_{xy}B_y.P_y = 0. \tag{13}$$

Here we used the notation $X.Y$ to denote the element by element multiplication of two vectors X and Y to yield a third vector of the same length. With this notation, Eq. (13) represents a linear relation between the four columns of D_o in Eq. (5).

The proof that the relations (12) generate closed-form solutions is constructive (i.e., the proof produces at the same time an algorithm to compute the solution), and is given in the following section.

3. Closed-form solution

With the three relations derived in the previous section, Eqs (5) can be rewritten as

$$\bar{D}\bar{x} = \begin{pmatrix} d_1 & d_2 & d_5 & d_7 & d_8 \end{pmatrix} \bar{x} = c, \tag{14}$$

with d_i the ith column of D, and

$$\bar{x}_1 = l_x^0 - P_{xx} l_x^{0'} - P_{yx} l_y^{0'} \tag{15}$$

$$\bar{x}_2 = l_y^0 - P_{xy} l_x^{0'} - P_{yy} l_y^{0'} \tag{16}$$

$$\bar{x}_3 = r_{xx} + \frac{P_{yx}}{P_{xx}} r_{xy} \tag{17}$$

$$\bar{x}_4 = r_{yx} + \frac{P_{yy}}{P_{xx}} r_{xy} \tag{18}$$

$$\bar{x}_5 = r_{yy} - \frac{P_{xy}}{P_{xx}} r_{xy}. \tag{19}$$

In deriving these equations, we used the following steps: (i) we replaced the third and fourth columns in D by the relations (12), and (ii) we replaced the sixth column by the linear combination of the fifth, seventh and eight column deduced from (13). (The choice of eliminating the sixth column is arbitrary.)

So, the 5×8 system in Eq. (5) is reduced to the 5×5 system in Eq. (14). In general, the coefficient matrix $\bar{D}$ is of full rank; if not, the design is *architecturally singular*, Ma and Angeles (1991), i.e., it has always at least one motion degree of freedom that the actuators in the prismatic legs cannot avoid. (An example was already pointed out in Lee and Roth (1993): if base and end effector are similar and *regular* hexagons, the resulting manipulator is architecturally singular.) This singular case is easily detected by checking the rank of $\bar{D}$, and solved by choosing either another base platform design (B_x and B_y), or other dependency constants $P_{xx} \dots P_{yy}$. In the general case of a non-singular design, Eq. (14) can be solved for the unknowns $\bar{x}$ using linear operations only; e.g., by some sort of Gaussian elimination. Hence, one gets three linear relations between four elements of the rotation matrix R, i.e., Eqs (17)–(19). The full matrix R can now be solved for in closed form, using also the orthonormality conditions in Eqs (9)–(11). Indeed, Eqs (17)–(19) and (9)–(11) allow to eliminate all but one of the r_{ij}, and to obtain a *quartic* equation in this remaining variable. For example, solving for r_{xy} gives

$$c_4 r_{xy}^4 + c_3 r_{xy}^3 + c_2 r_{xy}^2 + c_1 r_{xy} + c_0 = 0, \tag{20}$$

with

$$\begin{aligned} c_4 &= b_1^2 - b_2 b_3 & c_1 &= 2a_6 a_1 - a_5 a_2 + a_4 a_3 \\ c_3 &= 2b_1 a_1 - a_2 b_3 - a_3 b_2 & c_0 &= a_6^2 + a_4 a_5 \\ c_2 &= a_1^2 + 2a_6 b_1 - a_5 b_2 + a_4 b_3 - a_2 a_3, & & \end{aligned} \tag{21}$$

and

$$\begin{aligned} a_1 &= \bar{x}_3 + q_{yy}\bar{x}_4 - q_{yx}\bar{x}_5 & a_4 &= 1 - (\bar{x}_3)^2 - (\bar{x}_4)^2 \\ a_2 &= -2q_{xx}\bar{x}_3 - 2q_{yx}\bar{x}_4 & a_5 &= (\bar{x}_5)^2 - 1 \\ a_3 &= 2q_{yy}\bar{x}_5 & a_6 &= \bar{x}_4\bar{x}_5, \\ & & & \\ b_1 &= -q_{xx} - q_{yx}q_{yy} & & \\ b_2 &= q_{xx}^2 + q_{yx}^2 & & \\ b_3 &= 1 + q_{yy}^2 & & \end{aligned} \tag{22}$$

and

$$q_{xx} = \frac{P_{yx}}{P_{xx}}, \quad q_{yx} = \frac{P_{yy}}{P_{xx}}, \quad q_{yy} = \frac{P_{xy}}{P_{xx}}. \tag{23}$$

The quartic equation (20) gives four solutions for r_{xy}. In general, however, not all of them are real. Then, Eqs (17)–(19) give r_{xx}, r_{yx} and r_{yy} as linear functions of r_{xy}. The z components of the first two columns of R are then found from Eqs (10)–(11): one of them is used to find *two* solutions for, let's say, r_{zx}, with Eq. (9) being used to determine r_{zy}. The third column of R is straightforwardly constructed from its first two columns by taking their vector product.

Equations (15) and (16) give two linear relations between the three position unknowns l_x^0, l_y^0 and l_z^0; the unknowns $l_x^{0'}$ and $l_y^{0'}$ depend linearly on the three unprimed unknowns since they are connected by the (now fully known) rotation matrix R:

$$\begin{pmatrix} l_x^{0'} \\ l_y^{0'} \\ l_z^{0'} \end{pmatrix} = R^T \begin{pmatrix} l_x^0 \\ l_y^0 \\ l_z^0 \end{pmatrix} = \begin{pmatrix} r_{xx} & r_{yx} & r_{zx} \\ r_{xy} & r_{yy} & r_{zy} \\ r_{xz} & r_{yz} & r_{zz} \end{pmatrix} \begin{pmatrix} l_x^0 \\ l_y^0 \\ l_z^0 \end{pmatrix}. \tag{24}$$

The resulting two equations for l_x^0 and l_y^0 are supplemented by the nonlinear equation expressing the square of the length of leg zero:

$$l_0^2 = (l_x^0)^2 + (l_y^0)^2 + (l_z^0)^2. \tag{25}$$

¿From the first two equations, one can eliminate two of the three unknowns. For example, choosing to eliminate l_x^0 and l_y^0 yields

$$l_x^0 = \alpha + \beta l_z^0, \qquad l_y^0 = \gamma + \delta l_z^0, \tag{26}$$

with

$$\begin{array}{ll} \alpha = \frac{\bar{x}_1 v_2 - \bar{x}_2 v_1}{N} & \beta = -\frac{w_1 v_2 - w_2 v_1}{N} \\ \gamma = -\frac{\bar{x}_1 u_2 - \bar{x}_2 u_1}{N} & \delta = \frac{w_1 u_2 - w_2 u_1}{N} \end{array} \quad \text{and} \quad N = u_1 v_2 - u_2 v_1, \tag{27}$$

and

$$\begin{array}{ll} u_1 = 1 - P_{xx} r_{xx} - P_{yx} r_{xy} & u_2 = -P_{xy} r_{xx} - P_{yy} r_{xy} \\ v_1 = -P_{xx} r_{yx} - P_{yx} r_{yy} & v_2 = 1 - P_{xy} r_{yx} - P_{yy} r_{yy} \\ w_1 = -P_{xx} r_{zx} - P_{yx} r_{zy} & w_2 = -P_{xy} r_{zx} - P_{yy} r_{zy}. \end{array} \tag{28}$$

Hence, one gets the following quadratic equation in l_z^0:

$$A\,(l_z^0)^2 + B\,l_z^0 + C = 0, \tag{29}$$

$$\text{with} \quad A = 1 + \beta^2 + \delta^2, \quad B = \alpha\beta + \gamma\delta, \quad C = \alpha^2 + \gamma^2 - (l^0)^2. \tag{30}$$

Hence, for each of the eight rotation matrix solutions of Section 3 one finds two solutions for the relative position of the end effector platform with respect to the base platform.

4. Examples

Figure 2 shows some designs with different parameters $P_{xx} \ldots P_{yy}$. In these designs, we kept the geometry of the base platform fixed, to make clear that the design parameters can be tuned to change the shape of the end effector platform in a flexible (but not arbitrary!) way.

Singularities. The shape of the base platform can be freely chosen by the designer, as long as he avoids two or more coinciding leg attachment points, or a design with a rotational symmetry. These designs lead to a rank-deficient set of equations in Eq. (14), and hence to an architectural singularity. Some choices for the design parameters $P_{xy} \ldots P_{yy}$ lead to a singular design also; for example: $P_{xx} = P_{yy} = k, P_{yx} = P_{xy} = 0$, or $P_{xx} = P_{xy} = 0$, or $P_{yx} = P_{yy} = 0$.

5. Conclusions

This paper has described a class of fully parallel manipulators with closed-form forward kinematics, i.e., using only linear, quadratic, and quartic polynomials. Their architecture is defined by four design parameters: the geometry of the base platform can be chosen almost arbitrarily (some choices can lead to singular designs, which are easily detected), while the geometry of the corresponding end effector platform is then completely determined by the four design constants. The previously reported architectures with similar base and end effector platforms are a special case of the designs described in this paper.

Acknowledgements

H. Bruyninckx is Post-Doctoral Researcher of the Belgian National Fund for Scientific Research (NFWO).

References

L. Baron and J. Angeles. The decoupling of the direct kinematics of parallel manipulators using redundant sensors. In *Proceedings of the IEEE International Conference on Robotics and Automation*, pages 974–979, San Diego, CA, 1994.

K. C. Cheok, J. L. Overholt, and R. R. Beck. Exact methods for determining the kinematics of a Stewart platform using additional displacement sensors. *Journal of Robotic Systems*, 10(5):689–707, 1993.

E. F. Fichter. A Stewart platform-based manipulator: General theory and practical construction. *The International Journal of Robotics Research*, 5(2):157–182, 1986.

M. Griffis and J. Duffy. A forward displacement analysis of a class of Stewart platforms. *Journal of Robotic Systems*, 6(6):703–720, 1989.

K. Han, W. K. Chung, and Y. Youm. Local structurization for the forward kinematics of parallel manipulators using extra sensor data. In *Proceedings of the IEEE International Conference on Robotics and Automation*, pages 514–520, Nagoya, Japan, 1995.

A. Hara and K. Sugimoto. Synthesis of parallel micromanipulators. *Transactions of the ASME, Journal of Mechanisms, Transmissions, and Automation in Design*, 111:34–39, 1989.

K. H. Hunt. Structural kinematics of in-parallel-actuated robot-arms. *Transactions of the ASME, Journal of Mechanisms, Transmissions, and Automation in Design*, 105:705–712, 1983.

M. Husain and K. J. Waldron. Direct position kinematics of the 3-1-1-1 Stewart platforms. *Transactions of the ASME, Journal of Mechanical Design*, 116:1102–1107, 1994.

C. Innocenti and V. Parenti-Castelli. Direct position analysis of the Stewart platform mechanism. *Mechanism and Machine Theory*, 25(6):611–621, 1990.

D. R. Kerr. Analysis, properties, and design of a Stewart-platform transducer. *Transactions of the ASME, Journal of Mechanisms, Transmissions, and Automation in Design*, 111:23–28, 1989.

H.-Y. Lee and B. Roth. A closed-form solution of the forward displacement analysis of a class of in-parallel mechanisms. In *Proceedings of the IEEE International Conference on Robotics and Automation*, pages 720–724, Atlanta, GA, 1993.

O. Ma and J. Angeles. Optimum architecture design of platform manipulators. In *Proceedings of the International Conference on Advanced Robotics*, pages 1130–1135, Pisa, Italy, 1991.

J.-P. Merlet. Singular configurations of parallel manipulators and Grassmann geometry. *The International Journal of Robotics Research*, 8(5):45–56, 1989.

J.-P. Merlet. Direct kinematics and assembly modes of parallel manipulators. *The International Journal of Robotics Research*, 11(2):150–162, 1992.

J.-P. Merlet. Closed-form resolution of the direct kinematics of parallel manipulators using extra sensor data. In *Proceedings of the IEEE International Conference on Robotics and Automation*, pages 200–204, Atlanta, GA, 1993.

R. Nair and J. H. Maddocks. On the forward kinematics of parallel manipulators. *The International Journal of Robotics Research*, 13(2):171–188, 1994.

P. Nanua and K. J. Waldron. Direct kinematic solution of a special parallel ro bot structure. In A. Morecki, G. Bianchi, and K. Jaworek, editors, *Proc. 8th CISM-IFToMM Symposium on Theory and Prac tice of Robots and Manipulators*, pages 134–142, Warsaw, Poland, 1990.

P. Nanua, K. J. Waldron, and V. Murthy. Direct kinematic solution of a Stewart platform. *IEEE Transactions on Robotics and Automation*, 6(4):438–444, 1990.

X. Shi and R. G. Fenton. A complete and general solution to the forward kinematics problem of platform-type robotic manipulators. In *Proceedings of the IEEE International Conference on Robotics and Automation*, pages 3055–3062, San Diego, CA, 1994.

S. V. Sreenivasan, K. J. Waldron, and P. Nanua. Closed-form direct displacement analysis of a 6-6 Stewart platform. *Mechanism and Machine Theory*, 29(6):855–864, 1994.

G. Wang. Forward displacement analysis of a class of the 6- 6 Stewart platforms. In *Proc. 1992 ASME Design Technical Conferences–22nd Biennal Mechanisms Conference*, pages 113–117, Scottsdale, AZ, sep 1992.

C.-D. Zhang and S.-M. Song. Forward kinematics of a class of parallel (Stewart) platforms with closed-form solutions. *Journal of Robotic Systems*, 9(1):93–112, 1992.

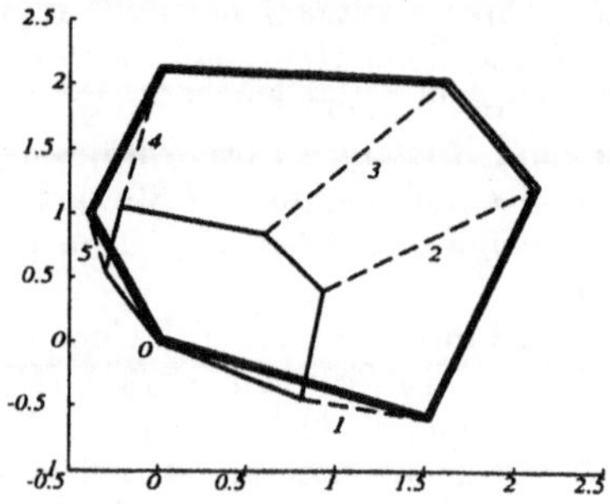

$P_{xx} = 0.5; P_{xy} = -0.1; P_{yx} = -0.1; P_{yy} = 0.5$

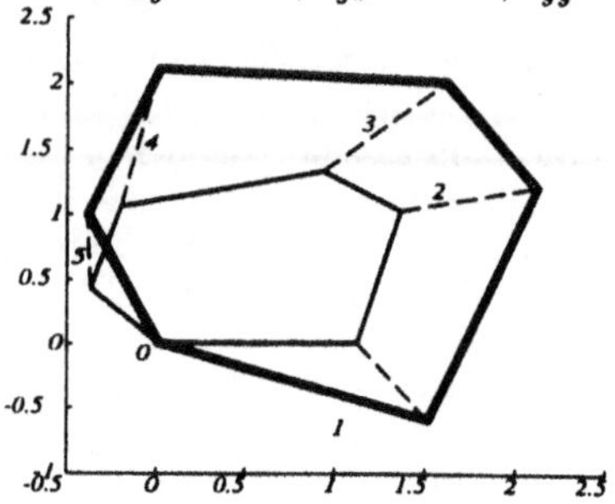

$P_{xx} = 0.7; P_{xy} = -0.1; P_{yx} = 0.2; P_{yy} = 0.5$

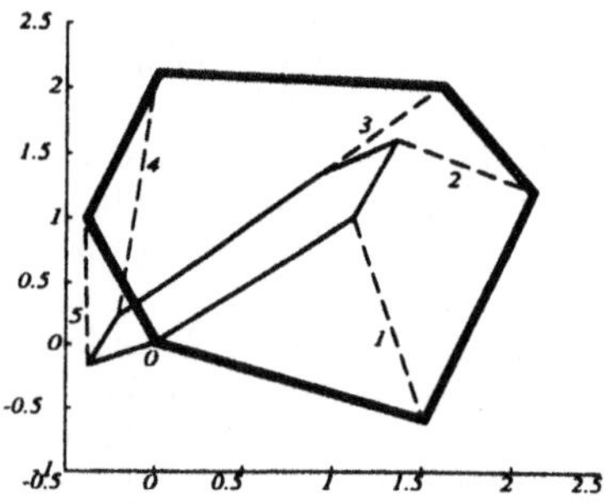

$P_{xx} = 0.7; P_{xy} = -0.1; P_{yx} = 0.7; P_{yy} = 0.1$

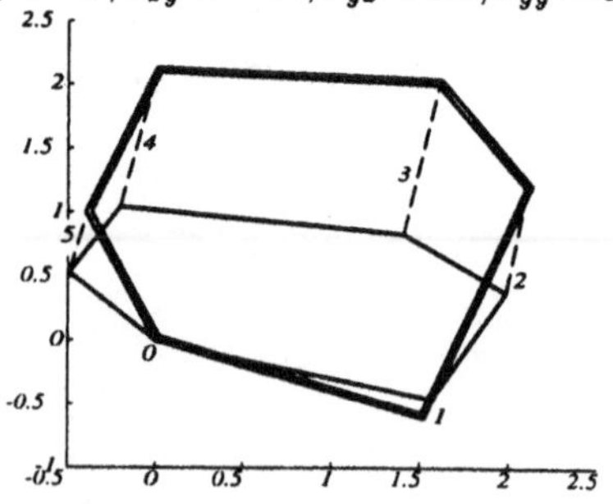

$P_{xx} = 1.0; P_{xy} = -0.1; P_{yx} = -0.1; P_{yy} = 0.5$

Figure 2. Some examples of base and end effector platform geometries for different values of the design parameters in Eqs (12). The base platforms are drawn with double lines; the legs are numbered from 0 to 5, and drawn in dashed lines. The manipulators are viewed from above, with the end effector platform parallel to, and straight above, the base platform (hence, the zeroth leg is seen as a point only).

KINEMATIC MAPPING OF 3-LEGGED PLANAR PLATFORMS WITH HOLONOMIC HIGHER PAIRS

M. J. D. HAYES AND P. J. ZSOMBOR-MURRAY
McGill University
Department of Mechanical Engineering
Centre for Intelligent Machines
817 r. Sherbrooke O., Rm 454
Montréal, Québec, H3A 2K6 Canada
Tel: (514) 398-6311 Fax: (514) 398-4476
email: <*johnh@cim.mcgill.ca*>, <*paul@cim.mcgill.ca*>

Abstract.

A planar parallel manipulator with holonomic higher pairs is introduced. The end effector is a circular disk which rolls with out slip along the straight lines of the non-grounded rigid links of each of three 2R legs. The contact points between the disk and legs are holonomic higher pairs. The forward kinematic problem of this manipulator is unlike that of planar Stewart-Gough type platforms because the initial assembly configuration must be included in the analysis. A procedure using kinematic mapping to solve the forward kinematics is discussed and a numerical example is given.

1. Introduction

Planar kinematic mappings map the set of all planar displacements onto the points of a three dimensional projective space with Cartesian homogeneous coordinates X_i $(i = 1, 2, 3, 4)$. It has recently been shown that this mapping has important applications in robotics, specifically, in the solution of the forward kinematics (FK) problem of planar and spatial Stewart-Gough (SG) type platforms [4].

In this paper, a kinematic mapping of planar displacements will be discussed. Its application will be demonstrated by an example wherein the FK problem of a planar parallel manipulator with holonomic higher pairs are solved. An algebraic approach was successfully used in [8] to obtain the FK

J. Lenarčič and V. Parenti-Castelli (eds.), Recent Advances in Robot Kinematics, 421-430.
© 1996 *Kluwer Academic Publishers. Printed in the Netherlands.*

solutions of the general planar three legged platform, but is computationally incompatible for manipulators with higher pairs of the type introduced in this paper.

2. A Kinematic Mapping of Planar Displacements

A general displacement in the plane requires three independent coordinates to fully characterise it. It is convenient to think of the relative planar motion between two rigid bodies as the motion of a Cartesian reference coordinate system, E attached to one of the bodies, with respect to the Cartesian coordinate system, Σ attached to the other, [1]. Without loss of generality, Σ may be considered as fixed while E is free to move. Then the position of a point in E relative to Σ can be given by

$$\begin{bmatrix} X' \\ Y' \end{bmatrix} = \begin{bmatrix} \cos\phi & -\sin\phi \\ \sin\phi & \cos\phi \end{bmatrix} \begin{bmatrix} x' \\ y' \end{bmatrix} + \begin{bmatrix} a \\ b \end{bmatrix}, \qquad (1)$$

where

i. (x', y') are the Cartesian coordinates of a point in E.
ii. (X', Y') are the Cartesian coordinates of the same point in Σ.
iii. (a, b) are the Cartesian coordinates of the origin of E measured in Σ, ie, the components of the position vector of the origin of E in Σ.
iv. ϕ is the rotation angle measured from the X'–axis to the x'–axis, the positive sense being counter–clockwise.

Equation (1) does not represent a linear transformation. This fact is computationally inconvenient, and can be remedied by the use of Cartesian homogeneous coordinates [7]

$$x' = \frac{x}{z}, \quad y' = \frac{y}{z}$$
$$X' = \frac{X}{Z}, \quad Y' = \frac{Y}{Z}.$$

Substituting these homogeneous coordinates in equation (1) and setting the homogenising coordinates to be equal, i.e. set $Z = z$, then multiplying through by z yields the following linear transformation

$$\begin{bmatrix} X \\ Y \\ Z \end{bmatrix} = \begin{bmatrix} \cos\phi & -\sin\phi & a \\ \sin\phi & \cos\phi & b \\ 0 & 0 & 1 \end{bmatrix} \begin{bmatrix} x \\ y \\ z \end{bmatrix}, \qquad (2)$$

which represents a displacement of E with respect to Σ.

All general planar displacements may be represented as a single rotation through a finite angle about a fixed axis normal to the plane. The coordinates of the piercing point of this axis is the *pole* of the displacement.

If E and Σ are initially coincident then after the displacement the pole has the same coordinates in both E and Σ. The location of the pole of a displacement along with the rotation angle convey sufficient information to characterise the displacement.

The value of the homogenising coordinate is arbitrary. If it is set equal to $\sin \phi/2$ it can be shown that the homogeneous coordinates of the pole, which are identical in each of the two coordinate systems Σ and E, in terms of the three displacement parameters a, b and ϕ are determined by the set of equations (3) [1, 6]

$$\begin{aligned} X_p = \quad x_p = \quad & \frac{1}{2} a \sin(\phi/2) - \frac{1}{2} b \cos(\phi/2) \\ Y_p = \quad y_p = \quad & \frac{1}{2} a \cos(\phi/2) + \frac{1}{2} b \sin(\phi/2) \\ Z_p = \quad z_p = \quad & \sin \phi/2. \end{aligned} \tag{3}$$

Many mappings can be defined that map a position (a, b, ϕ) of the moving coordinate system E with respect to the fixed system Σ in the plane to a point described by the homogeneous coordinates $(X_1 : X_2 : X_3 : X_4)$ of a three dimensional projective *image space*, Σ'. The mapping used here is as follows [1, 4, 6]:

$$(X_1 : X_2 : X_3 : X_4) \quad = \quad (X_p : Y_p : Z_p : \tau Z_p), \tag{4}$$

where

$$\begin{aligned} (X_1 : X_2 : X_3 : X_4) \; & \neq \; (0:0:0:0) \\ \tau \; & = \; \cot(\phi/2) \\ 0 \leq \; & \phi \; < 2\pi. \end{aligned}$$

$(X_p : Y_p : Z_p)$ depend on (a, b, ϕ) as given by the set of equations (3). The image of the pole coordinates under the kinematic mapping is called the *image point* of the displacement $D(a, b, \phi)$. The image point is given by

$$\begin{aligned} (X_1 : X_2 : X_3 : X_4) \quad = \quad & [(a \sin(\phi/2) - b \cos(\phi/2) : \\ & (a \cos(\phi/2) + b \sin(\phi/2) : \\ & 2 \sin(\phi/2) : 2 \cos(\phi/2)]. \end{aligned} \tag{5}$$

By virtue of the relationships expressed in (5), the linear transformation operator, the matrix from equation (2), may be expressed in terms of the homogeneous coordinates of the image space, Σ'. This means that we now have a linear transformation to express a displacement of E with respect to Σ in terms of the image point as given by (5):

$$\begin{bmatrix} X \\ Y \\ Z \end{bmatrix} = \begin{bmatrix} (X_4^2 - X_3^2) & -2X_3X_4 & 2(X_1X_3 + X_2X_4) \\ 2X_3X_4 & (X_4^2 - X_3^2) & 2(X_2X_3 - X_1X_4) \\ 0 & 0 & (X_4^2 + X_3^2) \end{bmatrix} \begin{bmatrix} x \\ y \\ z \end{bmatrix}. \quad (6)$$

Since equation (6) is a linear transformation, for each unique displacement described by (a, b, ϕ) there is a corresponding point in the image space. From equation (5), the inverse mapping is obtained. That is, for a given point of the image space, the displacement parameters are obtained from

$$\begin{aligned} \tan(\phi/2) &= X_3/X_4 \\ a &= 2(X_1X_3 + X_2X_4)/(X_3^2 + X_4^2) \\ b &= 2(X_2X_3 - X_1X_4)/(X_3^2 + X_4^2). \end{aligned} \quad (7)$$

Any image point with $X_3 = X_4 = 0$ can not be mapped to a displacement of the plane, and must be disregarded. It can be seen from equation (5) that this condition requires $\phi = 0°$ and $\phi = 180°$ simultaneously.

3. An Application for Planar Parallel Manipulators

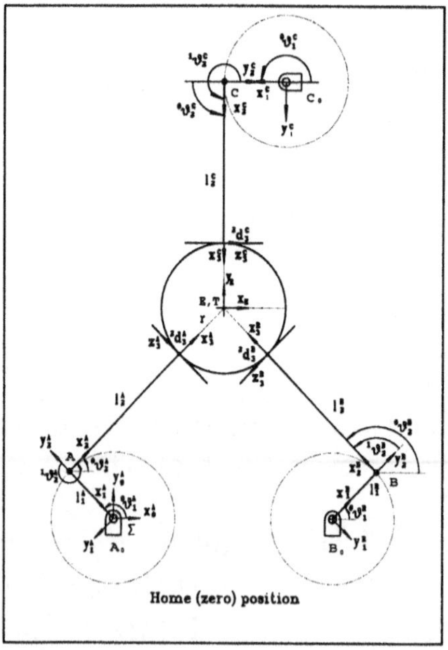

Figure 1. A planar manipulator with three DOF.

Consider the planar manipulator shown in Fig. 1. It consists of three closed kinematic chains. The disk, which is modelled as a pinion gear, rolls without slip on each of the three racks tangent to it. The rolling constraints are holonomic due to the pure rolling constraint and because the motion is planar, hence the constraint equations can be expressed in terms of displacement, i.e. in *integral form.* Each of the three legs connect a rack to a base point via two revolute (R) pairs. The leg links are rigid and the rack is welded at a right angle to the second leg link. The R–pairs connecting two links in a leg shall be referred to as *knee joints* A, B and C. The three base points A_0, B_0, C_0, are rigidly fixed.

Joint and fixed link parameters along with link reference frames are identified by left and right sub and superscripts. The generic parameter

$$ {}^{k}\Psi_{i}^{j} $$

is identified as follows:

1a. For a joint variable the right sub–script i, $i \in \{1,2,3\}$ identifies the joint number. For each manipulator leg, the joint number at the connection between the first link and the base is 1. Between the first and second links is 2. The higher pair between link 2 and the disk is 3.
1b. For a coordinate axis, the right sub–script i, $i \in \{0,1,2,3\}$ represents the link to which the coordinate system is attached. 0 is for the base, 1 is for the first link, etc..
2. The right super–script, j, $j \in \{A, B, C\}$ denotes a particular manipulator leg.
3. The left super–script, k, $k \in \{\Sigma, 0, 1, 2, E, T\}$ refers to the reference frame in which the variable is represented.
4. l_i^j is the length of link i in leg j.
5. r is the radius of the disk.

The link reference frames, save for E and T, were assigned using the Denavit and Hartenberg procedure [2], modified to accommodate the higher pairs. For reasons discussed below, it is convenient to have these two reference frames with origins incident on the disk centre. Both E and T translate with the disk, but only E rotates with it.

3.1. THE FORWARD KINEMATICS PROBLEM

The FK problem is conventionally expressed as a transformation of the position and orientation of the end effector from a joint space representation to a Cartesian space representation. That is, given a set of n joint variables, one for each n degrees of freedom, determine the position and orientation of the end effector with respect to a non–moving reference coordinate system. The pure rolling nature of the higher pairs make this manipulator markedly different from planar SG type platforms because the pure rolling condition renders FK solutions completely dependent on the initial assembly configuration (IAC). The FK analysis can not be reduced to the planar SG case because no equivalent mechanism exists which can exactly reproduce a rack–and–pinion motion, p.106 [3], hence the method in [8] can not be used. Furthermore, some displacements require a combination of two distinct types of rolling with respect to the fixed reference frame: 1) The disk rolls on a rack, or 2), a rack rolls on the disk. Each type of rolling produces a change in the location of the contact point but yields an entirely different displacement. As a result, conventional joint variable inputs can not be used.

We propose to modify the problem by using instead a set of *pseudo inputs* from which the position and orientation of the disk in the non–moving reference frame can be determined. The pseudo inputs are the positions of the knee joints in the rotating disk frame, E. These positions are

$$\begin{bmatrix} {}^{E}A \\ {}^{E}B \\ {}^{E}C \end{bmatrix}. \tag{8}$$

Each position is specified by a 2×1 array of Cartesian coordinates, having the form

$$\begin{bmatrix} {}^{E}x_2^j \\ {}^{E}y_2^j \end{bmatrix}.$$

Hence, six pseudo input variables are required. Because the knee joints are constrained to move on circles, the position and orientation of the disk in the non-moving frame Σ can be determined with the kinematic mapping discussed earlier and the procedure introduced by Husty [4].

The actual joint inputs are the change in variable joint lengths $\Delta^2 d_3^j = {}^2d_3^j - {}^2d_{30}^j$, $j \in \{A, B, C\}$, where the subscript '30' denotes the initial condition. These lengths are the change in distance of the contact point measured along the y_2^j coordinate axis, which is always parallel to the rack. This is why the solution is coupled with the ICA. They are related to the pseudo inputs in the following way:

$$\begin{bmatrix} {}^{E}x_2^j \\ {}^{E}y_2^j \end{bmatrix} = \begin{bmatrix} c^{\Sigma}\phi & s^{\Sigma}\phi \\ -s^{\Sigma}\phi & c^{\Sigma}\phi \end{bmatrix} \begin{bmatrix} {}^{T}x_2^j \\ {}^{T}y_2^j \end{bmatrix} = \begin{bmatrix} (l_2^j + r)c^{\Sigma}\vartheta_2^j - \Delta^2 d_3^j s^{\Sigma}\vartheta_2^j \\ (l_2^j + r)s^{\Sigma}\vartheta_2^j + \Delta^2 d_3^j c^{\Sigma}\vartheta_2^j \end{bmatrix} \tag{9}$$

where $c \equiv cos$, $s \equiv sin$, and ${}^{\Sigma}\phi$ is the orientation angle of the disk with respect to Σ. Since the reference frame T translates with the disk, ${}^{T}\phi = {}^{\Sigma}\phi$ and, of course, ${}^{T}\vartheta = {}^{\Sigma}\vartheta$. So, the pseudo inputs are theoretically *valid* as input parameters, except that the actual inputs can not be specified until the disk orientation is known. The higher pair variables along with the ICA must be specified or the disk orientation can not be determined. A cart–before–horse scenario, to be sure. While this approach to the FK problem is not necessarily practical, it is a start. To the best of our knowledge the FK of such a planar parallel platform with higher pairs have never been addressed.

Using the pseudo inputs, the FK problem of the manipulator shown in Fig. 1 can be stated in the following way: Given the coordinates of the three base points A_0, B_0, C_0 in an arbitrary fixed coordinate system, Σ, the coordinates of the knee joints ${}^EA, {}^EB, {}^EC$ expressed in an arbitrary coordinate system, E, which moves with the disk, the fixed lengths of each link, l_i^j, $i \in \{1,2\}$, $j \in \{A,B,C\}$, and the radius of the disk, find the positions and orientations of the disk such that the knee joints ${}^EA, {}^EB, {}^EC$ can be joined to the base points A_0, B_0, C_0 with legs of the given lengths.

To obtain the solutions for a given set of inputs and ICA, begin by removing the disk connections with legs B and C. Observe that the higher pairs are locked by virtue of the specified input parameter. The knee joint ${}^EA$ is constrained to move on a circle with A_0 as its centre and radius l_1^A. Furthermore, the rigid body comprised of link l_2^A and the disk can rotate about ${}^EA$. This two parameter motion corresponds to a two parameter set of points in the image space called a *constraint surface*, H. The equation of H is found using equation (6) and the fact that the moving point ${}^EA$ is bound to a circle, which gives

$$\begin{aligned} H: \quad & 0 = z^2(X_1^2 + X_2^2) + (1/4)[(x^2+y^2) - 2C_1xz - 2C_2yz + C_3z^2]X_3^2 + \\ & (1/4)[(x^2+y^2) + 2C_1xz + 2C_2yz + C_3z^2]X_4^2 + (C_1z - x)zX_1X_3 + \\ & (C_2z - y)zX_2X_3 - (y + C_2z)zX_1X_4 + (C_1z + x)zX_2X_4 + \\ & (C_2x - C_1y)zX_3X_4. \end{aligned} \tag{10}$$

It is shown in [1] that this constraint surface is a hyperboloid which contains the isotropic points $J_1(1 : i : 0 : 0)$ and $J_2(1 : -i : 0 : 0)$. Recall that any point with $X_3 = X_4 = 0$ can not be mapped to a displacement of the plane. When the other two points B and C are examined in turn, three hyperboloidal surfaces are generated, H_A, H_B, and H_C, which correspond to the complete range of possible displacements around the points still connected. The points of intersection of H_A, H_B and H_C represent the positions of the end–effector where its three knee joints are on their respective circles. Therefore, these points of intersection constitute the solutions to the FK problem. It is to be noted that three hyperboloids can intersect in at most eight points. However, all hyperboloidal constraint surfaces corresponding to planar displacements with one point that moves on a circle contain the points J_1 and J_2 and hence, these two points are always in the solution set, and must be disregarded. Therefore, there is a maximum of six real solutions to the FK problem for manipulators of this type.

3.2. EXAMPLE

Table 1 gives the coordinates of the base points A_0, B_0, C_0 in the fixed frame Σ with origin at A_0, the input variable coordinates of the knee joints $^EA, ^EB, ^EC$ in the moving frame E, with origin at centre of the disk, along with the ICA. The ICA parameters are distinguished by adjoining a '0' to the right subscript. The fixed link lengths, in generic units are $r = 4$, $l_1^j = 4$, $l_2^j = 10$, $j \in \{A, B, C\}$.

TABLE 1. Input parameters and ICA.

j_0	$^\Sigma x$	$^\Sigma y$	$^E j$	$^E x$	$^E y$	j	$^2d_{30}^j$	$^0\vartheta_{10}^j$	$^1\vartheta_{20}^j$
A_0	0	0	^EA	-9	-11	A	0	135°	270°
B_0	13	0	^EB	9	-11	B	0	45°	90°
C_0	10	26	^EC	9.5	10.5	C	0	180°	270°

Substituting the input data from Table 1 into equation (10) gives the following three constraint surfaces in the image space:

$$\begin{aligned} H_A: \quad & X_1^2 + X_2^2 + 46.5X_3^2 + 46.5X_4^2 + 9X_1X_3 + 11X_2X_3 + \\ & 11X_1X_4 - 9X_2X_4 = 0 \qquad (11) \\ H_B: \quad & X_1^2 + X_2^2 + 147.25X_3^2 + 30.25X_4^2 - 22X_1X_3 - 4X_2X_4 + \\ & 11X_2X_3 + 11X_1X_4 = 0 \qquad (12) \\ H_C: \quad & X_1^2 + X_2^2 + 424.125X_3^2 + 56.125X_4^2 - 19.5X_1X_3 - \\ & 36.5X_2X_3 + 15.5X_1X_4 - .5X_2X_4 - 142X_3X_4 = 0. \qquad (13) \end{aligned}$$

Since X_4 is the homogenising coordinate, its value is arbitrary, hence it is set $X_4 = 1$. Fig. 2 is a view of the resulting constraint hyperboloids where one of the intersections is visible. The set of three equations $H_A = 0, H_B = 0, H_C = 0$ can now be solved for the variables X_1, X_2, X_3. The following solutions are obtained:

$$\begin{aligned} S_1: \quad & X_1 = -5.35817508, X_2 = 1.69375244, X_3 = 0.18597447 \\ S_2: \quad & X_1 = -4.23444169, X_2 = 3.13635972, X_3 = 0.15325037 \\ S_3: \quad & X_1 = -4.71288212, X_2 = 2.25800666, X_3 = 0.20703047 \\ S_4: \quad & X_1 = -6.90743973, X_2 = 2.76957064, X_3 = 0.03229377 \\ S_5: \quad & X_1 = -4.306063 + 2.801994i, X_2 = 0.652824 + 0.102659i, \\ & X_3 = -0.043999 + 0.180029i \\ S_6: \quad & X_1 = -4.306063 - 2.801994i, X_2 = 0.652824 - 0.102659i, \\ & X_3 = -0.043999 - 0.180029i \end{aligned}$$

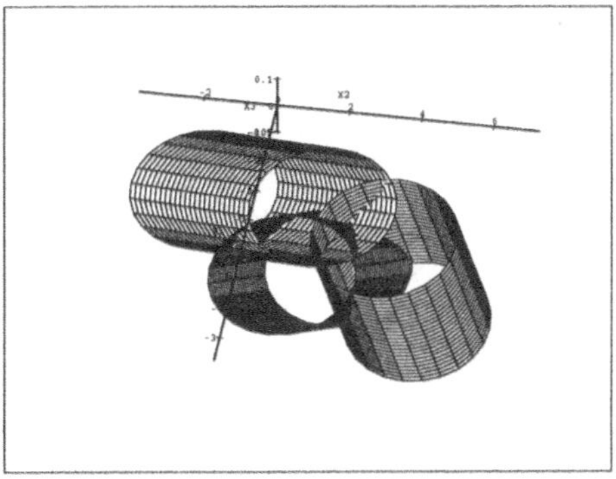

Figure 2. The constraint hyperboloids in the image space.

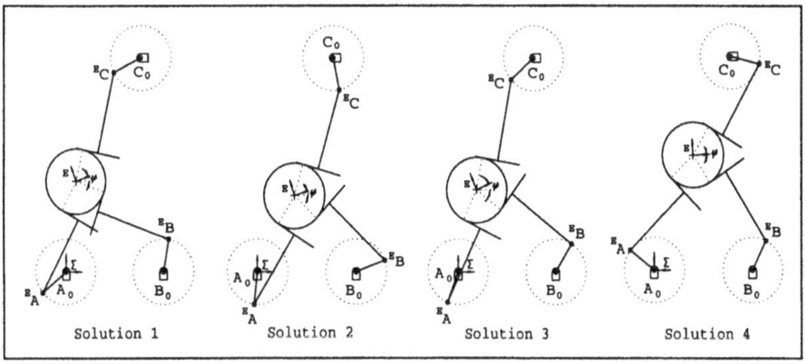

Figure 3. The four real solutions.

There are four real and one set of complex conjugate solutions for a total of six solutions, as expected. The position and orientation of the end–effector corresponding to each of these solutions in terms of the displacement parameters a, b, and ϕ can be found by substituting the solutions for X_1, X_2, X_3, along with $X_4 = 1$ into equations (7). The subsequent four sets of displacement parameters are given in Table 2. It is a simple matter of planar trigonometry to determine the the link parameters $\Delta^2 d_3^j$ and ${}^T\vartheta_2^j$, $j \in \{A, B, C\}$, also printed in Table 2, given the locations of the knee joints in E along with the fixed link lengths, disk radius and ICA. Recall that because the reference frame T translates with the disk but does not rotate, ${}^T\vartheta_2^j = {}^\Sigma\vartheta_2^j$. The four real solutions are illustrated in Figure 3.

As anticipated, the actual inputs, $\Delta^2 d_3^j$ are consistent in magnitude.

TABLE 2. Four real solutions.

	Sol'n 1	Sol'n 2	Sol'n 3	Sol'n 4
a	1.347918	4.860703	2.459188	5.087701
b	10.967028	9.213788	9.934891	13.979180
ϕ (deg.)	21.070388	17.425626	23.393454	3.699307
$\Delta^2 d_3^A$	2.449489	2.449489	2.449489	2.449489
$\Delta^2 d_3^B$	-2.449489	2.449489	2.449489	2.449489
$\Delta^2 d_3^C$	-2.121320	-2.121320	-2.121320	-2.121320
${}^T\vartheta_2^A$ (deg.)	241.856762	238.211999	244.179828	224.485681
${}^T\vartheta_2^B$ (deg.)	-19.715986	-43.209187	-37.241359	-56.935505
${}^T\vartheta_2^C$ (deg.)	77.548874	73.904112	79.871940	60.177794

There is one anomaly in the single negative $\Delta^2 d_3^B$ in the first solution. This anomaly appears to expose a flaw in the algorithm, but we are provided with three solutions where the magnitudes and sense of the actual inputs agree.

4. Conclusions

A kinematic mapping for planar displacements has been presented. An important application of this mapping is the solution of the forward kinematics problem of a planar parallel manipulator with holonomic higher pairs.

References

1. Bottema, O. & Roth, B., *Theoretical Kinematics*, Dover Publications, Inc., New York, 1979.
2. Denavit, J. & Hartenberg, R.S., *A Kinematic Notation for Lower-Pair Mechanisms Based on Matrices*, J. of Applied Mechanics, pp. 215-221, June, 1955.
3. Hunt, K.H., 1978, *Kinematic Geometry of Mechanisms*, Clarendon Press, Oxford, England.
4. Husty, M.L., *Kinematic Mapping of Planar Three–Legged Platforms*, Proc. 15th Canadian Congress of Applied Mechanics, CANCAM 1995, Vol. 2, pp. 876-877, 1995.
5. McCarthy, J.M., *An Introduction to Theoretical Kinematics*, The M.I.T. Press, Cambridge, Massachusetts, 1990.
6. De Sa, S., Roth, B., *Kinematic Mappings. Part 1: Classification of Algebraic Motions in the Plane*, ASME, J. of Mech. Design, Vol. 103, pp. 585-591, July, 1981.
7. Sommerville, D.M.Y., *Analytic Geometry of Three Dimensions*, Cambridge University Press, London, 1934.
8. Wohlahart, K., *Direct Kinematic Solution of the General Planar Stewart Platform*, Proc. of the Int. Conf. on Comp. Int. Manu., Zakopane, pp. 403–411, March, 1992.

A FORWARD ANALYSIS OF A TWO DEGREE OF FREEDOM PARALLEL MANIPULATOR

S.C. RIDGEWAY, C.D. CRANE, and J. DUFFY
Center for Intelligent Machines and Robotics
300 MEB, University of Florida
Gainesville, Florida, U.S.A. 32611

Abstract. The forward analysis of a parallel planar manipulator (PPM) with two degrees of freedom is developed. The two degree of freedom PPM has been applied to an articulated mobile robot and the forward analysis is one of the tools developed to aid in the design and control of the articulated mobile robot. The forward analysis is derived from the loop closure equation for a quadrilateral applied to two quadrilaterals that make up the PPM. One side and two angles are common between the two quadrilaterals. An input-output equation is derived that generates six closures for the mechanism. The forward analysis is applied to the geometry of the PPM utilized in the articulated mobile robot, and determines four real and two imaginary closures.

1. Introduction

An articulated mobile robot is a type of robot that is characterized as a serial chain of segments connected by one or more degree of freedom joints. Articulated mobile robots have been examined with respect to their suitability and usefulness in the environment expected in a nuclear power plant (Crane 1988, Hirose 1990). Several characteristics of the articulated structure make it suited for navigation and manipulation in an obstacle strewn environment with limited access. The payload capacity of an articulated robot relative to the robot's cross-sectional area would be relatively high if the payload were distributed between segments. This would enable the delivery of considerable resources to a work area within a restrictive environment. Some articulated robot concepts allow large changes in elevation and motion over obstacles (Ridgeway 1992).

The University of Florida, performing research for the U.S. Department of Energy's Advanced Technology Development Division, has initiated the development of an Articulated Transporter/Manipulator System (ATMS) for use in a nuclear reactor building. This effort began with a concept based on serially actuated segments. The concept consisted of 18 segments serially connected by two electro-mechanically actuated orthogonal rotational joints. Each segment was approximately 29 cm wide, 33 cm tall, and 70 cm long (Crane 1988). Significant resources were utilized in joint design and development. It eventually became apparent that the joint actuation scheme was not feasible based on

J. Lenarčič and V. Parenti-Castelli (eds.), Recent Advances in Robot Kinematics, 431-440.
© 1996 *Kluwer Academic Publishers. Printed in the Netherlands.*

currently available electro-mechanical actuators. An effort was directed at developing and implementing a joint actuation scheme that utilized currently available technology and met design criteria derived from performance requirements. Expected transportation and manipulation tasks of the ATMS require a light, stiff joint actuation scheme capable of generating large joint torques (4500 N-m for a distributed weight of 7 N per cm).

An effort to re-examine the design was begun with the intent of producing a design based on current technology that could be easily fabricated with facilities at hand. The apparent limiting factor was taken to be joint torque. The mechanical joint between articulated segments has typically been serial in nature, that is, one decoupled degree of freedom is associated with each prime mover of the joint. This paper details the forward analysis of a joint design methodology based on a novel parallel joint actuation scheme. The descriptor "parallel" is utilized to indicate that the degrees of freedom between two segments are coupled. Figure 1 illustrates a planar parallel manipulator that is utilized as the mechanical connection between segments of the design that was implemented. The planar parallel manipulator (PPM) has two degrees of freedom, represented here as ℓ_2 and θ. The two slider lengths can also be used to specify the PPM's configuration.

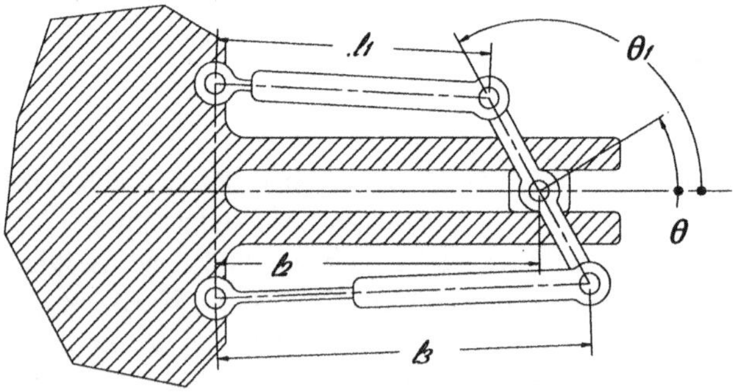

Figure 1. Parallel planar manipulator configuration.

The PPM utilized in the ATMS enables the reduction of mass per unit length by allowing each segment to change length. This feature is utilized in changing elevation, resulting in lower joint torque requirements. Figure 2 illustrates a simulation executing a change in elevation.

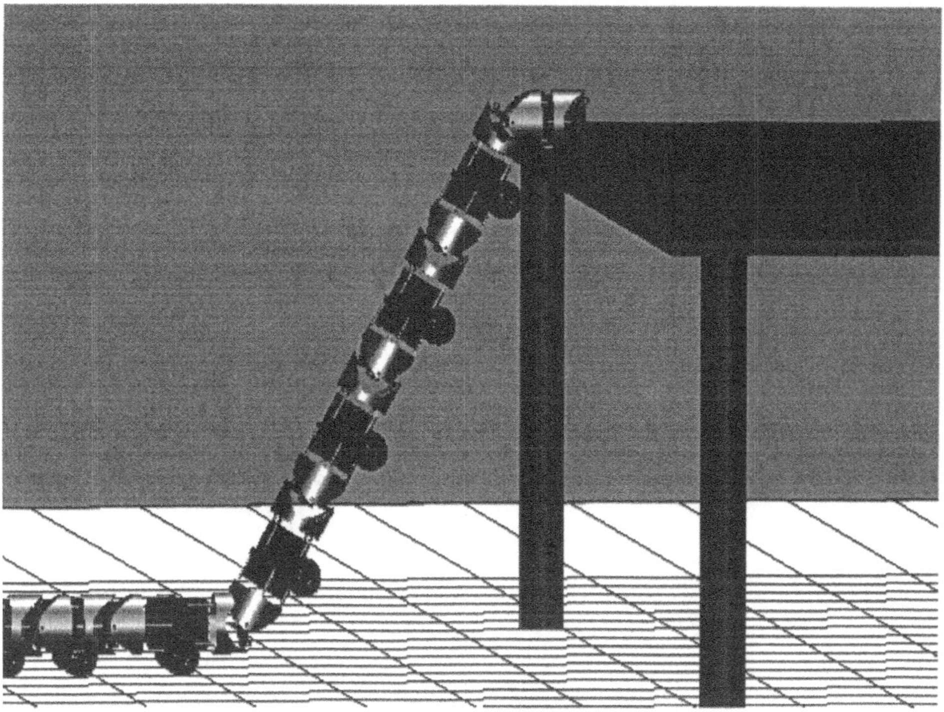

Figure 2 ATMS executing vertical navigation.

A forward analysis of the PPM utilized in the ATMS is necessary for both mechanism design and mechanism control. The work space of the ATMS is determined by the work space of the PPM's that make up the serial chain of segments. Typical linear actuators have limits on their range of motion. Actuator ranges impose limits on the work space of the PPM. A forward analysis defines the relationship between actuator length and PPM work space. The control of linear actuators is typically based on sensing and controlling actuator length. To control the PPM based on actuator length, a relationship between actuator length and PPM configuration is needed.

2. The Forward Analysis Problem

The formulation of the forward analysis is begun by developing the loop closure equation for a quadrilateral. This equation can then be applied to the mechanism in question. Figure 3 represents a quadrilateral with vertices labeled 1, 2, 3, 4 and sides a_{12}, a_{23}, a_{34}, and a_{41}. The relative orientation between sides is measured by the angles θ_1, θ_2, θ_3, and θ_4.

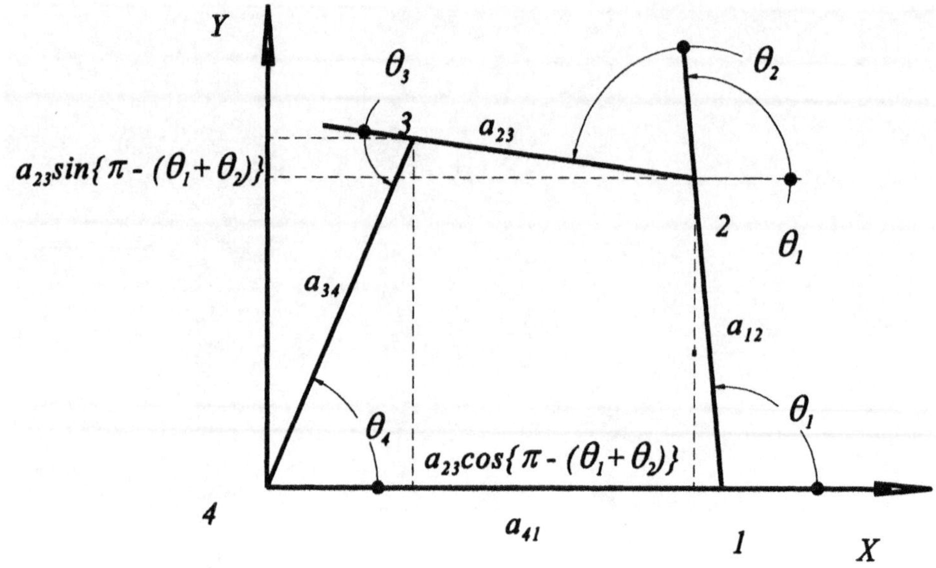

Figure 3 Quadrilateral.

Parallel projection on the x and y axes yields

$$a_{41} - a_{34}c_4 + a_{12}c_1 = - a_{23}c_{1+2} \tag{1}$$

$$a_{34}s_4 - a_{12}s_1 = a_{23}s_{1+2} \tag{2}$$

Adding the sums of the squares of (1) and (2), regrouping, and dividing by 2 gives

$$a_{12}a_{34}c_{4-1} + a_{41}a_{34}c_4 - a_{41}a_{12}c_1 + \frac{(a_{23}^2 - a_{12}^2 - a_{34}^2 - a_{41}^2)}{2} = 0 \tag{3}$$

Equation (3) is a general loop equation for a quadrilateral.

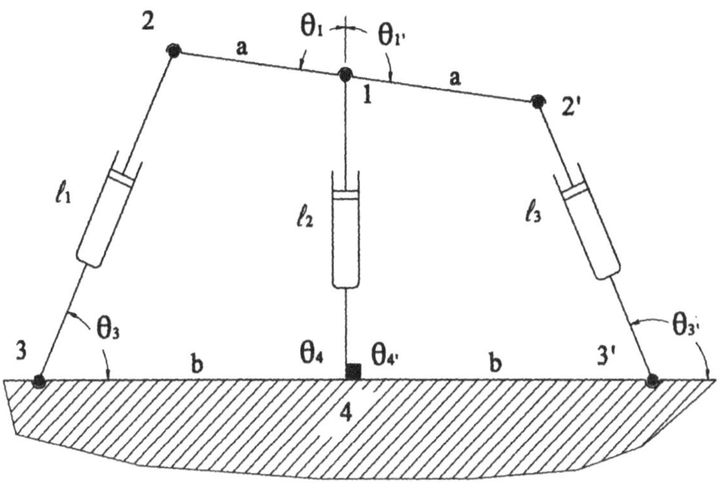

Figure 4 Planar parallel manipulator joint actuation scheme.

The forward analysis problem is stated as follows: given the mechanism constants a and b and the actuator lengths ℓ_1 and ℓ_3, find ℓ_2 and θ or θ_1. The two degree of freedom parallel planar manipulator can be modeled as two quadrilaterals with a common side and the angle associated with the common side fixed. Figure 4 illustrates this relationship with side ℓ_2 common and θ_4 and $\theta_{4'}$ fixed.

The quadrilateral loop equation applied to loop 4, 1, 2, 3 with the substitutions $a_{23} = \ell_1$, $a_{41} = \ell_2$, $a_{12} = a$, and $a_{34} = b$ and dividing through by ab yields

$$c_{4-1} + \frac{l_2 c_4}{a} - \frac{l_2 c_1}{b} + \frac{(l_1^2 - a^2 - b^2 - l_2^2)}{2ab} = 0 \tag{4}$$

The quadrilateral loop equation applied to loop 4, 1, 2', 3' with the substitutions $a_{2'3'} = \ell_3$, $a_{41} = \ell_2$, $a_{12} = a$, and $a_{34} = b$ and dividing through by ab yields

$$c_{4-1'} + \frac{l_2 c_{4'}}{a} - \frac{l_2 c_{1'}}{b} + \frac{(l_3^2 - a^2 - b^2 - l_2^2)}{2ab} = 0 \tag{5}$$

Substituting $\theta_{4'} = \pi - \theta_4$ and $\theta_{1'} = \pi - \theta_1$ into the previous equation leads to

$$c_{4-1} - \frac{l_2 c_4}{a} + \frac{l_2 c_1}{b} + \frac{(l_3^2 - a^2 - b^2 - l_2^2)}{2ab} = 0 \qquad (6)$$

The mechanism is constructed so that $\theta_4 = \theta = \pi/2$ is constant. This leads to the simplifications $c_4 = 0$ and $c_{4-1} = s_1$. Equations (4) and (6) can be expressed as

$$s_1 - \frac{l_2 c_1}{b} + \frac{(l_1^2 - a^2 - b^2 - l_2^2)}{2ab} = 0 \qquad (7)$$

and

$$s_1 + \frac{l_2 c_1}{b} + \frac{(l_3^2 - a^2 - b^2 - l_2^2)}{2ab} = 0 \qquad (8)$$

Equations (7) and (8) represent a coupled set of nonlinear equations in θ_1 and ℓ_2. It remains to eliminate one of the variables and develop an input-output equation in the remaining variable. A first step is subtracting the two equations, leading to

$$-l_2 c_1 + \frac{(l_1^2 - l_3^2)}{4a} = 0 \qquad (9)$$

Adding (7) and (8) gives

$$s_1 + \frac{(l_1^2 - 2(a^2 + b^2) - 2l_2^2 + l_3^2)}{4ab} = 0 \qquad (10)$$

Solving (9) and (10) for s_1 and c_1 yields.

$$c_1 = \frac{(l_1^2 - l_3^2)}{4a l_2} \qquad (11)$$

$$s_1 = \frac{-(l_1^2 - 2(a^2 + b^2) - 2l_2^2 + l_3^2)}{4ab} \tag{12}$$

Squaring equations (11) and (12) yields

$$c_1^2 = \frac{(l_1^2 - l_3^2)^2}{16a^2 l_2^2} \tag{13}$$

and

$$s_1^2 = \frac{(4l_2^4 + 4Kl_2^2 + K^2)}{16a^2 b^2} \tag{14}$$

$$\textit{where} \qquad K = -l_1^2 + 2(a^2 + b^2) - l_3^2 \tag{15}$$

Adding (13) and (14) and using the identity $s^2 + c^2 = 1$ gives a polynomial in ℓ_2

$$\frac{(4l_2^4 + 4Kl_2^2 + K^2)}{16a^2 b^2} + \frac{(l_1^2 - l_3^2)^2}{16a^2 l_2^2} = s_1^2 + c_1^2 = 1 \tag{16}$$

The input-output equation in ℓ_1 and ℓ_3 can be expressed as a sextic in ℓ_2

$$l_2^6 + Kl_2^4 + (\frac{K^2}{4} - 4a^2b^2)l_2^2 + \frac{(l_1^2 - l_3^2)^2 b^2}{4} = 0 \tag{17}$$

This equation has six roots, each one which can be used in equations (11) and (12) to find a unique $\theta_{1,}$ for six closures. The sextic can be further reduced by substituting $y = \ell_2^2$ and therefore

$$y^3 + Ky^2 + (\frac{K^2}{4} - 4a^2b^2)y + \frac{(l_1^2 - l_3^2)^2 b^2}{4} = 0 \tag{18}$$

Equation (18) yields three values for y and the six possible closures are represented by

$$l_2 = \pm \sqrt{y} \tag{19}$$

The three roots in y represent three pairs of reflected closures. Either one root or all three of the roots of (18) will be real, but none are necessarily positive. Either zero, one, two, or three pairs of real closures exist.

3. Numerical Example

The forward analysis is applied to the geometry of the PPM associated with the ATMS. The mechanism constants are a = b = 6.25 inches(15.9 cm). The actuator lengths ℓ_1 and ℓ_3 range between 18 and 28 inches (45.7 and 71.1 cm). The three roots of (18) generate three pair of reflected closures. All solutions satisfy the loop closure equations (7) and (8). Figure 5 illustrates the positive closures of the two real pairs of solutions for actuator lengths $\ell_1 = 20.0$ inches (50.8 cm) and $\ell_3 = 21.0$ inches (53.34 cm). For this geometry, two of these pairs are real solutions and the third pair is pure imaginary. The only closure pair the actual mechanism is capable of is illustrated in figure 5.1. The closure in figure 5.2 is unreachable due to mechanical constraints.

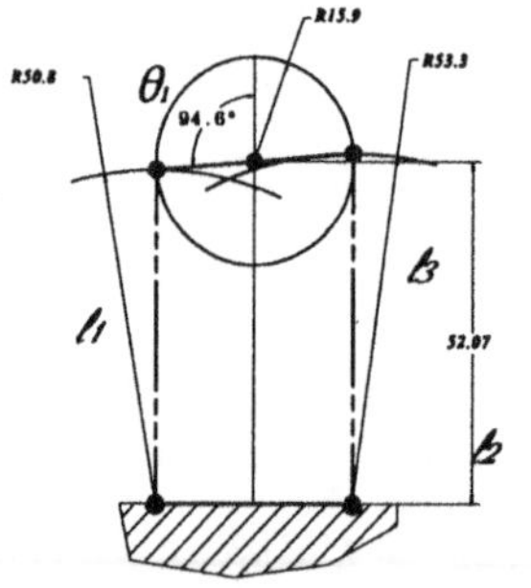

Figure 5.1

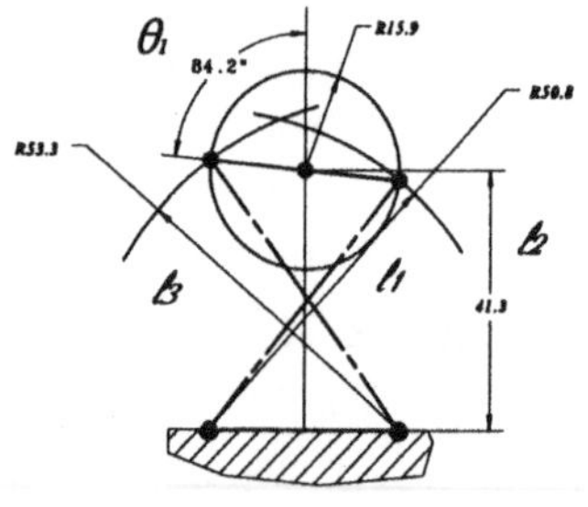

Figure 5.2

Positive real closures.

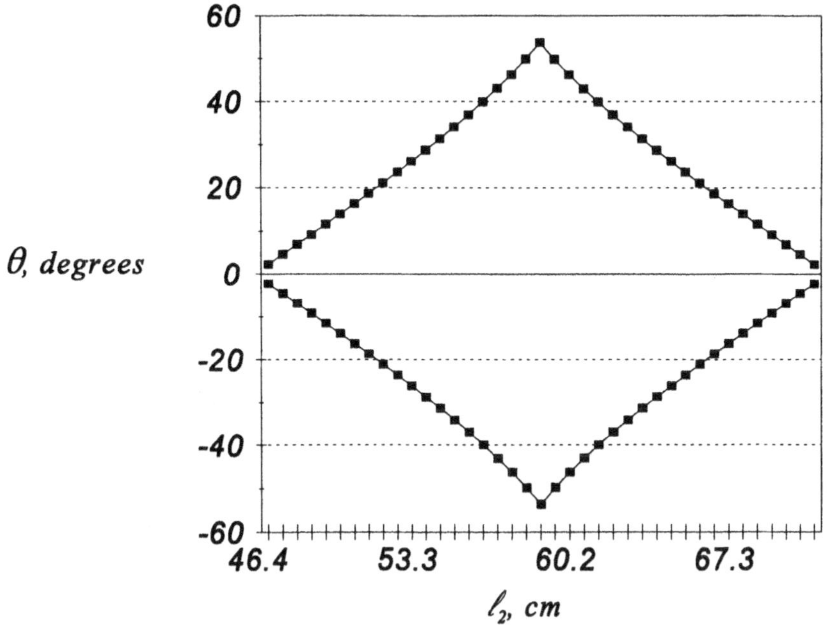

Figure 6 PPM work space.

Figure 6 illustrates the ℓ_2, $\theta = \theta_1 - \pi/2$ work space of the PPM utilized in the ATMS (see also figure 1). The variation in reachable work space is an important design and control issue.

4. Conclusions

A forward analysis is developed for a two degree of freedom planar parallel manipulator. The analysis results in an expression that generates six possible closures for the mechanism. The analysis is applied to a PPM utilized in the design of an articulated mobile robot. The real closures of the PPM are examined and the work space of the mechanism is defined based on mechanism parameters and actuator ranges.

5. Acknowledgments

The development of an articulated mobile robot for use in a nuclear power plant was supported by the United States Department of Energy as part of the Robotics for Advanced Reactors program under the grant # 4910450719612.

LIST OF REFERENCES

Crane, C., Chaing, S.C., Tulenko, J., and Carlton, R., (1988) Algorithm Development and Computer Graphics Simulation of an Articulated Transporter/Manipulator System, *ASME Computers in Engineering Conference*, San Francisco, CA, July 1988, vol 2, pp 253-257.

Hirose, S. and Morishima, A. (1990) Design and Control of a Mobile Robot with an Articulated Body, *The International Journal of Robotics Research*, April 1990, vol 9, No 2, pp 99-114.

Hirose, S. and Umetani, Y. (1981) An Active Cord Mechanism with Oblique Swivel Joints and its Control, *4th Symposium on Theory and Practice of Robots and Manipulators Proceedings*, Zaborow, Poland, 1981, pp 327-340.

Ridgeway, S., Crane, C., Adsit, P., and Harrell, R., (1992) The Mechanical Design of a Parallel Actuated Joint for an Articulated Mobile Robot, *ASME Mechanisms Conference*, Scottsdale, AZ, September 1992, vol. 45, pp 591.

CLOSED PLANE MECHANISMS AS A BASIS OF PARALLEL MANIPULATORS

L. SLUTSKI
Ben-Gurion University of the Negev
P.O.B. 653, 84105 Beer-Sheva, Israel

Abstract. The paper discusses closed plane mechanism applications to organize spatial parallel manipulators. It presents a classification of these mechanisms and a kinematic scheme is also extracted from this array that looks the most promising for multi-DOF robot creation. The structural features of this manipulator determine its properties that essentially simplify robot control. By using the graphic simulation method, this mechanism motion was examined.

1. Introduction

Closed or parallel manipulator mechanisms are a promising base on which to build new and effective constructions of robots. There are two principal directions in the organization of such machines. The first is based on the use of platform spatial manipulators. This approach has its origin in the Stewart platform (Stewart, 1965) and is rather widely used in present-day robot mechanics ((Hunt, 1983; Powell, 1982)). Platform manipulators are characterized by high rigidity and accuracy, but, at the same time, have a restricted workspace and some control difficulties because of not quite simple solutions of direct and inverse kinematics.

Therefore, another direction is now under development, based on the use of closed plane mechanisms as a foundation of spatial robots. The simplest example of a manipulator of this type is presented in Figure 1. Its basic mechanism has had many investigators (for example, (Bajpai and Roth, 1986; Djoldasbekov *et al.*, 1987; Smaili, 1995)). This mechanism is based on the use of a plane group of the 2nd class according to the classification of Assur-Artobolevski. Rotation about a vertical axis ensures its spatial movement. The mechanism's advantages are enhanced rigidity and

J. Lenarčič and V. Parenti-Castelli (eds.), Recent Advances in Robot Kinematics, 441-450.
© 1996 *Kluwer Academic Publishers. Printed in the Netherlands.*

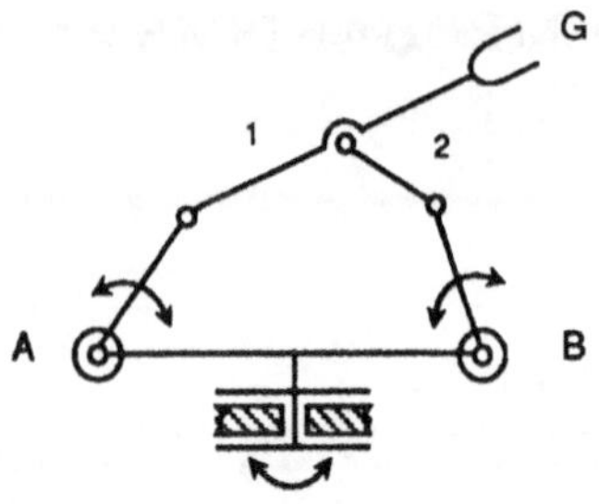

Figure 1. 3DOF-manipulator based on the 2nd class mechanism

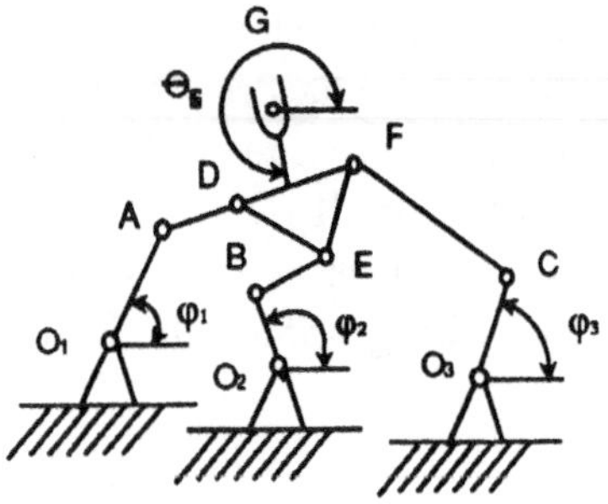

Figure 2. Plane manipulator based on the 3rd class mechanism

driving motor placement on a frame (the points A and B) and both these advantages are common properties of all closed mechanisms. However, only manipulators based on the closed plane mechanisms have the advantage of rather simple solutions of control problems. The scheme of Figure 1 has somewhat restricted workspace (Djoldasbekov *et al.*, 1987), determined by a distance between the joints A and B. Based on kinematic considerations, this distance may be chosen to be near zero. However, the problem is that some optimal value of this parameter ensures optimal rigidity and, respectively, accuracy (Kalinich and Slutski, 1990). Another feature of this mechanism is that if we need to implement a gripper orientation, it is also usually necessary to put an additional drive motor on a moving link.

A solution that allows one to solve this problem without putting a motor on the moving link involves the 3rd class group as a basic kinematic chain (Fig. 2). This mechanism has gained its reputation thanks to the papers of K.Hunt (for example, (1983)) and the following publications (Gosselin and Angeles, 1990; Pennok and Kassner, 1993). The 3rd class mechanisms have good prospects because of rather simple solution of both their kinematic problems. In this connection, a number of investigations were carried out to determine their characteristics (Djoldasbekov *et al.*, 1987; Gosselin and Angeles, 1990; Gosselin *et al.*, 1992; Pennok and Kassner, 1993). This paper develops this approach in creating closed structural schemes for robot mechanisms. Such layout possibilities are analyzed and a special variation of the discussed mechanisms, ensuring a high level of solution of the manipulation tasks to be performed, is proposed for development.

2. Structures of Robots Based on Closed Plane Mechanisms

A classification of basic plane mechanisms according to the class of the initial Assur group and the DOF number of mechanism is presented in Table 1. This classification is based on groups with only revolute kinematic pairs. It may be seen that conventional opened manipulators ($C_1 - N_2, C_1 - N_3$) are particular cases of this classification, as well as the simplest autooperator ($C_1 - N_1$).

Related to 2nd class mechanisms, except for those described in the introduction, they may also be effectively used with one drive link ($C_2 - N_1$) as a basis for implementation of autooperators with complicate trajectories of the working.

Fourth class mechanisms (Djoldasbekov *et al.*, 1985) might also be used, principally, as a basis for the discussed manipulators. For these mechanisms, not all possible structures are shown in Table 1. However, these have highly difficult solutions of direct and inverse kinematics; therefore it is not worth using them as a basis for multi-DOF robots. A positive side of these devices is their great number of kinematic parameters (i.e., link lengths); this determines their advantage as the base for autooperators ($C_4 - N_1$). Such a mechanism is usually synthesized (Djoldasbekov *et al.*, 1985) to solve a complicated trajectory problem and may give a better curve approximation than simpler mechanisms of the 2nd class (Sandor and Erdman, 1984).

As stated in the introduction, the 3rd class mechanism (of the type ($C_3 - N_3$) in Table 1, Fig. 2) can serve as a base on which to build spatial manipulators (for instance, as is carried out in the scheme of Fig. 1). Another type of such mechanisms uses a prismatic pair for linkage between Assur group and the frame or end link. It seems clear that the mechanisms with an additional kinematic pair connected to the frame (Fig. 1) are more practical because of their larger workspace, manipulability and so on. The most promising scheme among these will be discussed below. Before this, we discuss the common control possibilities of 3rd class mechanisms.

3. Control Possibilities of 3rd Class Mechanisms

For the closed plane mechanism (Fig. 2), there are 3 possible ways of setting up a control problem.

1) The first case of this manipulator control applies to the solving of trajectory tasks. This case is based on the ability of a closed mechanism

(C3 - N1, Table 1) to realize certain coupler-curves. For this, two of three driving links of the mechanism (Fig. 2) have to be stopped. In this case, as for all autooperator mechanisms, program curves are obtained by solving a corresponding synthesis problem (Sandor and Erdman, 1984).

2) Another form of the control problem is when the orientation angle Θ_G is not determined for the positioning task. The simplest way to solve this redundant control problem is to fix one of the generalized coordinates and to use only two others ($C_3 - N_2$, Table 1). Thus, control signals may be determined by solving inverse problems similar to those for the basic mechanism of Figure 1. Note that an optimal form of the redundant control problem may also occur when one introduces an optimality criterion. For example, a minimum volume of movements of the input links may be used for this criterion.

Another kind of manipulation problem, solved by using this mechanism, relates to curve generation but for two grippers attached to Assur group links (Smaili, 1995). A solution of this new problem is based on development of the known methods of synthesis (Sandor and Erdman, 1984).

3) Parallel manipulators of the general type ($C_3 - N_3$, Table 1) apply in this case. The usual problem of parallel manipulators is they have too many solutions of the direct kinematics problem. However, for the base plane robot under discussion (Fig. 2), direct kinematics has at most six possible solutions (Gosselin *et al.*, 1992; Peisakh and Nesterov, 1988), less for other types of platform manipulators. When solving inverse kinematics, the endpoint Cartesian coordinates and end link orientation angle are set (Fig. 2). The values of the coordinates ϕ_1, ϕ_2, ϕ_3 must be determined. The inverse kinematics solution is distinguished by extraordinary simplicity; it is discussed below.

Creation of a robot based on 3rd class mechanism is practical when this mechanism is specially constructed as is discussed in the next Section.

4. Especial Type of 3rd Class Mechanism

4.1. STRUCTURE OF ROBOT PROPOSED MECHANISM

This section considers a special kind of the basic 3rd class mechanism that develops the scheme shown in Figure 2. Concretly, its platform DEF is made in the form of a linear crankshaft and its input links are constructed as prismatic pairs; that is, sliders. These sliders 1, 2, 3 (Fig. 3) are placed in different parallel planes and are connected to the platform element. Since the platform element is in the form of a crankshaft, this prevents collisions of mechanism links. Such a structure leads to a considerable simplification of the control problem solution as compared with the initial mechanism (Fig. 2).

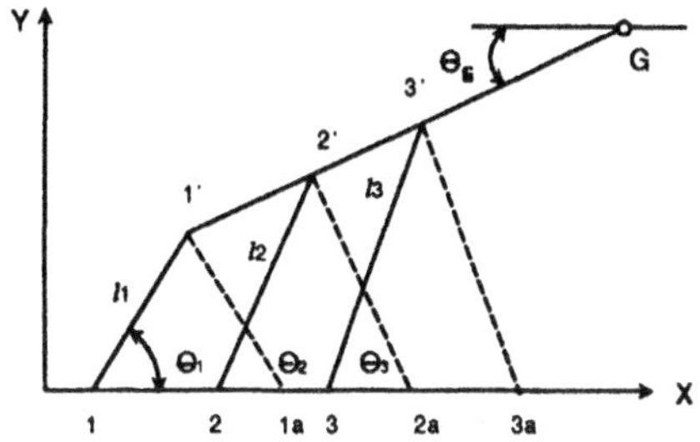

Figure 3. A scheme of the proposed manipulator mechanism

4.2. CONTROL OF ROBOT PROPOSED MECHANISM

4.2.1. *Inverse Kinematics*

To solve the inverse kinematics we can write corresponding equations for the end lever of the basic mechanism (Fig. 3):

$$\begin{vmatrix} x_{i'} \\ y_{i'} \end{vmatrix} = \begin{vmatrix} x_G \\ y_G \end{vmatrix} - l_i \begin{vmatrix} \cos\Theta_G \\ \sin\Theta_G \end{vmatrix}, \ i = 1,2,3. \tag{1}$$

For the points 1, 2, 3 of the input sliders, we write relationships:

$$\begin{aligned} x_i &= x_{i'} - l_i \cos\Theta_i, \\ y_{i'} &= l_i \sin\Theta_i, \ i = 1,2,3. \end{aligned} \tag{2}$$

We determine positions of the points $1', 2', 3'$ according to equations (1) and substitute these values into the equation

$$x_i = x_{i'} \pm \sqrt{l_i^2 - y_{i'}^2}, \tag{3}$$

obtained by eliminating the angle Θ_i from equations (2). It may be seen that the inverse kinematics for this mechanism has 8 possible configurations that consist of combinations of equation (3) solutions for different input links (points 1-1a, 2-2a, and 3-3a in Fig. 3). In order to simplify the control algorithms and to decrease the number of solutions it is worth using a velocity control of this mechanism.

4.2.2. *Velocity Control*

To solve the velocity control problem we differentiate the equations (1):

$$\begin{vmatrix} \dot{x}_{i'} \\ \dot{y}_{i'} \end{vmatrix} = \begin{vmatrix} \dot{x}_G \\ \dot{y}_G \end{vmatrix} + l_i \dot{\Theta}_G \begin{vmatrix} \sin\Theta_G \\ -\cos\Theta_G \end{vmatrix}, \ i = 1,2,3. \tag{4}$$

The equations (2) have also to be differentiated:

$$\begin{aligned} \dot{x}_i &= \dot{x}_{i'} + l_i \dot{\Theta}_i \sin\Theta_i, \\ \dot{y}_{i'} &= l_i \dot{\Theta}_i \cos\Theta_i, \ i = 1,2,3. \end{aligned} \tag{5}$$

When the gripper velocities $\dot{x}_G, \dot{y}_G$ and $\dot{\Theta}_G$ are set, the values $\dot{x}_{i'}$ and $\dot{y}_{i'}$ are found from equations (4). Substituting these values into equations (5) we then obtain values of the generalized velocities

$$\dot{x}_i = \dot{x}_{i'} + \dot{y}_{i'} \tan\Theta_i, \; i = 1,2,3. \tag{6}$$

One thing is quite clear from equations (6): to implement the velocity control algorithm, one must measure angle coordinates Θ_i and use these data for calculations according to these equations. A single solution exists in this case that may be used not only in differential form (6), but also in the form of difference equations.

4.2.3. *Control of Typical Motions*

A very original property of the discussed mechanism, that gives the most benefit from its use, is its ability to carry out simple motions either without performing any preliminary calculations, or by using some simple kinematic relationships. First of all we consider the case of

a) Horizontal shift of the gripper.

We consider a case when $\dot{y}_G = 0$ and $\dot{x}_G = V_G$. A result may be obtained by using the solution for velocity control (4). In the simplest case $\dot{\Theta}_G = 0$, we receive then relationships $\dot{x}_{i'} = \dot{x}_G = V_G$, $\dot{y}_{i'} = 0$, $i = 1,2,3$. By now using (5), it may be determined that for the studied motion ($\dot{\Theta}_i = 0$) the generalized velocities are equal $\dot{x}_i = \dot{x}_{i'} = V_G$, and we obtain a physically intelligible result that consists of simultaneous movement of all input sliders in the necessary direction with the same velocities.

b) Vertical shift of the gripper.

This is a case in which $\dot{x}_G = 0$ and $\dot{y}_G = V_G$. The simplest solution may also be obtained when $\dot{\Theta}_G = 0$. Thus, from (4), we determine $\dot{y}_{i'} = \dot{y}_G = V_G$, $\dot{x}_{i'} = 0$. After this, we obtain from (6) the formulas

$$\dot{x}_i = V_G \tan\Theta_i, \; i = 1,2,3. \tag{7}$$

To use the expressions (7) for calculations, it is necessary to implement the measurement of the angles $\Theta_i, i = 1,2,3$.

c) Rotational motion of the gripper.

This is a case when $\dot{x}_G = 0, \dot{y}_G = 0$ and $\dot{\Theta}_G = \Omega_G$. To obtain a required gripper rotation about a certain point we must first calculate velocities of the platform points i' $i = 1,2,3$. These may be determined by using kinematic relationships for this link rotated about an immobile point G:

$$v_{i'} = \sqrt{\dot{x}_{i'}^2 + \dot{y}_{i'}^2} = l_{i'G}\Omega_G, \tag{8}$$

where $l_{i'G}, i = 1,2,3$ are the lengths of the platform element from the gripper G to its i' points (Fig. 4). After this, we determine velocities of the

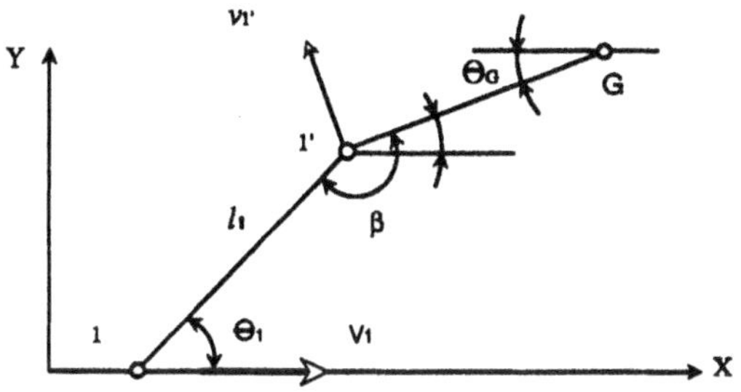

Figure 4. A scheme for velocity calculations

input links when solving a velocity problem for groups of links (Fig. 4). By considering these groups we obtain the expressions that connect velocities of the points i and i' ($i = 1, 2, 3$):

$$V_i = v_{i'} \frac{\sin(\Theta_i - \Theta_G)}{\cos \Theta_i}. \tag{9}$$

The final formulas may be obtained by joining expressions (8) and (9):

$$V_i = l_{i'G} \Omega_G \frac{\sin(\Theta_i - \Theta_G)}{\cos \Theta_i}, \; i = 1, 2, 3. \tag{10}$$

It may be seen that in order to carry out calculations according to (10), the values Θ_i and Θ_G have to be measured. The first values were already used for other calculations, but the latter angle has just to be measured for the solved problem. This may be more simply implemented by measuring a necessary angle in one of the joints i'. For example, it is possible to calculate the angle Θ_G through the measurement of the angle β between links $1, 1'$ and $1', G$ (when the angle Θ_1 is already known) by using the formula accounted for in the geometry of Figure 4: $\Theta_G = \beta + \Theta_1 - 180^o$.

5. Simulation of the Proposed Mechanism

Graphic simulation of the proposed mechanism was performed by using an advanced robotic package (ROBCAD) on the Silicon Graphic workstation. The work included the following stages: a) mechanism model creation; b) introduction of inverse kinematics ; c) programming of the positioning tasks and their simulation. The algorithm of velocity control was implemented while programming the robot model. Typical positioning tasks were simulated and successive spatial motions of the robot from one location to another one were checked. One picture of the simulation results is shown in Figure 5.

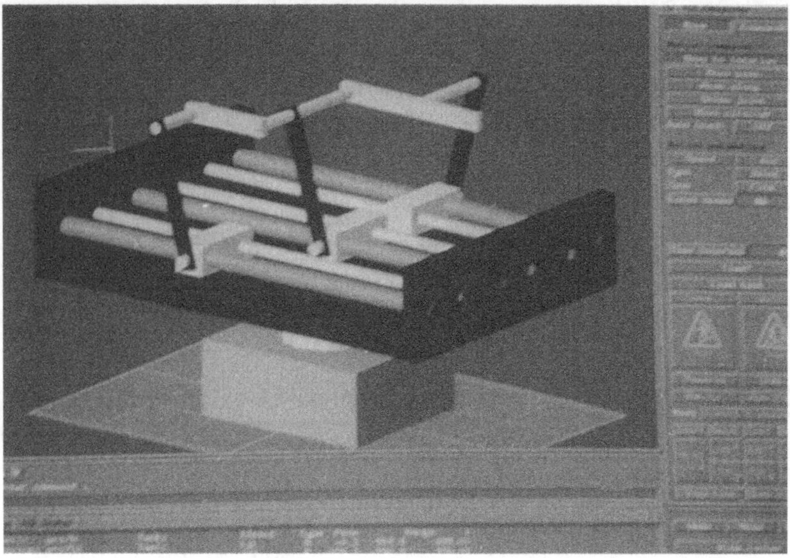

Figure 5. One picture of mechanism graphic simulations

6. Conclusion

Closed plane mechanisms are a very promising base to create multi-DOF robots with a number of original properties. This paper deals with such closed mechanisms, first, from the viewpoint of their common use for the above-noted purpose. Second, a particular scheme of the 3rd class mechanism is extracted from this array that looks to be the most promising for autooperator and robot creation. Its peculiarities are the platform link that is carried out in the form of a linear crankshaft (a linear shape of this link was also considered in (Gosselin *et al.*, 1992)) and input links that are performed in the form of the sliders placed in different parallel planes. While using this mechanism, its structural features essentially simplify robot software. We invented (Djoldasbekov *et al.*, 1987) a parallel manipulator which is a practical form of the construction recently called as H-Robot in paper (Herve, 1995). This manipulator allows us to get kinematic problem solutions that are either extraordinary simple or do not require any calculations.

By using graphic simulation, we examined the motions of the designed mechanism, enabling us to carry out evaluation of collision and reachability. It permitted to us to determine some peculiarities of this mechanism. For instance, it is necessary to note a loss of stability in certain poses of the mechanism that are related to its singular configurations (Gosselin and Angeles, 1990). However, this problem requires a separate presentation of obtained results and is not covered here as well as other important problems, for example, of robot workspace and Jacobian matrix determination.

In combination with a high accuracy and a rather great workspace, this device has good prospects to be used for solving a many industrial problems.

7. Acknowledgments

The author is grateful to the graduate student Ezra Boteach for his contribution to the discussed robot simulation. This work was partly supported by the Paul Ivanier Center for Robotic Research and Production Management of Ben-Gurion University of the Negev.

References

Bajpai, A. and Roth, B. (1986) Workspace and mobility of a closed-loop manipulator, *The International Journal of Robotic Research*, **Vol. 5 no. 2**, pp. 131-142.

Djoldasbekov, U.A., Slutski, L.I. and Baigunchekov, J.J. (1985) Manipulation devices based on high-class mechanisms, in *Theory and Pract. Rob. and Manipul., Proc. RoManSy'84, 5-th CISM - IFToMM Simp.*, Udine. London-Paris, pp. 313-319.

Djoldasbekov, U.A., Slutski, L.I. and Djumadullaev, S.K. (1987) Industrial robot, USSR author's certificate no. 1281401, *Bulletin of Inventions*, **no. 1** (in Russian).

Djoldasbekov, U.A., Slutski, L.I. and Izmambetov, M. (1989) Service characteristics investigation of manipulator closed plane mechanisms, in *Linkages and Manipulation Devices*: Collection of scientific papers. Kazakh State University, Alma-Ata, pp. 48-53 (in Russian).

Gosselin, C. and Angeles, J. (1990) Singularity analysis of closed loop kinematic chains, *IEEE Transactions on Robotics and Automation*, **Vol. 6 no. 3**, pp. 281-290.

Gosselen, C., Sefrioui, J. and Richard, M.J. (1992) Solutions polynomiales au probleme de la cinematique directe des manipulateurs paralleles plans a trois degres de liberte, *Mech. and Mach. Theory*, **Vol. 27 no. 2**, pp. 107-119.

Herve, J.M. (1995) Design of parallel manipulators via the displacement group, in *Proceedings of Ninth World Congress on the Theory of Machines and Mechanisms*, **Vol. 3**, Italy, pp. 2079-2082.

Hunt, K.H. (1983) Structural kinematics of in-parallel-actuated robot-arm, *ASME Journal of Mechanisms, Transmissions, and Automation in Design*, **Vol. 105**, pp. 705-712.

Kalinich, S.S. and Slutski, L.I. (1990) Positioning accuracy analysis of the manipulator closed mechanism with elastic links, in *Linkages and Manipulation Devices*: Collection of scientific papers. Kazakh State University, Alma-Ata, pp. 58-62 (in Russian).

Peisach, E.E. and Nesterov, V.A. (1988) *Design System of the Plane Linkages*. Mashinostroenie, Moscow (in Russian).

Pennok, G. and Kassner, D. (1993) The workspace of a general geometry planar three-degree-of-freedom platform-type manipulator, *ASME Journal of Mechanical Design*, **Vol. 115**, pp. 269-276.

Powell, I.L. (1982) The kinematic analysis and simulation of the parallel topology manipulator, *The Marconi Review*, **Vol. XLV no. 226**, pp. 121-138.

ROBCAD, Users Reference Guide. Tecnomatix Technologies, Herzelia, Israel.

Sandor, G.N. and Erdman, A.G. (1984) *Mechanical Design: Analysis and Synthesis*. Prentice Hall, Englewood Cliffs, N.J..

Smaili, A.A. (1995) Robomechs: a new class of linkage arm for multi-task applications, *Mech. and Mach. Theory*, **Vol. 30 no. 2**, pp. 169-176.

Stewart, D.A. (1965) Platform with six degree of freedom, *Proceedings of the Institute of Mechanical Engineering*, **66 Vol. 180 Part 1 no. 15**, pp. 371-386.

Table 1

Class of manipulator mechanism		I	II	III	IV	Notes
		C_1	C_2	C_3	C_4	
Number of DOF s						
1	N_1		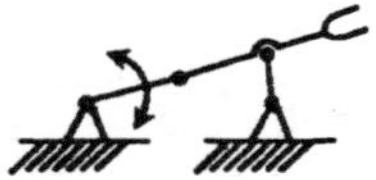		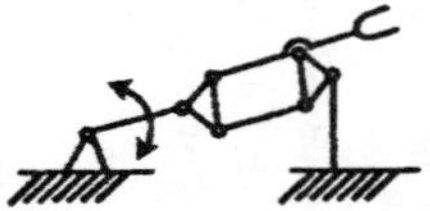	Autooperators
2	N_2				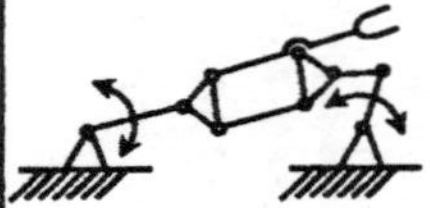	Position type manipulators
3	N_3		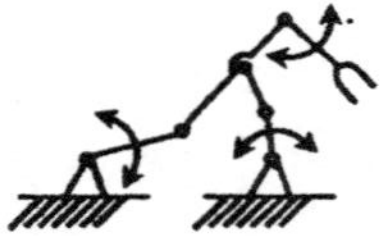			Position type manipulators with gripper orientation

A 3-D SENSOR FOR PARALLEL ROBOT CALIBRATION. A PARAMETER PERTURBATION ANALYSIS

G. FRIED, K. DJOUANI AND Y. AMIRAT
L.I.I.A, IUT de Créteil
61, av du Général de Gaulle 94010 CRETEIL, FRANCE
Phone/Fax: 33-1-45.17.16.90
email: djouani@univ-paris12.fr

AND

C. FRANCOIS
LGMPB, IUT de Cachan
9, av de la division Leclerc 94230 CACHAN, FRANCE
Phone/Fax:33-1-41.24.11.91

Abstract. High accuracy in robotic manipulation is an important requirement in order to ensure task specification and constraint. In this case, off line robot geometric parameter identification is very useful. In this paper, the authors present a new 3-D sensor for a 6 DOF parallel robot calibration. The forward and inverse geometric models for both the C5 parallel robot and the LGMPB-sensor are given. Then the parameter observability is studied by looking for the best minimum experimental configurations using some criteria based on the augmented Jacobian. A non-linear identification algorithm is proposed and first results are given.

1. Introduction

In performing accurate tasks using robot manipulator, one of the most important requirements is the minimizing of the positioning error in the operational space (Liegeois, 1984 and Aldon, 1986). For instance, when a specific task is defined in the operational space, the corresponding joint vector must be computed by using geometric and/or kinematic inverse models. The geometric parameters used within these models are not known very accurately and must be identified on the actual robot (Nahvi et al, 1994, Zhuang et

J. Lenarčič and V. Parenti-Castelli (eds.), Recent Advances in Robot Kinematics, 451-460.
© 1996 *Kluwer Academic Publishers. Printed in the Netherlands.*

al, 1994, Sayeh and Newman, 1994, and Doria et al, 1994). In static mode, the errors on the pose (position and orientation) of the robot end effector are absolute errors which may be both geometric and non geometric. On one hand, the geometric errors are due to the imprecise manufacturing of robot links and joints, or from the deviation of the encoder offsets. On the other hand, the non geometric errors may be due to friction, compliance, gear transmission and backlash (Liegeois, 1984, Aldon, 1986 and Payannet, 1985). The geometric errors may be also due to the poor estimation of the parameters defining the tool with respect to the terminal link (Kahlil et al, 1990). In this paper, the authors are concerned with the calibration of the geometric model of a C5 parallel robot (Dafaoui et al, 1994, 1996) for accurate compliant task performing, like assembly tasks and contour following, using a new 3-D sensor (Fried, 1995). This paper is organized in five parts. The first part is an overview of the C5 parallel robot geometric and kinematic modeling. In the second part, a perturbation parameter is studied, which justifies the robot calibration. Then the optimum measurement problem for parameter observability is treated. In the third part, a new 3-D sensor named LGMPB-sensor is presented, along with its geometric model and measurement system. Then the non-linear identification algorithm is given in part four, and simulation results are shown in part five.

2. Modeling of the C5 joint parallel robot

In this part, the models of the C5 joint parallel robot to calibrate are given. More details, can be found in Dafaoui et al, 1994, 1996.

2.1. MECHANISM DESCRIPTION

The C5 joint parallel robot consists of a static part and a mobile part connected together by six actuated links. Each segments is embedded to the static part at point A_i and linked to the mobile part through a C5 joint at point B_i (Fig. 1). Each C5 joint consists of a spherical joint tied to two cross sliding plates. Each segment is equipped with a ball screw linear actuator driven by a DC motor.

The notations used to describe the C5 joint parallel robot are as follows:
ρ_i is the lenght of the ith leg (length of vector $\overrightarrow{A_iB_i}$).

A_i is the ith imbedded point linking each actuator with the base and B_i the center of ith joint between the leg base and the mobile part.

Let $R_b\left(O, \vec{X}, \vec{Y}, \vec{Z}\right)$ and $R_p\left(O_p, \vec{X_p}, \vec{Y_p}, \vec{Z_p}\right)$ be respectively the fixed frame and the mobile frame, where O is the origine of the absolute coordinate system (tied on the base), which is the orthocenter of the base, and

Figure 1. Overview of the C5 parallel robot

O_p is the origine of the mobile coordinate system (tied to the mobile part), whose coordinates are $(x_c, y_c, z_c)^T$ in the absolute frame.

α, β, γ are the RPY (Roll Pitch and Yaw) angles i.e. the orientation parameters of the mobile coordinate system with respect to R_b.

The notation $\overrightarrow{X}_{/R_i}$ means that the components of vector $\overrightarrow{X}$ are expressed in the coordinate system R_i.

2.2. INVERSE GEOMETRIC MODEL

The inverse geometric model relates the joint variables which are the linear actuator lengths to the operational variables which define the pose of the end effector. This relation is given by the following equation ($i = 1, ..., 6$):

$$\rho_i = \|\overrightarrow{A_iB_i}_{/R_b}\| = \|\overrightarrow{A_iO}_{/R_b} + \overrightarrow{OO_p}_{/R_b} + [R]\overrightarrow{O_pB_i}_{/R_p}\| \tag{1}$$

where [R] is the rotation matrix. The expression of [R] is given in Dafaoui et al, 1994.

2.3. FORWARD GEOMETRIC MODEL

The forward geometric model is defined by the following equation:

$$\overrightarrow{X}\left(x_c, y_c, z_c, \alpha, \beta, \gamma\right) = F\left(\rho\right)$$

In C5 joint parallel robot case, an analytical solution is obtained through the following procedure: first, the orientation parameters (α, β, γ) are computed after trigonometric simplifications taking into account the C5 parallel robot geometric charasteristics. Then the (x_c, y_c, z_c) parameters are obtained by solving a simple linear system of three equations with three unknowns.

3. Parameter perturbation analysis and identifiability

Theoretical robot modeling (geometric and kinematic) doesn't take into account the robot design error. Thus, when performing very accurate tasks, the actual robot parameters have to be known. For this, the robot calibration using exteroceptive sensor and parameter identification is very helpful. In the following, a parameter perturbation analysis is given. When the 3-D sensor is mounted on the robot end effector, the joint displacements are given by:

$$\overrightarrow{q}_{robot} = F\left(\overrightarrow{q}_{sensor}, \overrightarrow{P}\right)$$

where $\overrightarrow{q}$ is the joint vector and $\overrightarrow{P}$ the parameter vector to be identified. In the parameter perturbation analysis, two hypothetic kinds of parameters, constructor parameters and supposed actual parameters, are used. By perturbing the constructor parameters, the new operational trajectory is computed. The table 1 shows the difference between the desired trajectory and the actual one; this proves the parameter identification requirement.

Table 1. Desired and actual trajectories

Desired trajectory			Actual trajectory		
X	Y	Z	X	Y	Z
2.5000	0.0000	2.0000	3.5078	0.9730	2.9655
2.1651	1.2500	2.0000	3.1730	2.2227	2.9657
1.2500	2.1651	2.0000	2.2580	3.1379	2.9657
0.0000	2.5000	2.0000	1.0081	3.4732	2.9655
-1.250	2.1651	2.0000	-0.242	3.1389	2.9651
-2.165	1.2500	2.0000	-1.157	2.2245	2.9648
-2.500	0.0000	2.0000	-1.492	0.9750	2.9644

3.1. OBSERVABILITY INDEX

The parameter observability was studied in the literature and some index are given in order to find the optimum robot posture for parameter identification (Driels and Pathre, 1990 and Bohrm and Menq, 1991).

In our study, two observability indices were used. They are defined as constraints imposed on the robot augmented jacobian matrix.

The first one is given by:

$$Cond\,(J) = \frac{\sigma_{\min}}{\sigma_{\max}}$$

where $\sigma_{\min}$ is the smallest singular value of J, and $\sigma_{\max}$ the largest one. J is the augmented jacobian matrix given by:

$$J = [J_1 \; J_2 \; ... \; J_m]^T$$

where J_i is the robot Jacobian for the ith configuration.
Obviously, $Cond\,(J) \leq 1$.
$Cond\,(J) = 1$ is the ideal case for observability of error parameters.
If $Cond\,(J) \to 0$, the error parameters are unobservable.
The second observability index is given by:

$$O\,(J) = \frac{\sqrt[L]{\prod_{i=1}^{L} \sigma_i}}{\sqrt{m}}$$

where L is the number of singular values of J and m the number of measurements. In this case, the error is unobservable when $O\,(J) = 0$. As $O\,(J)$ increases, the contribution of geometric error dominates the effects of non geometric errors and other unmodeled errors (Zhuang et al, 1994). For the case of the C5 joint parallel robot, simulation shows that 100 measurements are needed in order to satisfy the criteria on the error parameter observability requirement.

4. The three dimensional sensor

4.1. DESCRIPTION

The LGMPB-sensor consists of a mobile part, connected to the robot wrist, and a static part equipped with six dispacement gauges of LVDT type. The figure 2 shows the 3D-sensor where the mobile part is removed from the inside of the static part in order to show the displacement gauge tips. When the sensor is in operation, the contacts betweeen the displacement gauge tips and the mobile part are point joints (3 DOF in rotation and 2 DOF in translation). To derive the sensor geometric model, the same formalism as for the C5 parallel robot is used. In the following notations, the C5 robot and the LGMPB-sensor are differentiate by the sign (′).

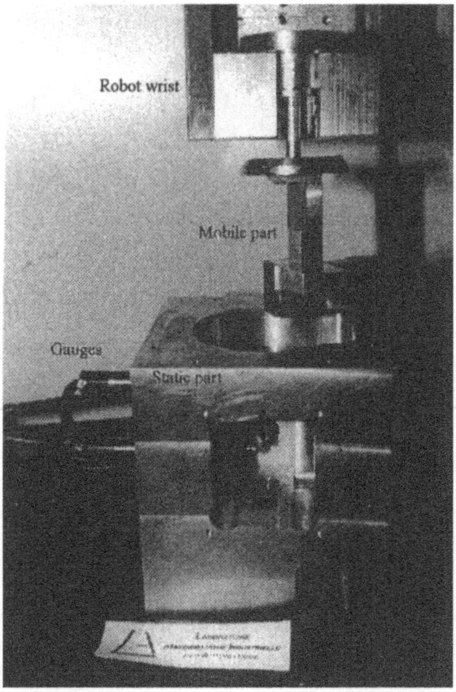

Figure 2. Overview of the 3D-sensor

Notations

A'_i is the ith imbedded point linking each displacement gauge with the static part.

B'_i is the center of ith point joint between the displacement gauge tip and the mobile part.

$R'_b\left(O',\overrightarrow{X'},\overrightarrow{Y'},\overrightarrow{Z'}\right)$ and $R'_p\left(O'_p,\overrightarrow{X'_p},\overrightarrow{Y'_p},\overrightarrow{Z'_p}\right)$ are respectively the fixed and the mobile frame.

In the fixed frame, the coordinates of A'_i and O'_p are:

$$\begin{array}{lll} A'_1 = (b_1 \quad 0 \quad c_{12})^T & A'_2 = (b_1 \quad 0 \quad 0)^T & A'_3 = (-c_{34} \quad -b_3 \quad 0)^T \\ A'_4 = (0 \quad -b_3 \quad 0)^T & A'_5 = (0 \quad c_{56} \quad -b_5)^T & A'_6 = (0 \quad 0 \quad -b_5)^T \\ O'_p = (x'_c \quad y'_c \quad z'_c)^T & & \end{array}$$

In the mobile frame, the coordinates of B'_i are:

$$\begin{array}{lll} B'_1 = \left(0 \quad 0 \quad z'_{b_1}\right)^T & B'_2 = \left(a_{12} \quad 0 \quad z'_{b_2}\right)^T & B'_3 = \left(x'_{b_3} \quad 0 \quad 0\right)^T \\ B'_4 = \left(x'_{b_4} \quad -a_{34} \quad 0\right)^T & B'_5 = \left(0 \quad y'_{b_5} \quad 0\right)^T & B'_6 = \left(0 \quad y'_{b_5} \quad -a_{56}\right)^T \end{array}$$

4.2. GEOMETRIC MODELING

The approach for the sensor geometric modeling is analogous to that used for the C5 joint parallel robot. When considering the gauges as six linear actuators of length ρ'_i $(i = 1, ..., 6)$ linking the mobile and the static part, the three dimensional sensor satisfies to criteria defining a parallel robot (Merlet, 1990). Thus, equation (1) becomes:

$$\overrightarrow{A'_iB'_i}_{/R'_b} = \overrightarrow{A'_iO'}_{/R'_b} + \overrightarrow{O'O'_p}_{/R'_b} + [R]\,\overrightarrow{O'_pB'_i}_{/R'_p} \tag{2}$$

The inverse geometric model is given by the following equation system:

$$\begin{cases} \rho'_1 = (x'_c - b_1) + \frac{r_{21}\ y'_c + r_{31}\ (x'_c - c_{12})}{r_{11}} \\ \rho'_2 = (x'_c - b_2) + \frac{r_{21}\ y'_c + r_{31}\ x'_c + a_{12}}{r_{11}} \\ \rho'_3 = (y'_c + b_3) + \frac{r_{12}\ (x'_c + c_{34}) + r_{32}\ z'_c}{r_{22}} \\ \rho'_4 = (y'_c + b_4) + \frac{r_{12}\ x'_c + r_{32}\ z'_c - a_{34}}{r_{22}} \\ \rho'_5 = (y'_c + b_5) + \frac{r_{13}\ x'_c + r_{23}\ (y'_c - c_{56})}{r_{33}} \\ \rho'_6 = (y'_c + b_6) + \frac{r_{13}\ x'_c + r_{23}\ y'_c - a_{56}}{r_{33}} \end{cases}$$

5. Parameter identification algorithm

The end effector location in the operationnal space is given by:

- the model: $\vec{X}_{model} = F\left(\vec{q}, \vec{P}\right)$
- the data from the LGMPB-sensor: $\vec{X}_{sensor} = F'\left(\vec{q}_{sensor}\right)$

Our aim is to identify the robot geometric parameter vector $\vec{P}$ by minimizing the norm:

$$Min_{\vec{P}} \sum_{t=0}^{N} \vec{\epsilon}\left(t, \vec{P}\right)^{T} [Q] \vec{\varepsilon}\left(t, \vec{P}\right)$$

where Q is a symmetrical positive definite matrix, and $\vec{\epsilon}$ the pose error between $\vec{X}_{model}$ and $\vec{X}_{sensor}$. Since the problem is non-linear, the identification algorithm proposed is developped in the following.

For each iteration k, the new parameters are given by:

$$\vec{P}^{k+1} = \vec{P}^{k} + \lambda_k \vec{d}_k \qquad (\lambda_k > 0)$$

where $\vec{d}_k$ is a stepest descent.

The identification algorithm is given by:

1- Initialisation:

$$\vec{P}^{0}, \beta_0, \vec{d}_0$$

2- While

$$\vec{P}^{k+1} - \vec{P}^{k} > \alpha \quad (required\ precision)$$

do:

$$\vec{d}_k = -\frac{\partial C\left(\vec{P}^{k}\right)}{\partial \vec{P}} + \beta_{k-1} \vec{d}_{k-1}$$

$$\beta_k = \frac{\left\| \nabla_{\vec{P}} C\left(\vec{P}^{k}\right) \right\|^2}{\left\| \nabla_{\vec{P}} C\left(\vec{P}^{k-1}\right) \right\|^2} \quad where \quad \nabla_{\vec{P}} C\left(\vec{P}^{k}\right) = \frac{\partial C\left(\vec{P}^{k}\right)}{\partial \vec{P}}$$

$$\vec{P}^{k+1} = \vec{P}^{k} + \lambda_k \vec{d}_k$$

6. Simulation results

In order to prove the reliability of the approach proposed for geometrical robot calibration, some simulations were conducted. The table 2 shows the

Table 2. Desired and actual trajectories after parameters identification

Desired trajectory			Actual trajectory		
X	Y	Z	X	Y	Z
2.5000	0.0000	2.0000	2.4981	-0.031	1.9983
2.1651	1.2500	2.0000	2.1624	1.2185	1.9985
1.2500	2.1651	2.0000	1.2468	2.1330	1.9987
0.0000	2.5000	2.0000	-0.0034	2.4674	1.9990
-1.250	2.1651	2.0000	-1.2532	2.1321	1.9990
-2.165	1.2500	2.0000	-2.1677	1.2197	1.9990
-2.500	0.0000	2.0000	-2.5019	-0.033	1.9988

desired and actual trajectories after parameter identification using the algorithm described above.

7. Conlusion

This paper presents a new 3-D sensor for robot calibration. This kind of sensor is used for our parallel robot calibration in the vicinity of the assembly task specification. In order to show identification requirement, a perturbation analysis is done. When investigating model parameters identification, the error parameter observability is a crucial problem. In our case, some criteria, based on the robot augmented jacobian, were used in order to find the optimal robot configurations which ensure the observability requirement, and first simulation results have been presented.

References

Liegeois, A. (1984) Analyse des Performances et CAO, *Les robots tome 7, Hermes publishing.*

Aldon, M.J. (1986) Identification des Paramètres Structuraux des Robots Manipulateurs, *conf Outils Mathématiques pour la Modélisation et la Commande des Robots*, Paris, pp 243-268.

Nahvi, A., Hollerbach, J.M., Hayward, V. (1994) Calibration of a Parallel Robot using Multiple Kinematic Closed Loop, *in Proc IEEE Intl. Conf Robot and Automation*, pp 407-412.

Zhuang, H., Roth, Z., Wang, K. (1994) Optimal Selection of Measurement Configuration using Simulated Annealing, *in Proc IEEE Intl. Conf Robot and Automation*, pp 393-397.

Sayeh, S., Newman, W. (1994) A New Technique for Solving Robot Calibration Equations with Partially Known Constaint, *in Proc IEEE Intl. Conf Robot and Automation*, pp 387-392.

Doria, A., Angrilli, F., De Marchi, S. (1994) Inverse Kinematic Robot Calibration by Spline Functions, *App. Math. Modelling, Vol 17*, pp 492-498.

Payannet, D. (1985) Modélisation et Correction des Erreurs Statiques des Robots Manipulateurs, *thèse de doctorat* Montpellier.

Kahlil, W., Gautier, M., Enguehard, C. (1990) Identifiable Parameters and Optimum Configurations for Robot Calibration, *Robotica Vol 9*, pp 63-70.

Dafaoui, M., Amirat, Y., Pontnau, J., François, C. (1994) Manipulateurs Parallèles 6 DDL: Modèles et Volume de Travail, *RAPA, n. 2 Vol 7*, pp 195-220.

Dafaoui, M., Amirat, Y., Pontnau, J., François, C. (1996) Analysis and Design of a six DOF Parallel Manipulator. Modeling, Singular Configurations and Workspace, *accepted for publication in the IEEE Trans. on Robotic and Automation.*

Fried, G. (1995) Modélisation Géométrique Inverse et Direct d'un Capteur Tridimentionnel, *intern report of LIIA.*

Driels, M., Pathre, U.S. (1990) Significance of Observation Strategy on the Design of Robot Calibration Experiments, *J. Robot System, 7(2)*, pp 197-223.

Bohrm, J.H., Menq, C.H. (1991) Determination of Optimal Selection of Measurement Configuration for Robot Calibration based on Observability Measure, *Int. J. Robotic Reasearch, Vol 10, n. 1*, pp 51-63.

Merlet, J.P. (1990), Les Robots Parallèles, *traité des nouvelles technologies, Hermes publishing.*

Author Index